Fast Non-destructive Detection Technology and Equipment for Food Quality and Safety

Fast Non-destructive Detection Technology and Equipment for Food Quality and Safety

Editors

Zhiming Guo
Zhao Zhang
Dong Hu

MDPI

Basel • Beijing • Wuhan • Barcelona • Belgrade • Novi Sad • Cluj • Manchester

Editors

Zhiming Guo	Zhao Zhang	Dong Hu
Jiangsu University	China Agricultural University	Zhejiang A&F University
Zhenjiang, China	Beijing, China	Hangzhou, China

Editorial Office
MDPI
St. Alban-Anlage 66
4052 Basel, Switzerland

This is a reprint of articles from the Special Issue published online in the open access journal *Foods* (ISSN 2304-8158) (available at: https://www.mdpi.com/journal/foods/special_issues/Fast_Non_Destructive_Detection_Technology_Equipment_Food_Quality_Safety).

For citation purposes, cite each article independently as indicated on the article page online and as indicated below:

Lastname, A.A.; Lastname, B.B. Article Title. *Journal Name* **Year**, *Volume Number*, Page Range.

ISBN 978-3-0365-9324-1 (Hbk)
ISBN 978-3-0365-9325-8 (PDF)
doi.org/10.3390/books978-3-0365-9325-8

Cover image courtesy of Zhiming Guo

Contents

About the Editors

Zhiming Guo

Dr. Zhiming Guo is a full professor in the School of Food and Biological Engineering at Jiangsu University, China. He is a principal investigator in food analytical and detection sciences. He was awarded as the first batch of "Shennong Young Talents" by the Ministry of Agriculture and Rural Affairs of China, the outstanding young teachers of "Qinglan Project" in Jiangsu universities, and the National Excellent Teacher Team in Universities by the Ministry of Education of China. He has published more than 100 peer-reviewed journal publications and holds 45 patents. He is the Vice Director of China Light Industry Key Laboratory of Food Intelligent Detection & Processing, the Deputy Secretary-General of Jiangsu Institute of Food Science and Technology (JIFST), the Deputy Secretary-General of the Agricultural Products Processing Machinery Branch in the Chinese Society of Agricultural Engineering (CSAE), and a member of the American Society of Agricultural and Biological Engineers (ASABE). His research interests focus on the fast and non-destructive detection of food quality and safety, optical sensing and automation for food quality evaluation, the Internet of Things (IOT) for food quality and safety, and advanced chemometrics methods. He has been funded by more than 20 national and provincial projects such as the National Key R&D Project, the National Natural Science Foundation of China, and the Jiangsu Provincial Key Research and Development Program. His research achievements have seen him be awarded second prize for the Natural Science of the Ministry of Education in 2020 (2/6), first prize for Science and Technology of Jiangsu Province in 2016 (6/10), first prize for Technological Invention of the China Light Industry Federation in 2020 (1/6), and a special prize for the China Business Federation in 2021 (4/14).

Zhao Zhang

Dr. Zhao Zhang earned his B.S. and M.S. from Northwest Agricultural and Forest University, majoring in Industrial Design and Agricultural Mechanization Engineering, respectively. He then earned his Ph.D. from Pennsylvania State University in the area of Agricultural Engineering, after which he joined USDA as a postdoc research fellow. He joined North Dakota State University as a member of the faculty, and focused on precision agriculture. He finally joined the College of Information and Electrical Engineering, China Agricultural University, as a member of the faculty, working in sensing and automation in agriculture. As a team member, he earned the Rain Bird Concept award of the American Society of Agricultural and Biological Engineering for his contribution to the development of a prototype of an in-field apple sorting system. The developed in-field apple grading and sorting system can work at a throughput rate of 12–15 apples per second, and the fruit bruising ratio is lower than 5%, thereby meeting industrial requirements Additionally, Dr. Zhang developed an apple-harvesting robot that can pick apples automatically, within which deep learning technology has been used for fruit detection and localization, and a newly developed end-effector has been used for grasping and detaching. The prototype has gone through multi-season field tests, and the technology currently being improved and readied for commercial application. Dr. Zhang also works on phenotyping, mainly focusing on developing autonomous technology for phenotyping information collection.

Dong Hu

Dong Hu, Ph.D., is an Associate Professor in the College of Optical Mechanical and Electrical Engineering at Zhejiang Agricultural and Forestry University, China. He was awarded the titles of one of the Young Excellent Talents under the Leading Talent Program of Zhejiang Province's

Universities, and one of the Outstanding Young Talents from Zhejiang A&F University. He has published more than 40 peer-reviewed journal publications and 2 chapters of English monographs, and has authorized more than 10 patents. He is a member of the American Society of Agricultural and Biological Engineers (ASABE), and a member of the Chinese Society of Agricultural Engineering (CSAE). His research interests include intelligent perception technology for agricultural and forestry information, optical detection technology, and equipment for agricultural and forestry products. He has been funded by the National Natural Science Foundation of China, and the Natural Science Foundation of Zhejiang Province. He has guided students to participate in various subject competitions and won multiple national awards.

foods

Editorial

Fast Nondestructive Detection Technology and Equipment for Food Quality and Safety

Zhiming Guo [1,2,3,*] and Heera Jayan [1,2,*]

1 School of Food and Biological Engineering, Jiangsu University, Zhenjiang 212013, China
2 International Joint Research Laboratory of Intelligent Agriculture and Agri-Products Processing, Jiangsu University, Zhenjiang 212013, China
3 China Light Industry Key Laboratory of Food Intelligent Detection & Processing, Jiangsu University, Zhenjiang 212013, China
* Correspondence: guozhiming@ujs.edu.cn (Z.G.); heerajayan93@outlook.com (H.J.); Tel.: +86-511-88780201 (Z.G.)

Citation: Guo, Z.; Jayan, H. Fast Nondestructive Detection Technology and Equipment for Food Quality and Safety. *Foods* **2023**, *12*, 3744. https://doi.org/10.3390/foods12203744

Received: 4 October 2023
Accepted: 8 October 2023
Published: 12 October 2023

Fast nondestructive detection technology in food quality and safety evaluation is a powerful support tool that fosters informatization and intelligence in the food industry, characterized by its rapid processing, convenient operation, and seamless online inspection. Over the past two decades, these technologies have found numerous successful applications in the field of food and agricultural product detection and processing. Owing to improvements in the development of photoelectric sensors and the ongoing progress in artificial intelligence and software algorithms, fast nondestructive detection technologies provide significantly enhanced accuracy, reliability, and stability, revolutionizing their role in food quality and safety detection and processing. Their seamless integration with the Internet of Things (IoT) and intelligent manufacturing is promoting a new wave of innovation in the food industry. The application of new sensing technology and equipment in the fast, nondestructive detection of food has always been at the forefront of scientific and technological research. The schematic diagram of the advance in research progress is shown in Figure 1. This Special Issue is dedicated to highlighting the latest research progress and jointly discussing the future directions of research and development in the field.

Raman spectroscopy is a fast and sensitive tool that has established itself as a valuable technique that has demonstrated successful applications in ensuring food safety and quality. Yin et al. [1] employed bimetallic core–shell nanoparticles and a specific redox reaction of carbimazole and chromium iron for the surface-enhanced Raman spectroscopy (SERS) detection of hexavalent chromium in tea. The developed techniques demonstrated excellent sensitivity, emphasizing the significant potential of rapid, non-destructive, and sensitive SERS detection in the field of food safety and quality analysis. Qiu et al. [2] developed an SERS-based method for the detection of polycyclic aromatic hydrocarbon (PAH) residues on the surface of fruits and vegetables. A flexible substrate (β-CD@AuNP/PTFE) was employed for enhancing the signals along with lightweight deep learning networks for data analysis. In addition, Zhang et al. [3] utilized a microfluidic chip for the capture of crop airborne disease spores for further detection using Raman spectroscopy. The use of support vector machine (SVM) and back-propagation artificial neural network (BPANN) ensured high accuracy in detection. Thus, the integration of deep learning in Raman spectroscopic data facilitates automated feature extraction, accommodates complex data relationships, and achieves high accuracy levels, thereby enabling its effective application in the field of food safety. Interestingly, Sun et al. [4] focused on theoretically calculating Raman spectra for five commonly used plasticizers, known as phthalic acid esters (PAEs). The density functional theory (DFT) calculations showed in the research have the potential to contribute to the development of Raman spectroscopic methods for the rapid detection of PAEs in the future, a crucial step in assessing their potential health risks. These innovative applications demonstrate the effectiveness of Raman spectroscopy in detecting contaminants and analyzing the quality of food products. Theoretical advancements in Raman spectral calculation help to gain insights into molecular structure, composition, and their interaction, which eventually has the potential to improve the accuracy and sensitivity of Raman-based analysis.

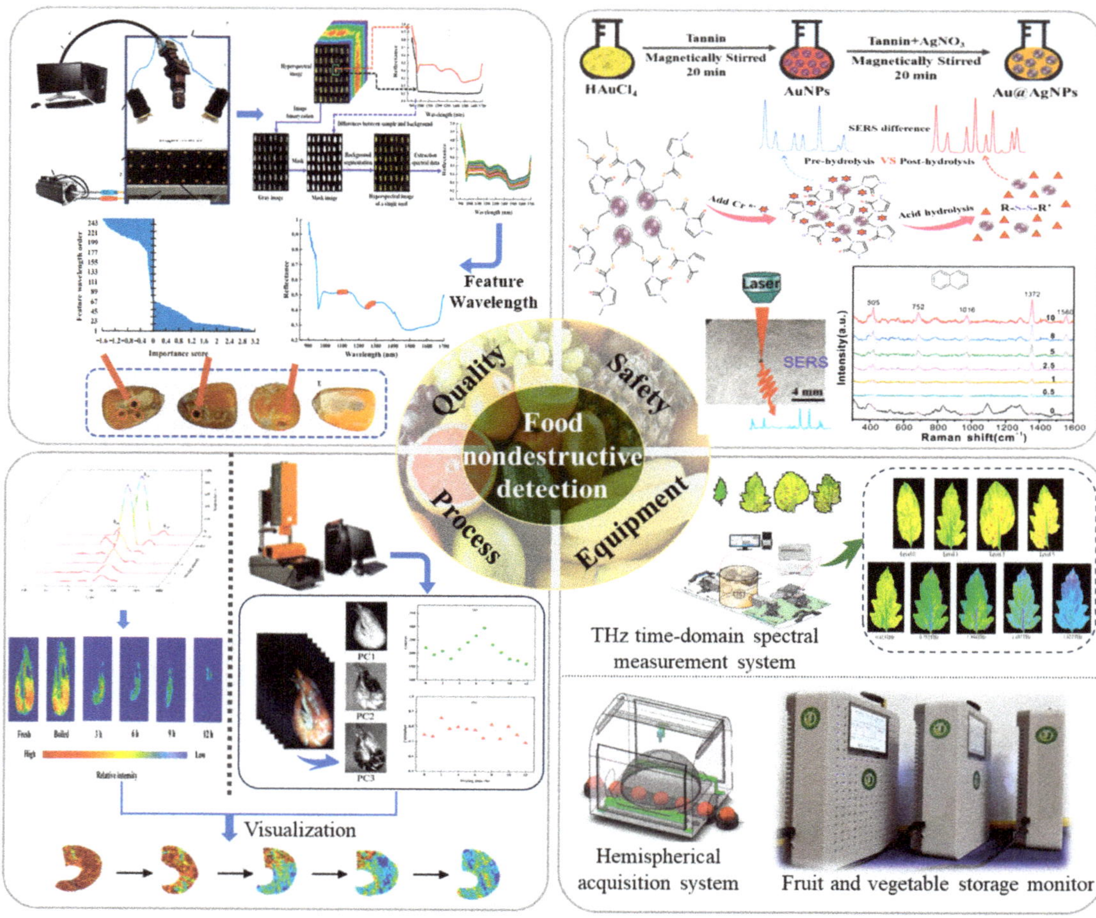

Figure 1. Schematic diagram of the advance research progress of fast nondestructive detection technology and equipment for food quality and safety.

Further, Sun et al. [5] employed visible/near-infrared (Vis/NIR) spectroscopy to detect the soluble solid content in fresh jujubes along with a least square support vector machine to develop a model. The proposed method yielded highly accurate prediction results, effectively tackling the demand for quality analysis of jujubes in the open fields. In addition, Jiang et al. [6] developed a calibration method for NIR spectroscopy to enhance the accuracy of the model for detecting the soluble solid content in apples of different sizes. The results hold high significance in advancing the development of dependable models for predicting the SSC in diverse fruits. While establishing a Vis/NIR spectroscopy detecting method for the stone cell content of Korla fragrant pears, Wang et al. [7] showed that the standardized normal variate (SNV) pre-processed successive projective algorithm–support vector regression (SPA-SVR) model effectively meets the requirements for intelligent evaluation, achieving high correlation coefficients for both calibration and validation sets. The effectiveness of Vis/NIR spectroscopy in food quality analysis was further proved by Wang et al. [8], where Vis/NIR spectroscopy was successfully employed to predict the anthocyanin content in purple Chinese cabbage with high accuracy. On the other hand, Migues et al. [9] developed a method for predicting the acceptability of Mandarin fruit based on the sugar and citric acid levels extracted from the NMR spectroscopic data. The study proved that the chemometric-based models facilitate data-driven decisions to optimize food quality, ensuring that the product meets consumer demands and regulatory standards. In addition, He et al. [10] evaluated the impact of ^{60}Co irradiation on turmeric essential oil composition using gas chromatography–ion mobility spectrometry (GC–IMS). The findings demonstrated that, even though compound composition remained constant, the peak intensities were altered, supporting a 5 kGy/min irradiation dose for preserving essential oil quality. The studies have effectively demonstrated that the integration of spectroscopic techniques along with advanced data analysis is a promising choice that cements the advancement in food safety and quality analysis. These approaches offer rapid and accurate predictions of key quality attributes in various food products.

The ability of machine learning to extract valuable information from high-dimensional spectral data was utilized to enhance the effectiveness and efficiency of hyperspectral imaging in analyzing the safety and quality of food products. Xu et al. [11] explored the relationship between water distribution and quality indicators in shrimp during hot air drying using hyperspectral imaging. The study revealed a positive association between shrimp moisture content and bound water, immobilized water, and free water. Conversely, attributes including hardness, stickiness, and chewiness showed negative correlations with bound water and free water. Likewise, Cao et al. [12] developed a rapid approach for assessing the texture profile analysis of common carp fillets, leveraging hyperspectral imaging and machine learning algorithms. The proposed method accelerated the assessment process and maintained the integrity of the product, making it a valuable alternative to traditional texture analysis methods. Additionally, Wang et al. [13] and Xu et al. [14] both employed hyperspectral imaging to assess the quality of the safety of maize seeds. However, Wang et al. [13] developed a method to detect mold growth in maize kernels by applying categorical analysis and data fusion to hyperspectral data. In contrast, Xu et al [14] developed a method to identify defective maize seeds by employing deep learning, particularly convolutional neural networks to hyperspectral images. Further, Zhang et al. [15] combined internal and external leaf features obtained from both near-infrared hyperspectral imaging and THz time-domain spectroscopy to assess the different grades of tomato leaf mildew infestation. The fusion of these sources of information allows for a high degree of accuracy in detection, preventing misdiagnosis associated with traditional disease detection methods. Hyperspectral imaging has brought about a transformative shift in food quality and safety assessment, providing an in-depth analysis of food products. The wide range of applications of hyperspectral imaging from understanding water distribution and assessing texture to detecting mold growth and defects emphasizes its versatility and reliability.

IoT plays a significant role in food safety and quality assurance by providing real-time monitoring, data accuracy, complete traceability, and early warning systems throughout the food supply chain. Yin et al. [16] developed a spoilage monitoring and early warning system based on the volatile component production during apple spoilage. The combination of a sensor prototype and multi-factor fusion early warning model provided the real-time evaluation of food spoilage. Thus, the development of novel sensors that have the capability to collect data from the environment is an integral part of the IoT to ensure miniaturization and energy efficiency. In this regard, a chemiresistive ethylene sensor, employing rGO/WSe2/Pd heterojunctions, has been developed for room-temperature (RT) ethylene detection. This sensor offers a practical solution to monitor ethylene concentration, improving fruit and vegetable quality control during transportation and reducing losses [17]. Similarly, Zhang et al. [18] developed a bi-layer containing an anthocyanin-loaded liposome that has the capability to indicate the freshness of shrimp products through visual color changes by monitoring the pH of the surrounding medium. These studies have great significance as they focus on sensor development, which is a key aspect of smart and intelligent agriculture.

Artificial intelligence and machine learning can provide a conceptual tool to transform food safety and quality data management, facilitating the early detection and prevention of food safety issues. The synergy between deep learning and image processing was harnessed by Liang et al. [19] to develop a real-time grading system for defective apples using an RGB camera machine vision system and a combination of semantic segmentation and a pruned YOLO V4 network. This approach ensured a high detection accuracy (92.42%) without compromising computational efficiency. In addition, Zhou et al. [20] found that the light penetration depth in apple tissues was around 2.2 mm when spatial frequency domain imaging (SFDI) was used to detect early stage bruises in apple tissue. These works proved the effectiveness of deep learning architecture in detecting early stage defects in thin-skinned fruits. Moreover, Chen et al. [21] focused on creating a methodology for assessing the degree of milling (DOM) in rice with digital image processing technology and deep learning. The research introduced an enhanced model that combines multi-scale information through the integration of the Inception-v3 structure and the residual network (ResNet) model, using the Bayesian optimization algorithm which achieved superior results. The method achieved an average detection accuracy of 96.9%. Similarly, Yu et al. [22] developed a model using YOLOv5 to identify small impurities in walnut kernels that showed a detection accuracy of 88.9%. The model achieved a faster detection time for single images using an improved YOLOv5 model. By replacing conventional Conv with Ghostconv, the detection time was reduced from 65.25 ms to 45.38 ms, ensuring the real-time detection of walnut impurities while maintaining detection performance.

Ensuring the safety and quality of food products is of paramount importance in the food supply chain. From production to distribution, rigorous food safety and quality inspections have to be conducted at every stage involving the monitoring and control of food processing methods, preventing contamination and maintaining the highest standard. Ongoing research and development efforts focus on creating non-destructive technologies and cutting-edge equipment to attain this goal. These innovations understandably play a crucial role in ensuring the safety and quality of food products without compromising the integrity and promoting public safety and confidence in the food supply chain. The application of artificial intelligence, big data, and the IoT has led to a transformative era for the food industry. This implementation has ushered in an improvement in quality and an increase in efficiency across the entirety of the food production and distribution chain. Predictive analytics powered by artificial intelligence help to optimize production and minimize postharvest loss, while IoT-associated sensors provide real-time data on various environmental factors, ensuring food safety and quality. Big data analytics provides insight into consumer preferences and market trends, leading to more informed decision making. These digital transformations promote the transformation of and upgrade the food industry, making it more sustainable, innovative, and responsive to the evolving needs of consumers.

In summary, this Special Issue explores a wide range of innovative research at the intersection of technology development, artificial intelligence, and IoT. From sensor development and emerging techniques to machine learning and chemometric analysis, the studies included in this Special Issue showcase the incredible progress in safety and quality analysis in the food industry. Moving forward, these advances hold great importance in revolutionizing, early detection, quality assessment, and safety evaluation, ultimately benefiting both consumers and the food industry.

Author Contributions: Conceptualization, Z.G.; resources, Z.G.; writing—original draft preparation, H.J.; writing—review and editing, Z.G.; supervision, Z.G.; project administration, Z.G.; funding acquisition, Z.G. All authors have read and agreed to the published version of the manuscript.

Funding: This research was funded by the National Key R&D Program of China (No. 2023YFE0107100), the National Natural Science Foundation of China (No. 31972151), the Outstanding Young Teachers of Blue Project in Jiangsu province, Key R&D Project of Jiangsu Province (No. BE2022363), Jiangsu Agriculture Science and Technology Innovation Fund (No. CX(22)3069).

Data Availability Statement: Referred datasets link at https://www.mdpi.com/journal/foods/special_issues/Fast_Non_Destructive_Detection_Technology_Equipment_Food_Quality_Safety (accessed on 3 October 2023).

Acknowledgments: We acknowledge Zhao Zhang and Dong Hu for organizing, editing, and promoting this Special Issue.

Conflicts of Interest: The authors declare no conflict of interest

References

1. Yin, L.; Jayan, H.; Cai, J.; El-Seedi, H.R.; Guo, Z.; Zou, X. Development of a Sensitive SERS Method for Label-Free Detection of Hexavalent Chromium in Tea Using Carbimazole Redox Reaction. *Foods* **2023**, *12*, 2673. [CrossRef] [PubMed]
2. Qiu, M.; Tang, L.; Wang, J.; Xu, Q.; Zheng, S.; Weng, S. SERS with Flexible β-CD@AuNP/PTFE Substrates for In Situ Detection and Identification of PAH Residues on Fruit and Vegetable Surfaces Combined with Lightweight Network. *Foods* **2023**, *12*, 3096. [CrossRef]
3. Zhang, X.; Bian, F.; Wang, Y.; Hu, L.; Yang, N.; Mao, H. A Method for Capture and Detection of Crop Airborne Disease Spores Based on Microfluidic Chips and Micro Raman Spectroscopy. *Foods* **2022**, *11*, 3462. [CrossRef] [PubMed]
4. Sun, T.; Wang, Y.; Li, M.; Hu, D. Raman Spectroscopic Study of Five Typical Plasticizers Based on DFT and HF Theoretical Calculation. *Foods* **2023**, *12*, 2888. [CrossRef] [PubMed]
5. Sun, H.; Zhang, S.; Ren, R.; Xue, J.; Zhao, H. Detection of Soluble Solids Content in Different Cultivated Fresh Jujubes Based on Variable Optimization and Model Update. *Foods* **2022**, *11*, 2522. [CrossRef] [PubMed]
6. Jiang, X.; Zhu, M.; Yao, J.; Zhang, Y.; Liu, Y. Calibration of Near Infrared Spectroscopy of Apples with Different Fruit Sizes to Improve Soluble Solids Content Model Performance. *Foods* **2022**, *11*, 1923. [CrossRef] [PubMed]
7. Wang, T.; Zhang, Y.; Liu, Y.; Zhang, Z.; Yan, T. Intelligent Evaluation of Stone Cell Content of Korla Fragrant Pears by Vis/NIR Reflection Spectroscopy. *Foods* **2022**, *11*, 2391. [CrossRef]
8. Wang, Y.Q.; Liu, G.M.; Hu, L.P.; Zhao, X.Z.; Zhang, D.S.; He, H.J. Prediction of Anthocyanidins Content in Purple Chinese Cabbage Based on Visible/Near Infrared Spectroscopy. *Foods* **2023**, *12*, 1922. [CrossRef] [PubMed]
9. Migues, I.; Rivas, F.; Moyna, G.; Kelly, S.D.; Heinzen, H. Predicting Mandarin Fruit Acceptability: From High-Field to Benchtop NMR Spectroscopy. *Foods* **2022**, *11*, 2384. [CrossRef] [PubMed]
10. He, Y.; Yin, L.; Zhou, W.; Wan, H.; Lei, C.; Li, S.; Huang, D. Evaluation of ^{60}Co Irradiation on Volatile Components of Turmeric (Curcumae Longae Rhizoma) Volatile Oil with GC–IMS. *Foods* **2023**, *12*, 2489. [CrossRef]
11. Xu, W.; Zhang, F.; Wang, J.; Ma, Q.; Sun, J.; Tang, Y.; Wang, J.; Wang, W. Real-Time Monitoring of the Quality Changes in Shrimp (*Penaeus vannamei*) with Hyperspectral Imaging Technology during Hot Air Drying. *Foods* **2022**, *11*, 3179. [CrossRef]
12. Cao, Y.M.; Zhang, Y.; Yu, S.T.; Wang, K.K.; Chen, Y.J.; Xu, Z.M.; Ma, Z.Y.; Chen, H.L.; Wang, Q.; Zhao, R.; et al. Rapid and Non-Invasive Assessment of Texture Profile Analysis of Common Carp (*Cyprinus carpio* L.) Using Hyperspectral Imaging and Machine Learning. *Foods* **2023**, *12*, 3154. [CrossRef]
13. Wang, W.; Huang, W.; Yu, H.; Tian, X. Identification of Maize with Different Moldy Levels Based on Catalase Activity and Data Fusion of Hyperspectral Images. *Foods* **2022**, *11*, 1727. [CrossRef]
14. Xu, P.; Sun, W.; Xu, K.; Zhang, Y.; Tan, Q.; Qing, Y.; Yang, R. Identification of Defective Maize Seeds Using Hyperspectral Imaging Combined with Deep Learning. *Foods* **2023**, *12*, 144. [CrossRef] [PubMed]
15. Zhang, X.; Wang, Y.; Zhou, Z.; Zhang, Y.; Wang, X. Detection Method for Tomato Leaf Mildew Based on Hyperspectral Fusion Terahertz Technology. *Foods* **2023**, *12*, 535. [CrossRef] [PubMed]
16. Yin, L.; Jayan, H.; Cai, J.; El-Seedi, H.R.; Guo, Z.; Zou, X. Spoilage Monitoring and Early Warning for Apples in Storage Using Gas Sensors and Chemometrics. *Foods* **2023**, *12*, 2968. [CrossRef] [PubMed]
17. Li, X.; Xu, C.; Du, X.; Wang, Z.; Huang, W.; Sun, J.; Wang, Y.; Li, Z. Assembled Reduced Graphene Oxide/Tungsten Diselenide/Pd Heterojunction with Matching Energy Bands for Quick Banana Ripeness Detection. *Foods* **2022**, *11*, 1879. [CrossRef] [PubMed]
18. Zhang, J.; Yang, Y.; Zhang, J.; Shi, J.; Liu, L.; Huang, X.; Song, W.; Li, Z.; Zou, X.; Povey, M. High-Stability Bi-Layer Films Incorporated with Liposomes@Anthocyanin/Carrageenan/Agar for Shrimp Freshness Monitoring. *Foods* **2023**, *12*, 732. [CrossRef] [PubMed]
19. Liang, X.; Jia, X.; Huang, W.; He, X.; Li, L.; Fan, S.; Li, J.; Zhao, C.; Zhang, C. Real-Time Grading of Defect Apples Using Semantic Segmentation Combination with a Pruned YOLO V4 Network. *Foods* **2022**, *11*, 3150. [CrossRef] [PubMed]
20. Zhou, T.; Hu, D.; Qiu, D.; Yu, S.; Huang, Y.; Sun, Z.; Sun, X.; Zhou, G.; Sun, T.; Peng, H. Analysis of Light Penetration Depth in Apple Tissues by Depth-Resolved Spatial-Frequency Domain Imaging. *Foods* **2023**, *12*, 1783. [CrossRef]

21. Chen, W.; Li, W.; Wang, Y. Evaluation of Rice Degree of Milling Based on Bayesian Optimization and Multi-Scale Residual Model. *Foods* **2022**, *11*, 3720. [CrossRef] [PubMed]
22. Yu, L.; Qian, M.; Chen, Q.; Sun, F.; Pan, J. An Improved YOLOv5 Model: Application to Mixed Impurities Detection for Walnut Kernels. *Foods* **2023**, *12*, 624. [CrossRef] [PubMed]

foods

MDPI

Article

SERS with Flexible β-CD@AuNP/PTFE Substrates for In Situ Detection and Identification of PAH Residues on Fruit and Vegetable Surfaces Combined with Lightweight Network

Mengqing Qiu [1,2], Le Tang [3], Jinghong Wang [3], Qingshan Xu [1], Shouguo Zheng [1,4,*] and Shizhuang Weng [3,*]

1. Hefei Institutes of Physical Science, Chinese Academy of Sciences, Hefei 230031, China; qmq_study@126.com (M.Q.); qshxu@aiofm.ac.cn (Q.X.)
2. Science Island Branch of Graduate School, University of Science and Technology of China, Hefei 230026, China
3. National Engineering Research Center for Agro-Ecological Big Data Analysis & Application, Anhui University, Hefei 230601, China; tang_ahut@126.com (L.T.); hongwang321@163.com (J.W.)
4. Anhui Institute of Innovation for Industrial Technology, Hefei 230088, China

* Correspondence: zhengsg@hfcas.ac.cn (S.Z.); weng_1989@126.com (S.W.); Tel.: +86-18709836209 (S.Z.); +86-13695601875 (S.W.)

Abstract: The detection of polycyclic aromatic hydrocarbons (PAHs) on fruit and vegetable surfaces is important for protecting human health and ensuring food safety. In this study, a method for the in situ detection and identification of PAH residues on fruit and vegetable surfaces was developed using surface-enhanced Raman spectroscopy (SERS) based on a flexible substrate and lightweight deep learning network. The flexible SERS substrate was fabricated by assembling β-cyclodextrin-modified gold nanoparticles (β-CD@AuNPs) on polytetrafluoroethylene (PTFE) film coated with perfluorinated liquid (β-CD@AuNP/PTFE). The concentrations of benzo(a)pyrene (BaP), naphthalene (Nap), and pyrene (Pyr) residues on fruit and vegetable surfaces could be detected at 0.25, 0.5, and 0.25 μg/cm^2, respectively, and all the relative standard deviations (RSD) were less than 10%, indicating that the β-CD@AuNP/PTFE exhibited high sensitivity and stability. The lightweight network was then used to construct a classification model for identifying various PAH residues. ShuffleNet obtained the best results with accuracies of 100%, 96.61%, and 97.63% for the training, validation, and prediction datasets, respectively. The proposed method realised the in situ detection and identification of various PAH residues on fruit and vegetables with simplicity, celerity, and sensitivity, demonstrating great potential for the rapid, nondestructive analysis of surface contaminant residues in the food-safety field.

Keywords: surface-enhanced Raman spectroscopy; flexible substrate; polycyclic aromatic hydrocarbons; in situ detection; deep learning

Citation: Qiu, M.; Tang, L.; Wang, J.; Xu, Q.; Zheng, S.; Weng, S. SERS with Flexible β-CD@AuNP/PTFE Substrates for In Situ Detection and Identification of PAH Residues on Fruit and Vegetable Surfaces Combined with Lightweight Network. *Foods* **2023**, *12*, 3096. https://doi.org/10.3390/foods12163096

Academic Editor: Dapeng Peng

Received: 31 July 2023
Revised: 12 August 2023
Accepted: 15 August 2023
Published: 17 August 2023

1. Introduction

Polycyclic aromatic hydrocarbons (PAHs) are persistent organic pollutants that are non-degradable, highly toxic, mutagenic, and carcinogenic [1,2]. Human exposure to PAHs can occur via the inhalation of polluted air, food intake, and skin contact, of which food intake accounts for more than 90% of cases [3–6]. In particular, fruit and vegetable surfaces tend to attract large deposits of PAHs owing to their long-term exposure to the atmosphere [7,8]. Therefore, it is of great scientific and practical significance to detect PAH residues on fruit and vegetable surfaces because of their strong carcinogenicity and teratogenicity.

In recent years, spectroscopic methods—such as colorimetry, fluorescence spectroscopy, near-infrared spectroscopy, and surface-enhanced Raman spectroscopy (SERS)—have been widely used in PAH analysis because of their efficiency, sensitivity, and automation [9–12]. Although PAHs exhibit macromolecular fluorescence, conventional fluorescence spectra are easily limited by the broadening of emission bands and it can be difficult to distinguish similar molecules because of their low specificity. SERS is a vibrational spectroscopy technique that provides information on the structural characteristics of molecules, enhances Raman scattering, and has been widely used in fast-trace analysis [13–15]. The key to SERS applications is the fabrication of nanostructures with local surface plasmon resonance as SERS-active substrates, and the interaction between the substrates and targets [16]. However, the adsorption of most PAH molecules onto the surface of metal nanoparticles (NPs) is low and resonance Raman scattering does not occur, hindering the effective SERS detection of PAHs. To solve this problem, many researchers have attempted to prepare functionalised plasma nanostructures by modifying the surfactants, antigens, antibodies, and supramolecules on the NP surfaces to promote target binding to SERS substrates [17,18]. However, these strategies are subject to interference from functionalised molecules during SERS detection.

Our previous studies [19] showed that β-cyclodextrins (β-CDs) modified on the surface of gold NPs (AuNPs) can effectively trap PAHs to form host–guest compounds because of their hydrophobic inner cavities (which exhibit a cyclooligosaccharide structure); β-CDs also exhibit weak Raman scattering properties that can reduce interference. Moreover, the surfaces of most fruit and vegetables are irregular and uneven. To improve the application of SERS technology for the detection of irregular sample surfaces, many researchers have attempted to construct flexible SERS-active substrates by assembling metal NPs on flexible materials—such as polymethylmethacrylate (PMMA), tape, and poly(ethylene terephthalate) (PET), which can be easily wrapped or formed to collect analytes from irregular sample surfaces [20–22]. Although these methods could achieve in situ detection, their sensitivity and stability required further optimisation because of the viscosity of some of the flexible materials that destroyed nanoarray structures during the stripping process. The superhydrophobic film could effectively narrow the gap between NPs under the action of hydrophobicity to generate a large number of hot spots, which could enhance the SERS signal, and was an effective method for introducing molecules into hot spot regions [23,24]. However, most reported superhydrophobic SERS substrates required various nanofabrication techniques—such as electron beam lithography, optical lithography, and reactive ion etching—thus increasing the cost of actual SERS applications [25,26]. Inexpensive polytetrafluoroethylene (PTFE) films, with the advantage of having a low surface tension, could be combined with lubricants to prepare flexible and hydrophobic platforms that contributed to the generation of hot spots and eliminated the effects of coffee rings and viscosity [27]. Consequently, PTFE films exhibited strong practical application potential for the in situ, sensitive, and stable SERS detection of PAH residues on the irregular surfaces of fruits and vegetables.

To achieve rapid, intelligent, and automated analysis, SERS spectra can be combined with deep learning (DL) methods to build a determination model [28,29]. In particular, lightweight DL networks—such as SqueezeNet, Xception, MobileNet, and ShuffleNet developed based on the representative convolutional neural network (CNN)—have been widely used because of small parameters, low computational overhead, and high precision, showing a higher specificity and sensitivity compared to typical chemometric analysis [30–33]. For example, Weng et al. [34] used SqueezeNet to develop regression models for the analysis of chlormequat chloride and acephate; excellent performance was obtained with coefficients of determination (R^2) of 0.9836 and 0.9826 and root-mean-square errors (RMSEs) of 0.49 and 4.08, respectively. Wang et al. [35] proposed a novel regression model, a lightweight one-dimensional CNN, for predicting the nicotine content in tobacco leaves with R^2 and RMSE values of 0.95 and 0.14, respectively. These results demonstrated that lightweight networks were suitable for the rapid, accurate analysis of SERS spectra. Consequently, SqueezeNet, MobileNet, and ShuffleNet were used to build classification models for the analysis of various PAH residues on fruit and vegetable surfaces.

In summary, this study aims to develop a method for the in situ detection and identification of various PAH residues on fruit and vegetable surfaces using flexible β-CD@AuNP/PTFE substrates and lightweight DL networks (Figure 1). The β-CD@AuNP/PTFE was prepared by assembling β-CD@AuNPs on a flexible PTFE film coated with perfluorinated liquid and the PAHs were detected based on the flexible substrate. SqueezeNet, MobileNet, and ShuffleNet were used to construct an intelligent analysis model combined with SERS spectra to classify various PAH residues on the fruit and vegetable surfaces.

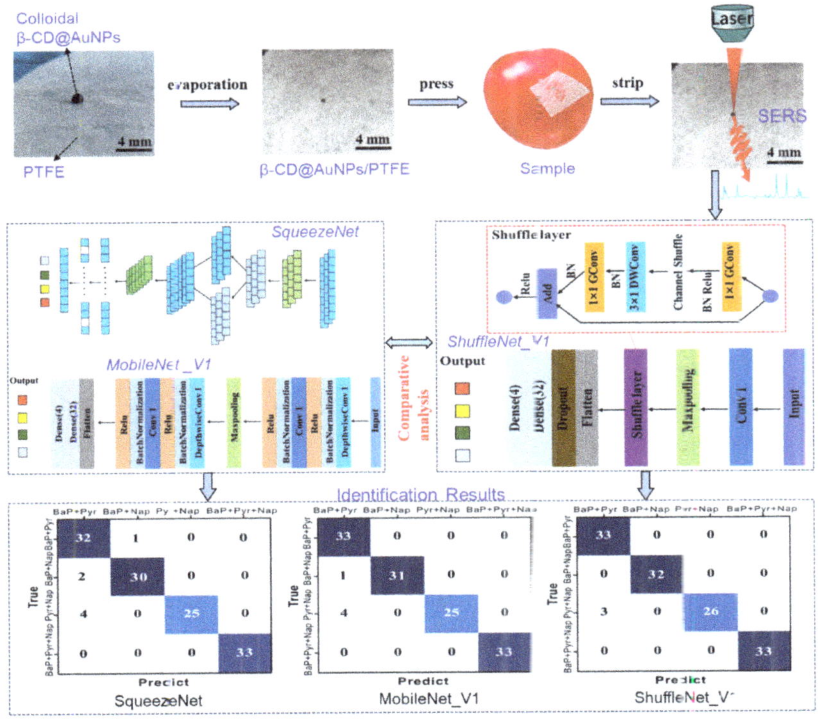

Figure 1. Schematic diagram of the flexible β-CD@AuNP/PTFE combined with lightweight networks to detect PAH residues on fruit and vegetable surfaces.

2. Materials and Methods

2.1. Materials

$HAuCl_4 \cdot 3H_2O$ (99%), β-CD (97%), BaP (96%), Pyr (98%), and Nap (99%) were purchased from Sigma-Aldrich. Na_2HPO_4, NaCl, acetone, methanol, and cyclohexane were obtained from Sinopharm Chemical Reagent Co., Ltd.(Shanghai, China), PTFE (pore size 0.1 μm, thickness 70 μm) film was bought from Whatman and polydimethylsiloxane (PDMS) was acquired from Anhui Zhongke Material Co., Ltd.(Hefei, China). Adhesive tape, apples, tomatoes, peaches, and cucumbers were purchased from local supermarkets Ultrapure water (18.25 MΩ) was used in all experiments.

2.2. Preparation of Flexible SERS Substrate

Synthesis of β-CD@AuNPs: β-CD@AuNP sol was prepared using β-CD as a reducing agent and stabiliser, according to Zhao et al. [36]. In brief, 5 mL of 0.1 M phosphate buffer (PB), 1 mL of 0.01 M chlorauric acid solvent, 10 mL of 0.01 M β-CD solution were successively added to 35 mL ultrapure water and stirred vigorously until fully mixed. The mixture was heated to a set boiling temperature and maintained at this temperature for 60 min. The scanning electron microscopy (SEM) images (Figure S1 in the Supporting Information) of the prepared β-CD@AuNPs showed that the particle size was uniform at approximately 20–25 nm.

Preparation of β-CD@AuNP/PTFE: The detailed steps for assembling β-CD@AuNPs on a hydrophobic smooth PTFE film were as follows. First, the PTFE film was glued to a 5 × 2 cm glass slide with double-sided adhesive. The slide was then adsorbed on the machine and 0.45 mL of a perfluorinated liquid was dispersed by spin coating at a low speed of 600 rpm for 30 s and a high speed of 1500 rpm for 1 min. The coated film was heated for 30 min to obtain a spare film. Finally, the concentrated 10 uL β-CD@AuNP colloidal solution was poured onto a hydrophobic PTFE film. During the drying process, the contact line shrank because of the low surface friction of the PTFE film. Eventually, the initial droplet was concentrated in the cell domain with a diameter of 0.5–1 mm (Figure S2 in the Supporting Information).

Preparation of β-CD@AuNP/tape: A two-dimensional NP array of β-CD@AuNPs was obtained using a simple liquid–liquid interface self-assembly method [19] and was subsequently transferred onto silicon wafers. A piece of adhesive tape (clipped using scissors) was then used to cover the nanoarray of β-CD@AuNPs, being pressed firmly for 3–5 s. The tape was gently peeled off from the surface of the silicon wafer and β-CD@AuNP/tape was formed by transferring the β-CD@AuNP array onto the tape.

Preparation of β-CD@AuNP/PDMS: Similarly, flexible β-CD@AuNP/PDMS was prepared by transferring monolayer NP arrays of β-CD@AuNPs using PDMS films instead of silicon wafers and air-drying at 25–30 °C.

2.3. Preparation of SERS Sample

A solution of 100 μg/mL was obtained by dissolving 10 μg BaP, Pyr, and Nap solid powders in a 0.1 L ethanol solution. Standard solutions of BaP, Pyr, and Nap at different concentrations (10, 8, 5, 2.5, 1, 0.5, 0.1, and 0.05 μg/mL) were prepared by diluting the 100 μg/mL solution with ethanol. Twenty samples were prepared at each concentration. A standard solution was then used to evaluate the effect of flexible β-CD@AuNP/PTFE on SERS detection. To simulate the actual environment, the spiked samples were prepared by spraying 10 μL of BaP, Pyr, and Nap standard solutions with different concentrations on a fixed area (1 × 1 cm^2) of the fruit and vegetable surfaces, after which they were air-dried at 25–30 °C. With 20 samples at each concentration, 5 spectra were collected for each.

The PAH samples comprised four classes—that is, BaP + Pyr, BaP + Nap, Pyr + Nap, and BaP + Pyr + Nap—with 20 samples in each class, covering a concentration range of 10 μg/mL to 0.05 μg/mL. Similarly, the 10 μL sample solution was sprayed onto a fixed area of the fruit and vegetable surfaces before being air-dried, followed by 10 μL ethanol being sprayed onto the fixed area to dissolve and extract various PAHs. Finally, the prepared SERS substrate was pasted onto the fruit and vegetable surfaces, gently pressed and lifted, the process being repeated two to three times to realise peel surface sampling. After sampling, the substrate was placed on a slide for SERS detection. Five spectra were collected for each sample.

2.4. Spectral Measurement

The morphology and structure of the substrates were characterised using scanning electron microscope (SEM, Zeiss, LSM 710, Oberkochen, Germany) and transmission electron microscopy (TEM, JEOL, JEM-2100F, Tokyo, Japan). The SERS signal of the sample was measured using a portable Raman spectrometer (BWTEK, i-Raman785 Plus, Newark, DE, USA) with a 785 nm He-Ne laser and an excitation light source of 150 mW. The integration time was 10 s, the laser power was 10%, and the spectral range was 300–1800 cm^{-1}.

2.5. Spectral Analysis Methods

SqueezeNet, MobileNet, and ShuffleNet were used to construct a classification model for the rapid, intelligent identification of various PAHs. SqueezeNet is a lightweight network based on a model-compression strategy [37]. The structure of the SqueezeNet used in this study is shown in Figure S3 in the Supporting Information. The first convolution layer and pooling layer are first used for the initial feature extraction; then, a 1 × 1 convolution layer (squeeze layer) is added, followed by a 1 × 1 convolution and a 3 × 1 convolution extended width (expand layer). The features of the two convolution layers are connected and sent to the flatten, dropout, and dense layers. Notably, the pooling operation of SqueezeNet is delayed, which ensures that a larger feature map is convolved, retaining more feature information, thereby effectively improving network performance. The parameter settings for SqueezeNet are listed in Table S1.

The core idea of MobileNet is to use depthwise separable convolution (DSC) instead of general convolution [38]. The MobileNet structure designed in this study is illustrated in Figure S4 in the Supporting Information. The DSC is implemented using DepthwiseConv and the common 1 × 1 convolution module, both of which are followed by batch normalisation and a rectified linear unit (ReLU) for batch normalisation and nonlinearisation. MobileNet comprises two DSC modules with an additional maximum pooling layer for dimensionality reduction, followed by a flatten layer and two dense layers. The parameter settings for MobileNet are listed in Table S1.

The ShuffleNet network includes group convolution and channel shuffling [39], the structure of which is shown in Figure S5 in the Supporting Information. Initial feature extraction is performed using a common convolution layer and a maximum pooling layer, followed by group convolution and channel shuffle using two shuffle layers. In the shuffle layer, the input is first convolved using a 1 × 1 group convolution, after which the channel shuffle module is used to shuffle the feature graphs of each group. The input is then convolved using a 3 × 1 DepthwiseConv and 1 × 1 group. Finally, the obtained input is added to the initial input to realise group convolution with channel shuffle. After extracting features through the two shuffle layers, the entire network can be completed through the flatten, dropout, and two dense layers. The parameter settings for ShuffleNet are listed in Table S1.

2.6. Model Evaluation

There were 400 spectra for the four classes of mixed samples; 30% of the spectra were randomly selected as the prediction dataset, the remaining 70% being divided into training and validation datasets in a 3:1 ratio, which were then used to adjust the network hyperparameters. The accuracy of the training, validation, and prediction datasets (ACC_T, ACC_V, and ACC_P), as well as the *Precision*, *Recall*, and F1-*score* of the prediction datasets, were used to evaluate the model performance. *Precision* is the percentage of true positives in all predicted positives; *Recall* is the percentage of predicted true positives in all positives; F1-*score* is the weighted harmonic average of *Precision* and *Recall*. The *ACC*, *Precision*, *Recall*, and F1-*score* may be conveniently calculated using the following expressions:

$$ACC = \frac{TP + TN}{TN + FP + FN + TP} \quad (1)$$

$$Precision = \frac{TP}{TP + FP} \quad (2)$$

$$Recall = \frac{TP}{TP + FN} \quad (3)$$

$$F1 - scores = \frac{Precision \times Recall}{Precision + Recall} \quad (4)$$

where *TP* (resp. *TN*) stands for true positive (resp. negative) and *FP* (resp. *FN*) for false positive (resp. negative).

3. Results and Discussion

3.1. Influence of Different Flexible Substrates on SERS Activity

Owing to the interference of several inherent flexible-film characteristics—such as transparency, fluorescent background, viscosity, and impurities—the morphology of the nanoarray structure transferred to its surface was affected, greatly affecting the SERS activity. To select an ideal flexible substrate and realise the in situ detection of trace PAH residues on irregular surfaces in a real-world environment, the influence of three flexible substrates on SERS activity was explored (Figure 2). Figure 2a, c, and e shows the SERS spectra of β-CD@AuNP/PDMS, β-CD@AuNP/tape, and β-CD@AuNP/PTFE, respectively, while the SERS spectra of 10 μg/mL BaP detected by them are shown in Figure 2b, d, and f. As is evident from the figure, β-CD@AuNP/PTFE obtained the best SERS activity (Figure 2f), the characteristic peak of BaP at 524, 607, 1231, and 1376 cm^{-1} being greatly improved. The vibration modes corresponding to each characteristic peak are listed in Table S2 in the Supporting Information. The SERS activity of the β-CD@AuNP/tape is the weakest (Figure 2d), owing to the viscosity and surface roughness of the tape, which easily destroys the two-dimensional nanoarray structure of the β-CD@AuNPs. The PDMS surface is not affected by the viscosity, but the PDMS hydrophobicity is lower than that of the PTFE, so the SERS activity of β-CD@AuNP/PDMS (Figure 2b) is also weaker. The surface hydrophobicity of the PTFE film coated with the perfluorinated liquid limits the diffusion of NPs, which is conducive to reducing the gap between the particles and generating a large number of hot spots. This result can be demonstrated using the SEM image of β-CD@AuNP/PTFE in Figure S6 of the Supporting Information, the gap between the aggregated β-CD@AuNPs being less than 10 nm, which generates abundant hot spots, resulting in a large SERS enhancement with Enhancement Factor of 10^6~10^7 (Figure S7 in Supporting Information). Therefore, β-CD@AuNP/PTFE was used as the flexible SERS substrate in subsequent experiments.

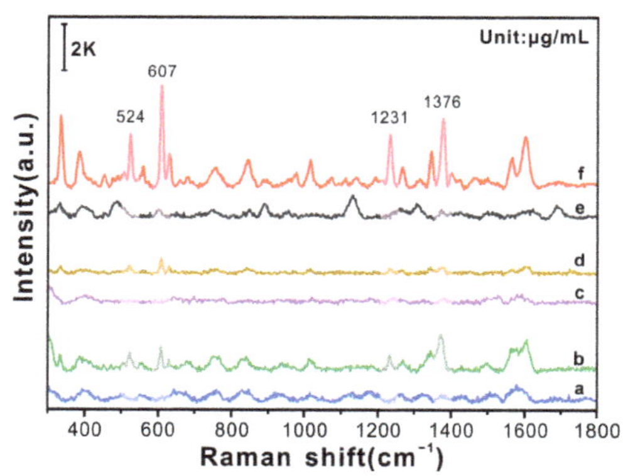

Figure 2. SERS spectra of 10 μg/mL BaP detected based on different flexible substrates: (a) β-CD@AuNP/PDMS, (b) 10 μg/mL BaP detected by β-CD@AuNP/PDMS; (c) β-CD@AuNP/tape, (d) 10 μg/mL BaP detected by β-CD@AuNP/tape; (e) β-CD@AuNP/PTFE, (f) 10 μg/mL BaP detected by β-CD@AuNP/PTFE.

3.2. SERS Detection of PAHs Based on Flexible β-CD@AuNP/PTFE

Nap, Pyr, and BaP were selected as targets because they are the most representative and widely distributed of the two-, four-, and five-aromatic PAHs, respectively; their structures are shown in Figure 3. SERS detection of BaP, Nap, and Pyr was conducted based on flexible β-CD@AuNP/PTFE, and its sensitivity, reproducibility, and stability were explored. The SERS spectra of BaP, Nap, and Pyr at different concentrations are shown in Figure 3A–C, respectively. The intensities of the characteristic peaks (highlighted in purple) decrease with decreasing concentration. The vibration modes corresponding to the characteristic peaks are listed in Table S2. When the concentration is as low as 0.05, 0.5, and 0.1 μg/mL, BaP at 607 cm^{-1}, Nap at 1372 cm^{-1}, and Pyr at 587 cm^{-1} still have weak Raman signals, indicating that the flexible β-CD@AuNP/PTFE substrate has strong sensitivity and good universality.

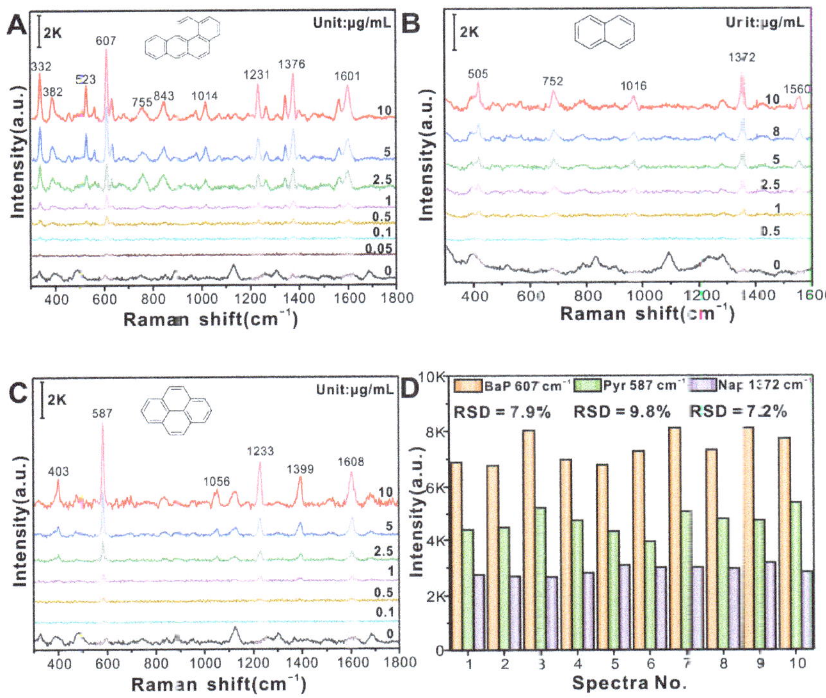

Figure 3. Detection of PAHs by β-CD@AuNP/PTFE; SERS spectra of (**A**) BaP, (**B**) Nap, and (**C**) Pyr standard samples, insets showing their molecular structures; (**D**) RSD values of characteristic peak intensities of 10 μg/mL BaP, Pyr, and Nap at 607, 587, and 1372 cm^{-1}, respectively.

The stability of β-CD@AuNP/PTFE was further evaluated by calculating the relative standard deviation (RSD) of the characteristic peak intensities of the 10 SERS spectra for 10 μg/mL PAHs, as shown in Figure 3D. The RSDs of BaP at 607 cm^{-1}, Nap at 1372 cm^{-1}, and Pyr at 587 cm^{-1} are 7.9%, 9.8%, and 7.2%, respectively. All RSD values are less than 10%, indicating that the SERS detection of PAHs based on the flexible β-CD@AuNP/PTFE constructed in this study exhibits good stability.

3.3. *In Situ Detection of PAHs on Fruit and Vegetable Surfaces*

The SERS spectra of the spiked samples prepared by spraying BaP, Pyr, and Nap onto the surfaces of apples, tomatoes, and peaches are shown in Figure 4. When BaP concentrations on the surfaces of apples (Figure 4(A1)), tomatoes (Figure 4(B1)) and peaches (Figure 4(C1)) are as low as 0.1, 0.25, and 0.25 μg/cm^2, respectively, there is still a weak SERS signal at 607 cm^{-1}, especially at 0.25 μg/cm^2, which can be attributed to the C-C and C-H bending vibration modes. The characteristic peaks at 587 cm^{-1} and 1233 cm^{-1} can still be observed at Pyr concentrations as low as 0.25, 0.5, and 0.25 μg/cm^2 on the surfaces of apples (Figure 4(A2)), tomatoes (Figure 4(B2)), and peaches (Figure 4(C2)), corresponding to C-C stretching and C-H bending vibration modes. The difference in spectral signal intensity between the different peels may be related to the roughness and chemical composition of the peels. When the concentration of Nap on the surface of the apple (Figure 4(A3)), tomato (Figure 4(B3)), and peach (Figure 4(C3)) is as low as 0.5, 1, and 0.5 μg/cm^2, respectively, the characteristic peak at 1372 cm^{-1} has a weak signal, while the signal at 505 cm^{-1} is almost invisible, which is related to the stretching and bending vibrations of the C-C bond.

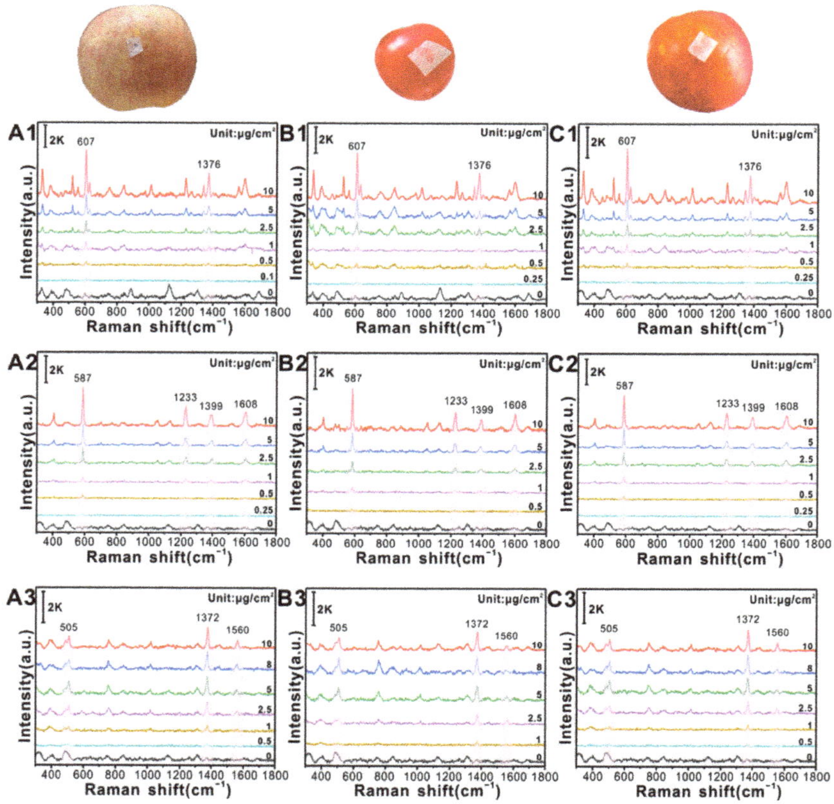

Figure 4. Detection of different concentrations of PAHs on fruit and vegetable surfaces by using a flexible β-CD@AuNP/PTFE substrate. (**A1–A3**), (**B1–B3**), and (**C1–C3**) are the SERS spectra of BaP, Pyr, and Nap with different concentrations on the surface of apple, tomato, and peach, respectively.

Although sensitive SERS detection of Nap on the fruit and vegetable surfaces is realised, the sensitivity is still weaker than that of BaP and Pyr because the C-H bond bending vibration mode of BaP and Pyr has higher polarisability and exhibits stronger peak intensity compared with the C-C bending vibration mode of Nap. The above results indicate that using β-CD@AuNP/PTFE as a flexible SERS substrate can realise the sensitive and in situ detection of PAH residues on the surface of fruits and vegetables, which is of great significance for food safety assessment.

3.4. Identification of Various PAHs on Fruit and Vegetable Surfaces

Because there is more than one PAH residue on the surface of fruits and vegetables in the real environment, the main purpose of this work is to realise the identification of various PAHs, without considering a single class. And the competitive adsorption of SERS can also lead to different contributions of PAHs to the SERS signal at the same concentration and can even cover the signals of other PAHs, resulting in low efficiency and accuracy in identifying various PAHs through the manual analysis of spectra. Consequently, in this study, a lightweight network combined with the SERS spectra of various PAHs was used to construct a classification model for intelligent, accurate identification.

The SERS spectra of various PAHs were obtained using the flexible β-CD@AuNP/PTFE substrate, as shown in Figure 5. Figure 5A shows the SERS spectra of Pyr, Nap, and BaP. It is evident that the characteristic peak of Nap at 505 cm^{-1} is different from those of

Pyr and BaP, and the characteristic peaks of Pyr at 587 and 1399 cm^{-1} are unique without overlapping. Similarly, the characteristic peak of BaP at 607 cm^{-1} is unique. These three targets have unique characteristic peaks that provide a basis for the identification of subsequent detection.

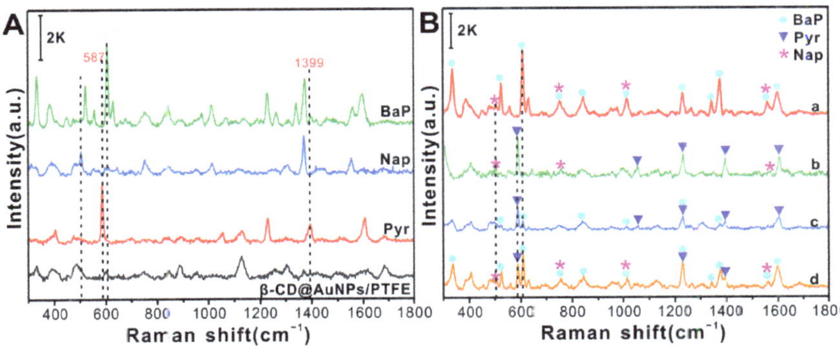

Figure 5. SERS spectra of BaP, Nap, and Pyr at a concentration of 10 μg/mL (**A**); SERS spectra of various PAHs (**B**), from top to bottom: (a) BaP + Nap, (b) Pyr + Nap, (c) BaP + Pyr, (d) BaP + Nap + Pyr.

Figure 5B shows the SERS spectra obtained after mixing Pyr, Nap, and BaP. From spectra d of BaP + Pyr + Nap, it is evident that the number of characteristic peaks is more than those of BaP + Nap, Pyr + Nap, and BaP + Pyr (Figure 5B a, b, c). Therefore, the identification performance for BaP + Pyr + Nap may be higher than that for the other types of spectra in subsequent identifications. Additionally, the spectra of a, b, and c in Figure 5B all contain the characteristic peaks of the two PAHs; moreover, there are many overlapping characteristic peaks that make quick and intuitive manual identification difficult to achieve. Consequently, the combination of the SERS spectra of PAHs with lightweight networks is an effective and robust method for constructing recognition models.

The identification results of the SERS spectra of various PAH residues on fruit and vegetable surfaces using the model constructed with three lightweight networks (SqueezeNet, MobileNet, and ShuffleNet) are shown in Table 1. The results obtained from SqueezeNet, with ACC_T, ACC_V, and ACC_P values of 99.57%, 93.22%, and 94.48%, respectively, were unsatisfactory. Based on the *Precision*, *Recall*, and *F1-score* of the prediction dataset, SqueezeNet is the best at identifying the mixed spectra of BaP + Pyr + Nap, primarily because of the distinct features of the mixed spectra. But the identification of the BaP + Pyr and Pyr + Nap spectra by this network is poor, indicating that the extracted features are not sufficiently rich, which is consistent with the confusion matrix predicted by the SqueezeNet model (Figure 6A). MobileNet performs better than SqueezeNet, with ACC_T, ACC_V, and ACC_P values of 100%, 94.92%, and 96.06%. The *Precision*, *Recall*, and *F1-score* of BaP + Pyr are 100%, 93.94%, and 96.88%, respectively, which are considerably higher than those of SqueezeNet, indicating that MobileNet can effectively capture BaP + Pyr features. Detailed prediction results were obtained from the confusion matrix of the MobileNet model (Figure 6B). Unfortunately, the ability of the network to recognise BaP + Pyr + Nap is low. ShuffleNet achieves the best identification results, with ACC_T = 100%, ACC_V = 96.61%, and ACC_P = 97.63%. These conclusions are also evident from the confusion matrix of the ShuffleNet model shown in Figure 6C, with the identification results of BaP + Pyr + Nap and BaP + Nap by the ShuffleNet model all being correct, and only three Pyr + Nap samples being misclassified as BaP + Pyr. The reason for this result can be found in spectra b and c in Figure 5B, the main characteristic peaks of the spectra of Pyr + Nap and BaP + Pyr being provided by Pyr, and the other weak characteristic peaks from BaP and Nap showing little difference. In general, the results indicate that a lightweight network combined with SERS provides a fast, accurate, and intelligent method for identifying various PAH residues on fruit and vegetable surfaces.

Table 1. Identification results of SERS combined with a lightweight network for various PAH residues on fruit and vegetable surfaces.

Methods	Classes	Accuracy (%)	Prediction Dataset		
			Precision (%)	*Recall* (%)	*F1-Score* (%)
Squeezenet	BaP + Pyr	$ACC_T = 99.57$	96.97	84.21	90.14
	BaP + Nap	$ACC_V = 93.22$	93.75	96.77	95.24
	Pyr + Nap	$ACC_P = 94.48$	86.21	100	92.59
	BaP + Pyr +Nap		100	100	100
Mobilenet_V1	BaP + Pyr	$ACC_T = 100$	100	93.94	96.88
	BaP + Nap	$ACC_V = 94.92$	96.88	100	98.42
	Pyr + Nap	$ACC_P = 96.06$	86.21	100	92.59
	BaP+Pyr+Nap		100	86.84	92.96
Shufflenet_V1	BaP + Pyr	$ACC_T = 100$	100	91.67	95.65
	BaP + Nap	$ACC_V = 96.61$	100	100	100
	Pyr + Nap	$ACC_P = 97.63$	89.66	100	94.55
	BaP + Pyr + Nap		100	100	100

Abbreviations: *ACC*, accuracy of correct classification; ACC_T, *ACC* of the training dataset; ACC_V, *ACC* of the validation dataset; ACC_P, *ACC* of the prediction dataset.

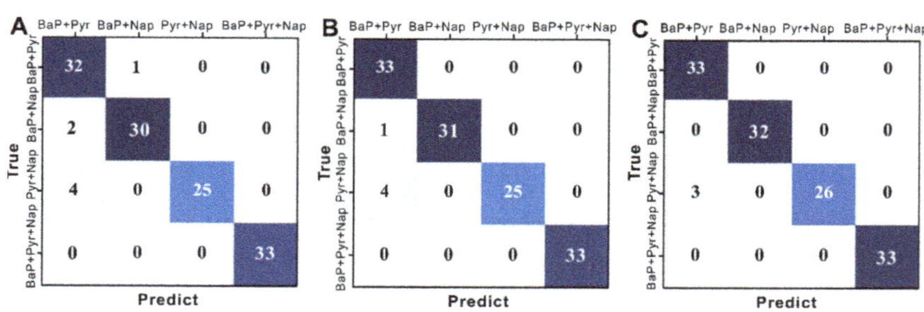

Figure 6. Confusion matrix of (**A**) SqueezeNet, (**B**) MobileNet_V1, and (**C**) ShuffleNet_V1.

3.5. Discussion

In recent years, many studies on in situ detection of targets by using flexible SERS substrates have been widely reported. Alyami et al. [20] fabricated novel AgNP/PDMS composites by self-assembly of organic AgNP solutions on flexible PDMS surfaces; CV and thiram concentrations as low as 1×10^{-7} M and 1×10^{-5} M were measured on contaminated fish skin and orange peel, respectively. Chen et al. [40] detected three-pesticide residues on tomato peel based on the SERS and flexible tape. Although these methods achieved in situ detection, the sensitivity was low due to the coffee ring effect caused by the weak hydrophobicity of the PDMS and tape surface. Moreover, with the adhesive tape it was easy to destroy the structure of the nanoarray during the "paste and peel off", resulting in low stability and reproducibility. In this study, we designed the flexible SERS substrate by assembling β-CD@AuNPs on PTFE film coated with perfluorinated liquid, effectively reducing the coffee ring effect and generating a large number of hot spots. The sensitivity and stability of SERS in situ detection were competitive with the strongest results reported by the above work, and the detection process is faster and more convenient, within 1 min.

In addition, DL methods, such as CNNs, recurrent neural networks (RNNs), and generative adversarial networks (GANs), with their strong self-learning ability and excellent fitting ability, were gradually used in spectral analysis to obtain fast and intelligent quantitative or qualitative analysis [28,41]. In particular, CNNs are widely used in the modelling of spectral data by virtue of their advantages with less preprocessing and easy expansion of network architecture. Erzina et al. [42] proposed the advanced route for express and precise recognition of normal and cancer cells by using SERS combined with a CNN, with 100% prediction accuracy. Yu et al. [43] obtained the accurate identification of six representative Vibrio species by combining label-free SERS technology with a CNN, achieving a high accuracy rate of 99 7%. However, these higher accuracy rates were obtained on the basis of building deeper and more complex networks, resulting in an increase in the number of parameters and memory footprint. In this study, the lightweight network developed based on a CNN was used for the first time to construct identification models of various PAHs and the accuracy rate was as high as 97.6%, indicating that this method could improve the computing speed and reduce the memory consumption while ensuring the model accuracy.

4. Conclusions

In this study, a flexible SERS substrate of β-CD@AuNP/PTFE combined with a lightweight network was designed to achieve the in situ detection and identification of various PAH residues on fruit and vegetable surfaces. AuNPs were modified with β-CD to enhance adsorption of PAHs. The flexible β-CD@AuNP/PTFE substrate was prepared by assembling β-CD@AuNPs on a PTFE film coated with a perfluorinated solution, contributing to the generation of a large number of hot spots and realising convenient in situ detection. The concentrations of BaP, Pyr, and Nap residues on fruit and vegetable surfaces can still be detected at 0.25, 0.5, and 0.25 μg/cm^2, and all the RSD values were less than 10%. Subsequently, SqueezeNet, MobileNet, and ShuffleNet networks were used to establish recognition models for various PAH residues on fruit and vegetable surfaces. ShuffleNet obtained the best recognition results with ACC_T = 100%, ACC_V = 96.61%, and ACC_P = 97.63%. These results demonstrated that the proposed method could achieve simple, sensitive, stable, and intelligent in situ detection and identification of various PAH residues on fruit and vegetable surfaces. This method offers great potential for the practical application of rapid, non-destructive analysis of surface contaminant residues in the food industry. However, owing to the large variety and low content of residual PAHs, it is necessary to optimise the SERS substrate to achieve highly sensitive PAH detection in complex matrices. Meanwhile, the SERS spectra of additional types of PAHs should be collected to further improve the recognition performance of the model.

Supplementary Materials: The following supporting information can be downloaded at: https://www.mdpi.com/article/10.3390/foods12163096/s1. Figure S1: SEM image (A) and Size (B) of the prepared β-CD@AuNPs; Figure S2: The preparation process of β-CD@AuNPs/PTFE; Figure S3: Structure of SqueezeNet; Figure S4: Structure of MobileNet; Figure S5: Structure of ShuffleNet; Figure S6: SEM image of (a) PTFE and (b) β-CD@AuNPs/PTFE; Figure S7: The field distribution of four AuNPs with A particle size of 24 nm with different gaps of (A) 8 nm, (B) 6 nm, (C) 5 nm, (D) 4 nm and (E) 3 nm; (F) Gap distribution of β-CD@AuNPs/PTFE; Table S1: Parameter setting of different models; Table S2: Raman shifts of BaP, Nap and Pyr and the corresponding vibration modes.

Author Contributions: M.Q.: Methodology, Validation, Investigation, Writing—original draft, Writing—review and editing. L.T.: Software, Methodology, Data curation. J.W.: Visualisation, Investigation. Q.X.: Investigation, Resources. S.Z.: Formal analysis, Supervision, Funding acquisition. S.W.: Conceptualisation, Methodology, Supervision, Project administration, Funding acquisition. All authors have read and agreed to the published version of the manuscript.

Funding: This study was supported by National Key R&D Program of China (No. 2022YFD2401305), Key Research and Development Program of Anhui Province (Nos. 202104a06020025), and National Natural Science Foundation of China (Nos. 31971789, 32001421).

Data Availability Statement: All related data and methods are presented in this paper. Additional inquiries should be addressed to the corresponding author.

Conflicts of Interest: The authors declare that they have no known competing financial interest or personal relationships that could have appeared to influence the work reported in this paper.

References

1. Lawal, A.T. Polycyclic aromatic hydrocarbons. A review. *Cogent Environ. Sci.* **2017**, *3*, 1339841. [CrossRef]
2. Han, J.; Liang, Y.; Zhao, B.; Wang, Y.; Xing, F.; Qin, L. Polycyclic aromatic hydrocarbon (PAHs) geographical distribution in China and their source, risk assessment analysis. *Environ. Pollut.* **2019**, *251*, 312–327. [CrossRef]
3. Mojiri, A.; Zhou, J.L.; Ohashi, A.; Ozaki, N.; Kindaichi, T. Comprehensive review of polycyclic aromatic hydrocarbons in water sources, their effects and treatments. *Sci. Total Environ.* **2019**, *696*, 133971. [CrossRef]
4. Sampaio, G.R.; Guizellini, G.M.; da Silva, S.A.; de Almeida, A.P.; Pinaffi-Langley, A.C.C.; Rogero, M.M.; de Camargo, A.C.; Torres, E.A. Polycyclic aromatic hydrocarbons in foods: Biological effects, legislation, occurrence, analytical methods, and strategies to reduce their formation. *Int. J. Mol. Sci.* **2021**, *22*, 6010. [CrossRef] [PubMed]
5. Li, T.; Wang, Y.; Hou, J.; Zheng, D.; Wang, G.; Hu, C.; Xu, T.; Cheng, J.; Yin, W.; Mao, X.; et al. Associations between inhaled doses of PM2.5-bound polycyclic aromatic hydrocarbons and fractional exhaled nitric oxide. *Chemosphere* **2019**, *218*, 992–1001. [CrossRef]
6. Li, G.; Wu, S.; Wang, L.; Akoh, C.C. Concentration, dietary exposure and health risk estimation of polycyclic aromatic hydrocarbons (PAHs) in youtiao, a Chinese traditional fried food. *Food Control* **2016**, *59*, 328–336. [CrossRef]
7. Ashayeri, N.Y.; Keshavarzi, B.; Moore, F.; Kersten, M.; Yazdi, M.; Lahijanzadeh, A.R. Presence of polycyclic aromatic hydrocarbons in sediments and surface water from Shadegan wetland–Iran: A focus on source apportionment, human and ecological risk assessment and sediment-water exchange. *Ecotox. Environ. Saf.* **2018**, *148*, 1054–1066. [CrossRef]
8. Sun, N.; Yu, S.; Cai, Z.; Liu, J.; Wang, T.; Qi, B.; Wang, Z.; Wang, S.; Yang, A.; Zhu, G.; et al. Inhibition of polycyclic aromatic hydrocarbon (PAHs) release from sediments in an integrated rice and crab coculture system by rice straw biochar. *J. Clean. Prod.* **2022**, *367*, 133058. [CrossRef]
9. Ma, C.M.; Lin, L.C.; Chuang, K.J.; Hong, G.B. Colorimetric detection of polycyclic aromatic hydrocarbons by using gold nanoparticles. *Spectrochim. Acta A* **2022**, *268*, 120701. [CrossRef]
10. Pena, E.A.; Ridley, L.M.; Murphy, W.R.; Sowa, J.R.; Bentivegna, C.S. Detection of polycyclic aromatic hydrocarbons (PAHs) in raw menhaden fish oil using fluorescence spectroscopy: Method development. *Environ. Toxicol. Chem.* **2015**, *34*, 1946–1958. [CrossRef]
11. Dong, G.; Li, X.; Yang, R.; Yang, Y.; Liu, H.; Wu, N. Correction method of effect of soil moisture on the fluorescence intensity of polycyclic aromatic hydrocarbons based on near-infrared diffuse reflection spectroscopy. *Environ. Pollut.* **2021**, *269*, 116150. [CrossRef]
12. Eremina, O.E.; Sergeeva, E.A.; Ferree, M.V.; Shekhovtsova, T.N.; Goodilin, E.A.; Veselova, I.A. Dual-Purpose SERS Sensor for Selective Determination of Polycyclic Aromatic Compounds via Electron Donor-Acceptor Traps. *ACS Sens.* **2021**, *6*, 1057–1066. [CrossRef]
13. Pilot, R.; Signorini, R.; Durante, C.; Orian, L.; Bhamidipati, M.; Fabris, L. A review on surface-enhanced Raman scattering. *Biosensors* **2019**, *9*, 57. [CrossRef] [PubMed]
14. Schlücker, S. Surface-Enhanced raman spectroscopy: Concepts and chemical applications. *Angew. Chem. Int. Edit.* **2014**, *53*, 4756–4795. [CrossRef] [PubMed]
15. Lopez-Lorente, A.I. Recent developments on gold nanostructures for surface enhanced Raman spectroscopy: Particle shape, substrates and analytical applications. A review. *Anal. Chim. Acta* **2021**, *1168*, 338474. [CrossRef] [PubMed]
16. Zhang, D.; Pu, H.; Huang, L.; Sun, D.W. Advances in flexible surface-enhanced Raman scattering (SERS) substrates for nondestructive food detection: Fundamentals and recent applications. *Trends Food Sci. Tec.* **2021**, *109*, 690–701. [CrossRef]
17. Gao, Y.; Li, L.; Zhang, X.; Wang, X.; Ji, W.; Zhao, J.; Ozaki, Y. CTAB-triggered Ag aggregates for reproducible SERS analysis of urinary polycyclic aromatic hydrocarbon metabolites. *Chem. Commun.* **2019**, *55*, 2146–2149. [CrossRef] [PubMed]
18. Poonia, M.; Kuster, T.; Bothun, G.D. Organic Anion Detection with Functionalized SERS Substrates via Coupled Electrokinetic Preconcentration, Analyte Capture, and Charge Transfer. *ACS Appl. Mater. Interfaces* **2022**, *14*, 23964–23972. [CrossRef] [PubMed]
19. Qiu, M.; Zheng, S.; Li, P.; Tang, L.; Xu, Q.; Weng, S. Detection of 1-OHPyr in human urine using SERS with injection under wet liquid–liquid self-assembled films of β-CD-coated gold nanoparticles and deep learning. *Spectrochim. Acta A* **2022**, *290*, 122238. [CrossRef]
20. Alyami, A.; Quinn, A.J.; Iacopino, D. Flexible and transparent Surface Enhanced Raman Scattering (SERS)-Active Ag NPs/PDMS composites for in-situ detection of food contaminants. *Talanta* **2019**, *201*, 58–64. [CrossRef]
21. Chen, J.; Huang, Y.; Kannan, P.; Zhang, L.; Lin, Z.; Zhang, J.; Chen, T.; Guo, L. Flexible and adhesive surface enhance Raman scattering active tape for rapid detection of pesticide residues in fruits and vegetables. *Anal. Chem.* **2016**, *88*, 2149–2155. [CrossRef] [PubMed]
22. Nowicka, A.B.; Czaplicka, M.; Kowalska, A.A.; Szymborski, T.; Kamińska, A. Flexible PET/ITO/Ag SERS Platform for Label-Free Detection of Pesticides. *Biosensors* **2019**, *9*, 111. [CrossRef]
23. Xiao, L.; Zhang, M.; Liu, Z.; Bian, W.; Zhang, X.; Zhan, J. Hydrophobic silver nanowire membrane for swabbing extraction and in situ SERS detection of polycyclic aromatic hydrocarbons on toys. *Anal. Methods* **2017**, *9*, 1816–1824. [CrossRef]
24. Zhang, D.; You, H.; Yuan, L.; Hao, R.; Li, T.; Fang, J. Hydrophobic slippery surface-based surface-enhanced Raman spectroscopy platform for ultrasensitive detection in food safety applications. *Anal. Chem.* **2019**, *91*, 4687–4695. [CrossRef] [PubMed]
25. Kasani, S.; Curtin, K.; Wu, N. A review of 2D and 3D plasmonic nanostructure array patterns: Fabrication, light management and sensing applications. *Nanophotonics* **2019**, *8*, 2065–2089. [CrossRef]

26. Huang, C.Y.; Tsai, M.S. Fabrication of 3D nano-hemispherical cavity array plasmonic substrate for SERS applications. *Int. J. Optomechatron.* **2018**, *12*, 40–52. [CrossRef]
27. Gu, Y.; Li, Q.; Yin, M.; Yang, D.; Yang, Y. A super-hydrophobic perfluoropolyether coated polytetrafluoroethylene sheets substrate for detection of acetamiprid surface-enhanced Raman spectroscopy. *Spectrochim. Acta A* **2022**, *278*, 121373. [CrossRef]
28. Ciloglu, F.U.; Caliskan, A.; Saridag, A.M.; Kilic, I.H.; Tokmakci, M.; Kahraman, M.; Aydin, O. Drug-resistant Staphylococcus aureus bacteria detection by combining surface-enhanced Raman spectroscopy (SERS) and deep learning techniques. *Sci. Rep.* **2021**, *11*, 18444. [CrossRef]
29. Weng, S.; Zhu, W.; Li, P.; Yuan, H.; Zhang, X.; Zheng, L.; Zhao, J.; Huang, L.; Han, P. Dynamic surface-enhanced Raman spectroscopy for the detection of acephate residue in rice by using gold nanorods modified with cysteamine and multivariant methods. *Food Chem.* **2020**, *310*, 125855. [CrossRef]
30. Gomes, R.; Rozario, P.; Adhikari, N. Deep Learning Optimization in Remote Sensing Image Segmentation using Dilated Convolutions and ShuffleNet. In Proceedings of the 2021 IEEE International Conference on Electro Information Technology (EIT), Mt. Pleasant, MI, USA, 14–15 May 2021; pp. 244–249.
31. Zhou, Y.; Chen, S.; Wang, Y.; Huan, W. Review of Research on Lightweight Convolutional Neural Networks. In Proceedings of the 2020 IEEE 5th Information Technology and Mechatronics Engineering Conference, Chongqing, China, 12–14 June 2020; pp. 1713–1720. [CrossRef]
32. Asghar, M.Z.; Albogamy, F.R.; Al-Rakhami, M.S.; Asghar, J.; Rahmat, M.K.; Alam, M.M.; Lajis, A.; Nasir, H.M. Facial mask detection using depthwise separable convolutional neural network model during COVID-19 pandemic. *Front. Public Health* **2022**, *10*, 855254. [CrossRef]
33. Lussier, F.; Thibault, V.; Charron, B.; Wallace, G.Q.; Masson, J.F. Deep learning and artificial intelligence methods for Raman and surface-enhanced Raman scattering. *TrAC* **2020**, *124*, 115796. [CrossRef]
34. Weng, S.; Tang, L.; Qiu, M.; Wang, J.; Wu, Y.; Zhu, R.; Wang, C.; Li, P.; Sha, W.; Liang, D. Surface-enhanced Raman spectroscopy charged probes under inverted superhydrophobic platform for detection of agricultural chemicals residues in rice combined with lightweight deep learning network. *Anal. Chim. Acta* **2023**, *1262*, 341264. [CrossRef] [PubMed]
35. Wang, D.; Zhao, F.; Wang, R.; Guo, J.; Zhang, C.; Liu, H.; Wang, Y.; Zong, G.; Feng, W. A Lightweight convolutional neural network for nicotine prediction in tobacco by near-infrared spectroscopy. *Front. Plant Sci.* **2023**, *14*, 1138693. [CrossRef] [PubMed]
36. Zhao, Y.; Huang, Y.; Zhu, H.; Zhu, Q.; Xia, Y. Three-in-one: Sensing, self-assembly, and cascade catalysis of cyclodextrin modified gold nanoparticles. *J. Am. Chem. Soc* **2016**, *138*, 16645–16654. [CrossRef]
37. Karthik, T.S.; Krishna, R.V.V.; Ramakrishna Rao, T.K.; Manoranjithem, V.; Kalaiarasi, S.; Jegajothi, B. Evolutionary Optimization Algorithm on Content Based Image Retrieval System using Handcrafted Features with Squeeze Networks. In Proceedings of the 2022 Second International Conference on Artificial Intelligence and Smart Energy, Coimbatore, India, 23–25 February 2022; pp. 1425–1431. [CrossRef]
38. Wang, W.; Li, Y.; Zou, T.; Wang, X.; You, J.; Luo, Y. A novel image classification approach via dense-MobileNet models. *Mob. Inf. Syst.* **2020**, *2020*, 7602384. [CrossRef]
39. Hu, Y.; Jiang, Z.; Zhu, K. An Optimized CNN Model for Engagement Recognition in an E-Learning Environment. *Appl. Sci.* **2022**, *12*, 8007. [CrossRef]
40. Tao, M.; Fang, H.; Feng, X.; He, Y.; Liu, X.; Shi, Y.; Wei, Y.; Hong, Z. Rapid Trace Detection of Pesticide Residues on Tomato by Surface-Enhanced Raman Spectroscopy and Flexible Tapes. *J. Food Qual.* **2022**, *2022*, 6947775. [CrossRef]
41. Beeram, R.; Vendamani, V.S.; Soma, V.R. Deep learning approach to overcome signal fluctuations in SERS for efficient On-Site trace explosives detection. *Spectrochim. Acta A* **2023**, *289*, 122218. [CrossRef]
42. Erzina, M.; Trelin, A.; Guselnikova, O.; Dvorankova, B.; Strnadova, K.; Perminova, A.; Ulbrich, P.; Mares, D.; Jerabek, V.; Elashnikov, R.; et al. Precise cancer detection via the combination of functionalized SERS surfaces and convolutional neural network with independent inputs. *Sensor. Actuat. B-Chem.* **2020**, *308*, 127660. [CrossRef]
43. Yu, H.; Yang, Z.; Fu, S.; Zhang, Y.; Panneerselvamc, R.; Li, B.; Zhang, L.; Chen, Z.; Wang, X.; Li, J. Intelligent convolution neural network-assisted SERS to realize highly accurate identification of six pathogenic Vibrio. *Chem. Commun.* **2023**, *59*, 5779–5782. [CrossRef]

foods

MDPI

Article

Rapid and Non-Invasive Assessment of Texture Profile Analysis of Common Carp (*Cyprinus carpio* L.) Using Hyperspectral Imaging and Machine Learning

Yi-Ming Cao [1], Yan Zhang [1], Shuang-Ting Yu [1,2], Kai-Kuo Wang [3], Ying-Jie Chen [3], Zi-Ming Xu [3], Zi-Yao Ma [3], Hong-Lu Chen [1], Qi Wang [1], Ran Zhao [1], Xiao-Qing Sun [1] and Jiong-Tang Li [1,*]

1 Key Laboratory of Aquatic Genomics, Ministry of Agriculture and Rural Affairs, Beijing Key Laboratory of Fishery Biotechnology, Chinese Academy of Fishery Sciences, Beijing 100041, China; caoyiming@cafs.ac.cn (Y.-M.C.); zhangy@cafs.ac.cn (Y.Z.); styuwork@163.com (S.-T.Y.); chenhongluz@cafs.ac.cn (H.-L.C.); wangqi@cafs.ac.cn (Q.W.); zhaoran@cafs.ac.cn (R.Z.); sunxiaoqing@cafs.ac.cn (X.-Q.S.)

2 Chinese Academy of Agricultural Sciences, Beijing 100181, China

3 National Demonstration Center for Experimental Fisheries Science Education, Shanghai Ocean University, Shanghai 201306, China; 18631836881@163.com (K.-K.W.); cyjttkl@163.com (Y.-J.C.); xuziming0916@163.com (Z.-M.X.); zql3703700@163.com (Z.-Y.M.)

* Correspondence: lijt@cafs.ac.cn; Tel.: +86-010-68691136

Abstract: Hyperspectral imaging (HSI) has been applied to assess the texture profile analysis (TPA) of processed meat. However, whether the texture profiles of live fish muscle could be assessed using HSI has not been determined. In this study, we evaluated the texture profile of four muscle regions of live common carp by scanning the corresponding skin regions using HSI. We collected skin hyperspectral information from four regions of 387 scaled and live common carp. Eight texture indicators of the muscle corresponding to each skin region were measured. With the skin HSI of live common carp, six machine learning (ML) models were used to predict the muscle texture indicators. Backpropagation artificial neural network (BP-ANN), partial least-square regression (PLSR), and least-square support vector machine (LS-SVM) were identified as the optimal models for predicting the texture parameters of the dorsal (coefficients of determination for prediction (r_P) ranged from 0.9191 to 0.9847, and the root-mean-square error for prediction ranged from 0.1070 to 0.3165), pectoral (r_P ranged from 0.9033 to 0.9574, and RMSEP ranged from 0.2285 to 0.3930), abdominal (r_P ranged from 0.9070 to 0.9776, and RMSEP ranged from 0.1649 to 0.3601), and gluteal (r_P ranged from 0.8726 to 0.9768, and RMSEP ranged from 0.1804 to 0.3938) regions. The optimal ML models and skin HSI data were employed to generate visual prediction maps of TPA values in common carp muscles. These results demonstrated that skin HSI and the optimal models can be used to rapidly and accurately determine the texture qualities of different muscle regions in common carp.

Keywords: common carp; hyperspectral imaging; texture; machine learning; visualization

Citation: Cao, Y.-M.; Zhang, Y.; Yu, S.-T.; Wang, K.-K.; Chen, Y.-J.; Xu, Z.-M.; Ma, Z.-Y.; Chen, H.-L.; Wang, Q.; Zhao, R.; et al. Rapid and Non-Invasive Assessment of Texture Profile Analysis of Common Carp (*Cyprinus carpio* L.) Using Hyperspectral Imaging and Machine Learning. *Foods* **2023**, *12*, 3154. https://doi.org/10.3390/foods12173154

Academic Editors: Zhiming Guo, Zhao Zhang and Dong Hu

Received: 24 July 2023
Revised: 11 August 2023
Accepted: 14 August 2023
Published: 22 August 2023

1. Introduction

The textural traits of fish, including gumminess, springiness, cohesiveness, resilience, hardness, brittleness, adhesiveness, and chewiness, are the most important traits in the aquaculture industry, and they affect the production process and the commercial value of fish [1–3]. Developing the fillet textual assessment method is beneficial for measuring the textual traits of fish-processed products [4]. Traditional fish textual assessment methods include measurements using a texture analyzer [5,6]. However, these methods are laborious and might destroy the integrity of the products. Therefore, there is an immediate requirement to construct an efficient and non-destructive method to detect the muscle texture of processed fish.

Recently, hyperspectral imaging (HSI) has become an alternative analytical approach that provides the benefits of rapid and non-destructive detection [4,7–11]. HSI combines image and spectral techniques to obtain both "spatial" and "spectral" information containing the sample [9–12]. Another feature of HSI is the ability to generate visual distribution maps of measured indicators to allow for the prediction and quantification of the composition of the sample and to determine their position on the sample surface [6,13]. Moreover, artificial intelligence and machine learning (ML) models can be used for prediction and modeling in the food industry [14]. With spectral images, HSI has been widely applied to evaluate the traits of meat products, including color, surface defects, damage, texture, water-holding capacity, flavor, freshness, and ripeness [4,15–21]. Ma et al. used HSI based on 400–1000 nm wavelengths to predict the different textural parameters of grass carp fillets during vacuum freeze-drying [4]. They predicted the Warner–Bratzler shear force, hardness, gumminess, and chewiness of fillets with prediction coefficients ranging from 0.79 to 0.87. ElMasry et al. predicted beef tenderness using hyperspectral imaging with a model based on partial least squares (PLS), showing a detection coefficient of 0.83 and a cross-validation narrative of 0.75 [17]. Zhou et al. predicted six texture parameters of silver carp muscle using HSI and ML methods, with coefficients ranging from 0.83 to 0.95 [8]. In addition, He et al. found that the SPA-LS-SVM prediction model and HSI had a prediction coefficient of 0.905 for the tenderness of salmon fillets [22]. These studies demonstrate that HSI and ML methods provide reliable solutions to measure processed fish textures.

In fish breeding, high-quality textures can provide fillets that are suitable for downstream processing and satisfy the consumer's taste. The traditional textual method requires the cut of fish muscle and is lethiferous [5,6]. Compared with the traditional textual method, HSI and ML methods have a non-destructive advantage, as they allow for the detection of the texture of live fish muscle. In the current literature, the majority of researchers have investigated the quality of fillets rather than intact fish using HSI, meaning that the spectra were usually obtained from the meat mass [16,23,24]. However, the application of HSI and ML methods to measure the live fish muscle has been less studied.

Common carp (*Cyprinus carpio*), an allotetraploid fish [25], is one of the most important freshwater-farmed fish in the world. Therefore, the aim of this study was to develop a non-invasive method in which skin HSI and ML are combined to detect the textual parameters of live fish muscle. We first acquired the skin HSI data of 387 scaled and live common carp with a hyperspectral imaging system at 400–1000 nm. Then, we measured the texture profiles of four corresponding muscle regions of each fish. The specific objectives of this study were to (1) utilize preprocessing methods to achieve spectral preprocessing and characteristic wavelength selection; (2) determine the optimal wavelengths that are most useful for the prediction of texture profile analysis (TPA) within the muscle of common carp; (3) determine the optimal relationship between the skin HSI data and muscle texture parameters using six ML methods and incorporate the skin hyperspectral index; and (4) apply the optimal model for the visualization of the distribution of muscle texture parameters.

2. Materials and Methods

2.1. Ethics Statement and Sampling

We performed this study following the recommendations of the Animal Care and Use Committee of the Chinese Academy of Fishery Sciences. In 2022, 387 live common carp (one-year-old, ~607.09 g average weight) were collected from the Chinese Academy of Fishery Sciences experimental fish farm (Fangshan, Beijing, China). These samples were previously cultivated at one pond and fed the same commercial diet (Tongwei, China). The sources of protein and lipids in commercial diet are fish meal, soybean meal, cottonseed meal, rapeseed meal, and soybean oil. The proximate compositions of the commercial diet (% dry matter) consist of 30% protein, 5% lipid, 12.5% moisture, and 15% ash. For each common carp, we selected four regions, including the dorsal, pectoral, abdominal, and gluteal regions.

2.2. Acquiring the Skin Hyperspectral Images of Four Regions of Live Common Carp

For each region of one live common carp, the scales on one side were removed. We scanned the skin of four regions of one scaled common carp with a hyperspectral imager (FigSpec Hyperspectral Camera FS-13, FigSpec Technology (Zhejiang) Co., Ltd., Hang Zhou, China). The detected wavelengths ranged from 400 to 1000 nm with the following parameters: a resolution of 2.5 nm, an exposure period of 150 ms, and the lens type of C-Mount. The imaging speed was 128 Hz in the full-wavelength range, and the scanning speed was 30 row/s. Before acquiring the hyperspectral images of each sample, the distance and intensity of the illumination source were adjusted to ensure the clearness of the acquired images.

2.3. Processing the Skin HSI

HSI enables the collection of signals from samples, as well as the environment, instruments, and other non-sample factors. To eliminate the signals from the non-sample factors, all HSI data were input into the reflectance calibration procedure. Briefly, the raw data were calibrated with black and white correction. The white balanced image (W) was obtained by collecting the reflectance value from the Teflon white surface, while the dark image (D) was acquired by turning off the illumination source and collecting the hyperspectral data when the lens was completely covered with its cap. The calibration image (I) was calculated using the following equation:

$$I = \frac{I_0 - D}{W - D} \times 100$$

where I represents the corrected reflectance hyperspectral image in a unit of relative reflectance (%); I_0 represents the raw hyperspectral image; D stands for the dark image (0% reflectance); and W is the white reference image (100% reflectance) [26].

Then, we used the Savitzky–Golay (SG) smoothing method to preprocess the images and eliminate the putative effects from the sampling environment and instruments [27]. We selected a region of interest (ROI) to represent each skin region with the ROI function of Environment in the Visualizing Images software (ENVI v5.3, Exelis Visual Information Solutions, Inc., Boulder, CO, USA) [5]. The size of an ROI was 200 pixels × 200 pixels. For each wavelength, the average spectrum of an ROI was calculated by averaging the spectra of all pixels. The reflectance values of all pixels were averaged at each wavelength variable to obtain an average value representative of each sample.

2.4. Selecting the Optimal Wavelength

One HSI dataset contained the spectral information of samples from 400 nm to 1000 nm, simultaneously. However, certain wavelengths had redundant data, resulting in the time-consuming processing of HSI data [28]. Therefore, it is necessary to eliminate wavelengths containing redundant and irrelevant information to optimize the texture profiles for data analysis samples using the wavelength/variant selection of hyperspectral data [23]. Regarding wavelength selection for HSI analysis, the regression coefficient (RC) is commonly utilized [29,30]. We utilized the RC to determine the optimal wavelength that contributed the most to the prediction of TPA values in common carp muscle. In the calculation of RC, the optimal wavelength is chosen by computing the β-coefficient from the full-wavelength PLSR model. The wavelength with the highest absolute value of the β-coefficient is considered to be the optimal wavelength [30]. The program for RC was operated in MATLAB 2021a software (The MathWorks Inc., Natick, MA, USA).

2.5. Measuring the Texture Indicators of Common Carp Muscle

We extracted the muscle corresponding to each of the four skin regions. The muscle size was 20 mm × 20 mm × 15 mm. Eight texture indicators of the muscle, including gumminess, springiness, cohesiveness, resilience, hardness, brittleness, adhesiveness, and chewiness, were measured using a texture analyzer (TA.XTC-18, Baosheng, Shanghai, China) and a TA/36 cylindrical probe. The measurement speed was 2 mm/s, and the trigger force was 5 N. The compressive deformation of one sample was set to 60%. These eight texture indicators were derived from the TPA curves of each sample, and the TPA parameters listed above were calculated using Bourne's technique [31].

For each texture indicator, to examine whether there were significant differences among four muscle regions, we measured the distances with PCA analysis using Tassel 5.0 [32]. The Spearman correlation coefficient of the contents of any two indicators in four samples was calculated using the R 'cor. test' function in the R software (version 4.0.2).

2.6. Estimating the Muscle Texture Indicators with the Skin HSI Data

With the processed skin HSI data in each region, we used six machine learning (ML) methods to estimate the texture indicators of the muscle in the corresponding region. The methods included partial least-square regression (PLSR), the interval partial least-square method (iPLS), the synergy interval partial least-square method (SiPLS), backward interval partial least squares (BiPLS), least-square support vector machines (LS-SVM), and backpropagation artificial neural network (BP-ANN).

PLSR projects the predictor variables and observable variables into a new feature space to build a linear regression model [33]. PLSR decomposes the independent variable X and the dependent variable Y into several X-scores (T) and constructs the PLSR model. Herein, the observed variables were the cross-validation performed to minimize the error between the predicted and the observed response values.

In the iPLS algorithm, the full spectral region is divided into smaller equidistant subintervals, and a PLS regression model is generated based on each subinterval. The best intervals and principal component scores are selected based on the principle of the lowest root-mean-square error for the calibration (RMSEC) value [34].

The SiPLS algorithm is a modified iPLS where the full spectral region is divided equally into subintervals. The combination with the lowest RMSEC value is selected [34].

The BiPLS algorithm divides the whole spectral region into N subintervals of equal width and performs PLS regression, each interval is omitted in turn, and the worst RMSEC value is obtained in the modeling; the subintervals continue to be removed until the lowest RMSEC value is obtained [34].

LS-SVM uses the radial basis kernel function (RBF), a non-linear function that reduces the complexity of the training process [35]. The regularization parameter gamma (γ) and the kernel parameter (σ^2), which can reduce the complexity, represent the width of the RBF kernel. To achieve high prediction accuracy, we performed the simulations of these two parameters, the values of which ranged from 0 to 1000 [6].

In BP-ANN models, an error-reversal propagation algorithm is used to train multilayer feedforward neural networks [36]. A BP-ANN, with an input layer, a hidden layer, and an output layer was established. Moreover, the transfer function, learning function, and training function were employed. The maximum training step was set to 1000, the learning goal was e^{-5}, and the learning rate and momentum factor were 0.01.

2.7. Evaluating the Accuracies of Six ML Models

The predictive accuracy of each ML model was assessed with multiple parameters, including coefficients of determination for calibration (r_C) and prediction (r_P), RMSEC, and the root-mean-square error for calibration and prediction (RMSEP) [37]. The r_C and r_P values were calculated as follows:

$$r_C = \sqrt{\frac{\sum_{i=1}^{n_c}(\hat{y}_i - y_i)^2}{\sum_{i=1}^{n_c}(\hat{y}_i - y_c)^2}}$$

$$r_P = \sqrt{\frac{\sum_{i=1}^{n_p}(\hat{y}_i - y_i)^2}{\sum_{i=1}^{n_p}\left(\hat{y}_i - y_p\right)^2}}$$

where $\hat{y}_i$ and y_i represent the predicted and measured TPA values, respectively; n_c and n_p represent the number of samples in the calibration and prediction sets, respectively.

The RMSEC and RMSEP were calculated as follows:

$$\textbf{RMSEC} = \sqrt{\frac{1}{N-1-R} \times \sum_{i=1}^{N} \left(y_i^{ref} - y_i\right)^2}$$

where N is the number of samples, R is the number of factors of the model, y_i^{ref} is the reference value of the sample, and i and y_i are the predicted values of the sample

$$\textbf{RMSEP} = \sqrt{\frac{1}{N} \times \sum_{i=1}^{N} \left(y_i^{ref} - y_i\right)^2}$$

Herein, for each texture indicator, y_i^{ref} was the observed value in the common carp muscle, while y_i was the predicted value with one ML method and the reflectance values of corresponding skin HSI. The lower RMSEC and RMSEP values indicated a smaller difference between the predicted texture indicator and the observed indicator. A good ML model was expected to have high r_C and r_P but low RMSEC and RMSEP values [38]

2.8. Visualizing the Images of TPA Values

TPA values were distributed varyingly in different muscle regions of the fillets [39], which resulted from the irregular distribution of lipids and protein in different muscle regions. To examine the differences among the TPA values in different muscle regions of one fish, distribution maps of TPA values were constructed to improve insight into the muscle texture of common carp. The optimal calibration model constructed by applying the spectra of the optimal wavelengths following RC selection was employed to generate new distribution maps of TPA values. Linear color scales are presented in the figure by visualizing the distribution maps, and the different colors in the color scales represent the predicted TPA parameter values in the fillets, thus facilitating the identification and capture of the variations in muscle TPA values by observing different color distributions. All the calculation and visualization procedures were implemented in programs operating in ENVI 5.3 (Exelis Visual Information Solutions, Inc., Boulder, CO, USA) and MATLAB 2021a software (The MathWorks Inc., Natick, MA, USA) [5].

3. Results

3.1. Spectral Features of the Skin of Scaled Common Carp

The spectral features of the four skin regions of common carp were distinct (Figure 1). For all regions, the spectrum at 430 nm had the lowest reflectance values. The distributions of the reflectance values in the gluteal, pectoral, and ventral skins were different from that of the dorsal skin. In general, for the former three skin regions, the reflectance values gradually increased at 430–600 nm. The values reached the plateau phase at 600–780 nm and fell at 780–970 nm. Finally, the values increased after 970 nm However, the reflectance values in the dorsal skin gradually increased from 430 nm to 1000 nm. The reflectance value of the gluteal skin at each wavelength was higher than those of the other skins. The reflectance values of the pectoral and ventral skin ranked second and third. The values of the dorsal skin were the lowest. The distinct distributions and the levels of the reflectance values among the four skins might indicate the different features of the four skin regions or the affiliated tissues.

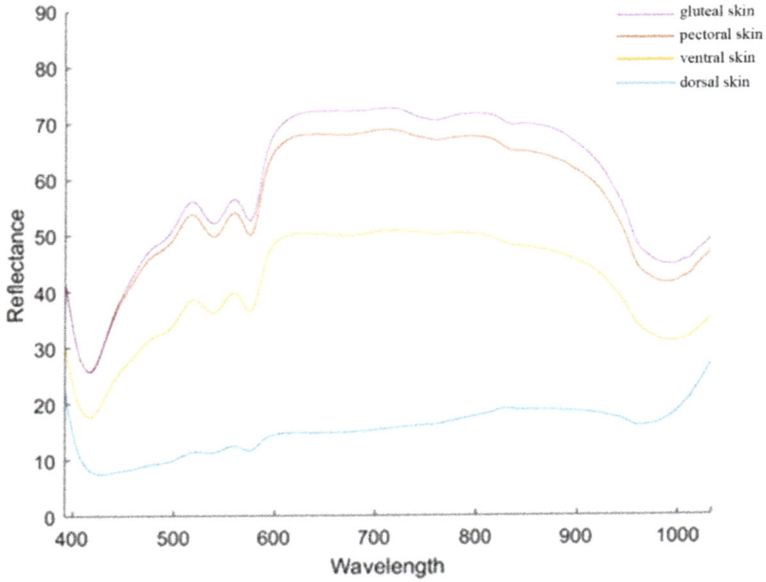

Figure 1. The spectral features of four skin regions of the common carp. The X-axis represents the different wavelengths, the Y-axis represents the reflectance values, and the curves of different colors represent the reflectance of four skin regions.

3.2. Texture Diversities of Common Carp Muscles

We obtained eight textural parameters of four muscle regions of 387 common carp (Supplementary Table S1). In the dorsal region, the first two principal components (PCs) explained 73.62% and 12.73% of all variances, respectively (Supplementary Figure S1). Intriguingly, the examined samples were grouped into two different clusters, suggesting different textural profiles among samples. Similar phenomena of two clusters were observed in the PCA analysis using the indicators of the pectoral, abdominal, and gluteal regions, respectively (Supplementary Figures S2–S4). The hardness of the gluteal region (median = 2010.53) was significantly higher than that of the other three regions (Figure 2 and Supplementary Table S2). The pectoral hardness was the lowest. Intriguingly, except for the cohesiveness (Supplementary Figure S5) and adhesiveness indicators (Supplementary Figure S6), the other five indicators in the gluteal region were also significantly higher than those of the other three regions (Supplementary Figures S7–S11). However, the adhesiveness parameter in this region was the lowest. These data revealed the different texture features of the four muscle regions.

The numbers of texture indicator pairs with a significant correlation were 17, 14, 9, and 7 in the abdominal, pectoral, dorsal, and gluteal regions, respectively (Supplementary Tables S3–S6). Only resilience had a significantly positive correlation with the springiness (coefficients ranging between 0.728 and 0.990) and cohesiveness (coefficients ranging between 0.148 and 0.473) in all four regions, respectively. The cohesiveness was also significantly positively correlated with chewiness (coefficients ranging between 0.12 and 0.951) in all four regions. Six texture indicator pairs had significant correlations among the three regions, including four positively correlated pairs and two negatively correlated pairs. These data suggested that the most significant correlation was not consistent in the four regions.

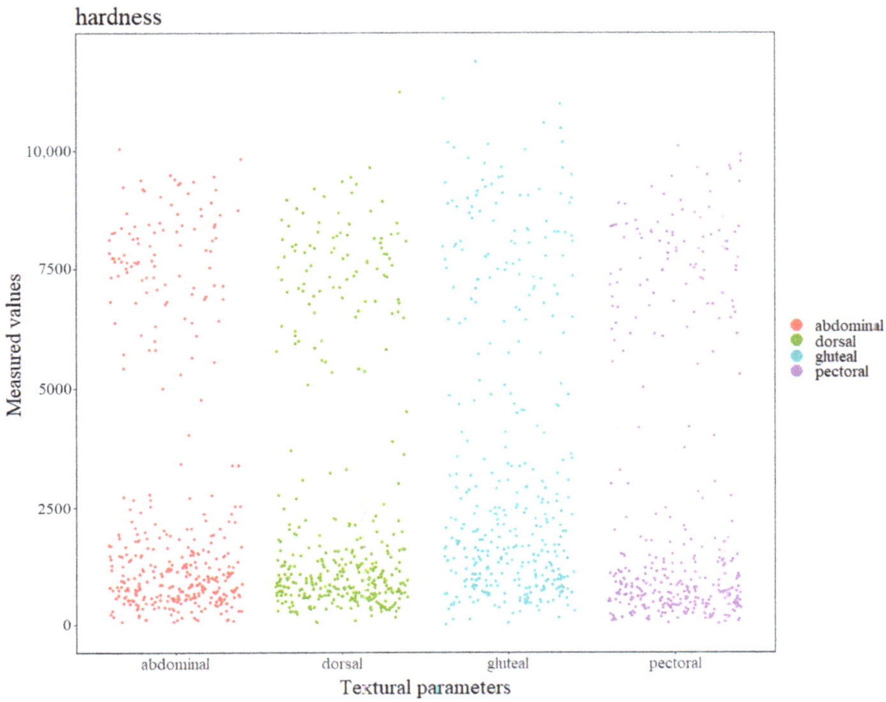

Figure 2. The hardness distributions in the four muscle regions of common carp.

3.3. Accurate Prediction of Muscle Texture Profiles Based on the Full Spectral Range

Since the texture profiles of the four muscle regions were different, we tried to determine whether it is possible to predict the muscle texture profiles. We used all reflectance values of the skin HSI data to predict the corresponding muscle texture indicators with different ML methods. For one texture indicator of one region, we only retained the prediction with the highest r_P for the downstream analysis (Supplementary Table S7). Predicting the chewiness of the four muscle regions had the highest r_P (from 0.9555 to 0.9836). The overall prediction accuracies of the gumminess (r_P from 0.9234 to 0.9863) and cohesiveness (r_P from 0.8952 to 0.9224) of the four muscle regions ranked second and third, respectively. The overall prediction accuracies of the hardness and adhesiveness indicators were also higher than 0.88 for all four regions. Among all the best predictions, the prediction accuracy of dorsal springiness was the lowest, with only 0.5612.

The BP-ANN, LS-SVM, and PLSR models were the best three methods to predict the muscle TPA parameters (Tables 1–4). The BP-ANN method had the best calibration results mainly for gumminess, chewiness, cohesiveness, hardness, and adhesiveness, including dorsal gumminess (0.9863), pectoral gumminess (0.9620), dorsal chewiness (0.9673), pectoral chewiness (0.9555), abdominal chewiness (0.9690), gluteal chewiness (0.9836), dorsal cohesiveness (0.9224), pectoral cohesiveness (0.9306), abdominal hardness (0.9401), and gluteal adhesiveness (0.9303). LS-SVM had the highest prediction accuracy for adhesiveness, including dorsal adhesiveness (0.9206) and abdominal adhesiveness (0.9206). PLSR had the optimal prediction effect mainly for gumminess, cohesiveness, and hardness, including abdominal chewiness (0.9318), gluteal chewiness (0.9836), abdominal gumminess (0.9318), gluteal gumminess (0.9234), and pectoral hardness (0.9033).

Table 1. Predicting the dorsal muscle texture profiles using the skin HSI with the reflectance values in the full-wavelength range.

Texture	Best ML Method	r_C	RMSEC	r_P	RMSEP
Gumminess	BP-ANN	0.9826	0.1439	0.9863	0.0685
Chewiness	BP-ANN	0.9852	0.1325	0.9673	0.1065
Cohesiveness	BP-ANN	0.9694	0.0842	0.9224	0.0735
Adhesiveness	LS-SVM	0.9556	0.1873	0.9206	0.1771
Brittleness	PLSR	0.9025	0.2577	0.8915	0.1978
Hardness	PLSR	0.9339	0.2573	0.8284	0.2031
Springiness	PLSR	0.8096	0.3924	0.7185	0.4938
Resilience	LS-SVM	0.8865	0.2792	0.5612	0.5218

Note: r_C: coefficients of determination for calibration. r_P: coefficients of determination for prediction. RMSEC: root-mean-square error for calibration. RMSEP: root-mean-square error for prediction. PLSR: partial least-square regression. LS-SVM: least-square support vector machines. BP-ANN: backpropagation artificial neural network.

Table 2. Predicting the pectoral muscle texture profiles using the skin HSI with the reflectance values in the full-wavelength range.

Texture	Best ML Method	r_C	RMSEC	r_P	RMSEP
Gumminess	BP-ANN	0.9812	0.1551	0.9620	0.1792
Chewiness	BP-ANN	0.9810	0.1578	0.9555	0.1709
Cohesiveness	BP-ANN	0.9648	0.0909	0.9306	0.1121
Hardness	PLSR	0.9457	0.2452	0.9033	0.2914
Adhesiveness	PLSR	0.9144	0.2684	0.8895	0.2227
Brittleness	PLSR	0.8958	0.2681	0.8933	0.2044
Resilience	PLSR	0.7934	0.3935	0.6329	0.5168
Springiness	BP-ANN	0.8461	0.3796	0.8140	0.4229

Table 3. Predicting the abdominal muscle texture profiles using the skin HSI with the reflectance values in the full-wavelength range.

Texture	Best ML Method	r_C	RMSEC	r_P	RMSEP
Chewiness	BP-ANN	0.9786	0.1710	0.9690	0.1356
Gumminess	PLSR	0.9712	0.2005	0.9318	0.1895
Hardness	BP-ANN	0.9682	0.1789	0.9401	0.2337
Adhesiveness	LS-SVM	0.9556	0.1873	0.9206	0.1771
Cohesiveness	PLSR	0.9499	0.1107	0.9213	0.1168
Resilience	PLSR	0.7890	0.4321	0.5798	0.5079
Springiness	PLSR	0.8441	0.3462	0.8294	0.3972
Brittleness	PLSR	0.9061	0.2413	0.8856	0.1891

Table 4. Predicting the gluteal muscle texture profiles using the skin HSI with the reflectance values in the full-wavelength range.

Texture	Best ML Method	r_C	RMSEC	r_P	RMSEP
Chewiness	BP-ANN	0.9910	0.1015	0.9836	0.1078
Gumminess	PLSR	0.9593	0.2153	0.9234	0.2366
Adhesiveness	BP-ANN	0.9338	0.223	0.9303	0.1907
Cohesiveness	PLSR	0.9269	0.1232	0.8952	0.0931
Hardness	PLSR	0.9069	0.2714	0.8990	0.2926
Brittleness	LS-SVM	0.8782	0.2667	0.8421	0.2171
Resilience	PLSR	0.7835	0.0118	0.6971	0.0124
Springiness	PLSR	0.7513	0.0738	0.6842	0.0865

3.4. Accurate Prediction of the Muscle Texture Profiles Based on the Optimum Wavelengths

Equivalent calibration results were obtained based on the optimal wavelengths compared with full wavelengths. This is because the optimal wavelengths carry the most important information relevant to the determination. Some peaks and valleys (positive and negative relationships with the TPA parameters) were selected at certain wavelengths, and the selection of optimal wavelengths was successfully conducted for the eight texture parameters using the RC method. We selected 60 to 114 optimal wavelengths of the dorsal, pectoral, abdominal, and gluteal skin regions to predict the muscle textures, respectively (Figure 3).

For one texture indicator of one region, we only retained the prediction with the highest r_P for the downstream analysis (Supplementary Table S8). In general, the prediction accuracies based on the optimal wavelengths were equal to those based on the full-wavelength range (Tables 5–8). Using the values in the range of the full wavelength, nine predictions had accuracies lower than 0.85. The accuracies of these regions were still lower than 0.85 using the optimal wavelengths. The remaining regions had accuracies over 0.85 using either the values of the full wavelength or the ones of the optimal wavelength. Moreover, the absolute prediction differences between the full wavelength values and the optimal wavelength values ranged from 5.93% to 15.20%, showing that the wavelength selection could make the reduced models more stable and robust.

Table 5. Predicting the dorsal muscle texture profiles using the reflectance values in the optimal wavelength range.

Texture	ML Method	No. of WLs	r_C	RMSEC	r_P	RMSEP
Gumminess	BP-ANN	88	0.9912	0.1361	0.9847	0.1070
Chewiness	BP-ANN	84	0.9755	0.2450	0.9469	0.2164
Cohesiveness	PLSR	88	0.9714	0.2535	0.9367	0.2836
Hardness	PLSR	86	0.9586	0.2972	0.9298	0.3165
Adhesiveness	BP-ANN	84	0.9432	0.3653	0.9191	0.3004
Brittleness	PLSR	97	0.9112	0.4423	0.8804	0.3848
Springiness	BP-ANN	110	0.8380	0.5171	0.7194	0.7554
Resilience	BP-ANN	80	0.8346	0.5676	0.6493	0.7074

Note: No. of WLs: number of wavelengths.

Table 6. Predicting the pectoral muscle texture profiles using the reflectance values in the optimal wavelength range.

Texture	ML Method	No. of WLs	r_C	RMSEC	r_P	RMSEP
Gumminess	BP-ANN	67	0.9883	0.1605	0.9574	0.2439
Chewiness	BP-ANN	60	0.9862	0.1771	0.9552	0.2285
Adhesiveness	LS-SVM	75	0.9424	0.3643	0.9370	0.2499
Cohesiveness	PLSR	88	0.9556	0.3081	0.9056	0.3765
Hardness	BP-ANN	69	0.9379	0.3561	0.9033	0.3930
Brittleness	PLSR	75	0.8835	0.5213	0.8846	0.3753
Springiness	LS-SVM	77	0.8866	0.4490	0.7376	0.5971
Resilience	LS-SVM	81	0.7095	0.6983	0.5964	0.7960

Table 7. Predicting the abdominal muscle texture profiles using the reflectance values in the optimal wavelength range.

Texture	ML Method	No. of WLs	r_C	RMSEC	r_P	RMSEP
Chewiness	BP-ANN	77	0.9858	0.1779	0.9776	0.1649
Gumminess	PLSR	81	0.9740	0.2418	0.9517	0.2360
Hardness	PLSR	92	0.9592	0.2860	0.9392	0.3158
Cohesiveness	PLSR	75	0.9631	0.2859	0.9070	0.3601
Adhesiveness	LS-SVM	88	0.9564	0.3245	0.8623	0.3358
Brittleness	BP-ANN	98	0.9309	0.4130	0.8617	0.3860
Springiness	PLSR	96	0.8954	0.4320	0.8322	0.5928
Resilience	LS-SVM	84	0.9367	0.3525	0.5609	0.7414

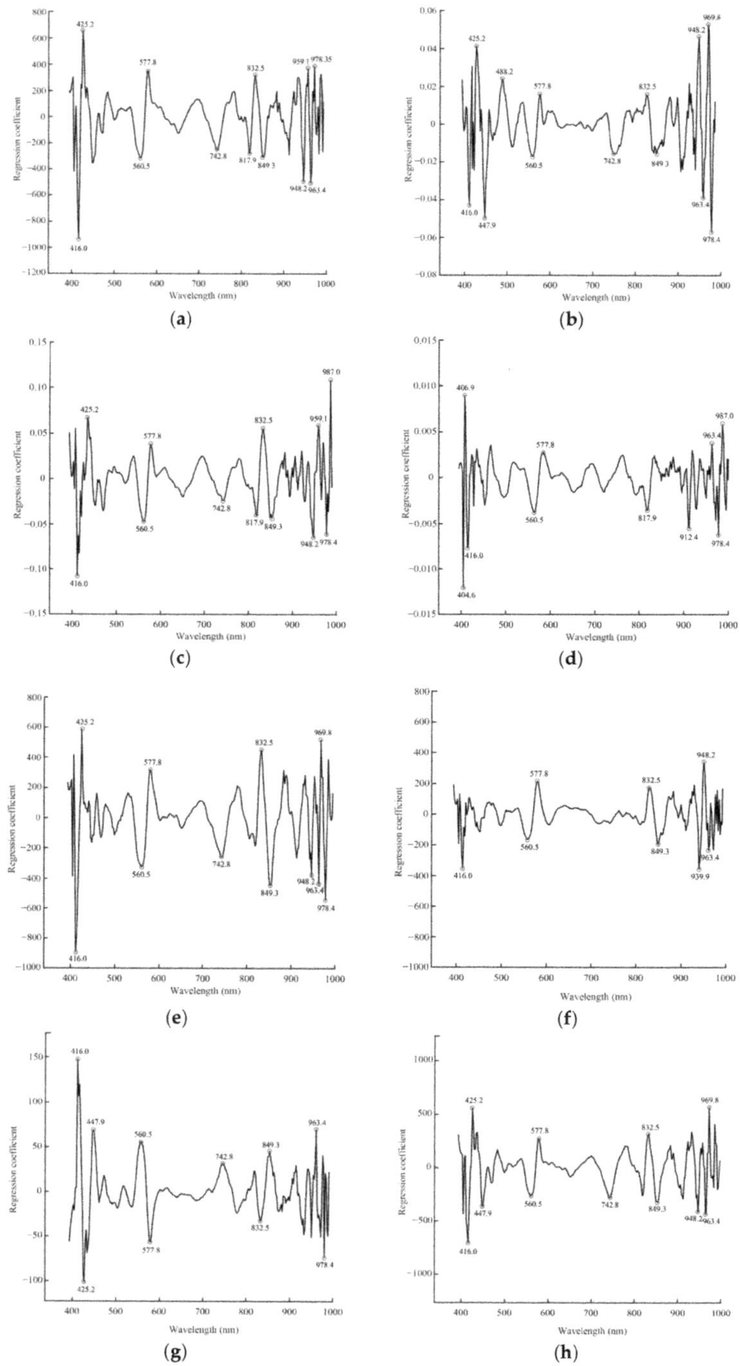

Figure 3. Selection of optimal wavelengths in the dorsal muscle. Regression coefficients method for (**a**) gumminess, (**b**) springiness, (**c**) cohesiveness, (**d**) resilience, (**e**) hardness, (**f**) brittleness, (**g**) adhesiveness, and (**h**) chewiness.

Table 8. Predicting the gluteal muscle texture profiles using the reflectance values in the optimal wavelength range.

Texture	ML Method	No. of WLs	r_C	RMSEC	r_P	RMSEP
Chewiness	BP-ANN	74	0.9768	0.2174	0.9768	0.1804
Gumminess	BP-ANN	70	0.9614	0.2804	0.9339	0.2856
Adhesiveness	BP-ANN	85	0.9421	0.3614	0.9304	0.2861
Cohesiveness	LS-SVM	79	0.9826	0.1891	0.8726	0.3938
Hardness	PLSR	80	0.9092	0.4186	0.8486	0.4933
Brittleness	PLSR	74	0.8883	0.4723	0.7613	0.4921
Springiness	PLSR	114	0.7668	0.6437	0.6396	0.7508
Resilience	PLSR	89	0.7066	0.7431	0.6316	0.7230

Compared with the prediction of TPA values based on the full-wavelength range, the r_P values of four TPA parameters (cohesiveness, hardness, springiness, and resilience) in the dorsal region were enhanced using the optimal wavelength, with the increase ranging from 0.0009 to 0.1014. The remaining four values slightly decreased. In the pectoral muscle, the r_P values of five TPA parameters, including adhesiveness, were increased by 0.0475. In the abdominal muscle region, three indicators (gumminess, springiness, and resilience) had improved r_P values. However, in the gluteal muscle region, only the accuracies of the gumminess raised by 0.0105. These results indicate that for, the former three regions, the prediction using the optimal wavelengths would be better than using the full-range wavelengths.

The BP-ANN, LS-SVM, and PLSR models were also the best three methods to predict the muscle TPA parameters. The BP-ANN method had the best calibration tool mainly for chewiness, gumminess, hardness, and adhesiveness, including dorsal chewiness (0.9469), pectoral chewiness (0.9552), abdominal chewiness (0.9776), gluteal chewiness (0.9768), dorsal gumminess (0.9847), pectoral gumminess (0.9574), pectoral hardness (0.9033), dorsal adhesiveness (0.9191), and gluteal adhesiveness (0.9304). LS-SVM had the highest prediction accuracy for gumminess and adhesiveness, including gluteal gumminess (0.9339) and pectoral adhesiveness (0.9370). PLSR had the optimal prediction effect mainly for gumminess, cohesiveness, and hardness, including abdominal gumminess (0.9517), dorsal cohesiveness (0.9367), pectoral cohesiveness (0.9056), abdominal cohesiveness (0.9070), dorsal hardness (0.9298), and abdominal hardness (0.9392).

3.5. Visualizing the Texture Parameters

The muscle texture parameters could be accurately predicted with the skin HSI spectra and the corresponding models. Therefore, the skin HSI spectra based on the optimal wavelengths and the above models were used to predict the predicted muscle parameters, which were further converted to the corresponding pixels in the tested samples, and the prediction maps were then generated. Figure 4 displays the visual prediction images of eight texture parameters in the dorsal region. The color variations presented in the test samples are automatically condensed in a linear color bar. The colors correspond to the different texture levels of the samples. The low values are highlighted in blue, and the high values are shown in orange.

In one distribution map, the spots with the same color were discretely distributed. Spots with high values were rare. Even in the same region, the prediction map of each parameter was in general different from the others. These maps reflected the minute texture difference in one region.

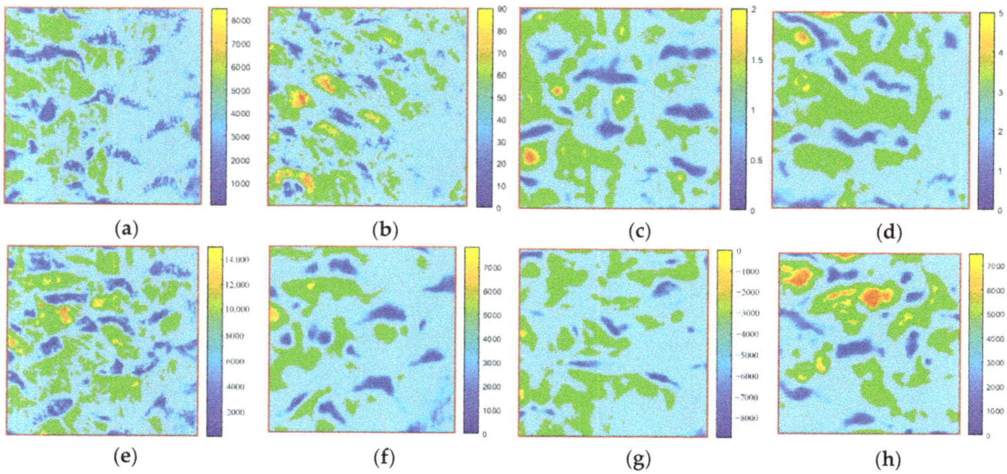

Figure 4. Visualizing the texture parameters of the dorsal muscle of scaled common carp: (**a**) gumminess, (**b**) springiness, (**c**) cohesiveness, (**d**) resilience, (**e**) hardness, (**f**) brittleness, (**g**) adhesiveness, and (**h**) chewiness. The numbers on the Y-axis represent the value of the texture measured, and the color from blue to orange represents the values from low to high.

4. Discussion

Previous works that focused on muscle texture prediction mainly utilized muscle HSI [4,5,8]. One highlight of our work was using the corresponding skin HSI of different muscle regions to estimate the texture features. Although we used the reflectance values of the skin, we found that the skin wavelength distributions were in agreement with the muscle wavelength distributions in a previous study [6]. Moreover, the high prediction accuracies of muscle texture profiles with skin HSI (prediction coefficients >0.9 for the majority of texture parameters) demonstrated that this strategy can be used in practice to detect muscle texture qualities. The prediction results were even higher than those already reported, so this method is feasible [5,8]. Another highlight of our work was performing multiple ML methods to predict muscle texture profiles, which is different from the methods used in previous studies [4]. Our results showed that BP-ANN, LS-SVM, and PLSR were the best three methods to predict the muscle TPA parameters. Each of these three methods was suitable for specific texture indicators in different muscle regions.

The distinct distributions and the levels of the reflectance values among the four skin regions, together with the different texture features of the four muscle regions, indicated the different features of these regions and the corresponding affiliated tissues. The differences were probably due to the differences in the primary chemical composition of the epidermis of the different muscle regions of common carp [40]. When the electromagnetic radiation emitted by light interacts with the internal structures of the sample, the various components of the sample exhibit distinct absorption properties at multiple particular wavelengths [17]. In the absorption, information in the 400–1000 nm spectral regions, overtones, and combinations of fundamental vibrations of functional bonds such as C-H, N-H, O-H, and S-H occur [24]. An interesting spectral trough was detected at around 430 nm, and comparable patterns were observed in the evaluation of total volatile basic nitrogen (TVB-N) and TPA in grass carp, although there have been few studies of this specific wavelength in common carp [4,6]. There was a noticeable and large absorption peak at around 500 nm, which might be associated with the residues of organic dietary items such as soybean meal [41]. Another local absorption at around 780 nm was mostly attributable to a third overtone O-H stretching [42]. The presence of water in fish caused absorption peaks at 980 nm (O-H stretching second overtone) [43].

The prediction coefficients for gumminess, chewiness, cohesiveness, and adhesiveness for different muscle regions ranged from 0.9206 to 0.9863. Wu et al. reported prediction coefficients for textural metrics of salmon fillets using full-wavelength spectroscopy ranging from 0.555 to 0.665 [5]. Chen et al. demonstrated a prediction coefficient of 0.80 for chewiness and RMSEP of 0.942 in beef [44]. Overall, the predictive ability was better than that of previous studies. A possible reason for this is the fact that different structures exhibit different characteristics of light scattering, projection, and reflection [45]. Tissues with denser muscle fibers and softer connective tissue result in better prediction [46]. These studies indicate the complexity of the elements that impact the prediction of meat quality characteristics.

Although the texture parameters were satisfactorily evaluated using full wavelengths, the volume of data and the amount of computation are enormous. Choosing the optimal wavelength can reduce the data dimensions and increase the computational speed of the model. Ma et al. applied the optimal wavelength to build a Warner–Bratzler shear force prediction model in which there was only a slight reduction from 0.8955 to 0.8913, while the number of variables both reduced significantly from 381 to 10 [4]. Our results indicate that the prediction accuracies using the optimal wavelength were not significantly different from those using the full wavelength, and similar results have been observed in previous studies [22,40]. Moreover, 81% of the wavelengths were excluded from the full-spectrum scope (114 compared to 600), indicating that RC was a valuable method of wavelength selection for identifying TPA values in the muscle of common carp. Decreasing the wavelength numbers while ensuring accuracy also reduced the runtime. Moreover, the prediction of TPA parameters such as gumminess, cohesiveness, and chewiness using the optimal wavelength obtained similar results, with r_P values ranging from 0.91 to 0.98. Using hyperspectral information from the skin and combining it with ML algorithms to predict the TPA of muscles had high accuracy. The prediction results were even higher than those already reported, so this method is feasible [5,8].

Although we obtained more optimized prediction results, the accuracy of the predictions depends on the quality and fitness of the calibration model. The prediction coefficients for springiness and resilience were lower in the present study (ranging from 0.5609 to 0.7613), and similar results were observed in a study of TPA of salmon fillets based on visible and near-infrared spectroscopy [5]. The reason for this unsatisfactory prediction may be due to the fact that springiness and resilience are subject to differences in muscle structure and connective tissue as well as differences in the water and myofibrillar protein content in muscle [47]. Therefore, there are still areas for improvement in HSI techniques, but the accuracy and efficiency of the method will continue to improve with advances in machine learning and spectroscopy.

The implementation of the visualization process is the ultimate but essential step in the HSI technique for texture prediction and will contribute to understanding the changes in TPA values in carp muscles that cannot be detected by the naked eye [4,5,13]. In one muscle region, different colored spots could be easily identified from the TPA distribution map, indicating that common carp muscles have a mixed composition and heterogeneous texture distribution [6,13]. The main reason for the color differences might be due to the distinct distribution of collagen and fat in the muscle [39,44]. The spatial distribution of textural features in common carp muscles can be conveniently observed by reference value distribution maps generated from the HSI of the samples [4]. Traditional methods can only detect a few specific points of the sample and are destructive and time-consuming. Hyperspectral imaging, on the other hand, with its superior spatial information, provides more detailed information for rapid, non-invasive measurement of TPA in common carp muscle.

The traditional method for measuring the muscle texture indicators requires a texture analyzer. The method is destructive and requires much time for muscle preparation and measurement. Our method, integrating skin hyperspectral imaging, the optimal ML method, and a visual prediction map, might provide a promising alternative tool to measure the muscle texture quality. First, this method does not require muscle preparation and is thus rapid and non-destructive. Second, the distribution maps of muscle TPA values are very useful for the meat industry to assess the sensory quality of common carp muscle by simply observing the color of the distribution map. Third, aside from the meat industry, the rapid, non-destructive, and visible features of our method are also helpful for screening common carp for food requirements.

5. Conclusions

The possibility of using HSI techniques (400–1000 nm) as a tool for determining the muscle texture profile in scaled common carp was evaluated. The optimal wavelength selected based on the RC data downscaling method with ML methods (BP-ANN, PLSR, and LS-SVM) performed most efficiently in predicting the TPA of different muscle regions in common carp. The results showed excellent performance in predicting gumminess, cohesiveness, adhesiveness, and chewiness. The r_P ranged from 0.8726 to 0.9847. Moreover, the visualization map of the distribution of TPA values was generated based on the optimal models, which provided further insight into the texture parameters in the common carp muscles. This study illustrated the tremendous potential of hyperspectral imaging technology as a robust and effective tool for the rapid and non-destructive measurement of TPA in different scaled common carp muscle regions. Despite the superior results of this study in predicting muscle texture parameters in common carp, it is still necessary to validate the developed models by applying numerous samples to ensure their reliability. In future studies, using hyperspectral imaging to acquire hyperspectral image data of other species of fish could be attempted for the rapid and non-destructive detection of meat quality.

Supplementary Materials: The following supporting information can be downloaded at: https://www.mdpi.com/article/10.3390/foods12173154/s1, Table S1: Eight textural parameters and HSI data of four muscle regions of 387 common carp, Table S2: Correlation analysis of eight textural parameters in four muscle regions, Table S3: Correlation analysis of the textural parameters in abdominal muscle of common carp (17 pairs/28), Table S4: Correlation analysis of the textural parameters in pectoral muscle of common carp (14 pairs), Table S5: Correlation analysis of the textural parameters in dorsal muscle of common carp (9 pairs), Table S6: Correlation analysis of the textural parameters in gluteal muscle of common carp (7 pairs), Table S7: Calibration and prediction results of TPA values for four muscle regions of common carp on the full spectral range, Table S8: Calibration and prediction results of TPA values for four muscle regions of common carp on the optimal spectral range, Figure S1: PCA plots clustering all common carp with eight textural indicators of dorsal muscles, Figure S2: PCA plots clustering all common carp with eight textural indicators of pectoral muscles, Figure S3: PCA plots clustering all common carp with eight textural indicators of abdominal muscles, Figure S4: PCA plots clustering all common carp with eight textural indicators of gluteal muscles, Figure S5: Box plot of cohesiveness in the four muscle regions of common carp, Figure S6: Scatter plot of adhesiveness values in the four muscle regions of common carp, Figure S7: Scatter plot of chewiness values in the four muscle regions of common carp, Figure S8: Scatter plot of gumminess values in the four muscle regions of common carp, Figure S9: Scatter plot of resilience values in the four muscle regions of common carp, Figure S10: Scatter plot of brittleness values in the four muscle regions of common carp, Figure S11: Scatter plot of springiness values in the four muscle regions of common carp.

Author Contributions: Conceptualization, J.-T.L.; methodology, Y.-M.C., Y.Z. and J.-T.L.; software, Y.-M.C.; validation, Y.-M.C. and J.-T.L.; formal analysis, Y.-M.C.; investigation, Y.-M.C., Y.Z., S.-T.Y., K.-K.W., Y.-J.C., Z.-M.X., Z.-Y.M., H.-L.C., Q.W., R.Z. and X.-Q.S.; resources, J.-T.L.; data curation, Y.-M.C. and S.-T.Y.; writing—original draft preparation, Y.-M.C.; writing—review and editing, J.-T.L.; visualization, Y.-M.C.; supervision, J.-T.L.; project administration, Y.Z. and J.-T.L.; funding acquisition, Y.Z. and J.-T.L. All authors have read and agreed to the published version of the manuscript.

Funding: This research was funded by the National Key Research and Development Program (grant number 2021YFD1200804), Beijing Municipal Natural Science Foundation (grant number 6212033), the Special Scientific Research Funds for Central Non-profit Institutes, Chinese Academy of Fishery Sciences (grant number 2020TD24), the fisheries innovation team of Beijing Agriculture Innovation Consortium (grant number BAIC07-2023-03), the Special Scientific Research Funds for Central Non-profit Institutes, Chinese Academy of Fishery Sciences (grant number 2023A003 and 2023XT0102), and the National Freshwater Genetic Resource Centre (grant number FGRC: 18537). All authors have contributed significantly to the manuscript.

Institutional Review Board Statement: The study was conducted according to the guidelines of the Declaration of China and approved by the Animal Care and Use Committee of the Chinese Academy of Fishery Sciences (protocol code ACUC-CAFS-20191202 and date of approval is 27 December 2018).

Informed Consent Statement: Not applicable.

Data Availability Statement: The data presented in this study are available on request from the corresponding author. The data presented in this study are available in Supplementary Materials.

Conflicts of Interest: The authors declare no conflict of interest.

References

1. Linhart, O.; Gela, D.; Rodina, M.; Šlechtová, V.; Šlechta, V. Top-crossing with paternal inheritance testing of common carp (*Cyprinus carpio* L.) progeny under two altitude conditions. *Aquaculture* **2002**, *204*, 481–491. [CrossRef]
2. Vandeputte, M. Selective breeding of quantitative traits in the common carp (*Cyprinus carpio*): A review. *Aquat. Living Resour.* **2003**, *16*, 399–407. [CrossRef]
3. Yue, G. Recent advances of genome mapping and marker-assisted selection in aquaculture. *Fish Fish.* **2014**, *15*, 376–396. [CrossRef]
4. Ma, J.; Sun, D.; Qu, J.; Pu, H. Prediction of textural changes in grass carp fillets as affected by vacuum freeze drying using hyperspectral imaging based on integrated group wavelengths. *LWT-Food Sci. Technol.* **2017**, *82*, 377–385. [CrossRef]
5. Wu, D.; Sun, D.; He, Y. Novel non-invasive distribution measurement of texture profile analysis (TPA) in salmon fillet by using visible and near infrared hyperspectral imaging. *Food Chem.* **2014**, *145*, 417–426. [CrossRef]
6. Cheng, J.; Sun, D.; Han, Z.; Zeng, X.A. Texture and structure measurements and analyses for evaluation of fish and fillet freshness quality: A review. *Compr. Rev. Food Sci. Food Saf.* **2014**, *13*, 52–61. [CrossRef]
7. Özdoğan, G.; Lin, X.; Sun, D. Rapid and noninvasive sensory analyses of food products by hyperspectral imaging: Recent application developments. *Trends Food Sci. Technol.* **2021**, *111*, 151–165. [CrossRef]
8. Zhou, J.; Wu, X.; You, J.; Xiong, S. Rapid determination of the textural properties of silver carp (*Hypophthalmichthys molitrix*) using near-infrared reflectance spectroscopy and chemometrics. *LWT* **2020**, *129*, 109545. [CrossRef]
9. León-Ecay, S.; López-Maestresalas, A.; Murillo-Arbizu, M.; Beriain, M.; Mendizabal, J.; Arazuri, S.; Jarén, C.; Bass, P.; Colle, M.; García, D.; et al. Classification of Beef longissimus thoracis Muscle Tenderness Using Hyperspectral Imaging and Chemometrics. *Foods* **2022**, *11*, 3105. [CrossRef]
10. Liu, C.; Chu, Z.; Weng, S.; Zhou, G.; Han, K.; Zhang, Z.; Huang, L.; Zhou, Z.; Zheng, S. Fusion of electronic nose and hyperspectral imaging for mutton freshness detection using input-modified convolution neural network. *Food Chem.* **2022**, *385*, 132651. [CrossRef]
11. Pu, H.; Yu, J.; Sun, D.; Wei, Q.; Shen, X.; Wang, Z. Distinguishing fresh and frozen-thawed beef using hyperspectral imaging technology combined with convolutional neural networks. *Microchem. J.* **2023**, *189*, 108559. [CrossRef]
12. Liu, Y.; Sun, D.; Cheng, J.; Han, Z. Hyperspectral imaging sensing of changes in moisture content and color of beef during microwave heating process. *Food Anal. Methods* **2018**, *11*, 2472–2484. [CrossRef]
13. Zhang, J.; Ma, Y.; Liu, G.; Fan, N.; Li, Y.; Sun, Y. Rapid evaluation of texture parameters of Tan mutton using hyperspectral imaging with optimization algorithms. *Food Control* **2022**, *135*, 108815. [CrossRef]
14. Arabameri, M.; Nazari, R.; Abdolshahi, A.; Mohammad, A.; Solmaz, M.; Nabi, S.; Francisco, B.; Amin, M. Oxidative stability of virgin olive oil: Evaluation and prediction with an adaptive neuro-fuzzy inference system (ANFIS). *J. Sci. Food Agric.* **2019**, *99*, 5358–5367. [CrossRef]
15. Barbin, D.; Elmasry, G.; Sun, D.; Allen, P. Near-infrared hyperspectral imaging for grading and classification of pork. *Meat Sci.* **2012**, *90*, 259–268. [CrossRef]
16. Cheng, K.; Wagner, L.; Moazzami, A.; Gómez-Requeni, P.; Schiller Vestergren, A.; Brännäs, E.; Pickova, J.; Trattner, S. Decontaminated fishmeal and fish oil from the Baltic Sea are promising feed sources for Arctic char (*Salvelinus alpinus* L.)—Studies of flesh lipid quality and metabolic profile. *Eur. J. Lipid Sci. Technol.* **2016**, *118*, 862–873. [CrossRef]
17. ElMasry, G.; Sun, D.; Allen, P. Near-infrared hyperspectral imaging for predicting colour, pH and tenderness of fresh beef. *J. Food Eng.* **2012**, *110*, 127–140. [CrossRef]
18. Liu, J.; Liu, S.; Shin, S.; Liu, F.; Shi, T.; Lv, C.; Qiao, Q.; Fang, H.; Jiang, W.; Men, H. Detection of apple taste information using model based on hyperspectral imaging and electronic tongue data. *Sens. Mater.* **2020**, *32*, 1767–1784. [CrossRef]
19. Talens, P.; Mora, L.; Morsy, N.; Barbin, D.; ElMasry, G.; Sun, D. Prediction of water and protein contents and quality classification of Spanish cooked ham using NIR hyperspectral imaging. *J. Food Eng.* **2013**, *117*, 272–280. [CrossRef]
20. Tan, W.; Sun, L.; Yang, F.; Che, W.; Ye, D.; Zhang, D.; Zou, B. Study on bruising degree classification of apples using hyperspectral imaging and GS-SVM. *Optik* **2018**, *154*, 581–592. [CrossRef]
21. Wu, J.; Peng, Y.; Li, Y.; Wang, W.; Chen, J.; Dhakal, S. Prediction of beef quality attributes using VIS/NIR hyperspectral scattering imaging technique. *J. Food Eng.* **2012**, *109*, 267–273. [CrossRef]

22. He, H.; Wu, D.; Sun, D. Potential of hyperspectral imaging combined with chemometric analysis for assessing and visualising tenderness distribution in raw farmed salmon fillets. *J. Food Eng.* **2014**, *126*, 156–164. [CrossRef]
23. Zhou, J.; Wu, X.; Chen, Z.; You, J.; Xiong, S. Evaluation of freshness in freshwater fish based on near infrared reflectance spectroscopy and chemometrics. *LWT* **2019**, *106*, 145–150. [CrossRef]
24. Moosavi-Nasab, M.; Khoshnoudi-Nia, S.; Azimifar, Z.; Kamyab, S. Evaluation of the total volatile basic nitrogen (TVB-N) content in fish fillets using hyperspectral imaging coupled with deep learning neural network and meta-analysis. *Sci. Rep.* **2021**, *11*, 5094. [CrossRef] [PubMed]
25. Li, J.; Wang, Q.; Huang Yang, M.; Li, Q.; Cui, M.; Dong, Z.; Wang, H.; Yu, J.; Zhao, Y.; Yang, C.; et al. Parallel subgenome structure and divergent expression evolution of allo-tetraploid common carp and goldfish. *Nat. Genet.* **2021**, *53*, 1493–1503. [CrossRef]
26. Wu, L.; He, J.; Liu, G.; Wang, S.; He, X. Detection of common defects on jujube using Vis-NIR and NIR hyperspectral imaging. *Postharvest Biol. Technol.* **2016**, *112*, 134–142. [CrossRef]
27. Ruffin, C.; King, R.L.; Younan, N. A combined derivative spectroscopy and Savitzky-Golay filtering method for the analysis of hyperspectral data. *GIScience Remote Sens.* **2008**, *45*, 1–15. [CrossRef]
28. Yu, H.; Qing, L.; Yan, D.; Xia, G.; Zhang, C.; Yun, Y.; Zhang, W. Hyperspectral imaging in combination with data fusion for rapid evaluation of tilapia fillet freshness. *Food Chem.* **2021**, *348*, 129129. [CrossRef]
29. Lim, J.; Kim, G.; Mo, C.; Kim, M.; Chao, K.; Qin, J.; Fu, X.; Baek, I.; Cho, B.-K. Detection of melamine in milk powders using near-infrared hyperspectral imaging combined with regression coefficient of partial least square regression model. *Talanta* **2016**, *151*, 183–191. [CrossRef]
30. Liu, D.; Sun, D.-W.; Zeng, X.-A. Recent advances in wavelength selection techniques for hyperspectral image processing in the food industry. *Food Bioprocess Technol.* **2014**, *7*, 307–323. [CrossRef]
31. Bourne, M. *Food Texture and Viscosity: Concept and Measurement*, 2nd ed.; Academic Press: New York, NY, USA, 2002.
32. Bradbury, P.; Zhang, Z.; Kroon, D.; Casstevens, T.; Ramdoss, Y.; Buckler, E. TASSEL: Software for Association Mapping of Complex Traits in Diverse Samples. *Bioinformatics* **2007**, *23*, 2633–2635. [CrossRef] [PubMed]
33. Abdi, H. Partial least squares regression and projection on latent structure regression (PLS Regression). *Wiley Interdiscip. Rev. Comput. Stat.* **2010**, *2*, 97–106. [CrossRef]
34. Wu, D.; Sun, D. Advanced applications of hyperspectral imaging technology for food quality and safety analysis and assessment: A review-Part I: Fundamentals. *Innov. Food Sci. Emerg. Technol.* **2013**, *19*, 1–14. [CrossRef]
35. Bao, Y.; Liu, F.; Kong, W.; Sun, D.; He, Y.; Qiu, Z. Measurement of soluble solid contents and pH of white vinegars using VIS/NIR spectroscopy and least squares support vector machine. *Food Bioprocess Technol.* **2014**, *7*, 54–61. [CrossRef]
36. Dai, Q.; Cheng, J.; Sun, D.; Pu, H.; Zeng, X.; Xiong, Z. Potential of visible/near-infrared hyperspectral imaging for rapid detection of freshness in unfrozen and frozen prawns. *J. Food Eng.* **2015**, *149*, 97–104. [CrossRef]
37. Kamruzzaman, M.; Makino, Y.; Oshita, S. Parsimonious model development for real-time monitoring of moisture in red meat using hyperspectral imaging. *Food Chem.* **2016**, *196*, 1084–1091. [CrossRef]
38. Zheng, X.; Li, Y.; Wei, W.; Peng, Y. Detection of adulteration with duck meat in minced lamb meat by using visible near-infrared hyperspectral imaging. *Meat Sci.* **2019**, *149*, 55–62. [CrossRef]
39. Casas, C.; Martinez, O.; Guillen, M.; Pin, C.; Salmeron, J. Textural properties of raw Atlantic salmon (*Salmo salar*) at three points along the fillet, determined by different methods. *Food Control* **2006**, *17*, 511–515. [CrossRef]
40. Wang, H.; Wang, K.; Zhu, X.; Zhang, P.; Yang, J.; Tan, M. Integration of partial least squares regression and hyperspectral data processing for the nondestructive detection of the scaling rate of carp (*Cyprinus carpio*). *Foods* **2020**, *9*, 500. [CrossRef]
41. Kimiya, T.; Sivertsen, A.H.; Heia, K. VIS/NIR spectroscopy for non-destructive freshness assessment of Atlantic salmon (*Salmo salar* L.) fillets. *J. Food Eng.* **2013**, *116*, 758–764. [CrossRef]
42. He, H.; Wu, D.; Sun, D. Non-destructive and rapid analysis of moisture distribution in farmed Atlantic salmon (*Salmo salar*) fillets using visible and near-infrared hyperspectral imaging. *Innov. Food Sci. Emerg. Technol.* **2013**, *18*, 237–245. [CrossRef]
43. Wu, D.; Sun, D.-W. Potential of time series-hyperspectral imaging (TS-HSI) for non-invasive determination of microbial spoilage of salmon flesh. *Talanta* **2013**, *111*, 39–46. [CrossRef] [PubMed]
44. Chen, Q.; Zhang, Y.; Guo, Y.; Cheng, Y.; Qian, H.; Yao, W.; Xie, Y.; Ozaki, Y. Non-destructive prediction of texture of frozen/thaw raw beef by Raman spectroscopy. *J. Food Eng.* **2020**, *266*, 109693. [CrossRef]
45. Hughes, J.; Oiseth, S.; Purslow, P.; Warner, R. A structural approach to understanding the interactions between colour, water-holding capacity and tenderness. *Meat Sci.* **2014**, *98*, 520–532. [CrossRef]
46. Science, P. Near-infrared reflectance spectroscopy as a method to predict chemical composition of breast meat and discriminate between different n−3 feeding sources. *Poult. Sci.* **2005**, *84*, 128. [CrossRef]
47. Xiong, Z.; Sun, D.; Dai, Q.; Han, Z.; Zeng, X.; Wang, L. Application of Visible Hyperspectral Imaging for Prediction of Springiness of Fresh Chicken Meat. *Food Anal. Methods* **2015**, *8*, 380–391. [CrossRef]

foods

Article

Raman Spectroscopic Study of Five Typical Plasticizers Based on DFT and HF Theoretical Calculation

Tong Sun *, Yitao Wang, Mingyue Li and Dong Hu

College of Optical, Mechanical and Electrical Engineering, Zhejiang A&F University, Hangzhou 311300, China; 2020604021041@stu.zafu.edu.cn (Y.W.); lmy@stu.zafu.edu.cn (M.L.); 2018004?@zafu.edu.cn (D.H.)
* Correspondence: suntong@zafu.edu.cn

Abstract: Phthalic acid esters (PAEs) are the most commonly used plasticizers, and long-term or high levels of exposure to PAEs have a huge potential risk to human health. In this study, the theories of Hartree–Fock (HF) and density functional theory (DFT) with different hybrid methods and basis sets were used to calculate the theoretical Raman spectra of five PAEs, and the comparison of calculated spectra between different theories, hybrid methods, and basis sets was conducted to determine the suitable theory with hybrid method and basis set for PAEs. Also, the Raman vibrations were assigned to the Raman peaks of PAEs according to the theoretical and experimental Raman spectra. The results indicate that DFT is more suitable for the theoretical study of PAEs than HF. In DFT, the hybrid method of B3LYP is more applicable to the theoretical study of PAEs than B3PW91, and the basis set of 6-311G(d, p) obtains the most consistent theoretical Raman spectra with the experimental spectra for PAEs. This study finds the optimal combination of the theoretical method and basis set for PAEs, and it will contribute to the establishment of the Raman fingerprint and the development of rapid detection for PAEs in the future.

Keywords: PAEs; Raman; DFT; HF; theoretical study

Citation: Sun, T.; Wang, Y.; Li, M.; Hu, D. Raman Spectroscopic Study of Five Typical Plasticizers Based on DFT and HF Theoretical Calculation. *Foods* 2023, *12*, 2888. https://doi.org/10.3390/foods12152888

Academic Editor: Ana Teresa Sanches-Silva

Received: 30 June 2023
Revised: 22 July 2023
Accepted: 24 July 2023
Published: 29 July 2023

1. Introduction

Plasticizers are polymer additives that are commonly used in packaging materials to increase plasticity [1]. Phthalic acid esters (PAEs) are the most commonly used plasticizers, which can enter the body with people's breath, diet, and even skin contact. When PAEs accumulate to a certain extent in the body for a long time, they can be harmful to human health, and will cause feminization of men, increase the risk of breast cancer in women, and leads to deformity and cancer [2]. In recent years, the frequent occurrence of excessive PAEs in food has attracted widespread concern and great importance from the government and society. In 2011, Taiwan's Food and Drug Administration found high concentrations of Di(2-ethyl)hexyl phthalate in a batch of probiotic ingredients [3]. Since then, PAEs have become known to the public. In 2012, the Plasticizer incident of Chinese Baijiu caused a sensation [4]. The incidents of artificially added PAEs in milk tea in 2017 and excessive PAEs in Ladue Blue Joe walnut oil in 2019 are even more controversial [5]. Therefore, it is very necessary to effectively monitor PAEs in food to protect people's health.

In recent years, Raman spectroscopy has been widely used in the field of food safety [6–8] and gradually applied to the detection of PAEs due to its characteristics of no sample pretreatment, fast detection speed, and response to molecular fingerprint information. Wu et al. [9] prepared homogeneous AuNPs films for the detection of Di (2-ethyl)hexyl phthalate in sorghum wine. Zhou et al. [10] formed Ag@Fe3O4@PEI nanoparticles, then modified them with cyclodextrin (β-CD). Finally, 1.3 mg/kg of BBP in white wine was able to detect using this substrate. Cao et al. [11] prepared Au-Ag-S nanostructured substrates using a one-pot method and used them for the detection of Di(2-ethyl)hexyl phthalate in juice. Wang et al. [12] and Wu et al. [13] used 2D silver plate and AuNPs as enhanced substrates to detect PAEs in edible oils. However, few of these studies incorporate structural theory, whereas identifying PAEs requires the incorporation of structural theory. Structure theory includes the semi-empirical method, density functional theory (DFT), and ab Initio [14,15]. It can help to understand the experimental results [16–18]. Ji et al. [19] assigned the Raman vibrations to the eight Raman peaks of the Dimethyl phthalate by DFT 6-31+G(d) calculations. Liu et al. [20] simulated the theoretical Raman spectra of di(2-ethyl)hexyl phthalate, dibutyl phthalate, and diethyl phthalate using DFT 3-21G; then, the theoretical Raman spectra with the corresponding experimental Raman spectra were compared and analyzed. Qiu et al. [21] calculated the Raman spectra of dimethyl phthalate, dibutyl phthalate, Di-n-octyl phthalate and their derivatives in the gaseous environment using DFT B3LYP 6-31g(d), which contributed to the studies of PAEs. Xu et al. [22] used DFT 6-31G(d) to calculate the theoretical spectra of di(2-ethyl)hexyl phthalate, dibutyl phthalate, and butyl benzyl phthalate. The theoretical Raman spectra were consistent with the experimental spectra, and the Raman vibrations were assigned to Raman peaks. Zuo et al. [23] used molecular dynamics simulations and DFT to reveal inter-molecular interactions of phthalic acid esters.

There is no comparison of theoretical Raman spectra of PAEs calculated by different theoretical methods and basis sets in the existing studies. The explanation of basis sets can be found in the second paragraph of Section 2.3. The theoretical Raman spectra calculated by some theoretical methods will have many spurious peaks, which are not found in the experimental Raman spectra; this will interfere with the analysis of the experimental data and cause errors. In addition, existing studies have only assigned the Raman vibrations to individual Raman peaks, but not to all major Raman peaks. In this research, the theoretical Raman spectra of five typical PAEs were simulated by different theoretical methods and basis sets, and were compared with the experimental Raman spectra in order to obtain the most applicable theoretical method and basis set for PAEs, which can effectively reduce the influence of spurious peaks on the analysis of PAEs detection in Food or other products. Also, all Raman peaks of the five PAEs were assigned according to the theoretical and experimental Raman spectra.

2. Materials and Methods

2.1. Materials and Equipment

Dimethyl phthalate (DMP), diethyl phthalate (DEP), and dibutyl phthalate (DBP) reagents were purchased from Sinopharm Chemical Reagent Co. (Shanghai, China) Di(2-ethyl)hexyl phthalate (DEHP) and diisononyl phthalate (DINP) reagents were purchased from Aladdin Reagent Co. (Shanghai, China) DMP, DEP, DBP, DEHP, and DINP are all analytical pure reagents with a purity greater than 99.5%. Table 1 shows the details of the five typical PAEs.

Table 1. Details of five typical PAEs.

Name	Abbreviations	Chemical Formula	Molecular Weight	Density (g/cm^3)	Harm
Dimethyl phthalate	DMP	$C_{10}H_{10}O_4$	194.19	1.18	Suppression of the central nervous system
Diethyl phthalate	DEP	$C_{12}H_{14}O_4$	222.24	1.12	Headaches, dizziness, and vomiting
Dibutyl phthalate	DBP	$C_{16}H_{22}O_4$	278.34	1.05	Teratogenic effect on embryos
Di(2-ethyl)hexyl phthalate	DEHP	$C_{24}H_{38}O_4$	390.56	0.99	Carcinogenic to animals
Diisononyl phthalate	DINP	$C_{26}H_{42}O_4$	418.61	0.97	Some effects on reproduction, development, and cancer

PAEs: phthalic acid esters.

A handheld portable Raman spectrometer (Bruker, Germany) equipped with dual wavelength lasers of 785 nm and 852 nm was used in this study. The spectral range, spot diameter, and spectral resolution of the spectrometer are 3200~300 cm^{-1}, 1~2 mm, and 10~12 cm^{-1}, respectively.

2.2. Spectral Acquisition

In order to reduce the influence of other impurities on the experimental Raman spectra of five PAEs, the analytical pure PAE samples were used for collecting Raman spectra. For DMP spectral acquisition, 3 mL of analytical pure DMP sample was put in the Raman vial first, then the Raman vial was placed into the liquid measurement accessory of the Raman instrument. After that, the DMP sample is irradiated with laser wavelengths of 785 nm and 852 nm in sequence, and the generated Raman spectrum signal passes through a lens, dichroic mirror, and long pass filter, and is then detected by the Raman spectrometer to obtain DMP Raman spectra at two laser wavelengths. Due to the fact that the positions of the fluorescence peaks do not change with the incident laser wavelength, while the positions of the Raman peaks will change with the incident laser wavelength, matching the Raman spectra of DMP at two incident laser wavelengths through software of OPUS8.7.31 can effectively eliminate fluorescence interference. Finally, the Raman spectrum of DMP that had eliminated fluorescence signal was obtained. The Raman spectra of DEP, DBP, DEHP, and DINP were acquired in the same way. The integration time and number of scans were both set to 6 s and 3, respectively, for five PAEs. Figure 1 shows the schematic diagram of Raman spectrum acquisition.

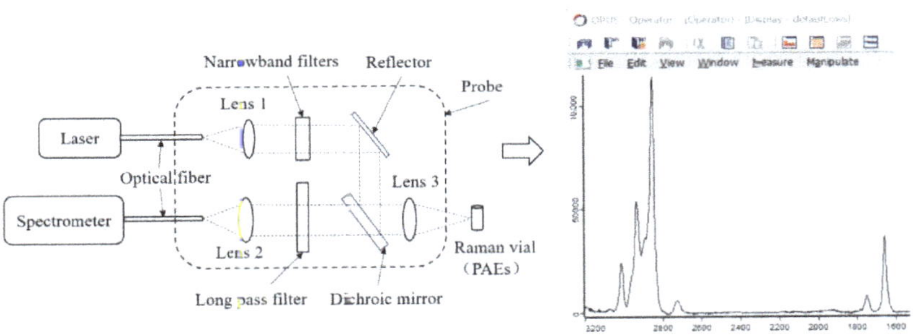

Figure 1. Schematic diagram of Raman spectrum acquisition. PAEs: phthalic acid esters.

2.3. Theoretical Calculation

The Hartree–Fock method (HF) is one of the ab initio methods, which is based on the Schrodinger equation [24]. DFT is a method for studying the electronic structure of multi-electron systems, which has a wide range of applications in the study of the properties of molecules and condensed matter. It is one of the most commonly used methods in the field of computational materials science and computational chemistry in condensed matter physics [25–27]. There are many hybrid methods in DFT. The hybrid methods of B3LYP (Becke-3 exchange with Lee–Yang–Parr gradient-corrected correlation functional) and B3PW91 (B3 exchange + PW91 correlation) are the most used in the calculation of organic matter [28]. Both B3LYP and B3PW91 are exchange-correlated general functions with similar calculated results [29], but the specific results are related to the studied substances.

The basis set is the second component of the theoretical calculation, and using a basis set means selecting a region of space where each electron is located [30]. For example, 6-311G+(2d, p): the first 6 refers to the six Gaussian functions describing the inner layer electrons, the latter 311 means that each valence orbit is represented by three basis functions, which are fitted by 3, 1, and 1 original functions, respectively; G means Gaussian basis set; d means one additional polarization function for each heavy atom (non-hydrogen atom); p means one additional polarization function for the hydrogen atom adds a polarization function; + means to add a dispersion function to the heavy atom; if the hydrogen atom also wants to add a dispersion function, then + is replaced by ++ [31]. The larger the basis set, the fewer the constraints imposed on the electrons and the more accurate the approximation of the true molecular wave function. The choice of the basis set depends on different accuracy requirements, theoretical approaches, and research object systems, etc. [32–34].

In this study, structural models of five PAEs(DMP, DEP, DBP, DEHP, and DINP) were constructed. Then, the two theories of DFT and HF with 6-31G(d) were used to calculate the theoretical Raman spectra of five PAEs, and the spectra of five PAEs calculated by DFT and HF were compared in order to obtain the suitable theory for PAEs. In DFT theory, the hybrid methods of B3LYP and B3WP91 were chosen, and the spectra calculated by B3LYP and B3WP91 of DFT were compared to determine which specific method would be more suitable. After that, different basis sets, (3-21G, 6-31G(d), 6-311G(d, p), and 6-311G+(d, p)), were used to simulate theoretical spectra of five PAEs, and the results were compared to choose the most applicable basis set for PAEs. Finally, the most applicable theoretical Raman spectra combined with experimental Raman spectra were analyzed to assign the Raman vibrations to the Raman peaks. All the theoretical calculations are prepared using the Gaussian09 (version 9.5), software.

3. Results

3.1. Molecular Structure of PAEs

The structural models of the five PAEs (DMP, DEP, DBP, DEHP, and DINP) and their molecular formulae are shown in Figure 2a–e. It can be found that the structure of DMP consists of a benzene ring, two carboxyl groups, and two methyl groups, and the structure of DMP is the simplest among these PAEs. The study of DMP has important reference values for other PAEs [35]. Also, the other four PAEs are relatively typical structures, which are important for the study of PAEs [36].

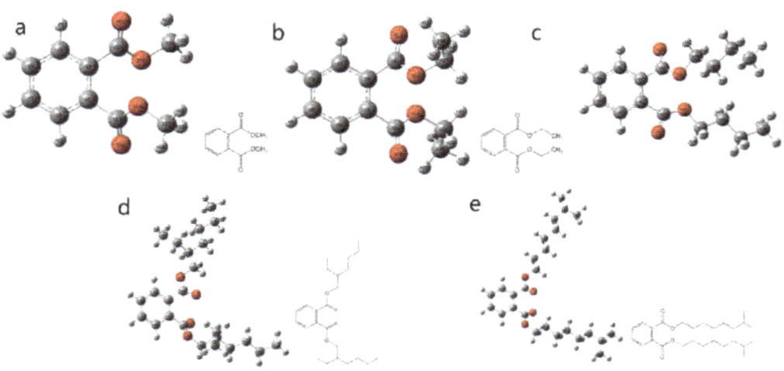

Figure 2. Optimized molecular structure diagram of PAEs and its molecular formula. (**a**): DMP; (**b**): DEP; (**c**): DBP; (**d**): DEHP; (**e**): DINP. DMP: dimethyl phthalate; DEP: diethyl phthalate; DBP: dibutyl phthalate; DEHP: di(2-ethyl)hexyl phthalate; DINP: diisononyl phthalate.

3.2. Experimental Raman Spectra of PAEs

Figure 3 shows the experimental Raman spectra of five PAEs. From Figure 3, it can be seen that the Raman peaks in the range of 2800~3200 cm^{-1} are very heterogeneous, and the peaks in this interval overlap with those of many solvents such as ethanol. Therefore, the range of 300~2000 cm^{-1} is chosen for this study. From Figure 3, it can be seen that the common Raman peaks of the five PAEs are 400, 650, 1040, 1120, 1160 1284, 1450, 1580, 1600, and 1726 cm^{-1}. The unique Raman peaks of DMP are 818 and 964 cm^{-1}; the unique Raman peaks of DEP are 352, 784, and 848 cm^{-1}; the unique Raman peaks of DBP are 810, 842, 940, and 962 cm^{-1}; the unique Raman peaks of DEHP are 834, 858, 894, and 956 cm^{-1}; and the unique Raman peaks of DINP are 822, 900, and 960 cm^{-1}. The partial experimental Raman peaks of PAEs in this study are consistent with the peaks of 650, 1040, 1580, 1600, and 1726 cm^{-1} for DEHP and DBP in the literature [22]. They are basically consistent with the peaks of 1038, 1120, 1578, 1599, and 1723 cm^{-1} for DEHP, DEP, and DBP in the literature [20], and the peaks of 403, 653, 1043, 1127, 1167, 1585, 1605, and 1731 cm^{-1} for eight PAEs in the literature [37].

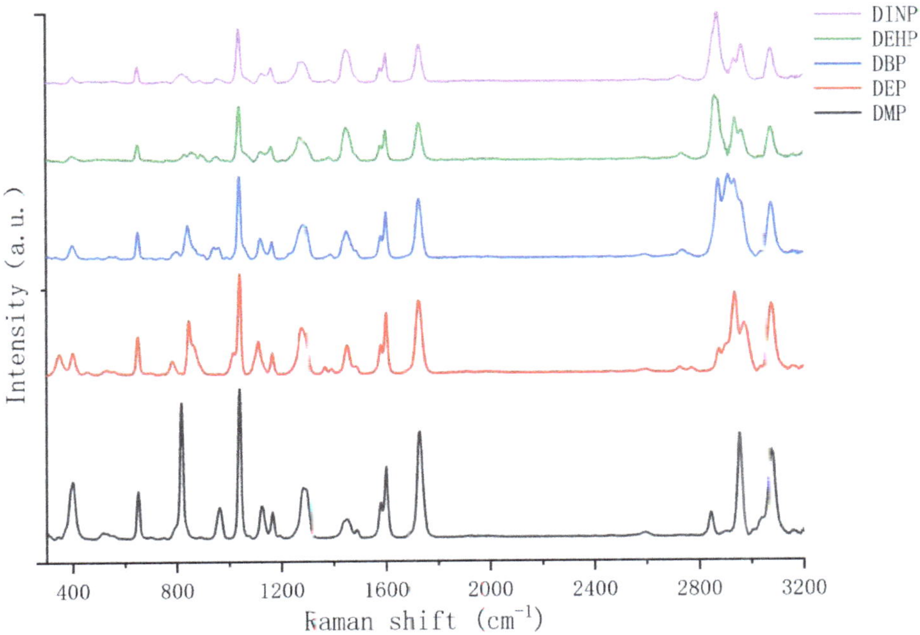

Figure 3. Experimental Raman spectra of phthalic acid esters. DINP: diisononyl phthalate; DEHP: di(2-ethyl)hexyl phthalate; DBP: dibutyl phthalate; DEP: diethyl phthalate; DMP: dimethyl phthalate.

3.3. Comparison of HF and DFT Methods

Different methods have different calculation accuracy for different substances, and the accuracy of the Raman shifts calculated by HF and DFT needs to be verified by comparison with experimental data. From the literature [20–22], it is clear that theoretical studies have been carried out on PAEs, but only one or two methods are selected for study, and no comparative studies have been conducted. The Raman spectra calculated by some theoretical methods will have many spurious peaks that are not present in the experimental spectra, which will cause interference and errors in the analysis of experimental data; so, it needs to find the suitable theoretical method for PAEs.

Figure 4 shows the theoretical Raman spectra of the five PAEs (DMP, DEP, DBP, DEHP, and DINP) calculated by HF and DFT with the 6-31G(d) basis set. Because many theoretical spectra have some offset errors from experimental spectra, it is necessary to use the scale factors from the database of frequency scale factors for electronic model chemistries [38] to correct the theoretical spectra in order to eliminate the offset error to the greatest extent [39,40]. From the scale factors database, it can be seen that the scale factors of HF 6-31G(d), DFT B3LYP 6-31G(d), and DFT B3PW91 6-31G(d) are 0.885, 0.952, and 0.947, respectively. In this study, the theoretical spectra are the spectra after correction.

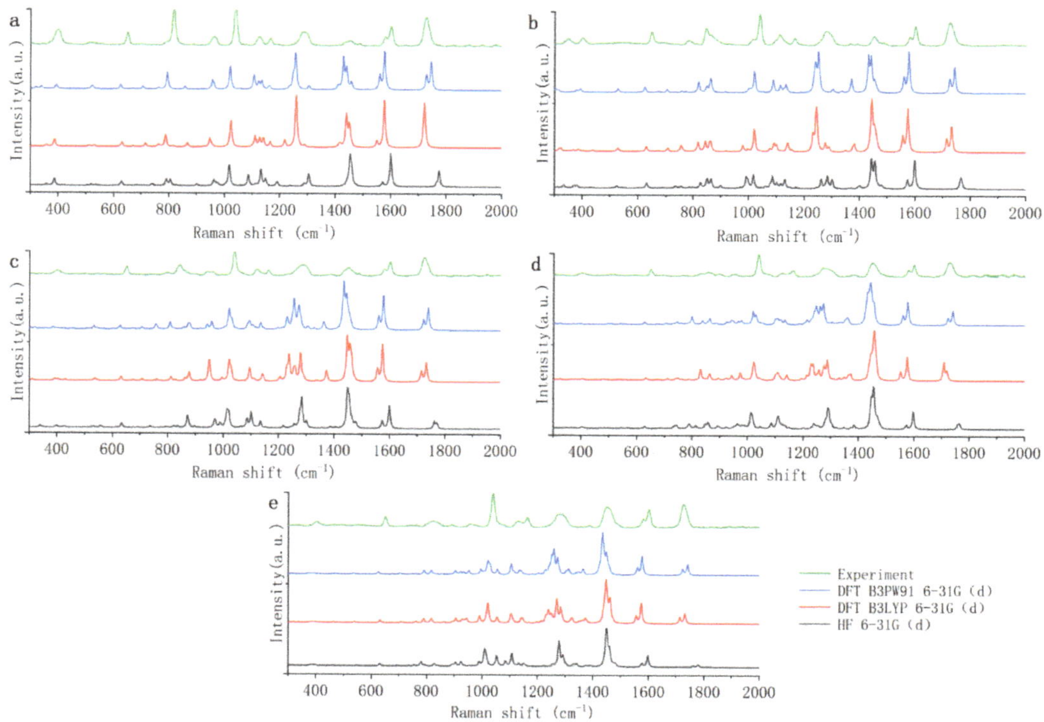

Figure 4. Theoretical Raman spectra of five phthalic acid esters based on HF and DFT: (**a**) DMP; (**b**) DEP; (**c**) DBP; (**d**) DEHP; (**e**) DINP. HF: Hartree–Fock method; DFT: density functional theory; DMP: dimethyl phthalate; DEP: diethyl phthalate; DBP: dibutyl phthalate; DEHP: di(2-ethyl)hexyl phthalate; DINP: diisononyl phthalate.

As shown in Figure 4, the theoretical Raman spectra calculated by HF and DFT show good agreement as a whole with the experimental Raman spectra, but there are large differences in individual Raman peaks. Comparing the theoretical Raman spectra calculated by HF with the experimental Raman spectra, it is found that the peaks of the five PAEs have common differences. The wide peaks of 1284 and 1450 cm^{-1} all become sharp, and the strong peaks of 1726 cm^{-1} become much weaker. The theoretical Raman peaks all have a red shift in the band of 300~800 cm^{-1}, while all have a blue shift in the band of 1500~2000 cm^{-1}. In addition to the above common differences, the theoretical Raman spectra of the five PAEs also individually have a lot of spurious Raman peaks. Among them, the theoretical Raman spectra of DMP, DEP, and DBP have more spurious peaks. In terms of DMP, the peak of 818 cm^{-1} is divided into peaks of 790 and 806 cm^{-1}, the peak of 1120 cm^{-1} is divided into peaks of 1130 and 1146 cm^{-1}, and there is a spurious peak of 1080 cm^{-1}. In terms of DEP, the peak of 1120 cm^{-1} is divided into peaks of 1086, 1102, and 1126 cm^{-1}, and there is a spurious peak of 992 cm^{-1}. In terms of DBP, the peaks of 940 and 962 cm^{-1} are shifted to 970 and 990 cm^{-1}, and the peak of 1120 cm^{-1} is divided into peaks of 1086 and 1110 cm^{-1}. Therefore, it can be seen that the theoretical Raman spectra calculated by HF 6-31G(d) have so many errors. This may be because HF ignores most of the electronic correlations [41], which makes the theoretical spectra of PAEs inaccurate.

There are also some errors in the theoretical Raman spectra calculated by DFT. However, the theoretical Raman spectra calculated by B3LYP and B3PW91 of DFT both have fewer spurious peaks than the theoretical Raman spectra calculated by HF. Therefore, DFT is more applicable to theoretical studies of PAEs. Then, the theoretical Raman spectra calculated by B3LYP and B3PW91 are further compared. It is found that the theoretical Raman spectra calculated by the B3PW91 have slightly more spurious peaks. In terms of DMP, there are two spurious peaks of 320 and 330 cm^{-1}. In terms of DEHP, there is a spurious peak of 800 cm^{-1}. Peak of 1726 cm^{-1} in five PAEs are all divided into two Raman peaks. In addition, the theoretical calculation time of the two methods is not significantly different. So, it can be concluded that the DFT B3LYP method is more applicable to the theoretical study of PAEs.

3.4. Different Basis Sets with DFT B3LYP

The above results show that the DFT B3LYP method is more suitable for the study of PAEs. Different basis sets of DFT B3LYP have different calculation accuracy, and further theoretical studies are needed to select the suitable basis set.

Figure 5 shows the theoretical Raman spectra of five PAEs (DMP, DEP, DBP, DEHP, and DINP) calculated by DFT B3LYP, with 3-21G, 6-31G(d), 6-311G(d, p), and 6-311G+(d, p) basis sets. From the scale factors database, it can be seen that the scale factors of DFT B3LYP 6-311G(d, p) is 0.9708. The scale factors of DFT B3LYP 6-31G(d) is as above, and the scale factors of 3-21G and 6-311G+(d, p) are not found in the scale factors database.

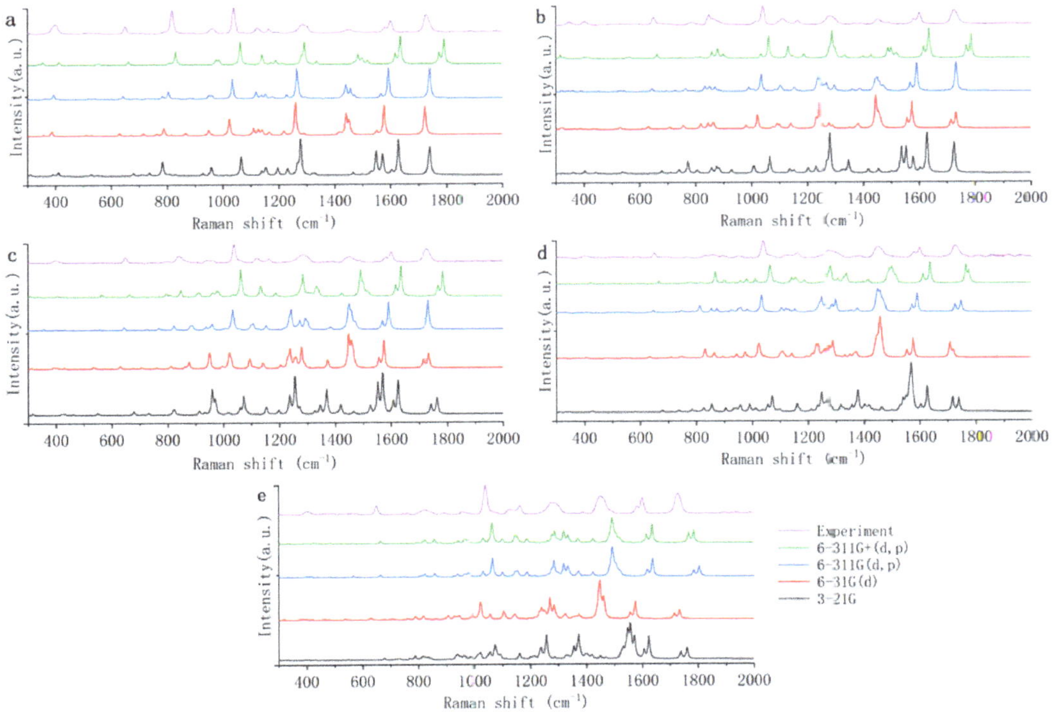

Figure 5. Theoretical Raman spectra of five phthalic acid esters calculated by DFT B3LYP with different basis sets: (**a**) DMP; (**b**) DEP; (**c**) DBP; (**d**) DEHP; (**e**) DINP. DFT: density functional theory; DMP: dimethyl phthalate; DEP: diethyl phthalate; DBP: dibutyl phthalate; DEHP: di(2-ethyl)hexyl phthalate; DINP: diisononyl phthalate.

As shown in Figure 5, compared with the other three basis sets, the theoretical Raman spectra of the 3-21G basis set have significantly more spurious peaks. Comparing the theoretical Raman spectra calculated by the 3-21G basis set with the experimental Raman spectra, it is found that the peaks of the five PAEs have common differences. The peaks of 650 cm^{-1} are all shifted to 680 cm^{-1}; the peaks of 1120 cm^{-1} are all divided into peaks of 1138 and 1156 cm^{-1}; the peaks of 1160 cm^{-1} are all shifted to 1190 cm^{-1}; and the wide peaks of 1450 cm^{-1} are all divided into peaks of 1548 and 1570 cm^{-1}. In addition to the above common differences, the theoretical Raman spectra of the five PAEs also individually have many spurious Raman peaks. This may be because the 3-21G basis set has only three original functions fitting per nuclear orbital basis function [42], which makes the theoretical spectra of PAEs inaccurate. Therefore, the 3-21G basis set is not applicable to the theoretical study of PAEs.

The theoretical Raman spectra calculated by the 6-31G(d) basis set have slightly more spurious peaks than the theoretical Raman spectra calculated by the 6-311G(d, p) and 6-311G+(d, p) basis sets. Comparing the theoretical Raman spectra calculated by 6-31G(d) with the experimental Raman spectra, it is found that the peaks of the five PAEs have common differences. The wide peaks of 1450 cm^{-1} all turn into a sharp peak and the intensity increases too much. The peaks of 1726 cm^{-1} are all divided into peaks of 1706 and 1720 cm^{-1}. In addition to the above common differences, the theoretical Raman spectra of the five PAEs also have some spurious peaks. It may be because the 6-31G(d) basis set is represented by two basis functions per valence orbit, which is one function less than the other two basis groups; so, the accuracy of the 6-31G(d) basis set is a bit worse for PAEs.

The difference between the theoretical spectra calculated by 6-311G(d, p) and 6-311G+(d, p) basis sets is extremely small, and theoretical spectra are both in good agreement with the experimental spectra. However, the scale factors of the 6-311G+(d, p) basis set is not found in the scale factors database. Therefore, compared with the Raman peaks of the experimental spectra, the Raman peaks of the theoretical spectra calculated by the 6-311G+(d, p) basis set are all blue shifted as a whole. In addition, because the 6-311G+(d, p) basis set has more plus dispersion functions on heavy atoms than 6-311G(d, p), the calculation of 6-311G+(d, p) takes nearly three times longer time than 6-311G(d, p). Therefore 6-311G(d, p) is more appropriate for the theoretical study of PAEs.

In summary, the DFT B3LYP 6-311G(d, p) is most suitable for the theoretical study of PAEs. However, the theoretical spectra calculated by DFT B3LYP 6-311G(d, p) still have some differences with experimental Raman spectra in some details. Therefore, the theoretical Raman spectra obtained by DFT B3LYP 6-311G(d, p) were further analyzed. Table 2 shows the common Raman peaks in the theoretical and experimental Raman spectra of the five PAEs. There are some differentiated peaks in theoretical Raman spectra. Compared with the results calculated using the DFT B3LYP 6-311G (d, p) method in the literature, the results of DEHP in this study are basically consistent with the results of DEHP in the literature [37]. In the literature, the experimental and theoretical Raman peaks of DEHP are 399, 653, 1043, 1127, 1167, 1585, 1605, 1731, and 385, 645, 1043, 1134, 1163, 1583, 1608, 1742, 1751 cm^{-1}, respectively.

Table 2. Common Raman peaks in theoretical and experimental Raman spectra of the five PAEs.

Experimental (cm^{-1})	DFT B3LYP 6-311G(d, p) (cm^{-1})				
	DMP	DEP	DBP	DEHP	DINP
400	392	390	356	384	400
650	642	646	646	646	664
1040	1034	1038	1032	1034	1064
1120	1118	1102	1106	1104	1102
1160	1140, 1152	1152	1152	1154	1145
1284	1264	1244, 1268	1242, 1272	1248, 1294	1284, 1320
1450	1440, 1456	1448	1448	1448	1538
1580	1564	1568	1568	1574	1620
1600	1592	1590	1592	1590	1636
1726	1740	1730	1732	1726, 1750	1786, 1800

PAEs: phthalic acid esters; DMP: dimethyl phthalate; DEP: diethyl phthalate; DBP: dibutyl phthalate; DEHP: di(2-ethyl)hexyl phthalate; DINP: diisononyl phthalate.

In addition to the peaks in the Table 2, there are still a few other peaks in the Raman spectra of PAEs. Compared with the experimental spectrum, the theoretical spectrum of DMP has two more peaks of 782 and 948 cm^{-1}, which are extremely weak and negligible. The theoretical spectrum of DEP has four more Raman peaks of 342, 834, 870, and 990 cm^{-1} than the experimental spectrum. The peak of 342 cm^{-1} can be regarded as the differentiated peak from the peak of 360 cm^{-1}, peaks of 834 and 870 cm^{-1} can be regarded as the differentiated peaks from the peak of 850 cm^{-1}, and peak of 990 cm^{-1} can be ignored because its peak strength is small. The theoretical spectrum of DEHP has two more Raman peaks of 924 and 982 cm^{-1} than the experimental spectrum, and peaks of 924 and 982 cm^{-1} can be regarded as the differentiated peaks from the peak of 958 cm^{-1}. The theoretical spectrum of DINP has one more Raman peak of 858 cm^{-1} than the experimental spectrum, which can be regarded as the differentiated peak from the peak of 822 cm^{-1}. The peaks of the theoretical spectrum of DINP are blue-shifted by nearly 40 cm^{-1}, while the theoretical Raman peaks of the other four PAEs are all shifted approximately 0~20 cm^{-1} relative to the experimental Raman peaks.

From the above results, it can be seen that the theoretical Raman spectra of PAEs still have some differences from the experimental Raman spectra. These differences may be caused by the following reasons. First, the Raman instrument has accuracy problems. Second, the DFT may take the electronic correlation too much into account, leading to calculation errors [20]. Third, theoretical studies generally calculate the structure of individual molecules, while the substances detected experimentally are multimolecular [23]. There are interactions between molecules, and this leads to errors between theory and experiment.

3.5. Vibration Mode Assignment of Raman Peaks

The above results show that the theoretical Raman spectra of five PAEs (DMP, DEP, DBP, DEHP, and DINP) calculated by DFT B3LYP 6-311G(d, p) are the best in agreement with the experimental Raman spectra, and have the least spurious peaks. Combining the experimental and theoretical spectra, the Raman peaks of the five PAEs were assigned. Table 3 shows the common Raman peaks and vibrational mode assignments of the five PAEs. Table 4 shows the unique Raman peaks and vibrational mode assignments of the five PAEs. The five PAEs can be identified by their unique Raman peaks.

Table 3. Common Raman peaks and vibrational mode assignments of the five PAEs.

PAEs Type	Theoretical (cm^{-1})	Experimental (cm^{-1})	Assignments	Strength
DMP, DEP, DBP, DEHP and DINP	390	400	γ(C-C of the benzene ring)	m
	634	650	β(Benzene)	s
	1020	1040	β(C-H of the benzene ring)	vs
	1104	1120	β(C-C-O) β(C-H of the benzene ring)	m
	1152	1160	β(C-H of the benzene ring)	s
	1264	1284	υ(C-C-O) γ(C-H) β(C-H of the benzene ring)	m w s
	1438	1450	γ(C-H)	m
	1546	1580	β(Benzene)	m
	1570	1600	β(Benzene)	s
	1696	1726	υ(C=O) γ(C-H)	s w

DMP: dimethyl phthalate; DEP: diethyl phthalate; DBP: dibutyl phthalate; DEHP: di(2-ethylhexyl phthalate; DINP: diisononyl phthalate;υ: telescopic vibration; β: in-plane bending vibration; γ: out-of-plane bending vibration; δ: deformation vibration; vs: very strong; s: strong; m: medium; w: weak.

Table 4. Unique Raman peaks and vibrational mode assignments of the five PAEs.

PAEs Type	Theoretical (cm^{-1})	Experimental (cm^{-1})	Assignments	Strength
DMP	802	818	δ(O=C-O)	w
			γ(C-H of the -CH3)	m
	960	964	υ(O-CH3)	m
			γ(C-H of the benzene ring)	m
DEP	360	352	υ(C-H of -C2H5)	m
	764	784	γ(C-H of -C2H5)	s
	850	848	β(O=C-O)	w
			γ(C-H of -C2H5)	m
DBP	822	810	β(O=C-O)	w
			γ(C-H of -C4H9)	m
	882	842	γ(C-H of -C4H9)	m
	936	940	υ(O-CH3)	s
	958	962		
DEHP	812	834	β(O=C-O)	w
	824	858	β(O=C-O)	m
			γ(C-H of -C2H3(C2H5)C4H9)	m
	874	894	γ(C-H of -C2H3(C2H5)C4H9)	m
	958	956	υ(C-O-C)	m
DINP	822	822	β(O=C-O)	m
	940	900	γ(C-H of -C7H13(CH3)2)	m
	978	960	υ(C-O-C)	m

DMP: dimethyl phthalate; DEP: diethyl phthalate; DBP: dibutyl phthalate; DEHP: di(2-ethyl)hexyl phthalate; DINP: diisononyl phthalate;υ: telescopic vibration; β: in-plane bending vibration; γ: out-of-plane bending vibration; δ: deformation vibration; vs: very strong; s: strong; m: medium; w: weak.

4. Conclusions

In this study, the theoretical Raman spectra of five PAEs (DMP, DEP, DBP, DEHP, and DINP) were calculated using different theoretical methods and basis sets, and the best theoretical method was determined by comparing with the experimental spectra. Also, the common and unique Raman peaks the of five PAEs were identified, and the vibration modes were assigned to these peaks. The results indicate that DFT is more suitable for the theoretical study of PAEs than HF. In the DFT, the B3LYP method is more accurate than the B3PW91 method to calculated the theoretical spectra of PAEs, and 6-311G (d, p) is most suitable for the theoretical study of PAEs among these four basis sets. So, DFT B3LYP 6-311G(d, p) is the most applicable method for the theoretical calculation of the Raman spectra of PAEs, which can reduce the influence of spurious peaks and help to identify the Raman characteristic peaks of PAEs. This will be beneficial for the detection of trace PAEs and the discrimination of PAEs in food products or human blood in the future. Also, the results of this study will help us to establish a Raman fingerprint for PAEs. In the future, further studies must be considered to detect the trace PAEs in food products or human blood by Raman spectroscopy combined with DFT calculation.

Author Contributions: Conceptualization, Y.W. and T.S.; methodology, Y.W. and T.S.; software, Y.W.; validation, Y.W. and M.L.; formal analysis, Y.W. and T.S.; investigation, Y.W. and T.S.; resources, T.S. and D.H.; data curation, Y.W.; writing—original draft preparation, Y.W. and T.S.; writing—review and editing, T.S. and D.H.; visualization, Y.W. and T.S.; supervision, T.S. and D.H.; project administration, T.S. and D.H.; funding acquisition, T.S. and D.H. All authors have read and agreed to the published version of the manuscript.

Funding: This research was funded by the Fundamental Research Funds for the Provincial Universities of Zhejiang (No. 2021TD002) and Talent launch project of Zhejiang A & F University Scientific Research Development Fund (No. 2019FR033).

Data Availability Statement: The data used to support the findings of this study can be made available by the corresponding author upon request.

Conflicts of Interest: The authors declare no conflict of interest.

References

1. Zhao, E.; Xu, Z.; Xiong, X.; Hu, H.; Wu, C. The impact of particle size and photoaging on the leaching of phthalates from plastic waste. *J. Clean. Prod.* **2022**, *367*, 133109. [CrossRef]
2. He, L.; Gielen, G.; Bolan, N.S.; Zhang, X.; Qin, H.; Huang, H.; Wang, H. Contamination and remediation of phthalic acid esters in agricultural soils in China: A review. *Agron. Sustain. Dev.* **2015**, *35*, 519–534. [CrossRef]
3. Li, J.H.; Ko, Y.C. Plasticizer incident and its health effects in Taiwan. *Kaohsiung J. Med. Sci.* **2012**, *28*, S17–S21. [CrossRef]
4. Ruan, H.; Rong, W.; Ma, Y.; Ji, W.; Liu, H. Recent progress of detection methods for phthalate esters. *Chin. J. Food Hygi* **2014**, *26*, 193–198.
5. Yu, S. Detection and toxicity analysis of plasticizer in plastic bottled vinegar. *China Condiment* **2019**, *44*, 171–175.
6. Li, J.; Hu, X.; Zhou, Y.; Zhang, L.; Ge, Z.; Wang, X.; Xu, W. β-Cyclodextrin-stabilized Au nanoparticles for the detection of butyl benzyl phthalate. *ACS Appl. Nano Mater.* **2019**, *2*, 2743–2751. [CrossRef]
7. Rong, Y.; Ali, S.; Ouyang, Q.; Wang, L.; Li, H.; Chen, Q. Development of a bimodal sensor based on upconversion nanoparticles and surface-enhanced Raman for the sensitive determination of dibutyl phthalate in food. *J. Food Compost. Anal.* **2021**, *100*, 103929. [CrossRef]
8. Xu, S.; Li, H.; Guo, M.; Wang, L.; Li, X.; Xue, Q. Liquid-liquid interfacial self-assembled triangular Ag nanoplate-based high-density and ordered SERS-active arrays for the sensitive detection of dibutyl phthalate (DBP) in edible oils. *Analyst* **2021**, *146*, 4858–4864. [CrossRef] [PubMed]
9. Wu, L.; Wang, W.; Zhang, W.; Su, H.; Liu, Q.; Gu, J.; Deng, T.; Zhang, D. Highly sensitive, reproducible and uniform SERS substrates with a high density of three-dimensionally distributed hotspots:gyroid-structured Au periodic metallic materials. *NPG Asia Mater.* **2018**, *10*, e462. [CrossRef]
10. Zhou, Y.; Li, J.; Zhang, L.; Ge, Z.; Wang, X.; Hu, X.; Xu, T.; Li, P.; Xu, W. HS-beta-cyclodextrin-functionalized $Ag@Fe_3O_4@Ag$ nanoparticles as a surface-enhanced Raman spectroscopy substrate for the sensitive detection of butyl benzyl phthalate. *Anal. Bioanal. Chem.* **2019**, *411*, 5691–5701. [CrossRef] [PubMed]
11. Cao, Q.; Che, R. Tailoring Au-Ag-S composite microstructures in one-pot for both SERS detection and photocatalytic degradation of plasticizers DEHA and DEHP. *ACS Appl. Mater. Interfaces* **2014**, *6*, 7020–7027. [CrossRef] [PubMed]
12. Wang, H.; Wang, C.; Huang, J.; Liu, Y.; Wu, Y.; You, R.; Zhang, J.; Lu, Y.; Shen, H. Preparation of SERS substrate with 2D silver plate and nano silver sol for plasticizer detection in edible oil. *Food Chem.* **2023**, *409*, 135363. [CrossRef] [PubMed]
13. Wu, X.; Ma, R.; Xu, B.; Wang, Z.; Du, Z.; Zhang, X.; Niu, Y.; Gao, S.; Liu, H.; Zhang, Y. Qualitative and quantitative studies of plasticizers in extra virgin olive oil by surface-enhanced Raman spectroscopy combined with chemometrics. *Vib. Spectrosc.* **2023**, *126*, 103527. [CrossRef]
14. Alparone, A. Density functional theory Raman spectra of cyclic selenium clusters Se-n (n = 5–12) *Comput. Theor. Chem.* **2012**, *988*, 81–85. [CrossRef]
15. Arici, K.; Yilmaz, R. Infrared Spectra, Density Functional Theory and Hartree-Fock Theoretical Calculations of 2-Methyl-8-quinolinol. *Chem. Asian J.* **2013**, *25*, 7106–7114. [CrossRef]
16. Ma, Y.; Hu, W.; Song, X.; Wang, C. Density Functional theory study on Raman spectra of rhodamine molecules in different forms. *Chin. J. Chem. Phys.* **2014**, *27*, 291–296. [CrossRef]
17. Weck, P.F.; Gordon, M.E.; Greathouse, J.A.; Bryan, C.R.; Meserole, S.P.; Rodriguez, M.A.; Payne, C.; Kim, E. Infrared and Raman spectroscopy of alpha-ZrW_2O_8: A comprehensive density functional perturbation theory and experimental study. *J. Raman Spectrosc.* **2018**, *49*, 1373–1384. [CrossRef]
18. Zhang, Z.; Wang, W.; Liu, S.; Chen, D. Experimental and density functional theory calculation studies on Raman and infrared spectra of 1,1′-Binaphthyl-2,2′-diamine. *Chin. J. Chem. Phys.* **2017**, *30*, 7–15. [CrossRef]
19. Ji, L.; Sun, Y.; Xie, Y.; Wang, H.; Qian, H.; Wang, L.; Yao, W. Density functional theory and surface-enhanced Raman spectroscopy studies on endocrine-disrupting chemical, dimethyl phthalate. *Vib. Spectrosc.* **2015**, *79*, 44–51. [CrossRef]
20. Liu, Y.; Jiang, X.; Fan, Y.; Shi, J.; Song, W. Comparative study of Raman spectroscopy of three kinds of the commonly used plasticizer. *J. Light. Scatt.* **2015**, *27*, 219–224.
21. Qiu, Y.; Xin, M.; Li, Y. Derivatization enhanced Raman characteristic vibration spectrum of PAEs based on pharmacophore model. *Spectrosc. Spect. Anal.* **2018**, *38*, 441–447.
22. Xu, X.; Shen, X.; Yang, X.; Chen, J.; Zhao, M.; Liu, J.; Zhao, Y.; Cui, F.; Li, X. Rapid analysis of phthalate esters in plastic toys by laser Raman technology. *Spectrosc. Spect. Anal.* **2020**, *40*, 1929–1933.
23. Zuo, X.; Yi, P.; Chen, Q.; Wu, M.; Zhang, L.; Pan, B.; Xing, B. Inter-molecular interactions of phthalic acid esters and multi-stage sorption revealed by experimental investigations and computation simulations. *Chem. Eng. J.* **2022**, *431*, 134018. [CrossRef]
24. Chernyshev, V.A.; Agzamova, P.A.; Arkhipov, A.V. Phonon spectrum of $Eu_2Sn_2O_7$: Ab initio calculation. *Opt. Spectrosc.* **2020**, *128*, 1800–1808. [CrossRef]
25. Huff, G.S.; Gallaher, J.K.; Hodgkiss, J.M.; Gordon, K.C. No single DFT method can predict Raman cross-sections, frequencies and electronic absorption maxima of oligothiophenes. *Synth. Met.* **2017**, *231*, 1–6. [CrossRef]
26. Kuzmin, V.V.; Novikov, V.S.; Ustynyuk, L.Y.; Prokhorov, K.A.; Sagitova, E.A.; Nikolaeva, G.Y. Raman spectra of polyethylene glycols: Comparative experimental and DFT. *J. Mol. Struct.* **2020**, *1217*, 128331. [CrossRef]

27. Kazachenko, A.S.; Tomilin, F.N.; Pozdnyakova, A.A.; Vasilyeva, N.Y.; Malyar, Y.N.; Kuznetsova, S.A.; Avramov, P.V. Theoretical DFT interpretation of infrared spectra of biologically active arabinogalactan sulphated derivatives. *Chem. Pap.* **2020**, *74*, 4103–4113. [CrossRef]
28. Su, Y.Q.; Wu, D.Y.; Tian, Z.Q. Density functional theory calculations of the influence of weak hydrogen bonding interactions on the Raman spectra of thiourea in aqueous Solution. *Acta Phys.-Chim. Sin.* **2014**, *30*, 1993–1999.
29. Avci, D.; Bahceli, S. Quantum chemical insight into molecular structure, spectroscopic and nonlinear optical studies on methylene bis(dithiobenzoate). *Opt. Spectrosc.* **2021**, *129*, 1207–1218. [CrossRef]
30. Jin, R.Y.; Sun, X.H.; Liu, Y.F.; Long, W.; Lu, W.T.; Ma, H.X. Synthesis, crystal structure, IR, H-1 NMR and theoretical calculations of 1,2,4-triazole Schiff base. *J. Mol. Struct.* **2014**, *1062*, 13–20. [CrossRef]
31. Bielecki, J.; Lipiec, E. Basis set dependence using DFT/B3LYP calculations to model the Raman spectrum of thymine. *J. Bioinform. Comput. Biol.* **2016**, *14*, 1650002. [CrossRef]
32. Erdogdu, Y.; Manimaran, D.; Gulluoglu, M.T.; Amalanathan, M.; Joe, I.H.; Yurdakul, S. FT-IR, FT-Raman, NMR spectra and DFT simulations of 4-(4-fluoro-phenyl)-1H-imidazole. *Opt. Spectrosc.* **2013**, *114*, 525–536. [CrossRef]
33. Wang, M.; Geng, J.J.; Wei, Z.B.; Wang, Z.Y. FT-IR, Raman and NMR spectra, molecular geometry, vibrational assignments, ab initio and density functional theory calculations for diethyl phthalate. *Chin. J. Struc. Chem.* **2013**, *32*, 890–902.
34. Wattanathana, W.; Nootsuwan, N.; Veranitisagul, C.; Koonsaeng, N.; Suramitr, S.; Laobuthee, A. Crystallographic, spectroscopic (FT-IR/FT-Raman) and computational (DFT/B3LYP) studies on 4,4′-diethy1-2,2′-[methylazanediylbis (methylene)] diphenol. *J. Mol. Struct.* **2016**, *1109*, 201–208. [CrossRef]
35. Zhang, H.; Zhao, C.; Na, H. Stability enhancement of a plastic additive (Dimethyl Phthalate, DMP) with environment-friendly based on 3D-QSAR model. *Pol. J. Environ. Stud.* **2021**, *30*, 3885–3896. [CrossRef] [PubMed]
36. Xu, T.; Chen, J.; Wang, Z.; Tang, W.; Xia, D.; Fu, Z.; Xie, H. Development of prediction models on base-catalyzed hydrolysis kinetics of phthalate esters with density functional theory calculation. *Environ. Sci. Technol.* **2019**, *53*, 5828–5837. [CrossRef] [PubMed]
37. Ji, L.; Xie, Y.; Yao, W. Rapid analysis of plastic stabilizer phthalates by surface enhancement Raman spectroscopy. *Sci. Tech. Food Ind.* **2012**, *33*, 297–300.
38. Bao, J.L.; Zheng, J.; Alecu, I.M.; Lynch, B.J.; Zhao, Y.; Truhlar, D.G. Database of Frequency Scale Factors for Electronic Model Chemistries. Available online: https://comp.chem.umn.edu/freqscale/version3b2.htm (accessed on 10 June 2023).
39. Buevich, A.V.; Sauri, J.; Parella, T.; De Tommasi, N.; Bifulco, G.; Williamson, R.T.; Martin, G.E. Enhancing the utility of (1)J(CH) coupling constants in structural studies through optimized DFT analysis. *Chem. Comm.* **2019**, *55*, 5781–5784. [CrossRef]
40. He, Y.; Shi, X.; Wang, Y.; Shen, Y. A fine frequency estimation algorithm based on DFT samples and fuzzy logic for a real sinusoid. *IET Radar Sonar Navig.* **2022**, *16*, 1364–1375. [CrossRef]
41. Mamarakhmonov, M.K.; Belen'kii, L.I.; Djurayev, A.M.; Chuvylkin, N.D.; Askarov, I.R. Quantum chemical study of ferrocene derivatives 1. Arylation reactions with aminobenzoic acids. *Russ. Chem. Bull.* **2017**, *66*, 721–723. [CrossRef]
42. Zara, Z.; Iqbal, J.; Ayub, K.; Irfan, M.; Mahmood, A.; Khera, R.A.; Eliasson, B. A comparative study of DFT calculated and experimental UV/Visible spectra for thirty carboline and carbazole based compounds. *J. Mol. Struct.* **2017**, *1149*, 282–298. [CrossRef]

foods

MDPI

Article

Spoilage Monitoring and Early Warning for Apples in Storage Using Gas Sensors and Chemometrics

Limei Yin [1,2], Heera Jayan [2], Jianrong Cai [2], Hesham R. El-Seedi [3,4], Zhiming Guo [2,4,*] and Xiaobo Zou [2,4]

1 Key Laboratory of Modern Agricultural Equipment and Technology, Ministry of Education, School of Agricultural Engineering, Jiangsu University, Zhenjiang 212013, China; yinlm6@163.com
2 School of Food and Biological Engineering, Jiangsu University, Zhenjiang 212013, China; heerajayan93@outlook.com (H.J.); jrcai@ujs.edu.cn (J.C.); zou_xiaobo@ujs.edu.cn (X.Z.)
3 Pharmacognosy Group, Department of Pharmaceutical Biosciences, Biology Medical Center, Uppsala University, P.O. Box 591, SE-751 24 Uppsala, Sweden; hesham.el-seedi@fkog.uu.se
4 International Joint Research Laboratory of Intelligent Agriculture and Agri-Products Processing, Jiangsu University, Zhenjiang 212013, China
* Correspondence: guozhiming@ujs.edu.cn

Abstract: In the process of storage and cold chain logistics, apples are prone to physical bumps or microbial infection, which easily leads to spoilage in the micro-environment, resulting in widespread infection and serious post-harvest economic losses. Thus, development of methods for monitoring apple spoilage and providing early warning of spoilage has become the focus for post-harvest loss reduction. Thus, in this study, a spoilage monitoring and early warning system was developed by measuring volatile component production during apple spoilage combined with chemometric analysis. An apple spoilage monitoring prototype was designed to include a gas monitoring array capable of measuring volatile organic compounds, such as CO_2, O_2 and C_2H_4, integrated with the temperature and humidity sensor. The sensor information from a simulated apple warehouse was obtained by the prototype, and a multi-factor fusion early warning model of apple spoilage was established based on various modeling methods. Simulated annealing–partial least squares (SA-PLS) was the optimal model with the correlation coefficient of prediction set (R_p) and root mean square error of prediction (RMSEP) of 0.936 and 0.828, respectively. The real-time evaluation of the spoilage was successfully obtained by loading an optimal monitoring and warning model into the microcontroller. An apple remote monitoring and early warning platform was built to visualize the apple warehouse's sensors data and spoilage level. The results demonstrated that the prototype based on characteristic gas sensor array could effectively monitor and warn apple spoilage.

Keywords: gas sensor; spoilage monitoring; early warning; logistics control; simulated annealing; apple

Citation: Yin, L.; Jayan, H.; Cai, J.; El-Seedi, H.R.; Guo, Z.; Zou, X. Spoilage Monitoring and Early Warning for Apples in Storage Using Gas Sensors and Chemometrics. *Foods* **2023**, *12*, 2968. https://doi.org/10.3390/foods12152968

Academic Editor: Bcce Zhang

Received: 24 June 2023
Revised: 28 July 2023
Accepted: 3 August 2023
Published: 6 August 2023

1. Introduction

Apple has the characteristics of high nutritional value and storage resistance, but due to its high sugar and moisture, coupled with complex external environmental factors, it is susceptible to fungal spoilage [1,2]. Apple spoilage is the result of changes in tissue composition under the action of physical, chemical, microbial and other environmental factors. Physical aspects include mechanical damage, frostbite, etc.; chemical aspects include water imbalance and quality deterioration caused by environmental changes; microbiological aspects include fruit rot, penicillium and rot heart disease, etc. Among them, the spoilage caused by fungi is the most serious. With the increase in storage time, spoilage fungi on the surface of apples enter the fruit through stomata or calyx and other parts to generate mycelium, which eventually leads to the spoilage of apples. In addition, spoilage fungi may also produce mycotoxins that pose the risk of disease, which can seriously endanger the health of consumers [3]. By sampling and isolating the epiphytic microbiota of fresh apples, it was found that the proportion of fungi (79.0%) was much higher than that of bacteria (13.8%) [4], and the dominant spoilage fungi included *Aspergillus niger* (*A. niger*), *Penicillium expansum*, (*P. expansum*), *Penicillium chrysogenum* (*Penicillium chrysogenum*, *P. chrysogenum*) and *Alternaria alternata* (*A. alternata*), etc. [5–8]. Therefore, monitoring and providing early warning of apple spoilage have important practical significance, and can ensure the food safety of consumers and provide technical support for the healthy development of the apple industry.

At present, the traditional research on the detection of fruit and vegetable spoilage mainly includes polymerase chain reaction (PCR) [9] and gas chromatography-mass spectrometry (GC-MS) [10]. Although these detection methods are accurate, the operation is complicated and requires professional operators, which cannot meet the needs of rapid real-time detection of apple spoilage [11]. Therefore, it is of practical significance to explore a fast and effective apple spoilage monitoring technique.

Electronic nose technology is a powerful tool that simulates the olfactory system of animals [12]. The electronic nose converts chemical signals into electrical signals through the gas sensor array and combines chemometrics to process the data matrix to realize the qualitative and quantitative analysis of the detected samples. It integrates sensors, computers, mathematics and other disciplines, and has been widely used in food fields, such as medicine and the environment [13]. The aroma of fruit is an important indicator for evaluating fruit quality, and is mainly composed of various volatile components. The type and concentration of volatile components can be affected by a variety of factors, such as actual type, maturity and storage time. Electronic noses are widely used in the field of fruit and vegetable testing. By acquiring the gas data of the testing samples, non-destructive rapid testing of the quality of fruits and vegetables can be achieved.

With the development of sensor technology and chemometrics, the accuracy of gas detection has gradually improved, and electronic nose technology has been gradually applied in the field of fruit and vegetable spoilage detection [14]. When the internal quality of fruits and vegetables changes or spoilage occurs, the volatile gases of fruits and vegetables change accordingly. By analyzing the type and content of volatile gases in fruits and vegetables, the degree of spoilage of fruits and vegetables and the detection of spoilage microorganisms can be performed.

Nouri et al. [15] took pomegranate as the research object and used the electronic nose system to identify pomegranates infected with Alternaria, with an accuracy rate of 100%, and reviewed the application of gas sensors in the detection of pomegranates infected with *Alternaria*. Liu et al. [16] proposed a non-destructive testing method based on hyperspectral imaging and electronic nose, which could rapidly detect the microbial content and variety of attributes during the rotting process of strawberries. This research used PCA to extract feature information and established a quantitative prediction model for strawberry microbial content and quality traits, indicating that the changes in the appearance and internal components of fungal-infected strawberries during storage were highly correlated with microbial content. Previous studies have shown that the combination of hyperspectral imaging and electronic nose can help improve the evaluation of strawberry quality and safety.

The spoilage of apples is caused by physiological disorders or aging of apple tissues, and infection by decay-causing microorganisms. The respiration of apples and the catabolism of microorganisms will change the composition and proportion of gases in the storage micro-environment. By clarifying the evolution law of gas composition in an apple spoilage environment, real-time monitoring of indicative gas in the storage environment is an effective way to provide early warning of apple spoilage. Electronic noses are currently widely used in the field of gas monitoring; however, the information collected by traditional desktop electronic noses cannot be linked to the spoilage status of apples, and most of the research objects are single apple samples, and thus the data cannot be applied to a monitoring and early warning model for the whole environment of fruit warehouse. Therefore, this study designed and developed a prototype of apple spoilage storage monitoring and attempted to analyze the change law of gas in the process of batch apple spoilage and the spoilage influence mechanism of multi-factor coupling. An apple spoilage early warning model with various influencing factors was established, and at the same time, a remote monitoring and early warning platform was built to realize remote monitoring of spoilage early warning information.

2. Materials and Methods

2.1. Design of Monitoring Prototype

The monitoring prototype was designed to detect volatile profiles, temperature and humidity during apple storage. The prototype consisted of hardware, software and mechanical components. The hardware system measured the gas components and included a microcontroller, gas sensor array and display screen. The software system was developed to control the microcontroller and to interface with the computer. The mechanical components involved a gas delivery system that transported the gas components from the storage environment to the sensors array. The design characteristics of the monitoring prototype are further explained in the following sections.

2.1.1. Selection and Optimization of Sensors

The complexity of gas composition in the apple storage environment was fully considered in the selection of sensors. The current storage method in the warehouse mainly adopts a combination of controlled atmosphere and refrigeration, and the storage gas concentration changes in real time. The high sensitivity of the gas sensor helps in the early detection of apple spoilage. Meanwhile, the field layout prototype needs to be small and high-precision with low power consumption and high accuracy to meet the needs of spoilage monitoring. After extensive research and experiments, the main gases in the apple quality and spoilage process were determined, and C_2H_4, CO_2, volatile organic compounds (VOC) and O_2 were optimized to be the characteristic gases of the apple warehouse [17–20]. An infrared gas sensor was selected for the CO_2 sensor, and the remaining sensors were electrochemical gas sensors. This ensured that the prototype had low power consumption and high precision, which is convenient for long-term monitoring in warehouses [21]. Table 1 shows the detection range, resolution, sampling accuracy and repeatability parameters of each sensor.

Table 1. Detection range, resolution, sampling precision and repeatability of each sensor.

Sensor	Detection Range	Resolution	Precision	Repeatability
C_2H_4	0–100 ppm	0.1 ppm	±2% FS	±1% FS
O_2	0–30% VOL	0.1% VOL	±2% FS	±1% FS
VOC	0–50 ppm	0.001 ppm	±2% FS	±1% FS
CO_2	0–5000 ppm	1 ppm	±2% FS	±1% FS
Temperature	−20–80 °C	0.1 °C	±0.3 °C	±1% FS
Humidity	0–100% rh	0.1 rh	±0.3% rh	±1% FS

% FS: the percentage of accuracy and full scale.

2.1.2. Air Chamber and Air Path Design

The design requirements of the prototype need to ensure the portability of the equipment. To improve the speed and efficiency of the gas contact sensor surface, the design of the gas chamber is considered, with factors such as the size, structure, and material of the gas chamber [22]. By designing the shell with air holes, the gas sensor array was wrapped, and each gas sensor scatter arranged at the bottom. In order to ensure the circulation of the gas flow path, the external gas was sucked into the prototype by the fan and evenly passed through the surface of the sensor array to better obtain the gas information of the storage micro-environment.

2.1.3. Hardware System Integration

The hardware system included a microcontroller, gas sensor array, TF card module, micro vacuum pump, power supply and display screen [23]. Figure 1 shows the schematic representation of various components in the prototype. The microcontroller was mainly used to control the collection of gas sensor data and control other hardware devices, such as fans. The gas sensor array was used to obtain the data of the gas in the apple warehouse. The TF card module was used to store the gas data. The display screen revealed the real-time sensing data of each gas sensor.

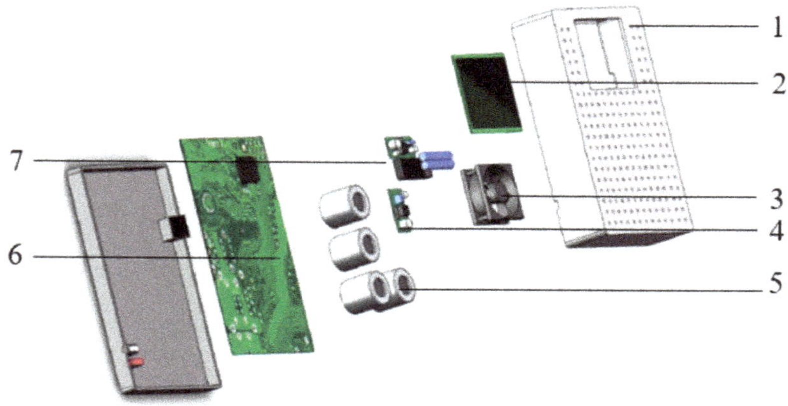

Figure 1. Schematic diagram of the structure of the apple spoilage monitoring prototype.1. Shell. 2. Display. 3. Fan. 4. Temperature/humidity sensor. 5. Gas sensor array. 6. Motherboard. 7. Battery.

The miniature air pump drove the flow of gas in the air chamber, and the one-way valve controlled the closure of the air path. The power module supplied power to the prototype, and the voltage was stabilized to 5V through the voltage regulator circuit to supply power to each sensor.

2.1.4. Software Structure Design

The software was developed under the Windows 10 operating system. The Windows operating system has been affirmed and welcomed by consumers and developers and has now been released to Windows 11. The Windows system provides many development interfaces and standards, and the maintenance difficulty is lower than other systems. The Windows 10 Professional operating system used in this study was based on the NT core, with good hardware support and higher development efficiency.

The microcontroller program was written in Keil uVision5 IDE using C language. Keil provides many library functions and development and debugging tools through the integrated environment, which is convenient for developers to call, and is currently the most popular microcontroller development tool.

Qt is a cross-platform C++ graphical user interface application development framework that enables rapid development of GUI programs and non-GUI programs. Through the visual graphical interface editor, the user can quickly and easily drag and drop controls, including buttons, radio boxes, check boxes, group boxes, tree views, table views and texts. Qt has the advantages of being cross-platform, object-oriented, easy to use and fast to run, and it is easy to transplant and can be quickly converted according to the operating system. It is widely used in the development of embedded products and device interfaces.

The dedicated prototype mainly included sensor signal acquisition, data display and data storage functions. To visualize the monitoring process, the prototype developed a special human–computer interaction interface, and the display of each sensor's data was mainly realized by the serial port screen. In order to visually display the data of each sensor, a dedicated display interface was designed.

2.1.5. Prototype System Integration

According to the software and hardware design scheme of the above-mentioned special-purpose prototype, the hardware and software systems were integrated, and the prototype assembly was finally completed. After debugging and optimization of the prototype, the repeatability and stability of the prototype were verified by acquiring apple sample information from the warehouse, and the batch test was carried out after reaching the expectation.

2.2. Apple Sample Preparation

2.2.1. Activated Culture and Inoculation of Spoilage Fungi

Aspergillus niger (CICC2089), the dominant spoilage fungi of apple, was purchased from China Industrial Microbial Species Preservation and Administration Center (CICC). Activation and culture procedures were performed in strict accordance with CICC instructions and guidelines.

The bacterial cells were recovered from lyophilization prior to inoculation. The top of the lyophilized tube with *Aspergillus niger* was placed on the alcohol lamp and heated evenly for 30 s. Then, 2–3 drops of sterile water were dropped onto the heated part. The tube wall was broken due to uneven heat, and the tear was knocked out with sterilized tweezers. The lyophilized powder was placed into a 1.5 mL centrifuge tube using an inoculum ring, and 200 μL of sterile water was added to dissolve it. The lyophilized powder solution was evenly coated on potato dextrose agar medium plates and placed in a constant temperature and humidity incubator at 28 °C. After seven days of culture, the spores of the third generation of fungi were scraped with a one-time inoculation ring and placed in sterile water, which was configured into fungal suspension. The fungal suspension was counted through a blood count plate and diluted with sterile water to a concentration of 10^6 cfu/mL.

Before inoculation, the apple skin was washed with distilled water, then wiped with 75% alcohol, and finally placed on a sterile workbench under ultraviolet light for half an hour. Apple samples were punctured with sterile syringe needles (diameter 3 mm, depth 5 mm) along the apple equator, with 3 puncture holes, each 120° apart. Then, 5 μL of fungal suspension was injected into each of the three holes and incubated in a constant temperature and humidity incubator (25 °C, 60% humidity) [5].

2.2.2. Micro-Environment Information Sensing

To simulate the conditions of apple storage in warehouses, nine simulated warehouses were set up in the laboratory [24]. Each simulated warehouse contained 30 fresh apple samples, and the gas sensing data and temperature and humidity data of the apple samples were collected for two days by the acquisition terminal prototype. Then, 10 apple samples were selected from each simulated warehouse to be inoculated with *Aspergillus niger*, and the inoculated apple samples were put back into the simulated warehouse. Data acquisitions were performed every 24 h for a total of 6 days. The data format detection system was stored in a two-dimensional table format. The collection time of a single sensor was 500 s, and the collection frequency was 1 s. The data of each simulated warehouse sample were collected as a 500 × 6 two-dimensional array based on 6 sensors. Then, the data were transformed from a 500 × 6 two-dimensional matrix into a 3000 × 1 one-dimensional array through flattening processing for subsequent model establishment.

2.3. Variable Selection Method

2.3.1. Genetic Algorithm

Genetic algorithm (GA) is an algorithm based on biological evolution rules, which automatically obtains and guides the optimized search space by simulating random search and optimization solving methods [25], which can quickly screen characteristic variables and eliminate the interference of irrelevant information [26,27], has the characteristics of simple operation and strong versatility, achieves the global optimum in a short time and can reduce the risk of falling into the local optimum search.

2.3.2. Simulated Annealing Algorithm

Simulated annealing (SA) is a probabilistic optimization algorithm for simulating the solid annealing process in metalwork [28]. SA has strict convergence characteristics following a Metropolis criterion, which effectively reduces the probability of falling into a local minimum. SA can quickly find the global optimal solution, and the final optimization result has nothing to do with the initial value [29]. It is a powerful tool for solving optimization and combination problems. SA has the characteristics of simplicity, flexibility and efficiency, which can effectively improve the generalization ability of the model.

2.3.3. Ant Colony Optimization Algorithm

Ant colony optimization (ACO) algorithm was inspired by the bionic intelligence of ant colony foraging behavior [28,30]. Ants use shared pheromones to quickly spread information in ant colonies, which helps to strengthen cooperation between ant colonies, improve global exploration capabilities and obtain better solution results [31]. The essence of ACO is based on its ability to optimize the creation of paths, and it has strong generality and robustness and is widely used in data optimization and fuzzy modeling.

2.3.4. Competitive Adaptive Reweighed Sampling

Competitive adaptive reweighed sampling (CARS) is a variable selection method suitable for high-dimensional data extraction [32]. In the sampling stage, CARS regards each variable as an independent individual, retains variables with larger weights, removes variables with smaller weights, and treats variables with significant weights as a new subset, which can effectively remove irrelevant variables and reduce collinear variables [33,34]. By selecting the optimized subset of variables, the algorithm can overcome the combinatorial explosion problem in variable selection to a certain extent, improve the prediction ability of the model, and reduce the prediction variance. CARS introduces an exponential decay function [35], which controls the retention rate of variables and improves the computational efficiency.

2.4. Apple Remote Monitoring and Early Warning Platform

The apple remote monitoring and early warning platform mainly included three parts: data upload module, remote monitoring module and spoilage early warning module, as shown in Figure 2. The platform development language was JAVA, which was developed through the SSM frameworks, including the SpringBoot, SpringMVC and MyBatis 3 frameworks [36,37]. The visualization of individual sensor data and spoilage levels was implemented by the Echarts visualization library [38].

The data upload module was mainly responsible for the upload of sensor and model data, which made it convenient for the subsequent monitoring module and spoilage early warning module to call data. The remote-monitoring module mainly displayed the trend and change of sensor data over time through a line graph and realized the visualization of each sensor's data. The schematic diagram of apple spoilage monitoring and early warning process is shown in Figure 3. The spoilage early warning module was mainly responsible for calling the data of sensors and spoilage models and realizing the visual display of spoilage levels through the dashboard.

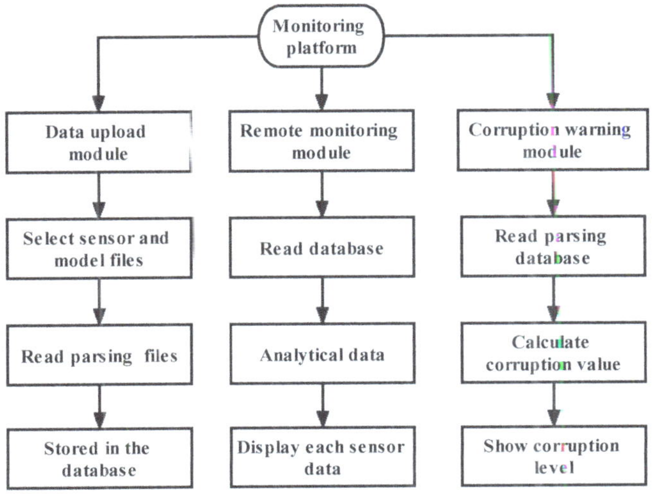

Figure 2. Flow chart of each module of apple remote monitoring and early warning platform.

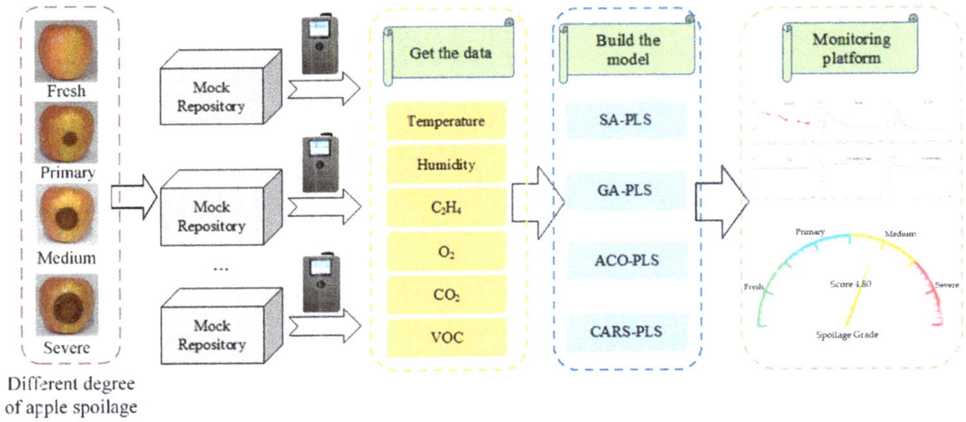

Figure 3. Schematic illustration of apple spoilage early warning model and remote monitoring and early warning.

3. Results

3.1. Analysis of Apple Volatile Gas

The volatile gas production during fungal spoilage of apples was recorded using the sensor and represented as sensor response value. The sensor response value indicated the response value of each sensor to the presence of certain gaseous chemicals over time.

The average response data of the VOC, CO_2, O_2 and C_2H_4 gas sensors of simulated warehouse apple samples are shown in Figure 4. The response value of the VOC sensor was 7 > 8 > 4 > 6 > 5 > 1 > 3 > 2, and the VOC content on the seventh and eighth days was much higher than that on the first and second days (Figure 4a). The results showed that the content of VOC gradually increased during the degradation of apples from fresh to severe spoilage. The microbial spoilage of apples led to changes in VOC emissions, which are classified as alcohols, terpenes, ketones, alkenes, benzenoids and sulfides [39].

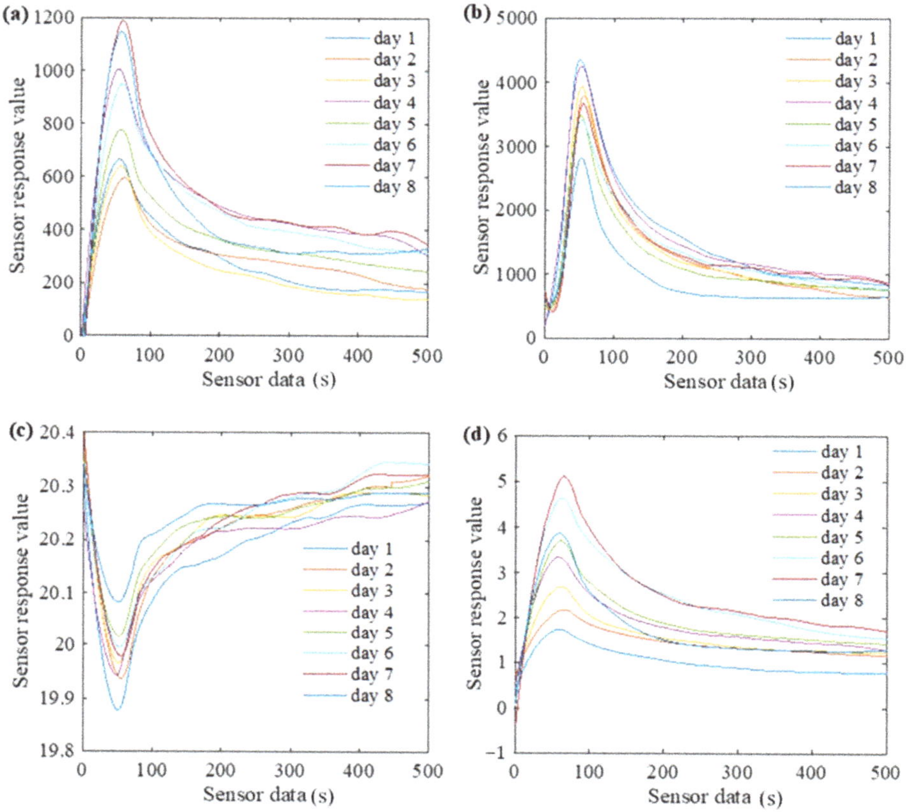

Figure 4. Response of each sensor for different days of spoilage. (**a**) VOC, (**b**) CO_2, (**c**) O_2 and (**d**) C_2H_4.

The responses of the CO_2 sensor indicated that the content on the eighth day was much higher than that on the first day: 1 < 4 < 3 < 2 < 7 < 5 < 6 < 8 in sequence (Figure 4b). The results showed that during the spoilage of apples, the CO_2 release decreased gradually. The responses of the O_2 sensor were 8 > 5 > 6 > 7 > 3 > 4 > 2 > 1 in sequence, and the content on the eighth day was much higher than that on the first day (Figure 4c). The results showed that the consumption of O_2 gradually decreased during apple spoilage process due to the natural senescence process that causes cell and tissue to breakdown.

The responses of the C_2H_4 sensor were 7 > 6 > 8 > 5 > 4 > 3 > 2 > 1 in sequence, with the highest content on the seventh day and the lowest content on the first day (Figure 4d). During the decaying process of apples from fresh to severe spoilage, the production of C_2H_4 increased gradually, except that the content decreased slightly on the eighth day of treatment [24].

Based on the results obtained [24,40], it was found that with the intensification of the degree of spoilage, the release of VOC and C_2H_4 gradually increased, while the release of CO_2 and the consumption of O_2 were generally reduced. This was because the metabolic capacity of apples decreases as they spoil, leading to a lower consumption of O_2 and a lower release of CO_2. With the increase in spoilage time, the apple samples consume O_2 and release CO_2, VOC and C_2H_4, but as the degree of spoilage progresses, the metabolic capacity of apples is reduced, leading to the low consumption of O_2 and the release of CO_2.

3.2. *Apple Spoilage Early Warning Model*

3.2.1. ACO-PLS Prediction Model of Apple Spoilage

The optimized parameters of the algorithm were set as follows: the initial population size was 50, the maximum number of cycles was 10, the maximum number of iterations was 50, the variable selection probability threshold P was 0.3 and the significance factor Q was 0.01. The characteristic variables of the sensor data were screened by the ACO method, and the screening results are shown in Figure 5a. Forty-eight characteristic variables were screened, and the ACO-PLS model results are shown in Figure 6a.

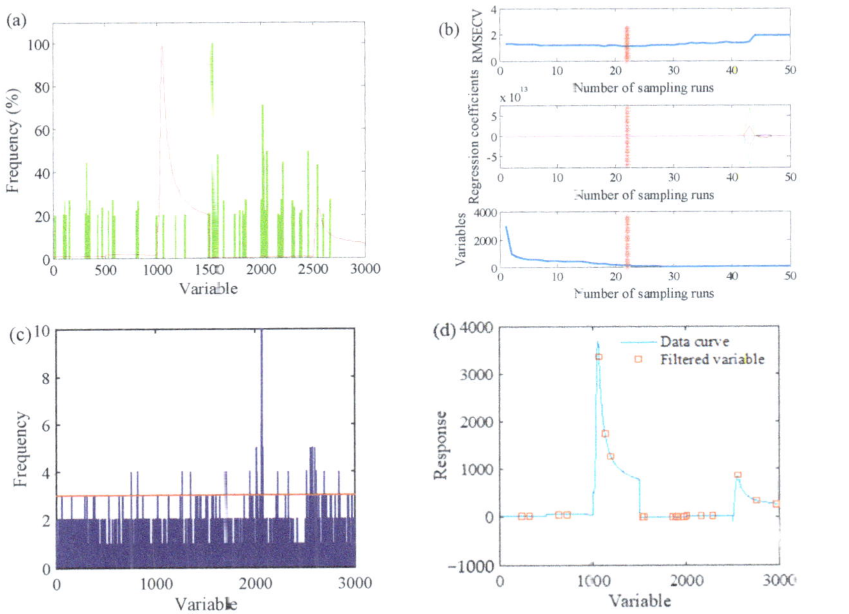

Figure 5. Variable selection results of apple spoilage days. (**a**) ACO-PLS, (**b**) CARS-PLS, (**c**) GA-PLS and (**d**) SA-PLS.

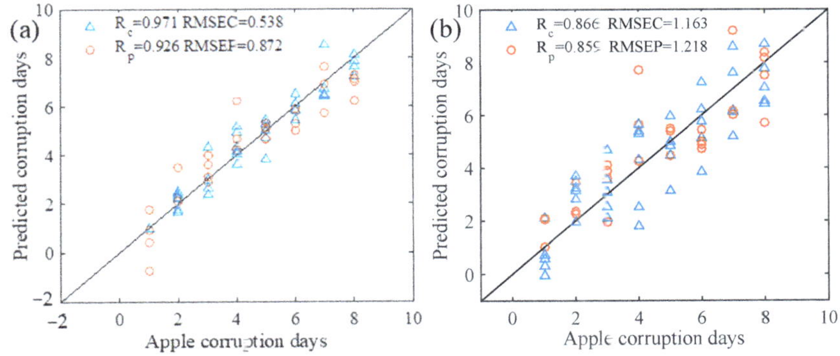

Figure 6. *Cont.*

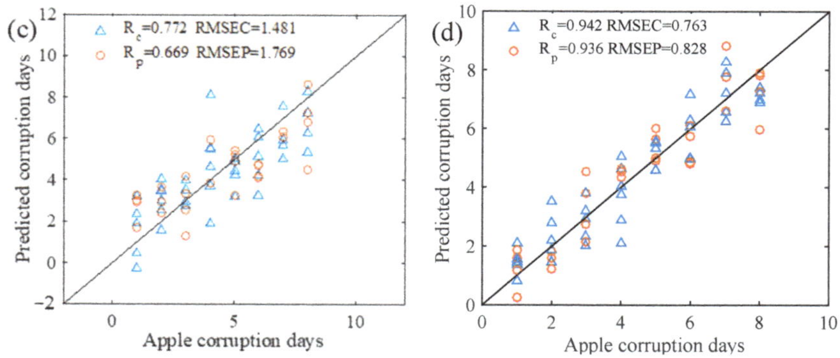

Figure 6. Prediction model of apple decay days. (**a**) ACO-PLS, (**b**) CARS-PLS, (**c**) GA-PLS and (**d**) SA-PLS. Blue triangle represents calibration set and red circle represents prediction set.

3.2.2. CARS-PLS Prediction Model of Apple Spoilage

The main parameters of CARS in this study were set as follows: the maximum number of principal components was 15, the number of interactive validation groups was 5 and the number of Monte Carlo sampling runs was 2000. Figure 5b shows the best calculation result of the CARS model of apple spoilage time. It can be seen from the figure that the RMSECV value was large at the beginning of the screening, the regression coefficient of each variable was small, and the number of variables was larger. With the increase in sampling times, the number of variables gradually decreased and the gap between the regression coefficients of each variable widened; 22 variables were screened out. The results of CARS-PLS are shown in Figure 6b.

3.2.3. GA-PLS Prediction Model of Apple Spoilage

The main parameters of GA were set as follows: the number of initial chromosomes was 30, the deletion group was 5, the mutation rate was 0.01, the crossover probability was 0.5 and the maximum number of iterations was set to 100. The cumulative frequency of the variables screened by the GA algorithm is shown in Figure 5c, and 32 variables were screened out. The PLS quantitative prediction model corresponding to the apple spoilage area was established through the screened characteristic variables. The specific scatter diagram of the model is shown in Figure 6c.

3.2.4. SA-PLS Prediction Model of Apple Spoilage

The main parameters of SA in this study were set as follows: the initial temperature was 10, the end temperature was 1, the Markov chain length was 10, the temperature cooling coefficient was 0.95, the window starting width was 10 and the window ending width was 20 and increasing. The step size was 1, the number of wavenumber points exchanged each time the Markov chain was 2 and the maximum number of principal components for modeling was 12. Figure 5d shows the results of variable screening of sensor data through SA, and 20 characteristic variables were screened out. The results of the established SA-PLS model are shown in Figure 6d.

3.3. Comparison and Analysis of Various Models

The variable selection method was selected to filter the characteristic variables of the sensor data, and the PLS prediction model of the number of days of apple spoilage was established. The specific results of the apple spoilage time prediction model established by the variable selection method are shown in Table 2. The scatter plot of the apple spoilage time prediction model is shown in Figure 6, in which the R_c and R_p of ACO-PLS are 0.971 and 0.926, respectively, the R_c and R_p of SA-PLS are 0.942 and 0.936, respectively, and ACO-PLS has the highest R_c, but R_p is low. In order to ensure the prediction accuracy of the apple spoilage model, the variables were numbered from 1 to 3000, and SA-PLS was used to establish an early warning model of apple spoilage. The characteristic variables were identified to be 1889, 1894, 1974, 2001, 2159, 2163, 2274, 2561, 2758 and 2965. A similar observation was made by Guo et al. [41], where they observed that the competitive adaptive reweighted sampling (CARS) algorithm combined with PLS effectively filtered irrelevant information and improved the accuracy of the model in predicting apple spoilage area from the electronic nose data. Table 3 showed the characteristic variables and original variable ranges screened by SA.

Table 2. Prediction model results of days before apple spoilage using C_2H_4, CO_2, VOC and O_2 sensor data.

Model	Calibration Set		Prediction Set	
	R_c	RMSEC	R_p	RMSEP
GA-PLS	0.772	1.481	0.669	1.769
SA-PLS	0.942	0.763	0.936	0.828
ACO-PLS	0.971	0.538	0.926	0.872
CARS-PLS	0.866	1.163	0.859	1.218

Table 3. Characteristic variables and original variable ranges of each sensor in the apple spoilage early warning model.

Sensor	Characteristic Variables	Original Variable Ranges
Temperature	228, 309	0–500
Humidity	622, 726	500–1000
CO_2	1064, 1126, 1188	1000–1500
C_2H_4	1526, 1538, 1861, 1889, 1894, 1974	1500–2000
O_2	2001, 2159, 2163, 2274	2000–2500
VOC	2561, 2758, 2965	2500–3000

According to the results of the SA-PLS spoilage early warning model, the dependent variables, independent variables, and coefficients of the model were derived. The model results are shown in Table 4. The apple spoilage early warning model is as follows: $Y = 0.3264\ X1 + 0.3708\ X2 + 0.0248\ X3 + 0.0363\ X4 - 0.0008\ X5 - 0.0005\ X6 - 0.0014\ X7 + 0.4734\ X8 + 0.3338\ X9 + 0.0248\ X10 - 0.0136\ X11 - 0.0118\ X12 - 0.0132\ X13 + 0.3407\ X14 - 1.9581\ X15 + 0.3719\ X16 + 0.5173\ X17 - 1.9010\ X18 + 0.0013\ X19 - 0.0009\ X20 + 38.9899$. Among these values, X1–X20 are the dependent variables, that is, the value of the sensor corresponding to the screening feature variable. When the value of Y is in the range of 1–8, the Y value from 1–2 indicates the freshness of the product, whereas a value between 3 and 4 indicates spoilage. Similarly, a value of 5–6 indicates that the spoilage grade is medium spoilage, and a value of 7–8 indicates that the spoilage grade is severe spoilage.

Table 4. Independent variables, dependent variables and coefficients of apple spoilage early warning model.

Number	Independent Variables	Dependent Variables	Number	Independent Variables	Dependent Variables
1	0.3264	228	11	−0.0136	1889
2	0.3708	309	12	−0.0118	1894
3	0.0248	622	13	−0.0132	1974
4	0.0363	726	14	0.3407	2001
5	−0.0008	1064	15	−1.9581	2159
6	−0.0005	1126	16	0.3719	2163
7	−0.0014	1188	17	0.5173	2274
8	0.4734	1526	18	−1.9010	2561
9	0.3338	1538	19	0.0013	2758
10	0.0248	1861	20	−0.0009	2965
Coefficient			38.9899		

4. Discussion

Apples are prone to spoilage due to physical damage and microbial infections, which can lead to widespread infection and post-harvest losses. The current study has developed a sensor prototype for apple spoilage monitoring and an early warning system for apples. The gas composition information collected by the sensor prototype from the simulated warehouse was used to develop a multi-factor fusion early warning model to predict apple spoilage during storage. Among the different models employed for the purpose, the ACO-PLS model showed the highest correlation coefficient in calibration as well as prediction set. However, it is worth mentioning that the SA-PLS model was suitable for identifying the characteristic variables. A similar outcome was obtained by Ren et al. [38] when a multilayer perceptron neural network (MLPN) model was employed on volatile components released by damaged apples upon mechanical injury. The model was able to classify the degree of damage with 100% accuracy based on the volatile components released from apples. However, the selection of a model for the prediction of the quality of fruits greatly varies depending on the specific situation and intention. Similarly, monitoring gas emitted by fruits and vegetables in combination with multivariate chemometric analysis has been proven to be effective in detecting spoilage inside refrigerators [42]. The use of artificial neural networks as a machine learning model is prevalent in developing a predictive model from electronic nose data [43]. However, compared to neural network models, PLS models are more interpretable, robust to noise and faster to train, making them suitable for real-world applications. With the involvement of too many predictor variables, the PLS model becomes unstable, and the prediction can be inaccurate. Thus, the selection of optimum parameters for modeling is an important task in the process of model building. In this study among the different variable selection methods (ACO, CARS, GA and SA) employed for the reduction of computational complexity, the SA method showed better results. Similarly, Zhao et al. [44] employed SSA to select optimal parameters for the BPNN network model developed for the prediction of fungal infection in apples using electronic nose data.

The developed prototype has a gas monitoring array to collect data on VOC, CO_2, O_2, C_2H_4, temperature and humidity. These data are then employed to build a multifactor fusion early warning model. Other works have reported that a mathematical model for shelf-life analysis could not be established based only on one specific parameter. The development of the model has shown very high efficiency in establishing shelf-life predictions in several industries, including dairy [45].

The developed model based on the gas sensor prototype has the potential to be employed in the early detection of fruit spoilage, which prevents postharvest losses and reduces economic loss significantly. There were similar attempts to predict the spoilage of bananas based on the 3D fluorescence data of the storage room gas. The spoilage benchmark for the stored banana was estimated as early as the 4th day, which demonstrates the reliability of the early warning systems [46]. However, it is worth mentioning that the early detection system was established based on the fusion of information from no less than five representative physical and chemical indicators from bananas. Moreover, Putnik et al. [47] have developed a mathematical model to compare the influence of modified atmosphere packaging in the prevention of apple browning, and the model was published online as a computer simulation that predicts shelf life given the basic quality parameters of apples, thus proving the practical applicability of model development in predicting apple spoilage. The models can be employed to answer questions about the economic benefits of using different treatments in industrial apple production and storage. This is indirectly helpful in extending the shelf-life of apples and providing economic benefits to producers, while also ensuring that consumers receive high-quality food.

Overall, this work presents a novel approach for monitoring and providing early warning of spoilage of apples in storage based on a gas sensor prototype. The developed system can help reduce post-harvest losses by preventing the spread of infection in the storage environment. Moreover, the developed method can assist in streamlining the process of monitoring and inspecting apples for fungal spoilage.

5. Conclusions

In view of the current problems and difficulties of spoilage monitoring in apple warehouses, we analyzed the overall framework of a sensor prototype for the detection of various gases produced during apple spoilage. The prototype was developed by integrating software and hardware development of acquisition terminals to continuously obtain sensor data from simulated warehouses and analyze the process of spoilage in batches of apples. The change in various gaseous composition and their influence on spoilage was identified to be a result of multi-factor coupling. The early warning model for apple spoilage was developed by considering multiple factors, including temperature, humidity and gas composition, and a variable selection method was employed to optimize the characteristic variable, which predicted the degree of spoilage. The developed remote monitoring platform had three major modules, including data upload module, remote monitoring module and spoilage early warning module. These modules enabled the upload of sensor data, visualization of the trends in the data over time and visual display of spoilage levels. Thus, the study proved that analyzing the change in gas composition in an apple spoilage micro-environment and real-time monitoring and warning of indicator gas is an effective way to reduce postharvest losses of fruits.

Author Contributions: Conceptualization, Z.G.; methodology, L.Y.; software, L.Y.; validation, L.Y. and H.J.; formal analysis, X.Z.; investigation, L.Y.; resources, Z.G.; data curation, J.C.; writing—original draft preparation, L.Y. and H.J.; writing—review and editing, H.R.E.-S.; visualization, L.Y.; supervision, Z.G. and J.C.; project administration, Z.G.; funding acquisition, Z.G. All authors have read and agreed to the published version of the manuscript.

Funding: This research was funded by the National Key R&D Program of China (No. 2022YFD2100604), the Outstanding Young Teachers of Blue Project in Jiangsu Province, Key R&D Project of Jiangsu Province (BE2022363), Jiangsu Agriculture Science and Technology Innovation Fund (CX(22)3069), Postgraduate Research & Practice Innovation Program of Jiangsu Province (SJCX21_3383), the Open Fund of Key Laboratory of Modern Agricultural Equipment and Technology of Ministry of Education (MAET202117) and the Youth Project of Faculty of Agricultural Equipment of Jiangsu University (NZXB20210205).

Institutional Review Board Statement: Not applicable.

Informed Consent Statement: Not applicable.

Data Availability Statement: The data used to support the findings of this study can be made available by the corresponding author upon request.

Conflicts of Interest: The authors declare no conflict of interest.

References

1. Butkeviciute, A.; Viskelis, J.; Liaudanskas, M.; Viskelis, P.; Janulis, V. Impact of Storage Controlled Atmosphere on the Apple Phenolic Acids, Flavonoids, and Anthocyanins and Antioxidant Activity In Vitro. *Plants* **2022**, *11*, 201. [CrossRef] [PubMed]
2. Shen, Y.; Zhang, J.; Nie, J.; Zhang, H.; Bacha, S.A.S. Apple microbial communities and differences between two main Chinese producing regions. *Food Qual. Saf.* **2022**, *6*, fyab033. [CrossRef]
3. Vidal, A.; Ouhibi, S.; Ghali, R.; Hedhili, A.; De Saeger, S.; De Boevre, M. The mycotoxin patulin: An updated short review on occurrence, toxicity and analytical challenges. *Food Chem. Toxicol.* **2019**, *129*, 249–256. [CrossRef]
4. Angeli, D.; Sare, A.R.; Jijakli, M.H.; Pertot, I.; Massart, S. Insights gained from metagenomic shotgun sequencing of apple fruit epiphytic microbiota. *Postharvest Biol. Technol.* **2019**, *153*, 96–106. [CrossRef]
5. Liu, C.-Q.; Hu, K.-D.; Li, T.-T.; Yang, Y.; Yang, F.; Li, Y.-H.; Liu, H.-P.; Chen, X.-Y.; Zhang, H. Polygalacturonase gene pgxB in *Aspergillus niger* is a virulence factor in apple fruit. *PLoS ONE* **2017**, *12*, e0173277. Erratum in *PLoS ONE* **2018**, *13*, e0191350. [CrossRef] [PubMed]
6. Žebeljan, A.; Vico, I.; Duduk, N.; Žiberna, B.; Krajnc, A.U. Dynamic changes in common metabolites and antioxidants during *Penicillium expansum*-apple fruit interactions. *Physiol. Mol. Plant Pathol.* **2019**, *106*, 166–174. [CrossRef]

7. Agustín, M.D.R.; Tarifa, M.C.; Vela-Gurovic, M.S.; Brugnoni, L.I. Application of natamycin and farnesol as bioprotection agents to inhibit biofilm formation of yeasts and foodborne bacterial pathogens in apple juice processing lines. *Food Microbiol.* **2023**, *109*, 104123. [CrossRef]
8. Shu, C.; Zhao, H.; Jiao, W.; Liu, B.; Cao, J.; Jiang, W. Antifungal efficacy of ursolic acid in control of *Alternaria alternata* causing black spot rot on apple fruit and possible mechanisms involved. *Sci. Hortic.* **2019**, *256*, 108636. [CrossRef]
9. Sikdar, P.; Okubara, P.; Mazzola, M.; Xiao, C.L. Development of PCR Assays for Diagnosis and Detection of the Pathogens *Phacidiopycnis washingtonensis* and *Sphaeropsis pyriputrescens* in Apple Fruit. *Plant Dis.* **2014**, *98*, 241–246. [CrossRef]
10. Kim, S.M.; Lee, S.M.; Seo, J.-A.; Kim, Y.-S. Changes in volatile compounds emitted by fungal pathogen spoilage of apples during decay. *Postharvest Biol. Technol.* **2018**, *146*, 51–59. [CrossRef]
11. Kim, Y.-H.; Kim, K.-H.; Szulejko, J.E.; Parker, D. Quantitative Analysis of Fragrance and Odorants Released from Fresh and Decaying Strawberries. *Sensors* **2013**, *13*, 7939–7978. [CrossRef] [PubMed]
12. Ali, M.M.; Hashim, N.; Aziz, S.A.; Lasekan, O. Principles and recent advances in electronic nose for quality inspection of agricultural and food products. *Trends Food Sci. Technol.* **2020**, *99*, 1–10. [CrossRef]
13. van der Sar, I.G.; Wijbenga, N.; Nakshbandi, G.; Aerts, J.G.J.V.; Manintveld, O.C.; Wijsenbeek, M.S.; Hellemons, M.E.; Moor, C.C. The smell of lung disease: A review of the current status of electronic nose technology. *Respir. Res.* **2021**, *22*, 246. [CrossRef] [PubMed]
14. Mota, I.; Teixeira-Santos, R.; Rufo, J.C. Detection and identification of fungal species by electronic nose technology: A systematic review. *Fungal Biol. Rev.* **2021**, *37*, 59–70. [CrossRef]
15. Nouri, B.; Mohtasebi, S.S.; Rafiee, S. Quality detection of pomegranate fruit infected with fungal disease. *Int. J. Food Prop.* **2020**, *23*, 9–21. [CrossRef]
16. Liu, Q.; Sun, K.; Zhao, N.; Yang, J.; Zhang, Y.; Ma, C.; Pan, L.; Tu, K. Information fusion of hyperspectral imaging and electronic nose for evaluation of fungal contamination in strawberries during decay. *Postharvest Biol. Technol.* **2019**, *153*, 152–160. [CrossRef]
17. Yumoto, M.; Kawata, Y.; Abe, T.; Matsuyama, T.; Wada, S. Non-destructive mid-IR spectroscopy with quantum cascade laser can detect ethylene gas dynamics of apple cultivar 'Fuji' in real time. *Sci. Rep.* **2021**, *11*, 20695. [CrossRef]
18. Du, M.; Liu, Z.; Zhang, X.; Li, H.; Liu, Z.; Li, X.; Song, J.; Jia, X.; Wang, L. Effect of pulsed controlled atmosphere with CO_2 on the quality of watercored apple during storage. *Sci. Hortic.* **2021**, *278*, 109854. [CrossRef]
19. Farneti, B.; Khomenko, I.; Cappellin, L.; Ting, V.; Costa, G.; Biasioli, F.; Costa, F. Dynamic volatile organic compound fingerprinting of apple fruit during processing. *LWT-Food Sci. Technol.* **2015**, *63*, 21–28. [CrossRef]
20. Park, S.H.; Park, C.W.; Na Kim, M.; Choi, S.Y.; Choi, D.S.; Kim, J.S.; Kim, Y.H.; Lee, S.J. The Characteristics of Gas Exchange for Purge-type in a Controlled Atmosphere Storage System. *Food Eng. Prog.* **2018**, *22*, 118–125. [CrossRef]
21. Hanafi, R.; Mayasari, R.D.; Masmui; Bakri, A.; Raharjo, J.; Nuryadi, R. Electrochemical Sensor for Environmental Monitoring System: A Review. In Proceedings of the 8th National Physics Seminar, Univeristas Negeri, Jakarta, Indonesia, 29–30 June 2019; 2169.
22. Kwiatkowski, A.; Drozdowska, K.; Smulko, J. Embedded gas sensing setup for air samples analysis. *Rev. Sci. Instrum.* **2021**, *92*, 074102. [CrossRef]
23. Konduru, T.; Rains, G.C.; Li, C. A Customized Metal Oxide Semiconductor-Based Gas Sensor Array for Onion Quality Evaluation: System Development and Characterization. *Sensors* **2015**, *15*, 1252–1273. [CrossRef] [PubMed]
24. Guo, Z.; Guo, C.; Sun, L.; Zuo, M.; Chen, Q.; El-Seedi, H.R.; Zou, X. Identification of the apple spoilage causative fungi and prediction of the spoilage degree using electronic nose. *J. Food Process. Eng.* **2021**, *44*, e13816. [CrossRef]
25. Arslan, M.; Xiaobo, Z.; Tahir, H.E.; Xuetao, H.; Rakha, A.; Basheer, S.; Hao, Z. Near-infrared spectroscopy coupled chemometric algorithms for prediction of antioxidant activity of black goji berries (*Lycium ruthenicum* Murr.). *J. Food Meas. Charact.* **2018**, *12*, 2366–2376. [CrossRef]
26. Du, X.-L.; Li, X.-Y.; Liu, Y.; Zhou, W.-H.; Li, J.-L. Genetic algorithm optimized non-destructive prediction on property of mechanically injured peaches during postharvest storage by portable visible/shortwave near-infrared spectroscopy. *Sci. Hortic.* **2019**, *249*, 240–249. [CrossRef]
27. Yuan, L.-M.; You, L.; Yang, X.; Chen, X.; Huang, G.; Chen, X.; Shi, W.; Sun, Y. Consensual Regression of Soluble Solids Content in Peach by Near Infrared Spectrocopy. *Foods* **2022**, *11*, 1095. [CrossRef] [PubMed]
28. Ren, G.; Sun, Y.; Li, M.; Ning, J.; Zhang, Z. Cognitive spectroscopy for evaluating Chinese black tea grades (*Camellia sinensis*): Near-infrared spectroscopy and evolutionary algorithms. *J. Sci. Food Agric.* **2020**, *100*, 3950–3959. [CrossRef] [PubMed]
29. Bin, J.; Fan, W.; Zhou, J.H.; Li, X.; Liang, Y.Z. Application of Intelligent Optimization Algorithms to Wavelength Selection of Near-Infrared Spectroscopy. *Spectrosc. Spectr. Anal.* **2017**, *37*, 95–102.
30. Hu, L.; Yin, C.; Ma, S.; Liu, Z. Rapid detection of three quality parameters and classification of wine based on Vis-NIR spectroscopy with wavelength selection by ACO and CARS algorithms. *Spectrochim. Acta Part A Mol. Biomol. Spectrosc.* **2018**, *205*, 574–581. [CrossRef]
31. Guo, Z.; Barimah, A.O.; Shujat, A.; Zhang, Z.; Ouyang, Q.; Shi, J.; El-Seedi, H.R.; Zou, X.; Chen, Q. Simultaneous quantification of active constituents and antioxidant capability of green tea using NIR spectroscopy coupled with swarm intelligence algorithm. *LWT-Food Sci. Technol.* **2020**, *129*, 109510. [CrossRef]

32. Ouyang, Q.; Yang, Y.; Wu, J.; Liu, Z.; Chen, X.; Dong, C.; Chen, Q.; Zhang, Z.; Guo, Z. Rapid sensing of total theaflavins content in black tea using a portable electronic tongue system coupled to efficient variables selection algorithms. *J. Food Compos. Anal.* **2019**, *75*, 43–48. [CrossRef]
33. Guo, Z.; Barimah, A.O.; Yin, L.; Chen, Q.; Shi, J.; El-Seedi, H.R.; Zou, X. Intelligent evaluation of taste constituents and polyphenols-to-amino acids ratio in matcha tea powder using near infrared spectroscopy. *Food Chem.* **2021**, *353*, 129372. [CrossRef]
34. Diallo, A.A.; Yang, Z.; Shen, G.; Ge, J.; Li, Z.; Han, L. Comparison and rapid prediction of lignocellulose and organic elements of a wide variety of rice straw based on near infrared spectroscopy. *Int. J. Agric. Biol. Eng.* **2019**, *12*, 166–172. [CrossRef]
35. Guo, Z.; Wang, M.; Agyekum, A.A.; Wu, J.; Chen, Q.; Zuo, M.; El-Seedi, H.R.; Tao, F.; Shi, J.; Ouyang, Q.; et al. Quantitative detection of apple watercore and soluble solids content by near infrared transmittance spectroscopy. *J. Food Eng.* **2020**, *279*, 109955. [CrossRef]
36. Ying, Y.; Lu, L.; Banerjee, S.; Xu, L.; Zhao, Q.; Wu, H.; Li, R.; Xu, X.; Yu, H.; Neculai, D.; et al. KVarPredDB: A database for predicting pathogenicity of missense sequence variants of keratin genes associated with genodermatoses. *Hum. Genom.* **2020**, *14*, 45. [CrossRef]
37. Yang, G.; Shi, X.; Feng, L.; He, S.; Shi, Z.; Chen, J. CEDAR: A Cost-Effective Crowdsensing System for Detecting and Localizing Drones. *IEEE Trans. Mob. Comput.* **2020**, *19*, 2028–2043. [CrossRef]
38. Yang, Y.; Chen, T. Analysis and Visualization Implementation of Medical Big Data Resource Sharing Mechanism Based on Deep Learning. *IEEE Access* **2019**, *7*, 156077–156088. [CrossRef]
39. Tiwari, S.; Kate, A.; Mohapatra, D.; Tripathi, M.K.; Ray, H.; Akuli, A.; Ghosh, A.; Modhera, B. Volatile organic compounds (VOCs): Biomarkers for quality management of horticultural commodities during storage through e-sensing. *Trends Food Sci. Technol.* **2020**, *106*, 417–433. [CrossRef]
40. Qiao, J.; Su, G.; Liu, C.; Zou, Y.; Chang, Z.; Yu, H.; Wang, L.; Guo, R. Study on the Application of Electronic Nose Technology in the Detection for the Artificial Ripening of Crab Apples. *Horticulturae* **2022**, *8*, 386. [CrossRef]
41. Guo, Z.; Guo, C.; Chen, Q.; Ouyang, Q.; Shi, J.; El-Seedi, H.R.; Zou, X. Classification for *Penicillium expansum* Spoilage and Defect in Apples by Electronic Nose Combined with Chemometrics. *Sensors* **2020**, *20*, 2130. [CrossRef]
42. Caya, M.V.C.; Cruz, F.R.G.; Fernando, C.M.N.; Lafuente, R.M.M.; Malonzo, M.B.; Chung, W.Y. Monitoring and Detection of Fruits and Vegetables Spoilage in the Refrigerator Using Electronic Nose Based on Principal Component Analysis. In Proceedings of the 2019 IEEE 11th International Conference on Humanoid, Nanotechnology, Information Technology, Communication and Control, Environment, and Management, HNICEM 2019, Laoag, Philippines, 29 November–1 December 2019. [CrossRef]
43. Wang, Y.; Fei, C.; Wang, D.; Wei, Y.; Qing, Z.; Zhao, S.; Wu, H.; Zhang, W. Quantitative analysis and early detection of postharvest soft rot in kiwifruit using E-nose and chemometrics. *J. Food Meas. Charact.* **2023**, 1–11. [CrossRef]
44. Zhao, C.; Ma, J.; Jia, W.; Wang, H.; Tian, H.; Wang, J.; Zhou, W. An Apple Fungal Infection Detection Model Based on BPNN Optimized by Sparrow Search Algorithm. *Biosensors* **2022**, *12*, 692. [CrossRef]
45. Mataragas, M.; Dimitriou, V.; Skandamis, P.; Drosinos, E. Quantifying the spoilage and shelf-life of yoghurt with fruits. *Food Microbiol.* **2011**, *28*, 611–616. [CrossRef] [PubMed]
46. Li, M.; Yin, Y.; Yu, H.; Yuan, Y.; Liu, X. Early Warning Potential of Banana Spoilage Based on 3D Fluorescence Data of Storage Room Gas. *Food Bioprocess Technol.* **2021**, *14*, 1946–1961. [CrossRef]
47. Putnik, P.; Kovačević, D.B.; Herceg, K.; Levaj, B. Influence of Cultivar, Anti-Browning Solutions, Packaging Gasses, and Advanced Technology on Browning in Fresh-Cut Apples During Storage. *J. Food Process. Eng.* **2016**, *40*, e12400. [CrossRef]

foods

MDPI

Article

Evaluation of ^{60}Co Irradiation on Volatile Components of Turmeric (Curcumae Longae Rhizoma) Volatile Oil with GC–IMS

Ye He [1,2], Lu Yin [1,2], Wei Zhou [1], Hongyan Wan [1,2], Chang Lei [1], Shunxiang Li [1,2,3,*] and Dan Huang [1,2,3,*]

1 State Key Laboratory of Chinese Medicine Powder and Medicine Innovation in Hunan (Incubator), Science and Technology Innovation Center, Hunan University of Chinese Medicine, Changsha 410208, China; hy070915@126.com (Y.H.); yl024083@163.com (L.Y.); fxxx99w@163.com (W.Z.); wanhongyan2001@163.com (H.W.); leichang1231@126.com (C.L.)

2 Hunan Engineering Technology Research Center for Bioactive Substance Discovery of Chinese Medicine, School of Pharmacy, Hunan University of Chinese Medicine, Changsha 410208, China

3 Hunan Province Sino-US International Joint Research Center for Therapeutic Drugs of Senile Degenerative Diseases, Changsha 410208, China

* Correspondence: lishunxiang@hnucm.edu.cn (S.L.); huangdan110@hnucm.edu.cn (D.H.)

Abstract: ^{60}Co irradiation is an efficient and rapid sterilization method. The aim of this work is to determine the changes in essential-oil composition under different irradiation intensities of ^{60}Co and to select an appropriate irradiation dose with GC–IMS. Dosages of 0, 5, and 10 kGy of ^{60}Co were used to analyze turmeric (Curcumae Longae Rhizoma) volatile oil after ^{60}Co irradiation (named JH-1, JH-2, and JH-3). The odor fingerprints of volatile organic compounds in different turmeric volatile oil samples were constructed by headspace solid-phase microextraction and gas chromatography–ion mobility spectrometry (GC–IMS) after irradiation. The differences in odor fingerprints of volatile organic compounds (VOCs) were compared by principal component analysis (PCA). The results showed that 97 volatile components were detected in the volatile oil of Curcuma longa, and 64 components were identified by database retrieval. With the change in irradiation intensity, the volatile compounds in the three turmeric volatile oil samples were similar, but the peak intensity was significantly different, which was attributed to the change in compound composition and content caused by different irradiation doses. In addition, the principal component analysis showed that JH-2 and JH-3 were relatively correlated, while JH-1 and JH-3 were far from each other. In general, different doses of ^{60}Co irradiation can affect the content of volatile substances in turmeric volatile oil. With the increase in irradiation dose, the peak area decreased, and so the irradiation dose of 5 kGy/min was better. It is shown that irradiation technology has good application prospects in the sterilization of foods with volatile components. However, we must pay attention to the changes in radiation dose and chemical composition.

Keywords: Curcumae Longae Rhizoma; volatile oil; ^{60}Co; GC–IMS

Citation: He, Y.; Yin, L.; Zhou, W.; Wan, H.; Lei, C.; Li, S.; Huang, D. Evaluation of ^{60}Co Irradiation on Volatile Components of Turmeric (Curcumae Longae Rhizoma) Volatile Oil with GC–IMS. *Foods* **2023**, *12*, 2489. https://doi.org/10.3390/foods12132489

Academic Editor: Shahab A. Shamsi

Received: 1 June 2023
Revised: 20 June 2023
Accepted: 25 June 2023
Published: 26 June 2023

1. Introduction

Curcumae Longae Rhizoma, a rootstock plant called turmeric belonging to Fam. Zingiberaceae in China, is extensively cultivated in South China, such as in the Sichuan, Guangxi, Guangdong, Yunnan, and Zhejiang provinces. Turmeric has been widely used in traditional Chinese medicine to promote Qi circulation, dissolve blood sludge, induce menstruation, and relieve pain. It can be used to relieve pain in the chest and hypochondriac regions, as well as treat amenorrhea, mass formation in the abdomen, rheumatic pain of the shoulders and arms, traumatic swelling, and pain (Committee for the Pharmacopoeia of P.R. China, 2020). Modern research shows that the abundant volatile oil and curcumin in turmeric (up to about 4%) have anti-tumour, anti-inflammatory, and antioxidant effects [1].

The quality of traditional Chinese medicine will have an impact on its efficacy. With the development of modern medicine, the quality standards of traditional Chinese medicine are also constantly improving. However, the source of traditional Chinese medicine is complex, and it is easy to be contaminated by bacteria in the process of production and storage. Traditional dry-heat sterilization, damp-heat sterilization and ethylene oxide gas sterilization technologies easily destroy the effective components of drugs, affecting the efficacy or residual organic solvents. The sterilization operation should remove microorganisms and ensure the quality and efficacy of drugs [2]. ^{60}Co-γ irradiation ray sterilization method, widely used in the sterilization of Chinese medicinal materials, is a new disinfection and sterilization process in the 20th century, which is based on the high-energy rays produced by X-ray, γ ray, and other ionizing radiation to inhibit the continuation of pests and diseases and to effectively kill insects. Sterilization at room temperature and high efficiency, with simple operation and other characteristics, dosages of 5, and 10 kGy are most widely used [3,4]. However, the influence of the irradiation dose on the component of traditional Chinese medicine remains unknown. Therefore, it is of great significance to explore the sterilization dose of irradiation sterilization of different traditional Chinese medicine varieties using ^{60}Co, which can be of great reference value to the quality problems of traditional Chinese medicine and its preparation products.

The traditional detection method of volatile compounds in turmeric uses gas chromatography-mass spectrometry (GC–MS) [5], but it takes a long time and has low sensitivity, and GC-MS needs complicated pre-treatments and is constrained in distinguishing isomeric molecules [6]. In recent years, a novel and powerful device, gas chromatography (GC) coupled with ion mobility spectrometry (IMS), has been chosen to accurately test flavor compounds, especially the volatile oil components in various foods. Gas chromatography–ion mobility spectrometry (GC–IMS) which has the advantages of simple operation, strong separation ability, short detection cycle, and retaining the original flavor of samples to the greatest extent [7], has been successfully applied to food [8], biological and aquatic odor analyses [9], quality detection [10], and other fields. However, there is no report on the study of volatile organic compounds in turmeric volatile oil by GC–IMS technology.

In this study, the volatile oil of turmeric was extracted by steam distillation and irradiated, with ^{60}Co rays of different intensities (Dosages of 0, 5, and 10 kGy). A total of 64 components were detected using GC–IMS technology to compare the changes before and after irradiation. Finally, the composition changes of volatile oil under different irradiation intensities were given to provide a certain reference for the production and sterilization of turmeric and its preparations. This study provides a sound basis for the use of ^{60}Co-γ ray irradiation sterilization technology during the preparation of medicinal herbs. GC–IMS, which has the advantages of simple operation, strong separation ability, short detection cycle, and retaining the original flavor of samples to the greatest extent, can be successfully applied to foods.

2. Materials and Methods

2.1. Materials

Turmeric was collected from Baise, Guangxi Province, China, and identified by Prof. Zhaoming Xie at the Hunan Academy of Traditional Chinese Medicine. A voucher specimen (HNATCM2022-006) was deposited in the herbarium of the Hunan Academy of Traditional Chinese Medicine.

2.2. Isolation of the Essential Oils

The essential oil was extracted through steam distillation by referring to the Chinese Pharmacopoeia 2020 edition (part 4) volatile oil determination method (General Principle 2204) for determination. We added 500 g of dried turmeric with an appropriate amount of water and a few glass beads into a 1000 mL round-bottom flask, treating the solution by slowly heating it to a boil, followed by keeping it slightly boiling for 6 h. The upper oil phase (the crude essential oil of turmeric) was gathered and sealed and stored away from light at 4 °C for further use.

2.3. Extraction Yield

After extraction, the extraction rate of volatile oil is calculated according to the following formula:

$$\text{Extraction yield (\%)} = m_1/m_0 \times 100 \quad (1)$$

where m_1 is the total mass of the extracted oil, and m_0 is the initial mass of the turmeric used in each extraction.

2.4. ^{60}Co-γ Irradiation

The resulting curcumin EOs was dehydrated with anhydrous Na_2SO_4 and then divided into three equal parts for ^{60}Co irradiation. The dose rates were 0, 5, and 10 kGy/min. The ^{60}Co γ radiation source was located at Hunan Radiological Technology Application Research Center (Changsha, China).

2.5. Analysis by GC–IMS

In the experiment, the volatiles were concentrated and separated by headspace solid-phase microextraction, with reference to Wang [11] and other methods and appropriate adjustments. Precisely-measured 50 μL of turmeric volatile oil sample was transferred into a 20 mL headspace bottle with Teflon spacer seal. The headspace bottle was heated at 80 °C and incubated for 10 min at 500 RPM. Then 100 μL of the sample was injected in non-shunt mode, and the temperature of the injection needle was kept at 85 °C.

The components of volatile compounds were identified by chromatography–ion mobility spectroscopy (GC–IMS; FlavourSpec®, G.A.S., Berlin, Germany). Gas chromatography (GC) was performed under the following conditions: carrier gas, nitrogen (99.99%); column, mxt-5 (15.0 m length × 0.53 mm ID × 1 μm thickness); running time, 50 min; flow rate, initial 2.0 mL/min, holding for 2 min, linearly increasing to 100 mL/min within 18 min, and holding for 20 min. Ion mobility spectroscopy (IMS) was carried out under the following conditions: drift gas, nitrogen (99.99%); flow rate, 150 mL/min; IMS detector temperature, 45 °C.

Three parallel samples are set for each irradiation intensity for volatile oil, and the difference in the spectrum of volatile organic compounds in the sample can be given after analysis. The NIST database and IMS database built into the software can conduct a qualitative analysis of substances.

2.6. Statistical Analysis

The analysis software Vocal matched with the instrument is used to view the qualitative and quantitative analysis spectrum and data. The NIST database and IMS database built into the application software can be used for qualitative analysis of substances. The porter plug-in directly compares the spectrum differences between samples (three-dimensional spectrum, two-dimensional top view, and difference spectrum). We compared the fingerprint of the gallery plot plug-in to intuitively and quantitatively compare the differences in volatile organic compounds between different samples. A dynamic PCA plug-in was used for dynamic principal component analysis, cluster analysis of samples, and rapid determination of unknown and unknown samples.

3. Results

3.1. Flavor Differences in Turmeric Volatile Oil under Different Irradiation Intensities Detected by GC–IMS

Three-dimensional topographic plots of turmeric irradiated by three different doses are shown in Figure 1A. We assessed the volatile organic compounds (VOC) in different samples from three perspectives: retention time, drift time, and peak intensity. The number of VOCs and the signal intensity of the peaks differed slightly among the three samples.

With the increase in irradiation intensity, only the signal intensity of the peaks changes slightly, and almost no new compounds form.

A two-dimensional topographic spectrum (planform of the 3D plot) was also obtained for its difficulty to observe the differences between three-dimensional groups (Figure 1B). In this plot, the red vertical line at 1.0 on the left is the reaction ion peak (RIP), and the background image is blue. Each point on both sides of the reaction ion peak represents a VOC, the color depth represents the volatile-compound content, the white area represents the low compound content, and the red area represents the high compound content. Using the difference comparison mode, we select the spectrum (JH-1) of one sample as the reference, and we deduct the spectrum of other samples from the reference to obtain Figure 1C. If the volatile organic compounds of the two samples are consistent, then the background after deduction is white, while red signifies that the concentration of the substance is higher than the reference, and blue implies that the concentration of the substance is lower than the reference. It can be seen from Figure 1B,C that signals are concentrated in areas A, B, and C. The color of some compounds in areas A and C is deepened, and the color of compounds in area B is lighter, which suggests that with the increase in irradiation dose, the compound content in regions A and C increases and in the regioqualitative analysis of volatile organic compounds.

Figure 1D shows the binary spectra of all volatile substances of turmeric volatile oil under three irradiation intensities. The volatile compounds in the volatile oil of turmeric were analyzed by GC. The NIST database and IMS database built into the IMS library were according to the retention index, retention time, and ion migration time. We identified 64 volatile components shown in Table 1 from the three sample varieties in Figure 1D. The substance numbers in Figure 1D are consistent with those in Table 1.

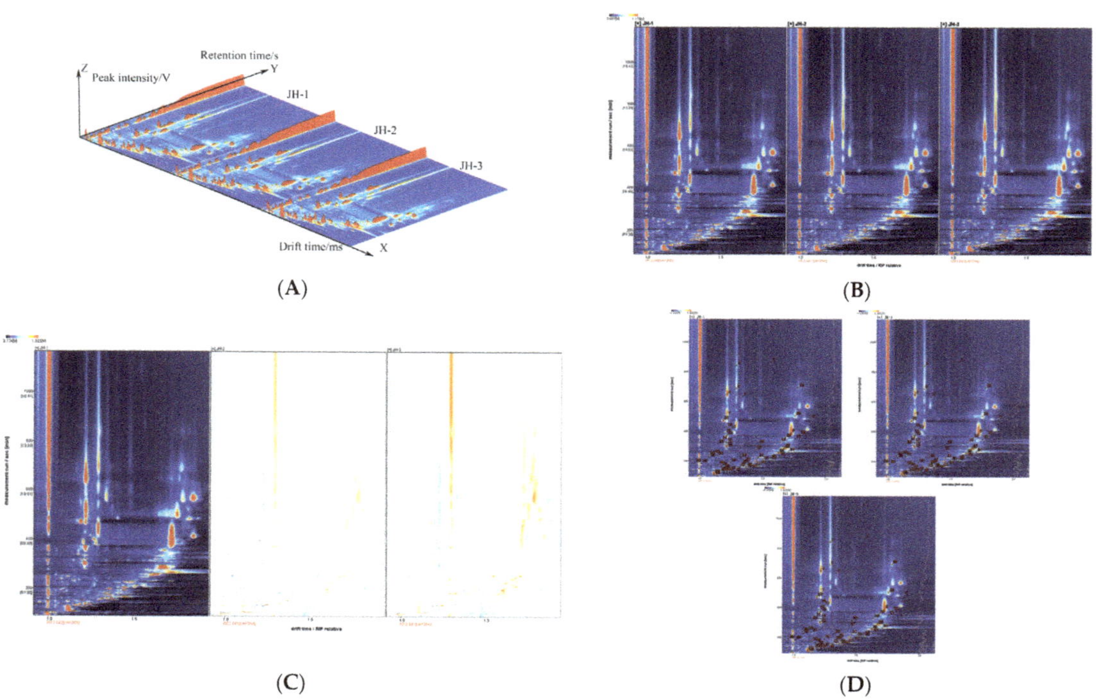

Figure 1. *Cont.*

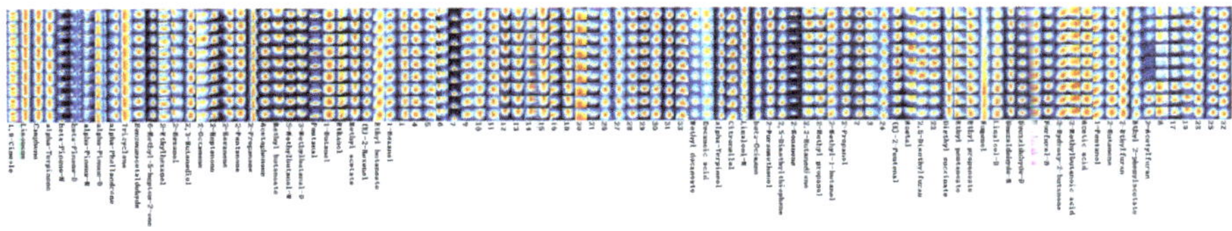

(E)

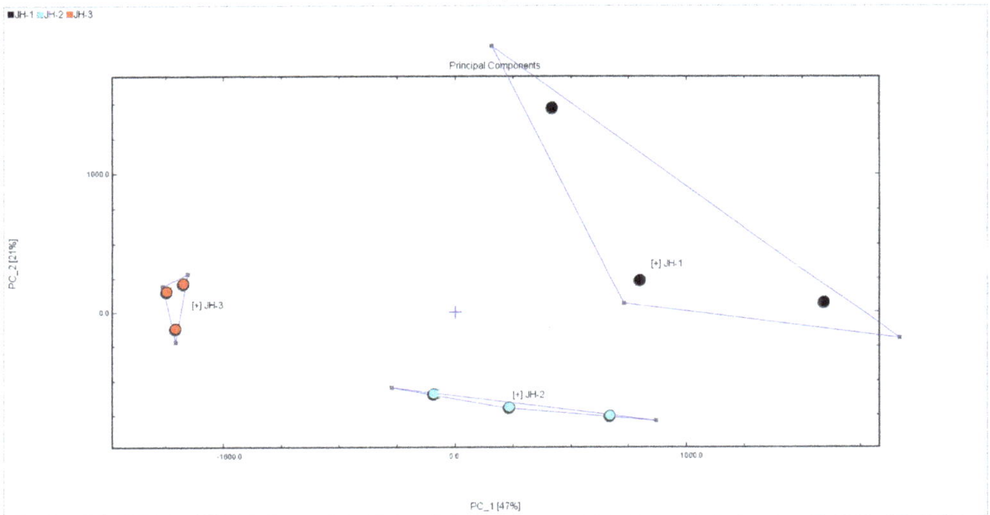

(F)

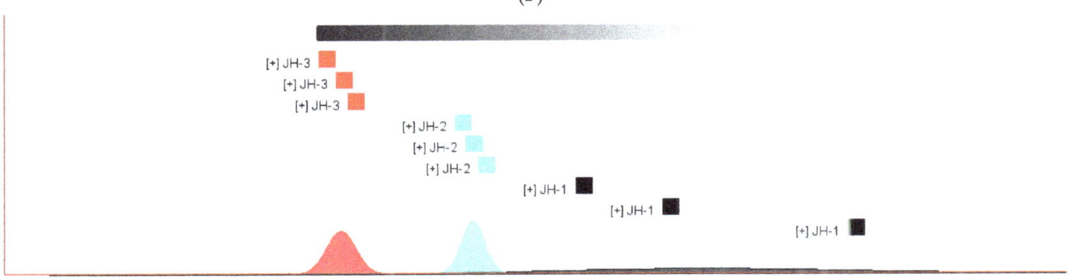

(G)

Figure 1. Volatile compounds analysis by GC-IMS. (**A**) 3D-topographic; (**B**) topographic plots; (**C**) topographic subtraction plots; (**D**) post-irradiation topographic plots; (**E**) volatile compounds fingerprint comparisons. Each row represents all the signals selected in a sample. Each column represents the signal of the same volatile compound. (**F**) Principal component analysis; (**G**) Fingerprint similarity analysis.

According to Tables 1 and 2, there are 64 monomers and dimers of volatile substances identified in turmeric volatile oil, including 17 alcohols and phenols, 11 aldehydes, 11 ketones, 9 terpenes, 8 esters, 4 carboxylic acids and their derivatives, 3 furans, and 1 thiophene. The chemical formula of the identified monomer and dimer is the same as the CAS number, but the form is different. The results are shown in Tables 1 and 2.

Table 1. Volatile substances contained in turmeric volatile oil.

No	Compounds	Molecular Formula	RI	Rt/s	Dt/ms (RIPrel)
1	Methyl decanoate	$C_{11}H_{22}O_2$	1508.0	1079.39	1.54878
2	Decanoic acid	$C_{10}H_{20}O_2$	1360.2	867.201	1.57124
3	Eugenol	$C_{10}H_{12}O_2$	1341.6	840.393	1.29711
4	alpha-Terpineol	$C_{10}H_{18}O$	1206.8	646.87	1.22248
5	Diethyl succinate	$C_8H_{14}O_4$	1234.7	686.937	1.29422
6	Citronellol	$C_{10}H_{20}O$	1197.5	633.591	1.34963
7	Linalool	$C_{10}H_{18}O$	1106.9	503.438	1.2229
8	2-Nonanone	$C_9H_{18}O$	1091.9	481.9	1.88944
9	1,8-Cineole	$C_{10}H_{18}O$	1027.6	389.61	1.73145
10	beta-Ocimene	$C_{10}H_{16}$	1054.5	428.22	1.21822
11	Benzeneacetaldehyde	C_8H_8O	1045.1	414.765	1.2491
12	Limonene	$C_{10}H_{16}$	1025.6	386.685	1.20792
13	alpha-Terpinene	$C_{10}H_{16}$	1015.0	371.475	1.22263
14	6-Methyl-5-hepten-2-one	$C_8H_{14}O$	990.5	339.885	1.17704
15	beta-Pinene	$C_{10}H_{16}$	976.0	327.6	1.21969
16	beta-Pinene	$C_{10}H_{16}$	976.0	327.6	1.64175
17	Camphene	$C_{10}H_{16}$	944.9	301.275	1.21822
18	alpha-Pinene	$C_{10}H_{16}$	931.8	290.16	1.22116
19	alpha-Pinene	$C_{10}H_{16}$	931.1	289.575	1.67557
20	2-Ethylhexanol	$C_8H_{18}O$	1018.3	376.21	1.79856
21	2-Octanone	$C_8H_{16}O$	999.2	348.832	1.7651
22	Benzaldehyde	C_7H_6O	959.8	313.914	1.15225
23	Benzaldehyde	C_7H_6O	959.4	313.517	1.47206
24	Ethyl pentanoate	$C_7H_{14}O_2$	901.2	264.315	1.71158
25	2-Heptanone	$C_7H_{14}O$	891.6	256.379	1.62862
26	2-Furanmethanol	$C_5H_6O_2$	873.3	246.856	1.1188
27	2-Acetylfuran	$C_6H_6O_2$	911.1	272.648	1.45333
28	2,5-Dimethylthiophene	C_6H_8S	856.5	238.127	1.07866
29	Furfural	$C_5H_4O_2$	828.3	223.446	1.08401
30	Furfural	$C_5H_4O_2$	826.7	222.652	1.3329
31	2-Methylbutanoic acid	$C_5H_{10}O_2$	827.5	223.049	1.46938
32	2-Hexanol	$C_6H_{14}O$	793.1	205.193	1.56974
33	2-Hexanone	$C_6H_{12}O$	791.9	204.546	1.50912
34	2,3-Butanediol	$C_4H_{10}O_2$	778.3	198.14	1.3627
35	1-Pentanol	$C_5H_{12}O$	760.5	190.908	1.25055
36	Acetal	$C_6H_{14}O_2$	748.9	186.155	0.96913
37	(E)-2-Pentenal	C_5H_8O	747.3	185.535	1.36062
38	3-Hydroxy-2-butanone	$C_4H_8O_2$	733.1	179.749	1.33778
39	2,5-Dimethylfuran	C_6H_8O	739.9	182.503	1.0218
40	Methyl butanoate	$C_5H_{10}O_2$	716.5	172.993	1.15331
41	Ethyl propanoate	$C_5H_{10}O_2$	695.4	164.398	1.44128
42	2-Pentanone	$C_5H_{10}O$	686.6	161.106	1.37552
43	2-Ethylfuran	C_6H_8O	674.9	157.998	1.3211
44	3-Methylbutanal	$C_5H_{10}O$	650.6	151.597	1.4118
45	1-Butanol	$C_4H_{10}O$	656.2	153.06	1.37552
46	3-Methylbutanal	$C_5H_{10}O$	645.1	150.134	1.20319
47	2-Butanone	C_4H_8O	591.7	136.053	1.24854
48	2-Methyl propanal	C_4H_8O	555.1	126.36	1.28596
49	2-Propanol	C_3H_8O	565.4	129.104	1.22814
50	2,3-Butanedione	$C_4H_6O_2$	582.1	133.493	1.18732
51	Acetic acid	$C_2H_4O_2$	565.4	129.104	1.16918
52	2-Propanone	C_3H_6O	539.2	112.462	1.11816
53	Ethanol	C_2H_6O	1508	105.513	1.12723
54	Methyl acetate	$C_3H_6O_2$	1360.2	125.263	1.20319
55	Pentanal	$C_5H_{10}O$	1341.6	163.667	1.18165
56	Ethyl 2-phenylacetate	$C_{10}H_{12}O_2$	1206.8	692.941	1.78609
57	Acetophenone	C_8H_8O	1234.7	471.206	1.19245
58	Linalool	$C_{10}H_{18}O$	1197.5	503.498	1.76126
59	alpha-Phellandrene	$C_{10}H_{16}$	1106.9	356.925	1.68764
60	Tricyclene	$C_{10}H_{16}$	1091.9	283.788	1.67365
61	(E)-2-Hexenol	$C_6H_{12}O$	1027.6	234.035	1.15052
62	1-Hexanol	$C_6H_{14}O$	1054.5	232.537	1.63849
63	Ethyl butanoate	$C_6H_{12}O_2$	1045.1	216.141	1.57005
64	2-Methyl-1-butanol	$C_5H_{12}O$	1025.6	178.89	1.48673

Table 2. Area of turmeric volatile oil.

No	Compounds	Molecular Formula	[+] JH-1	[+] JH-2	[+] JH-3
1	Methyl decanoate	$C_{11}H_{22}O_2$	853.04	807.50	706.77
2	Decanoic acid	$C_{10}H_{20}O_2$	343.61	425.17	323.73
3	Eugenol	$C_{10}H_{12}O_2$	3892.53	4916.40	5396.33
4	alpha-Terpineol	$C_{10}H_{18}O$	19,657.51	18,387.17	17,492.96
5	Diethyl succinate	$C_8H_{14}O_4$	4842.53	5279.30	5497.18
6	Citronellol	$C_{10}H_{20}O$	806.72	785.28	774.24
7	Linalool	$C_{10}H_{18}O$	8368.12	8339.72	8003.81
8	2-Nonanone	$C_9H_{18}O$	1129.50	845.90	803.89
9	1.8-Cineole	$C_{10}H_{18}O$	17,383.17	17,589.74	17,990.91
10	beta-Ocimene	$C_{10}H_{16}$	730.79	652.68	599.08
11	Benzeneacetaldehyde	C_8H_8O	461.94	443.61	419.23
12	Limonene	$C_{10}H_{16}$	1525.97	1546.91	1559.40
13	alpha-Terpinene	$C_{10}H_{16}$	3982.81	3847.04	3737.19
14	6-Methyl-5-hepten-2-one	$C_8H_{14}O$	764.36	841.66	779.81
15	beta-Pinene	$C_{10}H_{16}$	1892.73	1854.60	1873.25
16	beta-Pinene	$C_{10}H_{16}$	6168.35	6338.65	6214.42
17	Camphene	$C_{10}H_{16}$	5470.41	5424.69	5375.12
18	alpha-Pinene	$C_{10}H_{16}$	1211.24	1190.02	1129.20
19	alpha-Pinene	$C_{10}H_{16}$	6032.87	6101.62	6333.75
20	2-Ethylhexanol	$C_8H_{18}O$	196.51	191.32	202.42
21	2-Octanone	$C_8H_{16}O$	734.45	737.80	727.27
22	Benzaldehyde	C_7H_6O	248.36	256.24	260.37
23	Benzaldehyde	C_7H_6O	302.21	255.45	291.10
24	Ethyl pentanoate	$C_7H_{14}O_2$	16,363.09	16,853.35	16,297.10
25	2-Heptanone	$C_7H_{14}O$	6768.49	6831.28	6604.10
26	2-Furanmethanol	$C_5H_6O_2$	693.97	694.50	633.27
27	2-Acetylfuran	$C_6H_6O_2$	43.49	19.02	56.13
28	2,5-Dimethylthiophene	C_6H_8S	355.34	398.57	404.40
29	Furfural	$C_5H_4O_2$	151.02	163.77	172.80
30	Furfural	$C_5H_4O_2$	148.19	162.08	192.78
31	2-Methylbutanoic acid	$C_5H_{10}O_2$	97.08	98.83	103.59
32	2-Hexanol	$C_6H_{14}O$	1552.99	1696.63	1741.51
33	2-Hexanone	$C_6H_{12}O$	702.91	717.69	727.46
34	2,3-Butanediol	$C_4H_{10}O_2$	2616.14	2590.83	2552.24
35	1-Pentanol	$C_5H_{12}O$	98.02	97.41	107.12
36	Acetal	$C_6H_{14}O_2$	43.00	48.17	45.78
37	(E)-2-Pentenal	C_5H_8O	430.50	467.99	387.28
38	3-Hydroxy-2-butanone	$C_4H_8O_2$	794.53	799.28	854.01
39	2,5-Dimethylfuran	C_6H_8O	125.36	141.91	111.45
40	Methyl butanoate	$C_5H_{10}O_2$	89.02	93.33	91.42
41	Ethyl propanoate	$C_5H_{10}O_2$	2091.53	2078.08	2150.97
42	2-Pentanone	$C_5H_{10}O$	508.33	528.88	573.27
43	2-Ethylfuran	C_6H_8O	1519.15	1480.31	1638.41
44	3-Methylbutanal	$C_5H_{10}O$	149.00	147.46	136.26
45	1-Butanol	$C_4H_{10}O$	2718.89	2744.27	2798.98
46	3-Methylbutanal	$C_5H_{10}O$	786.67	816.78	863.78
47	2-Butanone	C_4H_8O	1163.04	1258.54	1331.09
48	2-Methyl propanal	C_4H_8O	116.66	127.12	118.02
49	2-Propanol	C_3H_8O	538.81	443.21	534.13
50	2,3-Butanedione	$C_4H_6O_2$	715.88	646.28	653.28
51	Acetic acid	$C_2H_4O_2$	899.26	900.87	978.94
52	2-Propanone	C_3H_6O	11,632.68	12,363.82	12,586.06
53	Ethanol	C_2H_6O	2364.98	2789.25	2848.56
54	Methyl acetate	$C_3H_6O_2$	355.35	368.09	363.62
55	Pentanal	$C_5H_{10}O$	59.13	56.64	56.97
56	Ethyl 2-phenylacetate	$C_{10}H_{12}O_2$	3291.02	3610.35	3732.50
57	Acetophenone	C_8H_8O	6600.73	6488.00	6267.98
58	Linalool	$C_{10}H_{18}O$	6085.73	6678.48	7028.94
59	alpha-Phellandrene	$C_{10}H_{16}$	1909.42	1986.81	1925.54
60	Tricyclene	$C_{10}H_{16}$	4633.64	4665.30	4772.88
61	(E)-2-Hexenol	$C_6H_{12}O$	623.21	621.26	613.55
62	1-Hexanol	$C_6H_{14}O$	330.60	337.76	336.93
63	Ethyl butanoate	$C_6H_{12}O_2$	267.79	278.89	273.73
64	2-Methyl-1-butanol	$C_5H_{12}O$	318.35	309.08	301.05

3.2. Odor Fingerprint of Volatile Substances

The fingerprint of VOCs corresponding to each sample with different irradiation intensities is displayed in Figure 1E. The same row represents the signal peaks of volatile compounds in the same turmeric volatile oil sample, and the same column represents the signal peaks of the same volatile compound in different turmeric volatile oil samples. The color from light to dark indicates the content of volatile compounds from low to high. The fingerprints of volatile organic compounds collected from three kinds of samples were divided into three regions with different colors. The volatile organic compounds in turmeric volatile oil varied substantially with different irradiation intensities.

The substances in the green box in Figure 1E, such as citronellol, alpha terpineol, decanoic acid, methyl decanoate, and linalool-m, have the highest content in JH-1 sample and gradually decrease with the increase in irradiation dose. The content of substances in the yellow area is the highest in JH-2 samples, such as 2,5-dimethylfuran, acetate, and (E)-2-pentanal. The content of substances in the red area is the highest in the JH-3 sample and the lowest in the JH-1 sample, such as 2,5-dimethylfuran, acetal, and (E)-2-glutaraldehyde. The JH-2 content of Yellow Zone 2 is the lowest among the three samples, such as 2-acetylfuran, benzaldehyde-d, benzaldehyde-m, furfural-d, furfural-m, and 2-methylbutanoic acid.

3.3. Similarity Analysis

Figure 1F is the PCA diagram of three volatile oil samples at different irradiation intensities, which can visually show the differences between different products. The volatile organic compounds in JH-1 and JH-2 are similar, and the distance between them is very close. The gap between JH-1 and JH-3 is the largest, and the distance between them is the farthest.

Figure 1G shows the fingerprint similarity analysis of three volatile oil samples with different irradiation intensities. It can also be seen from the figure that the volatile organic compounds of JH-2 and JH-3 are very similar, the distance between them is very close, and the results of the principal component analysis are consistent with those of fingerprint analysis. There are substantial differences in volatile organic compounds between volatile oil samples (JH-1, JH-3) with large differences in irradiation intensity.

4. Discussion

The main active components and volatile components of turmeric volatile oil exposed to different doses of ^{60}Co-γ irradiation were determined and analyzed. By using headspace sampling and GC–IMS technology, the compounds can be qualitatively analyzed according to the GC retention time and ion migration time of volatile substances. A total of 64 volatile compounds were identified by GC–IMS analysis and built-in NIST database retrieval. The results showed that the contents of various volatile substances were different under different irradiation intensities. The volatile oil in turmeric is responsible for the aroma of turmeric, while curcumin (curcumin and its analogues) is responsible for its bright yellow color [12,13]. Some literature studies on turmeric butter have identified sesquiterpenoids and monoterpenoids as the main components [14], including gingerone, curcumene, curcumin, sabinene, borneol, caryophyllene, and other compounds [15]. The results of this study show that ethyl-2-phenylacetate, 2-ethylfuran, 2-butanone, 1-pentanol, 2-methylbutanoic acid, 3-hydroxy-2-butanone, linalool-d, eugenol, ethyl propanoate, ethyl pentanoate, and diethyl succeed are the lowest in JH-1 and the highest in JH-3. 2,5-Dimethylfuran, acetal, (E)-2-pentanal, and three unmatched compounds have the highest content in JH-2, followed by JH-1, and JH-3. The contents of 2-methyl-1-butanol, 2-methylpropanal, 2,3-butanedione, 2-nonanone, 2,5-dimethylthiophene, 2-furanmethanol, beta-ocimene, linalool-m, citronellol, alpha-terpineol, decanoic acid, and methyl decanoate decreased gradually from JH-1 to JH-3. The content of 2-acetylfuran is higher in JH-3, followed by JH-1; and almost none is present in JH-2. The content of 2-propanol in JH-1 was the highest, JH-3 was the second, and JH-2 was the lowest. The rest of the ingredients did not change significantly. According to the literature, turmerin, turmerone, elemene, furanodiene, curdione, bisacurone, cyclocurcumin, calebin A, and germacrone and other compounds in turmeric volatile oil have anti-inflammatory and anti-cancer activities [16–18], anti-hyperlipidemic property [19–21], as well as used in the prevention of asthma [22], treatment of respiratory diseases, and anti-oxidation in vitro effect [23,24]. Most of these main components did not change much because of the influence of irradiation, and so ^{60}Co-γ ray irradiation did not have a great impact on the effectiveness of turmeric volatile oil.

Dosages of 0, 5, and 10 kGy of ^{60}Co were used to analyze turmeric (Curcumae Longae Rhizoma) volatile oil after ^{60}Co Irradiation (named JH-1, JH-2 and JH-3). With the increase in irradiation dose, the peak area decreased. It is of great significance to explore the sterilization dose of irradiation sterilization of different traditional Chinese medicine varieties using ^{60}Co. Dosages of 5 and 10 kGy are the most widely used. There is a maximum level and limit for radiation which should be discussed in the future. It could be better to do it for 0, 2.5, 5, 7.5, and 10.

In this study, the author found that gas chromatography–mass spectrometry can effectively identify volatile odor compounds such as alcohols, ketones, aldehydes, esters, and terpenes. Food [25,26], agriculture [27–29], and traditional Chinese medicine field are widely used [30]. Gas-phase ion mobility spectrometry widely used in food [25,26], agriculture [27–29], and traditional Chinese medicine fields can quickly and accurately conduct a qualitative analysis of turmeric volatile oil under different irradiation doses and elucidate the differences in the odor of volatile organic compounds between samples [30]. However, there are still some limitations. The author has not conducted a further quantitative analysis of each compound. The next step will be to do a more explicit quantitative analysis of the specific components of the volatile oil.

5. Conclusions

In this study, the volatile components of turmeric volatile oil samples with three different irradiation intensities were analyzed by GC–IMS. A total of 97 volatile substances were detected, and a total of 64 components were determined by database retrieval, such as dimers of some substances. The volatile organic compounds in three turmeric volatile oil samples, mainly including terpenes, esters, aldehydes, alcohols, and ketones, were found. The results showed that the chemical components of the three turmeric volatile oil samples were similar, but the contents were quite different, suggesting that the irradiation intensities might have an impact on the volatile organic compounds of turmeric volatile oil. JH-1 and JH-3 have great differences in irradiation strength, and the difference in the principal component analysis is also large, indicating that the difference in chemical composition is the largest. This study provides a scientific basis for the dose control of irradiation sterilization of turmeric and its volatile oil. With the increase in irradiation dose, the peak area decreased, and so the irradiation dose of 5 kGy/min was better for ^{60}Co irradiation of turmeric (Curcumae Longae Rhizoma). This study provides a sound basis for the use of ^{60}Co-γ ray irradiation sterilization technology during the preparation of medicinal herbs.

It is shown that irradiation technology has good application prospects in the sterilization of foods with volatile components, However, attention must be paid to the changes in radiation dose and chemical composition. GC–IMS which has the advantages of simple operation, strong separation ability, short detection cycle, and retaining the original flavor of samples to the greatest extent, can be successfully applied to foods.

Author Contributions: Conceptualization, L.Y. and Y.H.; methodology, W.Z.; software, H.W.; validation, L.Y., Y.H. and H.W.; formal analysis, L.Y.; investigation, Y.H.; resources, W.Z.; data curation, H.W.; writing—original draft preparation, S.L.; writing—review and editing, D.H.; visualization, C.L.; supervision, S.L.; project administration, S.L.; funding acquisition, D.H. All authors have read and agreed to the published version of the manuscript.

Funding: This research was funded by Hunan Provincial Natural Science Foundation of China (No. 2019JJ60018), Department of Science and Technology of Hunan Province (No. 2021CB1012), and the First-class Discipline Project on Traditional Chinese Medicine of Hunan University of Chinese Medicine (No. 2018), Department of Science and Technology of Xiangxi Tujiazu & Miaozu Autonomous Prefeture (No. 2022JSGG03). Hunan Province College Students Innovation and Entrepreneurship Training Program Project (No. 20212442).

Data Availability Statement: Data is contained within the article.

Conflicts of Interest: The authors declare no conflict of interest.

References

1. Hong, Z.-C.; Wu, H.-Z.; Yang, Y.-F.; Liu, Y.; Wang, X.-L.; Duan, X.-Y.; Chen, S.-H. Qualitative and quantitative analysis of volatile oil from Rhizoma Wenyujin Concisum by GC-MS and GC. *Cent. South Pharm.* **2019**, *17*, 426–434.
2. Feng, S.-J.; Wu, Z.-F.; Wang, Y.-Q.; Yue, P.-F.; Zhang, S.-J.; Yang, M. Current situation and problem analysis on sterilization process for Chinese materia medica. *Chin. Tradit. Herb. Drugs* **2015**, *46*, 2667–2673.
3. Wang, J.-Y. ^{60}Co-γApplication of irradiation sterilization in traditional Chinese medicine and its preparations. *Res. Pract. Chin. Med.* **2003**, *17*, 59–61.
4. Liu, C.-J.; Zhu, Q. Effect of ^{60}Co-γIrradiation on 3 Active Ingredients in Compound Hemostatic Capsules. *China Pharm.* **2018**, *21*, 992–996.
5. Jiang, J.-L.; Ding, H.-T.; Su, X.; Yuan, Y.-J. Identification of Anti-tumor Ingredients in Curcuma Volatile Oil Based on Composition-Activity Relationship. *Chin. J. Anal. Chem.* **2012**, *40*, 1488–1493. [CrossRef]
6. Li, Y.-H.; Gong, X.; Ren, F.; Cheng, Z.-H.; Zhou, W.; Li, J.-H. Flavor Changes of *Annona squamosa* L. under Different Storage Conditions by GC-IMS. *Sci. Technol. Food Ind.* **2019**, *40*, 263–266.
7. Meng, X.-T.; Qiao, X.; Pan, Y.; Zou, S.-P.; Zhang, T.; Zhang, Q. Characteristic Flavor Compounds Fingerprinting of Mutton from Different Producing Regions of Xinjiang, China by Gas Chromatography-Ion Mobility Spectrometry. *Food Sci.* **2020**, *41*, 218–226.
8. Chen, T.; Wu, Z.-Y.; Wang, Z.-Y.; Lu, D.-L.; Chen, B. Identification of Meat Species by Gas Chromatography-Ion Mobility Spectrometry and Chemometrics. *J. Chin. Inst. Food Sci. Technol.* **2019**, *19*, 221–226.
9. Liu, L.-L.; Yang, H.; Xiong, J.; Xu, C.; Zhang, Y.-F.; Yan, Z.-K.; Qi, Y.-H. Influence of Different Storage Containers on the Aroma Composition of Fengxiang-type Baijiu Analyzed by Gas Chromatography-Ion Mobility Spectroscopy and Electronic Nose. *Food Sci.* **2022**, *43*, 257–263.
10. Fang, W.-J.; Gong, Y.; Xu, M.-Z. Application of GC-IMS technology in rapid detection of perfume quality. *China Food Drug Adm. Mag.* **2019**, *2*, 57–61.
11. Wang, R.; Zhu, F.-F.; Sun, Y.-L.; Hu, H.-F. Analysis of fingerprint differences in volatile substances in leaves of different walnut varieties by GC-IMS. *J. Fruit Sci.* **2021**, *38*, 1930–1941.
12. Itokawa, H.; Shi, Q.; Akiyama, T.; Morris-Natschke, S.-L.; Lee, K.-H. Recent advances in the investigation of curcuminoids. *Chin. Med.* **2008**, *11*, 3. [CrossRef] [PubMed]
13. Chatterjee, S.; Variyar, P.-S.; Gholap, A.-S.; Bongirwar, D.-R. Effect of γ-irradiation on the volatile oil constituents of turmeric (*Curcuma longa*). *Food Res. Int.* **2000**, *33*, 103–106. [CrossRef]
14. Himaja, M.; Ranjitha, A.; Ramana, M.-V.; Anand, M.; Karigar, A. Phytochemical screening and antioxidant activity of rhizome part *Curcuma zedoaria*. *Int. J. Res. Ayurveda Pharm.* **2010**, *1*, 414–417.
15. Dosoky, N.S.; Setzer, W.N. Chemical Composition and Biological Activities of Essential Oils of *Curcuma* Species. *Nutrients* **2018**, *10*, 1196. [CrossRef]
16. Aggarwal, B.B.; Yuan, W.; Li, S.; Gupta, S.C. Curcumin-free turmeric exhibits anti-inflammatory and anticancer activities: Identification of novel components of turmeric. *Mol. Nutr. Food Res.* **2013**, *57*, 1529–1542. [CrossRef]
17. Lai, E.-Y.-C.; Chyau, C.-C.; Mau, J.-L.; Chen, C.-C.; Lai, Y.-J.; Shih, C.-F.; Lin, L.-L. Antimicrobial activity and cytotoxicity of the essential oil of *Curcuma zedoaria*. *Am. J. Chin. Med.* **2004**, *32*, 281–290. [CrossRef]
18. Lakshmi, S.; Padmaja, G.; Remani, P. Antitumour effects of isocurcumenol isolated from *Curcuma zedoaria* rhizomes on human and murine cancer cells. *Int. J. Med. Chem.* **2011**, *2011*, 253962. [CrossRef]
19. Ling, J.; Wei, B.; Lv, G.; Ji, H.; Li, S. Anti-hyperlipidaemic and antioxidant effects of turmeric oil in hyperlipidaemic rats. *Food Chem.* **2012**, *130*, 229–235. [CrossRef]
20. Singh, V.; Jain, M.; Misra, A.; Khanna, V.; Rana, M.; Prakash, P.; Malasoni, R.; Dwivedi, A.-K.; Dikshit, M.; Barthwal, M.-K. *Curcuma* oil ameliorates hyperlipidaemia and associated deleterious effects in golden Syrian hamsters. *Br. J. Nutr.* **2013**, *110*, 437–446. [CrossRef]
21. Lei, H. Studies on Anti-Hyperlipemia and Anti-Tumor Activities of Turmeric Oil. Master's Thesis, Nanjing University of Chinese Medicine, Nanjing, China, 2013.
22. Li, C.; Li, L.; Luo, J.; Huang, N. Effect of turmeric volatile oil on the respiratory tract. *China J. Chin. Mater. Med.* **1998**, *23*, 624–625.
23. Wu, B. Study on the Extraction and Antibacterial Activity of Turmeric Oil and the Purification of Curcumin. Master's Thesis, Central South University of Forestry and Technology, Changsha, China, 2009.

24. Lu, C.-H.; Mou, D.-H. Antiinflammatory, Analgesic and in Vitro Antioxidant Activities of Turmeric Oil. *Food Sci.* **2018**, *39*, 243–249.
25. Li, W.; Chen, Y.-P.; Blank, I.; Li, F.; Li, C.; Liu, Y. GC × GC-ToF-MS and GC-IMS based volatile profile characterization of the Chinese dry-cured hams from different regions. *Food Res. Int.* **2021**, *142*, 110222. [CrossRef] [PubMed]
26. Liu, H.; Xu, Y.; Wu, J.; Wen, J.; Yu, Y.; An, K.; Zou. B. GC-IMS and olfactometry analysis on the tea aroma of Yingde black teas harvested in different seasons. *Food Res. Int.* **2021**, *150*, 110784. [CrossRef] [PubMed]
27. Fan, X.; Jiao, X.; Liu, J.; Jia, M.; Blanchard, C.; Zhou, Z. Characterizing the volatile compounds of different sorghum cultivars by both GC-MS and HS-GC-IMS. *Food Res. Int.* **2021**, *140*, 109975. [CrossRef]
28. Yang, Y.; Wang, B.; Fu, Y.; Shi, Y.-G.; Chen, F.-L.; Guan, H.-N.; Liu, L.-L.; Zhang, C.-Y.; Zhu, P.-Y.; Liu, Y.; et al. HS-GC-IMS with PCA to analyze volatile flavor compounds across different production stages of fermented soybean whey tofu. *Food Chem.* **2021**, *346*, 128880. [CrossRef]
29. Zhang, K.; Zhang, C.; Gao, L.; Zhuang, H.; Feng, T.; Xu, G. Analysis of volatile flavor compounds of green wheat under different treatments by GC-MS and GC-IMS. *J. Food Biochem.* **2021**, *27*, e13875. [CrossRef]
30. Zhou, Q.; Dai, Y.-P.; Guo, W.; Wang, P.; Shi, D.-H. Analysis of volatile organic compounds (VOCs) fingerprint of raw and honey-fried licorice based on headspace-gas-chromatography ion-mobility spectrometry (HS-GC-IMS). *China J. Chin. Mater. Med.* **2020**, *45*, 3857–3862.

foods

MDPI

Article

Development of a Sensitive SERS Method for Label-Free Detection of Hexavalent Chromium in Tea Using Carbimazole Redox Reaction

Limei Yin [1], Heera Jayan [2], Jianrong Cai [1], Hesham R. El-Seedi [3,4], Zhiming Guo [2,4,*] and Xiaobo Zou [2,4]

1 Key Laboratory of Modern Agricultural Equipment and Technology of Ministry of Education, School of Agricultural Engineering, Jiangsu University, Zhenjiang 212013, China; yinlm6@163.com (L.Y.); jrcai@ujs.edu.cn (J.C.)
2 China Light Industry Key Laboratory of Food Intelligent Detection & Processing, School of Food and Biological Engineering, Jiangsu University, Zhenjiang 212013, China; heerajayan93@outlook.com (H.J.); zou_xiaobo@ujs.edu.cn (X.Z.)
3 Pharmacognosy Group, Department of Pharmaceutical Biosciences, BMC, Uppsala University, Box 591, SE 751 24 Uppsala, Sweden; hesham.el-seedi@fkog.uu.se
4 International Joint Research Laboratory of Intelligent Agriculture and Agri-Products Processing, Jiangsu University, Zhenjiang 212013, China
* Correspondence: guozhiming@ujs.edu.cn

Abstract: Tea plants absorb chromium-contaminated soil and water and accumulate in tea leaves. Hexavalent chromium (Cr^{6+}) is a very toxic heavy metal; excessive intake of tea containing Cr^{6+} can cause serious harm to human health. A reliable and sensitive surface-enhanced Raman spectroscopy (SERS) method was developed using Au@Ag nanoparticles as an enhanced substrate for the determination of Cr^{6+} in tea. The Au@AgNPs coated with carbimazole showed a highly selective reaction to Cr^{6+} in tea samples through a redox reaction between Cr^{6+} and carbimazole. The Cr^{6+} in the contaminated tea sample reacted with methimazole—the hydrolysate of carbimazole—to form disulfide, which led to the decrease in the Raman intensity of the peak at 595 cm^{-1}. The logarithm of the concentration of Cr^{6+} has a linear relationship with the Raman intensity at the characteristic peak and showed a limit of detection of 0.945 mg/kg for the tea sample. The carbimazole functionalized Au@AgNPs showed high selectivity in analyzing Cr^{6+} in tea samples, even in the presence of other metal ions. The SERS detection technique established in this study also showed comparable results with the standard ICP-MS method, indicating the applicability of the established technique in practical applications.

Keywords: SERS detection; chromium contamination; tea sample; carbimazole hydrolysate; Au@Ag nanoparticles

Citation: Yin, L.; Jayan, H.; Cai, J.; El-Seedi, H.R.; Guo, Z.; Zou, X. Development of a Sensitive SERS Method for Label-Free Detection of Hexavalent Chromium in Tea Using Carbimazole Redox Reaction. *Foods* 2023, *12*, 2673. https://doi.org/10.3390/foods12142673

Academic Editor: Lili He

Received: 30 May 2023
Revised: 7 July 2023
Accepted: 10 July 2023
Published: 11 July 2023

1. Introduction

Chromium has been widely employed in various industries and has become a major threat to the environment and human health [1,2]. The heavy metal chromium enters the metabolic and digestive system of the human body through contaminated foods and causes various health effects based on the valence state. Chromium exists stably in the environment as two different oxidation states, the trivalent chromium (Cr^{3+}) cation and the hexavalent chromium (Cr^{6+}) anion; however, their chemical properties are markedly different, and Cr^{6+} is far more dangerous than Cr^{3+} [3]. The presence of an appropriate amount of Cr^{3+} in the human body is non-toxic and beneficial to human metabolism and health. Cr^{3+} is a necessary trace element for glucose, fat, and protein metabolism in mammals [4]. However, the presence of Cr^{6+} in food increases the risk of cancer and mutation in the human body [4]. Cr^{6+} mainly exists in the industrial discharge of dyes and tanneries. If the tea plants were grown in chromium-contaminated soil or irrigated with chromium-contaminated water, the tea plants continued to absorb and enrich chromium pollutants, eventually leading to the accumulation of chromium pollutants in the tea. The safety assessment of agricultural products as a key important food source of heavy metals is of great significance to ensuring food safety.

Tea (*Camellia sinensis*) is the most popular flavored and functional beverage worldwide. Tea is particularly rich in polyphenols, amino acids, caffeine, and other effective components, with lowering blood lipids and blood sugars, anti-inflammatory, antibacterial, antioxidant and other health effects, which are thought to contribute to the health benefits [5]. In the international trade of tea, the primary problem is safety, in which respect the residue of risk substances, especially heavy metal pollutants, has aroused great concern. The maximum residual limit for chromium in tea is 5 mg/kg according to a standard issued by the Ministry of Agriculture of China [4]. Thus, a rapid, efficient, and accurate quantitative detection method is essential to prevent chromium-contaminated from entering the tea planting and production process from different sources. Nowadays, the determination methods of Cr^{6+} mainly include atomic absorption spectroscopy [6], liquid chromatography [7], immunoassay [8], fluorescence spectroscopy [9], and inductively coupled plasma mass spectrometry (ICP-MS) [10]. However, these techniques usually require sophisticated equipment, skilled personnel, and long and tedious sample preparation [11]. Therefore, a simple, sensitive, and reliable method for the determination of Cr^{6+} would be very useful to ensure the safety and quality of food products [12].

Surface-enhanced Raman spectroscopy (SERS) is a reliable, sensitive, and non-destructive method popular for the detection of trace amounts of contaminants from complex matrices. In the past decade, SERS detection has been extensively applied in several fields including environmental monitoring, food safety, and pharmaceutical analysis [13]. The increasing use of SERS substrates has prompted researchers to use green reducing agents to prepare nanoparticles [14,15]. Carbimazole (Ethyl 3-methyl-2-thioimidazoline-1-carboxylate) is the main functional component of an oral drug for the treatment of thyroid diseases. After entering the human body, it can be hydrolyzed into methimazole in the acidic environment of the stomach [16]. By simulating the human stomach environment, acidic conditions can be generated in an aqueous solution. Tannins, the green natural, tannins have a strong reducing ability and can be employed for the synthesis of gold-silver core–shell nanoparticles (Au@AgNPs) at room temperature. In the reaction solution, carboprazole was adsorbed on the surface of Au@AgNPs, and its Raman signal was greatly enhanced due to the "hot spot" between the nanoparticles [17].

In this study, a convenient and sensitive SERS method for the detection of Cr^{6+} in tea was developed by carbimazole redox reaction. The main objectives of this study are: (1) to synthesize a SERS substrate of bimetallic core–shell nanoparticles (Au@AgNPs) to acquire a higher and stable Raman signal; (2) to obtain the highest Raman enhancement factor by optimizing the tannin with different volumes and concentrations of $HAuCl_4$ and $AgNO_3$; (3) to elucidate the mechanism of the decrease in Raman characteristic peak intensity caused by the redox reaction between Cr^{6+} and methimazole, the hydrolysate of carbimazole; and (4) to establish the linear quantitative equation of Raman intensity and the concentration of Cr^{6+} and to analyze the specific selectivity of carbimazole to Cr^{6+}. The schematic diagram of SERS detection of Cr^{6+} is illustrated in Figure 1.

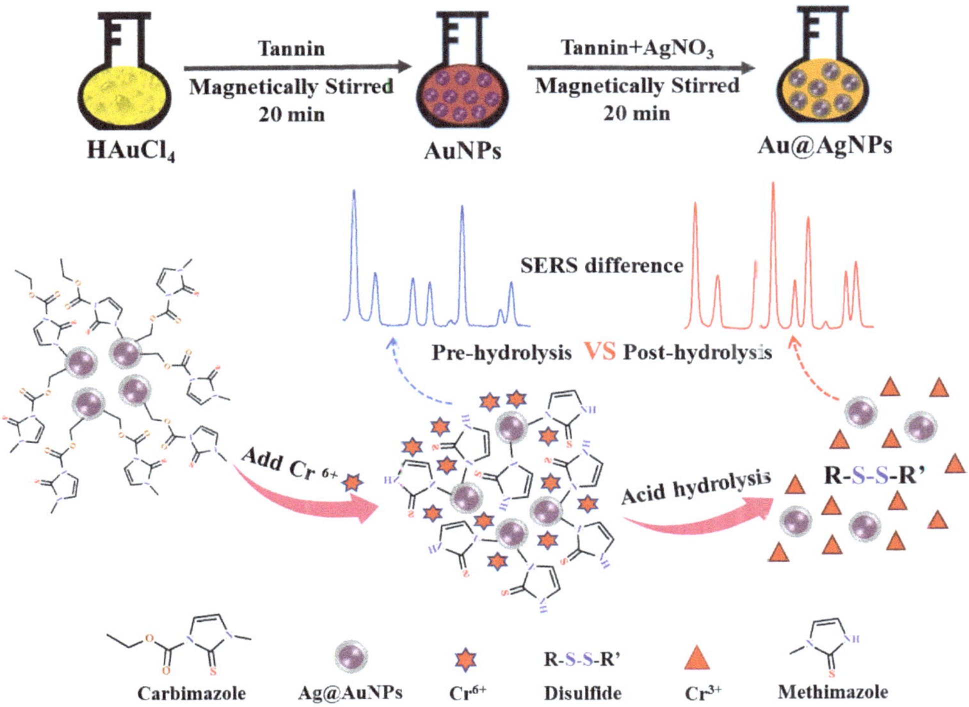

Figure 1. Schematic illustration of the mechanism of SERS detection of hexavalent chromium.

2. Materials and Methods

2.1. Materials

Tannin (92%), 4-mercaptobenzoic acid (4-MBA, 97%), chloroauric acid tetrahydrate ($HAuCl_4{\cdot}4H_2O$, 99%), silver nitrate ($AgNO_3$, 99.8%), carbimazole ($C_7H_{10}N_2O_2S$, 98%), sodium chloride (NaCl, 99%), potassium carbonate (K_2CO_3, 99%), and nitric acid (HNO_3, 68%) were obtained from National Pharmaceutical Group Chemical Reagents Company (Beijing, China). Standard solution of mercury (Hg), manganese (Mn), nickel (Ni), lead (Pb), copper (Cu), cadmium (Cd), iron (Fe), arsenic (As), sodium (Na), and chromium (Cr), were procured from Siyuan Chemical Glass Co., Ltd. (Zhenjiang, China), and all their purities were greater than 97%. Ultrapure water was used for all the experiments. All reagents used in the study were analytical grade unless stated otherwise.

2.2. Instruments

Au@AgNPs were characterized using Ultraviolet–Visible (UV–Vis) absorption spectroscopy (Agilent Technologies Inc., Palo Alto, CA, USA), Tecnai 12 transmission electron microscope (TEM) (Philips, Amsterdam, Holland,) and Fourier–transform infrared (FTIR) spectroscopy (Beijing Rayleigh Analytical Instrument Co., Ltd., Beijing, China). Microwave digestion instrument (MARS 6) (CEM, Charlotte, NC, USA) was used for the digestion of food samples. The confocal micro-Raman imaging spectrometer (XploRA PLUS, HORIBA, MPL, France) was employed to collect Raman spectra from the samples. The employed excitation wavelength and objective lens were 785 nm and 50× (Spot size: 1.28 μm), respectively. The ICP-MS (Thermo Fisher Scientific, X Series 2, Waltham, MA, USA) analysis was performed to detect Cr^{6+} in the spiked tea digestions.

2.3. Synthesis of SERS Substrate Au@AgNPs

Synthesis of AuNPs: Gold nanoparticles were prepared by utilizing tannins as a reducing agent [18]. Briefly, 1 mL of 10 mM tannin was added to a 50 mL beaker containing 18.5 mL of ultrapure water under stirring conditions (500 rpm), followed by the addition of 500 μL of 1% $HAuCl_4 \cdot 4H_2O$. The total volume of the reaction mixture was about 20 mL, and the reaction was allowed to continue for 20 min. The pH of the solution was measured at 6 due to the acidic nature of the tannin. Another batch of nanoparticles was also prepared by adding $HAuCl_4 \cdot 4H_2O$ (1%) after adjusting the pH of the solution to 7 with K_2CO_3 solution (0.2 M).

Synthesis of Au@AgNPs: The prepared AuNPs were coated adding a silver layer by reducing $AgNO_3$ using tannins as a reducing agent [19]. In a 25 mL round-bottom flask, 9.5 mL of prepared AuNPs colloidal solution was stirred at 500 rpm. Then, 250 μL of 10 mM tannin solution and 250 μL of 10 mM $AgNO_3$ solution were sequentially added to the solution with constant stirring for 30 min at room temperature. The formation of the core–shell structure was indicated by the change in colour of the solution from wine red to orange.

Synthesis optimization: It was found that the volume of $HAuCl_4$ and $AgNO_3$ during the preparation of Au@AgNPs greatly affects the size and concentration of nanoparticles formed in the solution, which will, in turn, affect the enhancement capabilities of the prepared substrate. In order to obtain high enhancement abilities, the volume of the two reactants ($HAuCl_4$ and $AgNO_3$) was optimized. The synthesis of AuNPs involved varying the volume of $HAuCl_4$ from 100 to 600 μL with an increment of 100 μL. The enhancement of the AuNPs was evaluated using 4-MBA (10^{-3} M) as a Raman signal probe. Meanwhile, the synthesis of Au@AgNPs involved varying the volume of $AgNO_3$ from 100 to 600 μL with an increment of 100 μL. The UV–Visible spectra of the synthesized Au@AgNPs substrates were collected to better understand the difference in surface plasmon resonance. Further, the enhancement of the Au@AgNPs was evaluated using 4-MBA (10^{-5} M) as a Raman signal probe.

2.4. Tea Sample Preparation

The black tea sample was purchased from the supermarket in Zhenjiang, Jiangsu. After grinding it into powder, the tea powder (0.2 g) was weighed and added to the microwave digestion tank. Concentrated HNO_3 (65%, 8 mL) and the Cr^{6+} standard solution was added to the tea sample before digestion (1 h). The digestion tanks were put in an acid extractor (130 °C, 60 min), cooled, and then digested in a microwave digester following the standard operating procedure with some modifications. Finally, a light green clarified solution was obtained without any solid residue [20]. The digested samples were degassed by sonication (100 °C, 10 min), and the inner cap was rinsed with a little water. Then the resulting digestion solution was filtered into 50 mL volumetric flasks with syringe filter (diameter 33 mm, pore size 0.22 μm) and filled up to the mark with ultrapure water to obtain the final solution to be assayed. A total of 8 samples spiked with various amounts of Cr^{6+}, including 100, 80, 60, 40, 20, 10, 5 μg L^{-1}, and blank, were used for the analysis. The prepared samples were also used for ICP-MS and SERS analysis.

2.5. Detection of Cr^{6+} in Tea Samples

The concentration of Cr^{6+} in tea samples was analyzed using Au@AgNPs in the presence of carbimazole solution. First, 40 μL of Au@AgNPs solution and 5 μL carbimazole solution (10 mM in chloroform) were mixed together for 10 min. Then, digested tea samples were added and mixed thoroughly. Finally, 5 μL NaCl (10 mM) was introduced to the solution to aggregate the nanoparticles in the solution which increases the local hotspot and improves the Raman signal intensity [21]. A piece of tin foil tape measuring approximately 5 cm in length (with a thickness of 0.09 mm and width of 20 mm) was carefully affixed flat onto a glass slide. The reaction solution with a total volume of 50 μL was allowed to sit undisturbed for 10 min. Then, 1 μL of the solution was gently placed onto the surface of the tin foil tape, shaping the droplet into a round shape as much as possible. It was left to air dry, and the SERS spectra were subsequently collected from within the dried droplet [22]. All spectra were collected using a confocal micro-Raman imaging spectrometer equipped with a 785 nm excitation laser (100% power). The total acquisition time was set at 5 s. Five spectra were randomly collected from the droplets of each sample, and the average values of the spectra of each concentration were taken as the final spectral data. The characteristic peaks for the Cr^{6+} were identified by comparing the obtained spectra with that of the Raman spectra of carbimazole powder. The standard curve for the quantitative analysis was obtained using digested tea samples containing different concentrations of Cr^{6+} and digested tea solution without Cr^{6+} was used as blank. The spectral intensity and intensity ratio of specific peak positions were taken into consideration to establish the calibration curve. Subsequently, the quantitative ability and accuracy were analyzed based on the calibration curve.

2.6. Specific Selectivity and Spike Recovery for Cr^{6+} Detection in Tea

According to the previous literature [23,24], common metal ion pollutants in tea include Hg^{2+}, Mn^{2+}, Ni^{2+}, Pb^{2+}, Cu^{2+}, Cd^{2+}, Fe^{3+}, As^{3+}, and Na^{+}. To validate the selectivity of carbimazole in detecting Cr^{6+}, the change of Raman intensity at the characteristic peak (595 cm^{-1}) was compared when these metal ions were added to the system at an equal concentration (100 μg/L). Additionally, the experiment was repeated 3 times for the spiked tea sample (5 μg/L), and the spectra of 15 different points were obtained on each prepared detection solution to analyze the repeatability of the method. The relative standard deviation (RSD) of the spectral intensity at the characteristic peak was calculated to analyze the repeatability of the detection method. the reproducibility of the method

The quantitative data obtained from the SERS analysis were compared against the standard ICP-MS method [25]. Tea samples with different concentrations of Cr^{6+} (2.5, 5, 10, 20 mg/kg) were used for ICP-MS analysis with triplicates, and the recovery rate and RSD values were calculated for each sample. Thus, the practical applicability of the developed method for the detection of Cr^{6+} was confirmed.

3. Results

3.1. Synthesis of Au@AgNPs and Optimization

Tannins are naturally occurring polyphenols that can act as an ideal reductant in the synthesis of nanoparticles. Tannins were responsible for the reduction of $HAuCl_4$, resulting in the formation of stable Au NPs which then serve as a seed to induce Ag NPs synthesis. The change in colour of the colloidal nanoparticle from wine red to orange indicated the formation of the core–shell structure. The surface plasmon resonance of the prepared nanoparticles with different amounts of precursors exhibited a change in surface plasmon resonance, as shown in Figure 2a. As the volume of $HAuCl_4$ increased, the colour of the nanoparticle solution gradually changed, and when the volume of $HAuCl_4$ reached 600 μL, the synthesized nanoparticle solution appeared turbid and exhibited aggregation and precipitation after 3 days of storage [26]. The UV–Vis spectra of AuNPs were shown in Figure 2a, and the peak at 520 cm^{-1} increased gradually and a slight red shift occurred as the volume of $HAuCl_4$ increased. This observation suggested that the particle size of Au NPs increased [27]. The enhancement effect of AuNPs on the Raman reporter molecule 4-MBA (10^{-3} M) was also used to optimize the amount of $HAuCl_4$. As shown in Figure 2c, the Raman intensity at 1074 cm^{-1} was the greatest when the amount of $HAuCl_4$ was 500 μL. It was attributed to AuNPs with larger particle size can produce strong localized surface plasmon resonance. However, as the amount of $HAuCl_4$ continued to increase (600 μL), the Raman intensity decreased. This is because the larger particle size leads to the instability of the Au NPs. Considering the stability of storage and Raman enhancement effect, the optimal volume of $HAuCl_4$ was selected as 500 μL.

To increase the Raman intensity, the growth of the Ag shell on the Au NPs was carried out. The amount of $AgNO_3$ was optimized because it affected the thickness of the Ag shell and thus affected the Raman intensity. As shown in Figure 2b, as the amount of $AgNO_3$ increased, the Ag absorption peak around 400 nm gradually intensified, while the Au absorption peak near 520 nm weakened rapidly or even vanished entirely. The Raman intensity of 4-MBA (10^{-6} M) at 1584 cm^{-1} was also used to determine the optimal amount of $AgNO_3$. As shown in Figure 2d, the Raman intensity reached its maximum when the volume of $AgNO_3$ was 500 μL. However, the Raman intensity weakened slightly when the amount of $AgNO_3$ increased to 600 μL, which was due to the 4-MBA signal transmission being hindered by a thicker Ag shell. Therefore, the optimal amount of $AgNO_3$ was 500 μL.

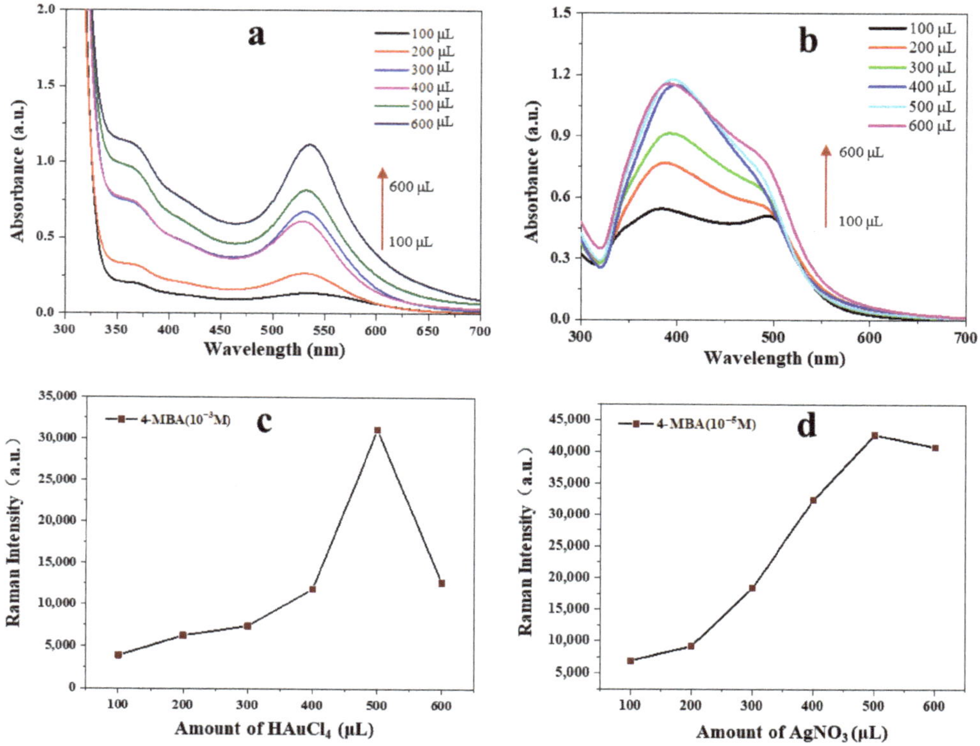

Figure 2. (**a**) UV spectra of AuNPs synthesized by tannin with different volumes of $HAuCl_4$; (**b**) UV spectra of Au@AgNPs reduced by tannin with different volumes of $AgNO_3$; (**c**) SERS enhancement of AuNPs synthesized using different volumes of $HAuCl_4$; (**d**) SERS enhancement of Au@AgNPs synthesized using different volumes of $AgNO_3$.

3.2. Characterization of Au@AgNPs

The morphology and distribution of nanoparticles were observed and measured by TEM. The size and morphology of the prepared nanoparticles changed based on the pH of the solution as shown in Figure 3a,b. The nanoparticles synthesized under pH 7 were aggregated and non-uniform compared to AuNPs synthesized under pH 6. The colloidal solution contained triangular nanoparticles, polygonal nanoparticles, nanorods, and irregular particles; however, the round nanoparticles account for the majority. Figure 3c shows the TEM image of Au@AgNPs synthesized under pH 6, which indicated that the formed particle had a particle size of around 10–30 nm. Since the shape and size of the nanoparticles formed under pH 6 conditions showed more uniformity, pH 6 was selected for further synthesis and applications.

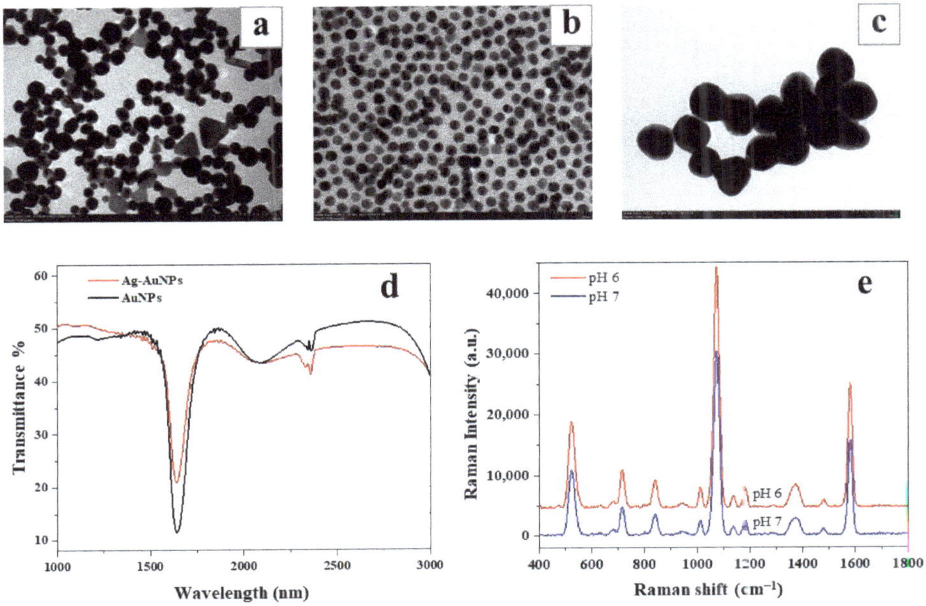

Figure 3. TEM image of AuNPs synthesized under pH 7 (**a**) and pH 6 (**b**); (**c**) TEM image of Au@AgNPs synthesized under pH 6; (**d**) FT-IR spectra of AuNPs and Au@AgNPs; (**e**) SERS enhancement of Au@AgNPs synthesized under pH 6 and pH 7 conditions.

For the FTIR characterization, spectra were collected from a 10 μL of drop cast nanoparticle solution, and the spectra displayed a characteristic peak at 1400 cm^{-1} for AuNPs (Figure 3d). Further, the silver coating caused a decrease in the peak intensity at 1400 cm^{-1} and a slight shift. The enhancement effects of Au@AgNPs synthesized under pH 6 and pH 7 conditions were also compared, and the obtained spectra are shown in Figure 3e. The calculated enhancement factors for Au@Ag nanoparticles synthesized under pH 6 and pH 7 conditions were 3.56×10^5 and 2.39×10^5, respectively. The Au@Ag further prepared by AuNPs synthesized at pH 6 had a better enhancement effect, which was closely related to the better uniform dispersion of AuNPs under this condition.

3.3. *Quantitative Detection of Cr^{6+} in Tea Samples*

The detection and quantitative analysis of Cr^{6+} in tea samples were obtained based on the redox reaction between carbimazole and Cr^{6+}. The hydrolytic product (methimazole) of carbimazole was adsorbed on the surface of Au@Ag via Ag-S and Ag-N bonds, and its Raman signal was greatly enhanced. At the same time, the aggregation of these functionalized substrates caused by NaCl also resulted in significant SERS enhancement [16]. In order to identify the characteristic peaks of Cr^{6+}, the spectra of high concentration of Cr^{6+} solution (100 μg/L) were compared with the blank tea sample (Cr^{6+} = 0 μg/L) (Figure 4a). Carbimazole was hydrolyzed to methimazole under acidic media that causes carbimazole to have a similar Raman spectrum to methimazole. The strongest characteristic peak was shown at 595 cm^{-1} due to the enhancement of Au@AgNPs on methimazole. The presence of Cr^{6+} causes a redox reaction between Cr^{6+} and methimazole, forming a disulfide compound and resulting in a decrease in Raman intensity at 595 cm^{-1}. Therefore, the quantitative analysis of Cr^{6+} can be realized by using the change of Raman intensity at 595 cm^{-1}. The main peaks of the methimazole contribution are listed in Table 1. As can be seen from Figure 4a, the characteristic peak at 595 cm^{-1} was attributed to the vibration of the C–N–S bend.

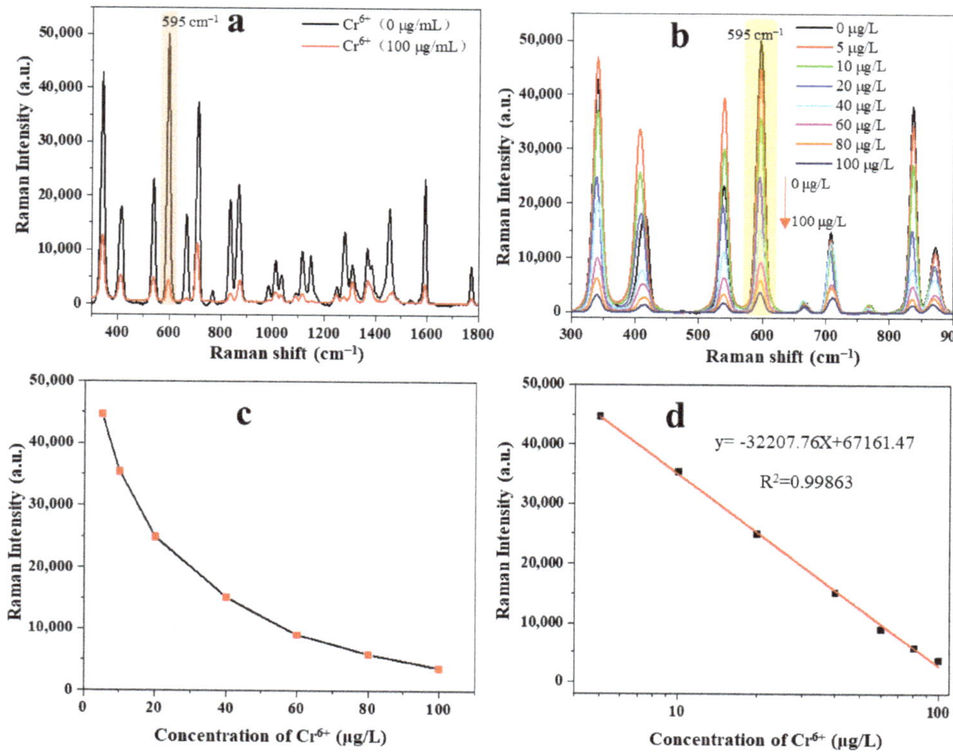

Figure 4. (**a**) SERS spectra of Cr^{6+} (0 μg/L) and Cr^{6+} (100 μg/L); (**b**) SERS spectra of Cr^{6+} in tea at different concentrations ranging from 0 to 100 μg/L; (**c**) the SERS intensity at 595 cm^{-1} at various concentration of Cr^{6+}; (**d**) calibration curve between the SERS intensity at 595 cm^{-1} and logarithm of the concentration of Cr^{6+}.

Table 1. Raman band assignments for peak obtained during Cr^{6+} detection.

Raman Shift (cm^{-1})	Band Assignment
412.63	Ring rotation, C–N–S bend
539.54	S=S stretching
595.85	C–N–S bend
708.48	Ring rotation, CH(NH) bend
872.67	Ring rotation, CH(NH) bend, C–N–S bend
1014.35	C–N stretching, CH(NH) bend
1370.14	C–N stretching, bend and rotation
1469.72	C–S stretching, CN stretching, NH bend
1593.76	C–C stretching, CH(NH) bend

The SERS spectra of different concentrations of Cr^{6+} ranging from 0 to 100 μg/L were obtained and are shown in Figure 4b. The Raman intensity at 595 cm^{-1} significantly decreased with increasing Cr^{6+} concentration in the tea samples (Figure 4c) as the methimazole was reduced to disulfide. The calibration curve of Cr^{6+} was obtained using the relationship between the Raman intensity at 595 cm^{-1} and the logarithm of the concentration of Cr^{6+} (Figure 4d). The calibration curve showed a good linear relationship in the range of 5~100 μg/L (R^2 = 0.99863). The linear quantitative relationship was described by the equation y = −32,207.76X + 67,161.47. Furthermore, the limit of detection (LOD) was defined as the concentration of Cr^{6+}, resulting in a 3% decrease in Raman intensity related to the blank tea sample (0 µg/L) [28]. According to the established equation, the LOD was calculated to be 3.78 µg/L, indicating that the immunosensor had a good sensitivity. Due to the dilution factor of 250 times in the pretreatment of tea samples, the detection range of Cr^{6+} in tea sample was 1.25~25 mg/kg, with a LOD of 0.945 mg/kg, which was much lower than the recommended tolerable level (5 mg/kg) of Cr^{6+} in tea.

3.4. Specific Selectivity and Recovery for Cr^{6+} Detection in Spiked Tea

For evaluating the specificity of the SERS method for Cr^{6+} analysis, some metal ions commonly found in tea, including Hg^{2+}, Mn^{2+}, Ni^{2+}, Pb^{2+}, Cu^{2+}, Cd^{2+}, Fe^{3+}, As^{3+}, Na^{+}, and Cr^{6}, were selected as interferences [10]. The Raman intensity of the peak at 595 cm^{-1} was shown in Figure 5a, and the results showed that only the Cr^{6+} (100 µg/L) have a specific redox reaction with the hydrolysate of the carbimazole, resulting in the decrease in the Raman intensity at the 595 cm^{-1} peak. Therefore, the presence of other ions does not interfere with the quantitative detection of Cr^{6+}.

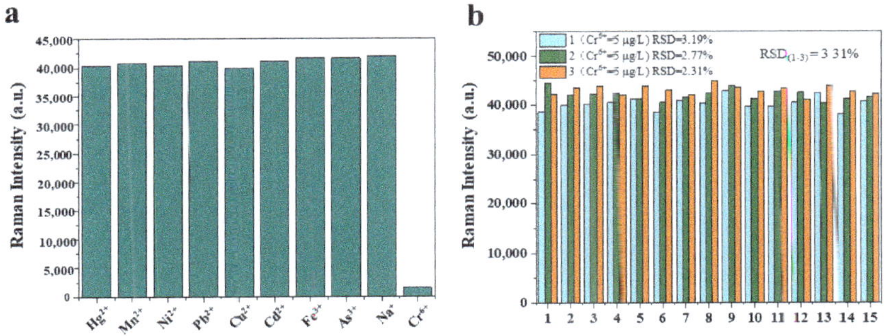

Figure 5. (**a**) Specificity of SERS detection for Cr^{6+} in tea when the concentration of all the ions was 100 µg/L; (**b**) relative standard deviation (RSD) of three repeated experiments and 15 random points in the detection solution at 595 cm^{-1}.

The experiment was repeated three times at the same concentration (Cr^{6+} = 5 µg/L) to verify the reproducibility. For each detection solution, 15 random points were chosen to obtain the Raman intensity at peak 595 cm^{-1}. As shown in Figure 5b, in each experiment, the RSD value of 15 points were all less than 5%, indicating that the detection solution had good uniformity. Similarly, the RSD value of the three repeated experiments was less than 5%, indicating that the detection method had a good reproducibility.

Further the quantitative results obtained from SERS detection were validated against standard ICP-MS, and the results are shown in Table 2. The recovery rates for the SERS method ranged from 91.62% to 104.84% (RSD $\leq$ 3.07%), while the recovery rates of the ICP-MS method ranged from 97.65% to 103.86% (RSD $\leq$ 1.53%). Comparative analysis indicates that the SERS method demonstrates higher accuracy and good reproducibility in detecting Cr^{6+} content in tea leaves.

Table 2. Comparison between the SERS method (this study) and ICP-MS quantitative detection of Cr^{6+}.

Spiked Value (mg/kg)	SERS			ICP-MS		
	Obtained Value (mg/kg)	Recovery Percentage (%)	RSD (%, n = 3)	Obtained Value (mg/kg)	Recovery Percentage (%)	RSD (%, n = 3)
2.5	2.621	104.84	2.06	2.528	101.12	1.07
5	4.686	93.72	1.39	4.8825	97.65	0.96
10	9.162	91.62	2.12	10.386	103.86	0.89
20	20.526	102.63	3.07	19.638	98.19	1.53

4. Discussion

The present study demonstrated the development of a sensitive and reliable method for the quantitative detection of Cr^{6+} in tea samples. The method relied on the redox reaction between Cr^{6+} and carbimazole, leading to a decrease in the intensity of the Raman peak at 595 cm^{-1}. This study provided valuable insights into the potential application of SERS technology as a rapid and label-free detection technique for Cr^{6+} in tea. The results indicated that the Raman intensity of the characteristic peak at 595 cm^{-1} was inversely correlated with the concentration of Cr^{6+}. Moreover, the relationship between the Raman intensity at 595 cm^{-1} and the logarithm of the concentration of Cr^{6+} was linear. The LOD (3.78 μg/L) of Cr^{6+} calculated using this approach indicated the sensitivity of the developed method. Compared with the previous study [29] using photoelectrochemical to detect Cr^{6+} (LOD = 0.01 μM) in the environment, the proposed method exhibited similar sensitivity and simpler detection operation.

Another key aspect investigated in this study was the selectivity of carbimazole towards Cr^{6+} in the presence of other metal ions. A specific selection of Cr^{6+} by carbimazole was demonstrated, highlighting the potential applicability of the developed method for real tea samples, which may contain various metal ions as contaminants. To validate the accuracy and reliability of the developed SERS method, a comparison was carried out with the standard method (ICP-MS). The reproducibility, accuracy, and recovery rate of the SERS technique were thoroughly analyzed. The results showed that the SERS method exhibited comparable performance to the ICP-MS method in terms of accuracy and recovery rate, indicating its suitability for practical applications.

The findings of this study exhibited the potential of SERS technology for the sensitive and label-free detection of Cr^{6+} in tea samples. The advantages of SERS, such as its high sensitivity, non-destructivity, and minimal sample preparation requirements, make it an attractive alternative to traditional analytical techniques. The established SERS method offers a convenient and sensitive approach for the determination of Cr^{6+}, which is of great importance considering the potential health hazards associated with its presence in tea. The results obtained in this study contribute to the growing interest in the application of SERS technology for the detection of contaminants in food, and it paves the way for further research.

5. Conclusions

In this study, a highly sensitive nano-SERS substrate was developed for the detection of Cr^{6+} in tea samples. The Au@AgNPs with uniform particle size distribution and good enhancement effect was synthesized using tannin. The combination of the specific redox reaction of carbimazole and Cr^{6+} and NaCl-induced aggregation of nanoparticles enhanced the EF value of Au@AgNPs to 3.56×10^5. Compared with AuNPs, the EF of Au@AgNPs was two orders of magnitude higher, thus improving the sensitivity of Cr^{6+} detection. Quantitative analysis showed a linear relationship between the Raman intensity of the characteristic peak at 595 cm^{-1} and the logarithm of the Cr^{6+} concentration. The LOD of this proposed SERS method was 3.78 μg/L, and the detection range was established to be 5~100 μg/L. Due to the 250-fold dilution during tea sample processing, the detection range of Cr^{6+} in tea sample was 1.25~25 mg/kg, with a LOD of 0.945 mg/kg. Additionally, the proposed method showed high specificity, even in the presence of other metal ions, and good reproducibility in detecting Cr^{6+} in the tea sample. In particular, it showed good accuracy (recovery rates ranged from 91.62% to 104.84%) and precision (RSD $\leq$ 3.07%) in the recovery experiment, and the obtained results were validated against ICP-MS. Conclusively, this method has great potential for rapid and label-free detection of Cr^{6+} in tea.

Author Contributions: Conceptualization, Z.G.; methodology, L.Y.; software, J.C.; validation, L.Y. and H.J.; formal analysis, L.Y.; investigation, L.Y.; resources, J.C. and X.Z.; data curation, L.Y.; writing—original draft preparation, L.Y. and H.J.; writing—review and editing, H.R.E.-S.; visualization, L.Y.; supervision, Z.G.; project administration, Z.G.; funding acquisition, Z.G. All authors have read and agreed to the published version of the manuscript.

Funding: This research was funded by National Natural Science Foundation of China (31972151), the Postgraduate Research and Practice Innovation Program of Jiangsu Province (KYCX21_3383), the Outstanding Young Teachers of Blue Project in Jiangsu province, Key R&D Project of Jiangsu Province (BE2022363), Jiangsu Agriculture Science and Technology Innovation Fund (CX(22)3069), the Open Fund of Key Laboratory of Modern Agricultural Equipment and Technology of Ministry of Education (MAET202117), and the Youth Project of Faculty of Agricultural Equipment of Jiangsu University (NZXB20210205).

Data Availability Statement: The data presented in this study are available on request from the corresponding authors.

Conflicts of Interest: The authors declare no conflict of interest.

References

1. Yu, X.-L.; He, Y. Challenges and Opportunities in Quantitative Analyses of Lead, Cadmium, and Hexavalent Chromium in Plant Materials by Laser-Induced Breakdown Spectroscopy: A Review. *Appl. Spectrosc. Rev.* **2017**, *52*, 605–622. [CrossRef]
2. Ao, M.; Chen, X.; Deng, T.; Sun, S.; Tang, Y.; Morel, J.L.; Qiu, R.; Wang, S. Chromium Biogeochemical Behaviour in Soil-Plant Systems and Remediation Strategies: A Critical Review. *J. Hazard. Mater.* **2022**, *424*, 127233. [CrossRef] [PubMed]
3. Katz, S.A.; Salem, H. The Toxicology of Chromium with Respect to Its Chemical Speciation: A Review. *J. Appl. Toxicol.* **1993**, *13*, 217–224. [CrossRef] [PubMed]
4. Ji, W.; Wang, Y.; Tanabe, I.; Han, X.; Zhao, B.; Ozaki, Y. Semiconductor-Driven "Turn-off" Surface-Enhanced Raman Scattering Spectroscopy: Application in Selective Determination of Chromium(VI) in Water. *Chem. Sci.* **2014**, *6*, 342–348. [CrossRef]
5. Guo, Z.; Barimah, A.O.; Yin, L.; Chen, Q.; Shi, J.; El-Seedi, H.R.; Zou, X. Intelligent Evaluation of Taste Constituents and Polyphenols-to-Amino Acids Ratio in Matcha Tea Powder Using near Infrared Spectroscopy. *Food Chem.* **2021**, *353*, 129372. [CrossRef]
6. Habila, M.; Unsal, Y.E.; Alothman, Z.A.; Shabaka, A.; Tuzen, M.; Soylak, M. Speciation of Chromium in Natural Waters, Tea, and Soil with Membrane Filtration Flame Atomic Absorption Spectrometry. *Anal. Lett.* **2015**, *48*, 2258–2271. [CrossRef]
7. Wu, J.; Lu, G.; Huang, X. Fabrication of Monolith-Based Solid-Phase Microextraction for Effective Extraction of Total Chromium in Milk and Tea Samples Prior to HPLC/DAD Analysis. *Microchem. J.* **2020**, *159*, 105549. [CrossRef]
8. Ling, S.; Zhao, Q.; Iqbal, M.N.; Dong, M.; Li, X.; Lin, M.; Wang, R.; Lei, F.; He, C.; Wang, S. Development of Immunoassay Methods Based on Monoclonal Antibody and Its Application in the Determination of Cadmium Ion. *J. Hazard. Mater.* **2021**, *411*, 124992. [CrossRef]
9. Zheng, M.; Xie, Z.; Qu, D.; Li, D.; Du, P.; Jing, X.; Sun, Z. On–off–on Fluorescent Carbon Dot Nanosensor for Recognition of Chromium(VI) and Ascorbic Acid Based on the Inner Filter Effect. *ACS Appl. Mater. Interfaces* **2013**, *5*, 13242–13247. [CrossRef]
10. Areo, O.M.; Njobeh, P.B. Risk assessment of heavy metals in rooibos (*Aspalathus linearis*) tea consumed in South Africa. *Environ. Sci. Pollut. Res.* **2021**, *28*, 59687–59695. [CrossRef]
11. Guo, Z.; Chen, P.; Yosri, N.; Chen, Q.; Elseedi, H.R.; Zou, X.; Yang, H. Detection of Heavy Metals in Food and Agricultural Products by Surface-Enhanced Raman Spectroscopy. *Food Rev. Int.* **2023**, *39*, 1440–1461. [CrossRef]
12. Guo, Z.; Wang, M.; Agyekum, A.A.; Wu, J.; Chen, Q.; Zuo, M.; El-Seedi, H.R.; Tao, F.; Shi, J.; Ouyang, Q.; et al. Quantitative Detection of Apple Watercore and Soluble Solids Content by near Infrared Transmittance Spectroscopy. *J. Food Eng.* **2020**, *279*, 109955. [CrossRef]
13. Pérez-Jiménez, A.I.; Lyu, D.; Lu, Z.; Liu, G.; Ren, B. Surface-Enhanced Raman Spectroscopy: Benefits, Trade-offs and Future Developments. *Chem. Sci.* **2020**, *11*, 4563–4577. [CrossRef]
14. Prakash, J. Fundamentals and Applications of Recyclable SERS Substrates. *Int. Rev. Phys. Chem.* **2019**, *38*, 201–242. [CrossRef]
15. Tian, S.; Hu, Y.; Chen, X.; Liu, C.; Xue, Y.; Han, B. Green Synthesis of Silver Nanoparticles Using Sodium Alginate and Tannic Acid: Characterization and Anti-*S. Aureus* Activity. *Int. J. Biol. Macromol.* **2022**, *195*, 515–522. [CrossRef] [PubMed]
16. Sultan, S.M. A Kinetic Method for the Determination of Carbimazole in Pharmaceutical Preparations by Oxidation with Bichromate in Sulfuric Acid. *Anal. Sci.* **1992**, *8*, 503–506. [CrossRef]
17. Bu, X.; Zhang, Z.; Zhang, L.; Li, P.; Wu, J.; Zhang, H.; Tian, Y. Highly Sensitive SERS Determination of Chromium(VI) in Water Based on Carbimazole Functionalized Alginate-Protected Silver Nanoparticles. *Sens. Actuators B Chem.* **2018**, *273*, 1519–1524. [CrossRef]
18. Huang, X.; Wu, H.; Pu, S.; Zhang, W.; Liao, X.; Shi, B. One-Step Room-Temperature Synthesis of Au@Pd Core–Shell Nanoparticles with Tunable Structure Using Plant Tannin as Reductant and Stabilizer. *Green Chem.* **2011**, *13*, 950–957. [CrossRef]
19. Gangwar, C.; Yaseen, B.; Kumar, I.; Singh, N.K.; Naik, R.M. Growth Kinetic Study of Tannic Acid Mediated Monodispersed Silver Nanoparticles Synthesized by Chemical Reduction Method and Its Characterization. *ACS Omega* **2021**, *6*, 22344–22356. [CrossRef]
20. Ren, Z.; Li, Z.; Chen, Z.; Zhang, Y.; Lin, X.; Weng, W.; Yang, H.; Li, B. Characteristics and Application of Fish Oil-in-Water Pickering Emulsions Structured with Tea Water-Insoluble Proteins/κ-Carrageenan Complexes. *Food Hydrocoll.* **2021**, *114*, 106562. [CrossRef]

21. Zuo, Q.; Chen, Y.; Chen, Z.P.; Yu, R.Q. Quantification of Cadmium in Rice by Surface-Enhanced Raman Spectroscopy Based on a Ratiometric Indicator and Conical Holed Enhancing Substrates. *Anal. Sci.* **2018**, *34*, 1405–1410. [CrossRef] [PubMed]
22. Hertaeg, M.J.; Rees-Zimmerman, C.; Tabor, R.F.; Routh, A.F.; Garnier, G. Predicting Coffee Ring Formation upon Drying in Droplets of Particle Suspensions. *J. Colloid Interface Sci.* **2021**, *591*, 52–57. [CrossRef] [PubMed]
23. Welna, M.; Szymczycha-Madeja, A.; Pohl, P. Novel ICP-OES-Based Method for the Reliable Determination of the Total Content of 15 Elements in Yerba Mate Drinks along with the Determination of Caffeine and the in Vitro Bioaccessibility of the Compounds. *Molecules* **2023**, *28*, 3374. [CrossRef] [PubMed]
24. Grembecka, M. Assessment of the Mineral Composition and the Selected Physicochemical Parameters of Dietary Supplements Containing Green Tea Extracts. *Foods* **2022**, *11*, 3580. [CrossRef]
25. Guo, Z.; Barimah, A.O.; Guo, C.; Agyekum, A.A.; Annavaram, V.; El-Seedi, H.R.; Zou, X.; Chen, Q. Chemometrics Coupled 4-Aminothiophenol Labelled Ag-Au Alloy SERS off-Signal Nanosensor for Quantitative Detection of Mercury in Black Tea. *Spectrochim. Acta Part A Mol. Biomol. Spectrosc.* **2020**, *242*, 118747. [CrossRef]
26. Huang, H.; Toit, H.D.; Besenhard, M.O.; Ben-Jaber, S.; Dobson, P.; Parkin, I.; Gavriilidis, A. Continuous Flow Synthesis of Ultrasmall Gold Nanoparticles in a Microreactor Using Trisodium Citrate and Their SERS Performance. *Chem. Eng. Sci.* **2018**, *189*, 422–430. [CrossRef]
27. Haiss, W.; Thanh, N.T.K.; Aveyard, J.; Fernig, D.G. Determination of Size and Concentration of Gold Nanoparticles from UV-Vis Spectra. *Anal. Chem.* **2007**, *79*, 4215–4221. [CrossRef]
28. Chen, P.; Yin, L.; El-Seedi, H.R.; Zou, X.; Guo, Z. Green reduction of silver nanoparticles for cadmium detection in food using surface-enhanced Raman spectroscopy coupled multivariate calibration. *Food Chem.* **2022**, *394*, 133481. [CrossRef]
29. Wu, W.; Tan, Z.; Chen, X.; Chen, X.; Cheng, L.; Wu, H.; Li, P.; Zhang, Z. Carnation-like Morphology of $BiVO_4$-7 Enables Sensitive Photoelectrochemical Determination of Cr(VI) in the Food and Environment. *Biosensors* **2022**, *12*, 130. [CrossRef]

foods

Article

Prediction of Anthocyanidins Content in Purple Chinese Cabbage Based on Visible/Near Infrared Spectroscopy

Ya-Qin Wang [1,2], Guang-Min Liu [1,2], Li-Ping Hu [1,2], Xue-Zhi Zhao [1,2], De-Shuang Zhang [3,*] and Hong-Ju He [1,2,*]

1 Institute of Agri-Food Processing and Nutrition, Beijing Academy of Agriculture and Forestry Sciences, Beijing 100097, China; wangyaqin@iapn.org.cn (Y.-Q.W.)
2 Key Laboratory of Vegetable Postharvest Processing of Ministry of Agriculture and Rural Areas, Beijing 100097, China
3 Beijing Vegetable Research Center, Beijing Academy of Agriculture and Forestry Sciences, Beijing 100097, China
* Correspondence: zhangdeshuang@nercv.org (D.-S.Z.); hehongju@iapn.org.cn (H.-J.H.)

Abstract: Purple Chinese cabbage (PCC) has become a new breeding trend due to its attractive color and high nutritional quality since it contains abundant anthocyanidins. With the aim of rapid evaluation of PCC anthocyanidins contents and screening of breeding materials, a fast quantitative detection method for anthocyanidins in PCC was established using Near Infrared Spectroscopy (NIR). The PCC samples were scanned by NIR, and the spectral data combined with the chemometric results of anthocyanidins contents obtained by high-performance liquid chromatography were processed to establish the prediction models. The content of cyanidin varied from 93.5 mg/kg to 12,802.4 mg/kg in PCC, while the other anthocyanidins were much lower. The developed NIR prediction models on the basis of partial least square regression with the preprocessing of no-scattering mode and the first-order derivative showed the best prediction performance: for cyanidin, the external correlation coefficient (RSQ) and standard error of cross-validation (SECV) of the calibration set were 0.965 and 693.004, respectively; for total anthocyanidins, the RSQ and SECV of the calibration set were 0.966 and 685.994, respectively. The established models were effective, and this NIR method, with the advantages of timesaving and convenience, could be applied in purple vegetable breeding practice.

Keywords: near infrared spectroscopy; vegetables; anthocyanidins; fast determination

Citation: Wang, Y.-Q.; Liu, G.-M.; Hu, L.-P.; Zhao, X.-Z.; Zhang, D.-S.; He, H.-J. Prediction of Anthocyanidins Content in Purple Chinese Cabbage Based on Visible/Near Infrared Spectroscopy. *Foods* **2023**, *12*, 1922. https://doi.org/10.3390/foods12091922

Academic Editors: Zhiming Guo, Zhao Zhang and Dong Hu

Received: 22 March 2023
Revised: 4 May 2023
Accepted: 5 May 2023
Published: 8 May 2023

1. Introduction

Chinese cabbage (*Brassica rapa* L. ssp. *pekinensis*) is the most widely cultivated and consumed vegetable in East Asia with the characteristics of high yield, good cold resistance, long supply period, and rich nutrition. The inner leaf color of Chinese cabbage is mainly white and yellow. Purple leaf Chinese cabbage (PCC) is mainly generated by the cross of common green Chinese cabbage with red leaf mustard (*Brassica juncea* Coss.), purple flowering Chinese cabbage, or red bok choy (*Brassica rapa* L. ssp. *chinensis*) [1,2]. It has become increasingly popular due to its beautiful color, special flavor, and high level of anthocyanidins [2]. Anthocyanidins, a class of flavonoid substances, exist in different colors in fruits, flowers, and vegetables, such as purple, blue, and red. They contain a C6-C3-C6 carbon skeleton and -OH or -OCH_3 groups and specific sugar or acylated sugar residues located at C3, C5, and C7 positions [3,4]. Based on the type and location of the substituents, anthocyanidins are generally classified into six major groups: cyanidin, delphinidin, petunidin, malvidin, peonidin and pelargonidin, and the main anthocyanidin in PCC is cyanidin accumulated in the vacuoles [2,3,5]. Anthocyanidins have a wide range of anti-inflammatory, cardioprotective, chemotherapy, and hepatoprotective effect for human disease prevention [6]. Studies have proved that anthocyanidins have a good role in the chemoprevention and treatment of breast cancer [7]. Blueberry anthocyanidins can effectively improve the solubility of lipids [8], and extracted anthocyanidins from apples have an inhibitory effect on gastric cancer cells [9]. Due to their human health benefits, anthocyanidins have received more and more attention from public in recent years [9,10]. Creating colorful leaf vegetables, such as PCC, which contain abundant anthocyanidins, is of significant commercial interest and the new trend of breeding.

Visible/near-infrared spectroscopy (NIR) is a widely used technique in the agriculture and food industry with the advantages of fast, non-destructive, environmentally friendly, and accurate analysis. NIR is a molecular vibrational spectrum with wavelengths ranging from 400–750 (visible) and 750–2500 nm (near-infrared), in which the absorption signals of the reflected chemical components are assigned mainly to overtone and octave vibrations of hydrogen-containing groups, including C-H, N-H, O-H, and S-H [11,12]. Theoretically, no two compounds produce the same visible/near-infrared spectra since their unique composition of atoms [12]. It has been widely used in the field of bioactive compound detection in vegetables and fruits, and its applicability has been proven. Prodromidis et al. have successfully used FT-IR and UV-Vis spectroscopy to measure the onion anthocyanidins during heating [13]. Johnson et al. used attenuated total reflection Fourier transform infrared spectroscopy to predict the total anthocyanidin content in ethanolic extracts of plum with an R^2 of 0.93 [14]. Additionally, using NIR spectroscopy in the prediction of anthocyanidins content and antioxidant activity in grape juice is feasible [15]. Tian et al. established a prediction model for the detection of water content and anthocyanidins content in purple potatoes by visible near-infrared hyperspectroscopy [16]. With the development of algorithms, chemometrics, and artificial intelligence, the application of NIR spectroscopy will be extended for fast screening and quantitative analysis of anthocyanidins.

Purple leaf Chinese cabbage has become a popular breeding interest; meanwhile fast and accurate determination of the anthocyanidins contents is an important task for improving its nutritional quality. The commonly used determination methods for anthocyanidins are based on ultrasonic or microwave-assisted liquid extractions and high-performance liquid chromatography (HPLC) and liquid chromatography-tandem mass spectrometry (LC-MS/MS) detection [17,18]. However, the extraction is complicated and time-consuming while the reagents used may be harmful to the environment and human health [19], and the equipment are more expensive and require experts for analysis. By comparing with chemical analysis techniques, spectroscopic techniques are relatively simple and do not require further expansion of sample preparation [20]. NIR spectroscopy could be a powerful tool to fulfill this task. To date, no studies have focused on the quantitative prediction of anthocyanidins in PCC by NIR spectroscopy. It is an urgent need to build a suitable NIR method for simple and fast prediction of anthocyanidins to help the breeders and producers since the prediction models established by different food matrices cannot be simply applied to PCC. Therefore, this study aims to develop an accurate quantitative prediction method for anthocyanidin content in PCC using NIR spectroscopy, which laid a foundation for the fast and convenient detection of the nutritional quality of agri-food and the rapid screening of purple vegetable breeding materials.

2. Materials and Methods

2.1. Sample Preparation

The purple leaf Chinese cabbage samples from different breeding backgrounds with distinct color phenotypes were collected from Beijing Vegetable Research Center (Beijing, China, 116°30′ E, 39°94′ N). Specifically, the purple color trait was from the variety of 15NG28, as previously described [21], and the green parents were different Chinese cabbages with distinct shapes of leaves, holding patterns, and maturity traits. Totally 106 PCC samples were harvested and transferred to the laboratory within half an hour on 19 November 2021. Then, the vegetable leaves of each sample were cut into 2.0 cm length pieces, uniformly mixed, and freeze-dried (BIOCOOL vacuum freeze dryer, Boyikang Co., Ltd., Beijing, China). The dry samples were ground into a fine powder, passed through an 80-mesh sieve, then stored at −40 °C for further anthocyanidins content determination and NIR spectral profiles acquisition.

2.2. HPLC Analysis of Anthocyanidins

Extraction and HPLC analysis of anthocyanidins in PCC were carried out according to the Agricultural Industry Standard of the People's Republic of China (NY/T 2640-2014, Determination of anthocyanidins in plant origin products-High performance liquid chromatography). Basically, accurately weighed 0.200 g powdered samples were placed in a 15 mL plastic tube, and 5.00 mL of extracting solution consisting of ethanol:water:hydrochloric acid = 2:1:1 (volume) was added to extract anthocyanidins. The extraction mixture was sonicated for 30 min at room temperature, then hydrolyzed under boiled water for one hour. Then, the cooled extraction mixture was centrifuged using a HITACHI high-speed refrigerated centrifuge (Katsuta, Japan) at 8000 rpm for 10 min. The supernatant was accurately fixed to 5.00 mL volume and filtered through a 0.45 μm polyvinylidene fluoride syringe filter before HPLC analysis.

The quantification of anthocyanidins was carried out on a reversed-phase HPLC system (LC-20AD, Shimadzu, Tokyo, Japan) coupled with a photodiode array (PDA) detector (SPD-M20A, Shimadzu, Tokyo, Japan). The column used was a Waters C18 (3.9 × 150 mm, 5 μm) kept at 35 °C. The gradient elution was carried out with a binary solvent system consisting of ultrapure water (A) and acetonitrile (B), both containing 1% formic acid, at a constant flow rate of 0.8 mL/min. The injection volume was 20 μL. Anthocyanidin compounds were detected at the wavelength of 530 nm. Individual anthocyanidins were quantified via comparison of the peak areas with those of the known standards. The anthocyanidins standards (delphinidin, cyanidin, petunidin, pelargonidin, peonidin, and malvidin) were purchased from Sigma-Aldrich (Darmstadt, Germany).

2.3. NIR Spectral Acquisition

The NIR spectrometer used in this study was a FOSS NIR Systems model 5000 (Foss NIRSystems Inc., Silver Spring, MD, USA). The NIR spectrometer was preheated for 30 min before the sample scanning, and the samples were only scanned when the spectra and noise tests were passed. The dried PCC powders were evenly spread in the sample round cups, respectively and compacted with the lid to ensure the sample powder was covered evenly. The spectra were scanned in the wavelength range of 400–1100 nm and 1100–2498 nm under diffuse reflection mode. Each sample was scanned three times. The scanned spectral curves were collected, and the data were processed using the Foss WinISI III calibration software throughout the whole process.

2.4. Data Processing

The PCC samples were divided into two sets by systematic sampling method; 86 of them were used as calibration sets to establish the prediction models, and 20 samples not involved in the calibration were used as validation sets for external validation of the effectiveness of the developed models. The chemical determination results of anthocyanidin content obtained by HPLC of the calibration set samples were imported into the chemometric software accompanying the instrument and processed for NIR spectroscopy to obtain a cal. file. The spectral data were preprocessed using a partial least squares regression (PLSR) method at three different wavelength bands. These three bands included 400–1100 nm, 1100–2498 nm (full band); 400–800 nm (visible band); 800–1100 nm, 1100–2498 nm (near infrared band). The pre-processing scattering model of the spectral data included no scattering processing (None), standard normal variables transformation + de-trending processing (SNV+Detrend), standard normal variation processing (SNV Only), de-trending processing (Detrend Only), standard multivariate scattering correction (Standard MSC), weighted multivariate scattering correction (Weighted MSC), and two different derivative treatments, namely, no derivative and first-order derivative were employed. The final prediction models built under different preprocessing methods were compared, and the model with the internal cross-validation correlation coefficient (1-VR) close to 1 and lower standard error of cross-validation (SECV) was selected as the best one. These two sets of data can basically reflect the prediction performance of the calibration model for unknown samples. Subsequently, samples of the validation set were analyzed to test the predictive ability of the proposed model. The criterion was that the higher the external correlation coefficient (RSQ) value and the lower the standard deviation of prediction (SEP), the more accurate the model.

3. Results

3.1. Anthocyanidins Contents in PCC Samples

Anthocyanidins contents of the PCC samples were analyzed by the HPLC method (the results are shown in Table S1), and the content distribution of the anthocyanidins fractions is shown in Table 1. Four kinds of anthocyanidins were detected, with cyanidin the most abundant one in PCC, which was a coincidence with the previous report [2]. Cyanindin had the largest range of content variation from 93.5 to 12,802.4 mg/kg, and the average content was 5741.2 mg/kg. In most samples, cyanidin accounted for more than 95% of the total anthocyanidins. The content distribution range of cyanidin in the selected samples was wide, which can well represent PCC samples with different contents of anthocyanidins; it meant that it was a suitable sample set for establishing of NIR model.

Table 1. Distribution of anthocyanidins contents in purple leaf Chinese cabbage (mg/kg).

Compound	Content Range	Average Content	Percentage of Total %
delphinidin	nd[1]~193.7	159.4	2.66
cyanindin	93.5~12,802.4	5741.2	95.71
pelargonidin	nd[1]~66.0	52.3	0.87
peonidin	nd[1]~63.0	45.4	0.76

nd[1]: not detected.

Delphinidin was detected in most PCC samples, with a content up to 193.7 mg/kg, and the average content was 159.4 mg/kg. Compared to cyaniding, pelargonidin and peonidin were much lower in PCC, whose average contents were 52.3 mg/kg and 45.5 mg/kg, respectively, accounted for less than 1% content to the total anthocyanidins in PCC.

3.2. Visible/NIR Spectral Analysis of PCC Samples

Using the software WinISS III, the chemically determined values were input to the corresponding spectral positions, and the spectral data were analyzed in combination with chemical analysis data. The raw spectra of the PCC samples obtained after visible/NIR spectroscopy scan (Figure 1A), in which the horizontal coordinate was the wavelength, and the vertical coordinate was the absorbance expressed as log 1/R, showed that several samples of PCC had a clear trend of decreasing absorption peaks in the wavelength range of 400 to 800 nm, which indicated that different samples had specific absorption characteristics in the visible wavelength band. The large variation in their spectrograms also indirectly indicated the different contents of each sample composition.

The raw NIR spectra contained comprehensive information on all chemical structures and a lot of irrelevant information and noise, so mathematical data pretreatment methods were applied to remove noise, compensate for baseline shifting, reduce the influence of non-target variation, and assist in smoothing the spectrum. The derivative transformation could partially compensate for baseline offset between samples and reduce instrument drift effects [22]. Figure 1B shows the spectral curve of the original spectrum after SNV+Detrend and first-order derivative pretreatment. The pretreated spectrum had more obvious undulations, the peaks became more and sharper, and the absorption peaks appeared in the originally smooth part. Figure 1C shows the spectral profile of the original spectrum after the SNV only and first-order derivative pretreatment, and the fitting phenomenon could be observed. On the processed spectrograms, we observed more clearly several characteristic peaks of the spectrum, with the peak at 672 nm associated with chlorophyll [23]. The peak at about 760 nm corresponds to the third overtone of the O-H vibration [24].

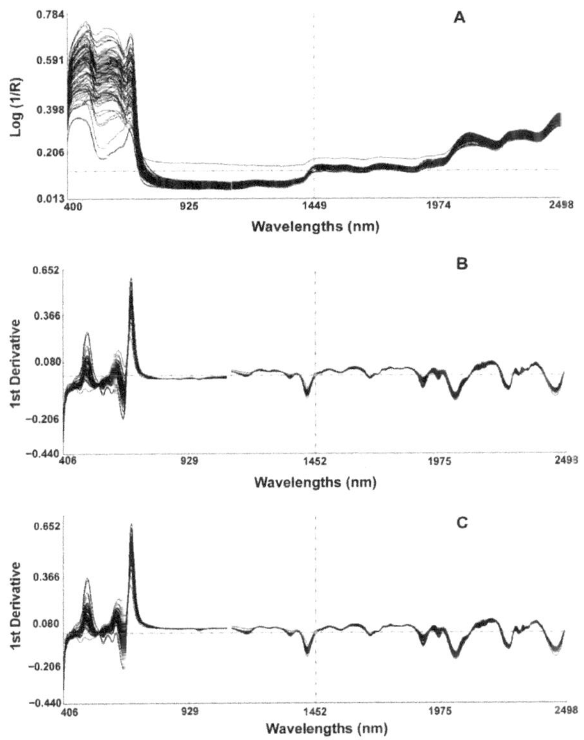

Figure 1. Visible/Near infrared spectra of purple leaf Chinese cabbages. (**A**): original spectra; (**B**): spectrum after SNV+Detrend and first derivative processing; (**C**) spectrum after SNV only and first derivative processing.

3.3. Establishment of Quantitative Models for Anthocyanidins Content in PPC

3.3.1. Model for Cyanidin Content Prediction

The spectral curves obtained from the scanned samples and the chemical analysis data were processed using PLSR to establish calibration models, and the calibration equation results are shown in Table 2. All spectral pre-treatment models performed well, with RSQ all above 0.91. Successful calibrations usually had a correlation coefficient of determination above 0.9. The 1-VR value of cyanidin in the full spectral band from 400 to 1100 nm and 1100 to 2498 nm after no scattering processing and first-order derivative pretreatment was 0.942 at the maximum, the SECV value was 693.004 at the smaller value, and the RSQ was 0.965. Figure 2A shows the cross-validation result of the prediction model established, the linear regression relationships between the NIR predicted values, and the chemically determined results (reference value). The slope of the line was 0.976, which is closed to 1; the samples were irregularly distributed on both sides of the line with the overall trend of discrete The model fits well and can achieve the purpose of good quantitative prediction. So, the model after no scattering processing (None) and first-order derivative pretreatment was chosen to be used in the rapid screening of high-quality PCC breeding materials. The highest 1-VR of delphinidin prediction models was 0.172, and the SECV was 12.030, obtained by SNV only (first-order derivative) in the 400–800 nm band. The values of correlation coefficients were small and could not accurately predict the content of delphinidin fraction in PCC. After no scattering processing and first-order derivative preprocessing in 400–800 nm visible light of pelargonidin, the 1-VR value was 0.467 at maximum, and SECV value was 3.887 at minimum, so its detection model was poorly predictive and could not accurately predict the content of pelargonidin. After the Detrend only and first-order derivative pretreatment under 400–800 nm visible light, the 1-VR value was 0.652 at maximum, and the SECV value was 3.557 for peonidin, so its detection model prediction was weakly correlated and could not accurately predict the content of peonidin fraction in PCC, which need further study. Considering that the contents of delphinidin, pelargonidin, and peonidin were relatively low, which accounted for less than 5% of the total anthocyanidins, it is negligible of their contribution to the quality of PCC.

Table 2. Calibration equations of cyanidin content in purple leaf Chinese cabbage using different pretreatment models.

Wave Band	Spectral Pre-Treatment Model	RSQ [1]	SEC [2]	1-VR [3]	SECV [4]
400~1100 nm 1100~2498 nm	None (no derivative)	0.922	808.339	0.908	887.788
	SNV+Detrend (no derivative)	0.928	772.348	0.894	948.462
	SNV only (no derivative)	0.913	852.994	0.866	1063.419
	Detrend only (no derivative)	0.942	685.539	0.923	801.505
	Standard MSC (no derivative)	0.923	784.604	0.896	924.328
	Weighted MSC (no derivative)	0.937	748.514	0.908	909.465
	None (first-order derivative)	0.965	531.591	0.942	693.004
	SNV+Detrend (first-order derivative)	0.959	576.934	0.931	754.230
	SNV only (first-order derivative)	0.956	602.184	0.924	799.911
	Detrend only (first-order derivative)	0.955	592.659	0.941	684.969
	Standard MSC (first-order derivative)	0.955	603.501	0.924	796.853
	Weighted MSC (first-order derivative)	0.952	622.972	0.917	825.123

[1] RSQ: external correlation coefficient; [2] SEC: standard error of calibration set; [3] 1-VR: internal cross-validation correlation coefficient; [4] SECV: standard error of cross-validation.

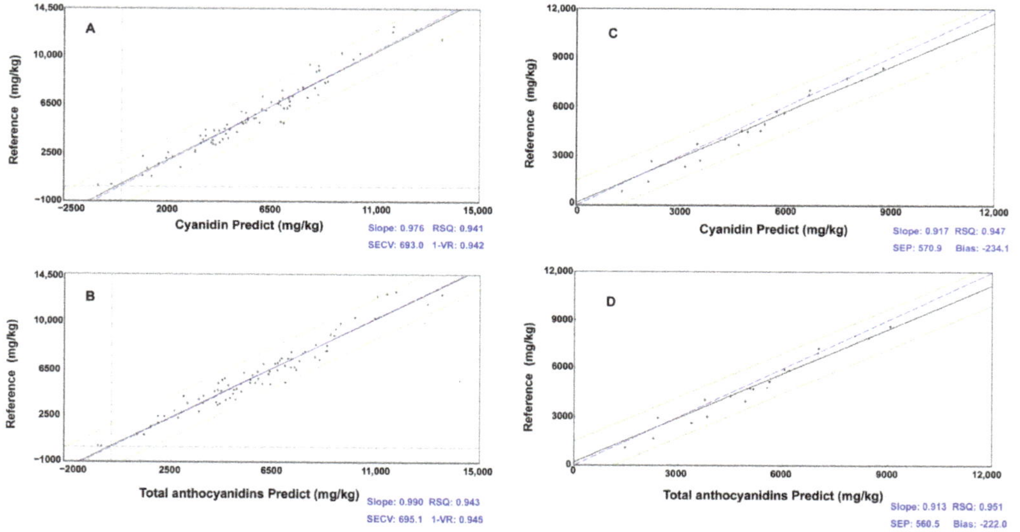

Figure 2. The cross-validation and external validation results of cyanidin and total anthocyanidin prediction models. (**A**): cross-validation of cyanidin prediction model; (**B**): cross-validation of total anthocyanidins prediction model; (**C**): external validation of cyanidin prediction model; (**D**): external validation of total anthocyanidins prediction model.

3.3.2. Model for Total Anthocyanidins Content Prediction

The performances of total anthocyanidin content prediction models were parallel with the models for cyanidin content prediction because cyaniding was the vast majority of anthocyanidin in PCC. As shown in Table 3, the 1-VR value of the total anthocyanidins in the visible/NIR spectral bands from 400 to 1100 nm and 1100 to 2498 nm after no scattering processing and first-order derivative preprocessing was 0.944, the minimum SECV value was 685.994, and the external correlation coefficient RSQ was 0.968, which meant that the cross-test effect was very satisfying. Combined with the cross-validation result shown in Figure 2B, the line slope was 0.990, and the samples scattered with no big deviation. The content of total anthocyanidins in PCC can be accurately predicted using the model after pretreatment of no scattering processing and first-order derivative.

Table 3. Calibration equations of total anthocyanidins content in purple leaf Chinese cabbage using different pretreatment models.

Wave Band	Spectral Pre-Treatment Model	RSQ [1]	SEC [2]	1-VR [3]	SECV [4]
400~1100 nm 1100~2498 nm	None (no derivative)	0.925	801.928	0.911	881.019
	SNV+Detrend (no derivative)	0.929	773.686	0.896	950.028
	SNV only (no derivative)	0.915	854.883	0.869	1062.558
	Detrend only (no derivative)	0.939	710.771	0.916	846.407
	Standard MSC (no derivative)	0.924	787.616	0.898	926.614
	Weighted MSC (no derivative)	0.938	749.215	0.910	910.829
	None (first-order derivative)	0.966	532.072	0.944	685.994
	SNV+Detrend (first-order derivative)	0.959	576.751	0.932	751.574
	SNV only (first-order derivative)	0.956	602.713	0.925	802.668
	Detrend only (first-order derivative)	0.956	592.196	0.941	691.249
	Standard MSC (first-order derivative)	0.956	595.245	0.930	761.400
	Weighted MSC (first-order derivative)	0.953	620.968	0.920	819.108

[1] RSQ: external correlation coefficient; [2] SEC: standard error of calibration set; [3] 1-VR: internal cross-validation correlation coefficient; [4] SECV: standard error of cross-validation.

3.3.3. External Validation of the Calibration Models

Using the mathematical model developed by WinISI III software, the samples not involved in the calibration were analyzed for external validation of the effectiveness of the developed model. The effectiveness of the validation was indicated by RSQ, SEP, and Bias. After validation, the RSQ were 0.947 and 0.951, respectively, for cyanindin and total anthocyanidins models at 400 to 1100 nm and 1100 to 2498 nm visible/NIR spectra, after no scattering processing and first-order derivative pretreatment (Figure 2C,D). The test deviation biases were small, which were −234.079 and −222.0, respectively. The slopes of the external validation prediction plots (the linear regression between the NIR predicted values and the chemically determined results) were 0.917 for cyanidin and 0.913 for total anthocyanidins. The validation samples were irregularly distributed on both sides of the line, and there was no big deviation, which meant that the models worked well; they could output accurate results for efficient and rapid screening of high anthocyanidins content materials. In addition, the validation results of delphinidin, pelargonidin, and peonidin prediction models showed very poor performance, as we could expect.

4. Discussion

As an osmoregulatory substance, anthocyanins are one of the most important pigments in plant leaves. It has an irreplaceable role in improving the cold, drought, and disease resistance of plants, and, therefore, monitoring the content of anthocyanidins in plants can help to understand the physiological state of plants [3]. Meanwhile, the benefits for human health of anthocyanidins have drawn a great deal of people's attention. At present, the detection of anthocyanidins contents in plants and plant products mainly uses HPLC or HPLC-MS method; the application of rapid and non-destructive detection using NIR is still in its initial stage, but high throughput, convenient operation, and no use of organic solvents will make NIR a powerful support tool in horticulture practice and agri-food industry. Huang et al. [25] proposed a NIR spectroscopic detection method based on an ant colony algorithm (ACO) combined with interval partial least squares (iPLS) in order to detect anthocyanidins content in flower tea quickly and accurately, indicating that NIR spectroscopy has promising applications in measuring total anthocyanidins in plants. NIR spectroscopy can be used to determine the anthocyanidin content of berries in completely satisfied results without breaking the composition of the berries [26]. In this study, we successfully developed suitable prediction models for cyanidin and total anthocyanin content in PCC, and they could be applied in the breeding practice of PCC to realize rapid and efficient screening of high-quality breeding materials.

NIR spectroscopy belongs to an indirect analysis technology; the accuracy of the prediction result relies on the quality of the calibration models. So, the establishment of a high-quality model, with accurate chemical analysis and spectrum scanning data, strong anti-interference capability, and broad enough representation, is vitally important. A large number and representative sample sets are essential factors for model building. Additionally, an appropriate algorithm to divide sample subsets is also critical [27]. In order to expand the application scope of our established models, further improvement using a larger number of PCC samples with different breeding backgrounds and distinct phenotypes is required. In terms of algorithms for NIR model establishment, there are several regression methods frequently used for the prediction/quantification of chemical content, including multiple linear regression (MLR), principal component analysis (PCA) for the exploration of the data, and partial least squares regression (PLSR) analysis to obtain a quantitative prediction of the parameters of interest [12,28]. Among them, PLSR is the most widely used multivariate statistical data analysis method for quantitative analysis of the NIR spectrum, with strong anti-interference ability. In this study, we used a PLSR method to process the spectral data, and the quantitative prediction results were satisfied.

Compared to chemical analysis methods, the sensitivity of NIR spectroscopy is relatively low and cannot be used for trace analysis, but its modeling is suitable for the detection of components with high content and a wide range of variation. In a previous report, a satisfied NIR prediction model was established to detect anthocyanidin content in flower teas with a content range of 0.17 to 1.60 mg/g [25]. In this study, the model prediction performance of cyanidin content and total anthocyanidin content with a wide range of variation was relatively good, which could be used for rapid screening of breeding materials and prediction of anthocyanidin content in PCC breeding practice. Meanwhile, the prediction model performance of delphinidin, pelargonidin, and petunidin with less abundant contents in PCC was very poor. Considering the contents of these three anthocyanidins were relatively low, their contribution to the phenotype and nutritional quality of PCC could be neglected. However, in other plant materials which contain a much higher proportion of these anthocyanidins, much more samples with a wide range of contents need to be included, and further optimization of their model-building methods is needed.

There are still some parts of NIR spectroscopy detection technology that need to be improved, but with the development of algorithm, spectroscopy, and artificial intelligence, the predictive ability, accuracy, and operability of this technology will continue to be improved on the original basis. With its obvious time-saving, high throughput, and non-destructive advantages, NIR spectroscopy will certainly have a broader development prospect in the agricultural, food industry, and market inspection.

5. Conclusions

Cyanidin was the most abundant anthocyanidin in purple leaf Chinese cabbage, with an average content of 5741.20 mg/kg, accounting for 95.7% of the total anthocyanidins. The prediction models established using visible/NIR spectroscopy on the basis of PLSR after no scattering processing and first-order derivative pretreatment method were suitable and effective for accurate and fast quantification of cyanidin and total anthocyanidin contents in PCC. The result laid a foundation for the application of NIR, with its obvious timesaving, convenience, and organic solvents free advantages, in the fast prediction of anthocyanidins in vegetables and rapid screening of purple vegetable breeding materials.

Supplementary Materials: The following supporting information can be downloaded at: https://www.mdpi.com/article/10.3390/foods12091922/s1, Table S1: Contents of anthocyanidins in purple Chinese cabbage analyzed by high-performance liquid chromatography (mg/kg).

Author Contributions: Conceptualization, D.-S.Z. and H.-J.H.; software, G.-M.L.; validation, Y.-Q.W. and L.-P.H.; formal analysis, G.-M.L. and X.-Z.Z.; investigation, Y.-Q.W. and X.-Z.Z.; data curation, G.-M.L. and L.-P.H.; writing—original draft preparation, Y.-Q.W.; writing—review and editing, Y.-Q.W., L.-P.H. and H.-J.H.; supervision, D.-S.Z. and H.-J.H.; project administration, H.-J.H.; funding acquisition, D.-S.Z. and H.-J.H. All authors have read and agreed to the published version of the manuscript.

Funding: This research was funded by the National Key Research and Development Program of China (No. 2022YFD1200805), Innovation and Capacity Building Project of Beijing Academy of Agriculture and Forestry Sciences (KJCX20200213, KJCX 20210437), National Natural Science Foundation of China (No. 31872094), and Beijing Innovation Consortium of Agriculture Research System (BAIC01-2022).

Data Availability Statement: Data is contained within the article or supplementary material.

Conflicts of Interest: The authors declare no conflict of interest.

References

1. Jiang, N.; Chung, S.-O.; Lee, J.; Ryu, D.; Lim, Y.P.; Park, S.; Lee, C.; Song, J.; Kim, K.; Park, J.-T.; et al. Increase of phenolic compounds in new Chinese cabbage cultivar with red phenotype. *Hortic. Environ. Biotechnol.* **2013**, *54*, 82–88. [CrossRef]
2. He, Q.; Zhang, Z.; Zhang, L. Anthocyanidins accumulation, antioxidant ability and stability, and a transcriptional analysis of anthocyanidins biosynthesis in purple heading Chinese cabbage (*Brassica rapa* L. ssp. *pekinensis*). *J. Agric. Food Chem.* **2006**, *64*, 132–145. [CrossRef]
3. Alvarez-Suarez, J.M.; Cuadrado, C.; Redondo, I.B.; Giampieri, F.; Gonzalez-Paramas, A.M.; Santos-Buelga, C. Novel approaches in anthocyanidins research-Plant fortification and bioavailability issues. *Trends Food Sci. Technol.* **2021**, *117*, 92–105. [CrossRef]
4. Zhang, N.; Jing, P. Anthocyanidinss in *Brassicaceae*: Composition, stability, bioavailability, and potential health benefits. *Crit. Rev. Food Sci. Nutr.* **2020**, *62*, 2205–2220. [CrossRef]
5. Harborne, J.B.; Williams, C.A. Advances in flavonoid research since 1992. *Phytochemistry* **2000**, *55*, 481–504. [CrossRef] [PubMed]
6. Putta, S.; Yarla, N.S.; Peluso, I.; Tiwari, D.K.; Reddy, G.V.; Giri, P.V.; Kumar, N.; Malla, R.; Rachel, V.; Bramhachari, P.V.; et al. Anthocyanidinss: Multi-target agents for prevention and therapy of chronic diseases. *Curr. Pharm. Des.* **2017**, *23*, 6321–6346. [CrossRef] [PubMed]
7. Wang, L.; Peng, C.; Li, Z.; Wei, W. Chemopreventive and therapeutic properties of anthocyanidinss in breast cancer: A comprehensive review. *Nutr. Res.* **2022**, *107*, 48–64.
8. Wang, P.; Liu, J.; Zhuang, Y.; Fei, P. Acylating blueberry anthocyanidinss with fatty acids: Improvement of their lipid solubility and antioxidant activities. *Food Chem. X* **2022**, *15*, 100420. [CrossRef] [PubMed]
9. Liu, Y.; Zhao, Y.; Zhuo, Y.; Li, Y.; Meng, J.; Wang, Y.; Li, H. Ultrasound-assisted extraction of anthocyanidinss from *Malus* 'Royalty' fruits: Optimization, separation, and antitumor activity. *Molecules* **2022**, *27*, 4299. [CrossRef] [PubMed]
10. Fakhri, S.; Khodamorady, M.; Naseri, M.; Farzaei, M.H.; Khan, H. The ameliorating effects of anthocyanidinss on the cross-linked signaling pathways of cancer dysregulated metabolism. *Pharmacol. Res.* **2020**, *159*, 104895. [CrossRef]
11. Kutsanedzie, F.Y.H.; Guo, Z.; Chen, Q. Advances in nondestructive methods for meat quality and safety monitoring. *Food Rev. Int.* **2019**, *35*, 536–562. [CrossRef]
12. Giovenzana, V.; Beghi, R.; Civelli, R.; Guidetti, R. Optical techniques for rapid quality monitoring along minimally processed fruit and vegetable chain. *Trends Food Sci. Technol.* **2015**, *46*, 331–338. [CrossRef]
13. Prodromidis, P.; Mourtzinos, I.; Biliaderis, C.G.; Moschakis, T. Stability of natural food colorants derived from onion leaf wastes. *Food Chem.* **2022**, *386*, 132750. [CrossRef] [PubMed]
14. Johnson, J.B.; El Orche, A.; Naiker, M. Prediction of anthocyanidins content and variety in plum extracts using ATR-FTIR spectroscopy and chemometrics. *Vib. Spectrosc.* **2022**, *121*, 103406. [CrossRef]

15. Beltrame, K.K.; Gonçalves, T.R.; Março, P.H.; Gomes, S.T.M.; Matsushita, M.; Valderrama, P. Pseudo-Univariate calibration based on NIR spectroscopy in the determination of anthocyanidinss and antioxidant activity in grape juices. *J. Braz. Chem. Soc.* **2021**, *32*, 1131–1136.
16. Tian, X.Y.; Aheto, J.H.; Bai, J.-W.; Dai, C.; Ren, Y.; Chang, X. Quantitative analysis and visualization of moisture and anthocyanidinss content in purple sweet potato by Vis-NIR hyperspectral imaging. *J. Food Process. Preserv.* **2021**, *45*, e15128. [CrossRef]
17. Krysa, M.; Szymańska-Chargot, M.; Zdunek, A. FT-IR and FT-Raman fingerprints of flavonoids—A review. *Food Chem.* **2022**, *393*, 133430. [CrossRef]
18. Watson, R.R. *Polyphenols in Plants: Isolation, Purification and Extract Preparation*; Academic Press: New York, NY, USA, 2014; pp. 187–199.
19. Manzoor, M.F.; Hussain, A.; Naumovski, N.; Ranjha, M.M.A.N.; Ahmad, N.; Karrar, E.; Xu, B.; Ibrahim, S.A. A narrative review of recent advances in rapid assessment of anthocyanidinss in agricultural and food products. *Front. Nutr.* **2022**, *9*, 901342. [CrossRef]
20. Geraldes, C.F.G.C. Introduction to infrared and Raman-based biomedical molecular imaging and comparison with other modalities. *Molecules* **2020**, *25*, 5547. [CrossRef]
21. Xin, X.; Zhang, D.; Zhao, H.; Su, T.; Zhao, X.; Wang, W.; Li, P.; Yu, Y.; Wang, J.; Yu, S.; et al. Identification of a monosomic alien chromosome addition line responsible for the purple color trait in heading Chinese cabbage. *Horticulturae* **2023**, *9*, 146. [CrossRef]
22. Ye, W. Application of Near-Infrared Spectroscopy for Determination of Nutrient Contents in Manure. Ph.D. Thesis, Iowa State University, Ames, IA, USA, 2003.
23. Gómez, A.H.; He, Y.; Pereira, A.G. Non-destructive measurement of acidity, soluble solids and firmness of Satsuma mandarin using Vis/NIR spectroscopy techniques. *J. Food Eng.* **2006**, *77*, 313–319. [CrossRef]
24. Golic, M.; Walsh, K.; Lawson, P. Short-wavelength near-infrared spectra of sucrose, glucose, and fructose with respect to sugar concentration and temperature. *Appl. Spectrosc.* **2003**, *57*, 139–145. [CrossRef]
25. Huang, X.; Zou, X.; Zhao, J.; Shi, J.; Zhang, X.; Holmes, M. Measurement of total anthocyanidinss content in flowering tea using near infrared spectroscopy combined with ant colony optimization models. *Food Chem.* **2014**, *164*, 536–543.
26. Ribera-fonseca, A.; Noferini, M.; Rombolá, A.D. Non-destructive assessment of highbush blueberry fruit maturity parameters and anthocyanidinss by using a visible/near Infrared (vis/NIR) spectroscopy device: A preliminary approach. *J. Soil Sci. Plant Nutr.* **2016**, *16*, 174–186.
27. Galvão, R.K.H.; Araujo, M.C.U.; José, G.E.; Pontes, M.J.C.; Silva, E.C.; Saldanha, T.C.B. A method for calibration and validation subset partitioning. *Talanta* **2005**, *67*, 736–740. [CrossRef] [PubMed]
28. Beć, K.B.; Huck, C.W. Breakthrough potential in near-infrared spectroscopy: Spectra simulation. A review of recent developments. *Front. Chem.* **2019**, *7*, 48. [CrossRef]

foods

MDPI

Article

Analysis of Light Penetration Depth in Apple Tissues by Depth-Resolved Spatial-Frequency Domain Imaging

Tongtong Zhou [1,†], Dong Hu [1,†], Dekai Qiu [1], Shengqi Yu [1], Yuping Huang [2], Zhizhong Sun [3], Xiaolin Sun [1], Guoquan Zhou [1], Tong Sun [1] and Hehuan Peng [1,*]

1 College of Optical Mechanical and Electrical Engineering, Zhejiang A&F University, Hangzhou 311300, China; zhoutt@stu.zafu.edu.cn (T.Z.); 20180047@zafu.edu.cn (D.H.)
2 College of Mechanical and Electronic Engineering, Nanjing Forestry University, Nanjing 210037, China
3 College of Chemistry and Materials Engineering, Zhejiang A&F University, Hangzhou 311300, China
* Correspondence: penghh@zafu.edu.cn
† These authors contributed equally to this work.

Abstract: Spatial-frequency domain imaging (SFDI) has been developed as an emerging modality for detecting early-stage bruises of fruits, such as apples, due to its unique advantage of a depth-resolved imaging feature. This paper presents theoretical and experimental analyses to determine the light penetration depth in apple tissues under spatially modulated illumination. Simulation and practical experiments were then carried out to explore the maximum light penetration depths in 'Golden Delicious' apples. Then, apple experiments for early-stage bruise detection using the estimated reduced scattering coefficient mapping were conducted to validate the results of light penetration depths. The results showed that the simulations produced comparable or a little larger light penetration depth in apple tissues (~2.2 mm) than the practical experiment (~1.8 mm or ~2.3 mm). Apple peel further decreased the light penetration depth due to the high absorption properties of pigment contents. Apple bruises located beneath the surface peel with the depth of about 0–1.2 mm could be effectively detected by the SFDI technique. This study, to our knowledge, made the first effort to investigate the light penetration depth in apple tissues by SFDI, which would provide useful information for enhanced detection of early-stage apple bruising by selecting the appropriate spatial frequency.

Keywords: light penetration depth; apple; spatial-frequency domain imaging; depth-resolved; bruise; scattering

Citation: Zhou, T.; Hu, D.; Qiu, D.; Yu, S.; Huang, Y.; Sun, Z.; Sun, X.; Zhou, G.; Sun, T.; Peng, H. Analysis of Light Penetration Depth in Apple Tissues by Depth-Resolved Spatial-Frequency Domain Imaging. *Foods* **2023**, *12*, 1783. https://doi.org/10.3390/foods12091783

Academic Editor: Corrado Costa

Received: 19 March 2023
Revised: 13 April 2023
Accepted: 23 April 2023
Published: 25 April 2023

1. Introduction

Optical sensing techniques, such as near-infrared spectroscopy and hyperspectral imaging, have been extensively researched and increasingly utilized for detecting multiple defects of agro-food products [1–3]. The past decade has witnessed the development of spatial-frequency domain imaging (SFDI) for detecting various surface and subsurface defects of fruits [4–7]. As one of the typical defect types, surface bruises in apples often occur during harvest, transportation, storage, and sorting processes. Slight early-stage bruises are invisible to our naked eyes and are challenging to be recognized by traditional imaging techniques under uniform or diffuse illumination, which are more sensitive to the obvious surface properties. Thanks to frequency-dependent light attenuation within tissues, as depicted in Figure 1, SFDI enables acquiring depth-resolved information regarding tissue constituents and structure. Based on this remarkable feature, SFDI is proven to be capable of detecting the early-stage bruises of apples beneath the peels [6]. As opposed to the traditional uniform light imaging techniques, such as machine vision and hyperspectral imaging, spatially modulated light in a sinusoidal waveform is used in SFDI to acquire pattern images from samples. Through image demodulation and inverse estimation processing, SFDI produces 2-D optical property mappings in a pixel-by-pixel fashion, i.e., absorption coefficient (μ_a) mapping and reduced scattering coefficient (μ'_s) mapping. The differences of optical properties between non-bruised apple tissues and bruised ones can be directly used for early-stage bruise detection. It is well known that there are quite a lot of optical property measuring methods, e.g., time-resolved, spatially resolved, and integrating sphere, which have been employed for measuring optical properties of diverse agro-food products [8–11]. However, they are generally limited to point measurement and cannot attain depth-resolved information, resulting in great challenges in the non-destructive detection of early-stage bruises of apples [12,13]. In the technique of SFDI under the spatially modulated illumination, high-frequency light is more sensitive to the shallower tissue, while the low-frequency component has a much larger light penetration depth (~mm) [14,15], which provides a theoretical basis for detecting early-stage bruises of apples.

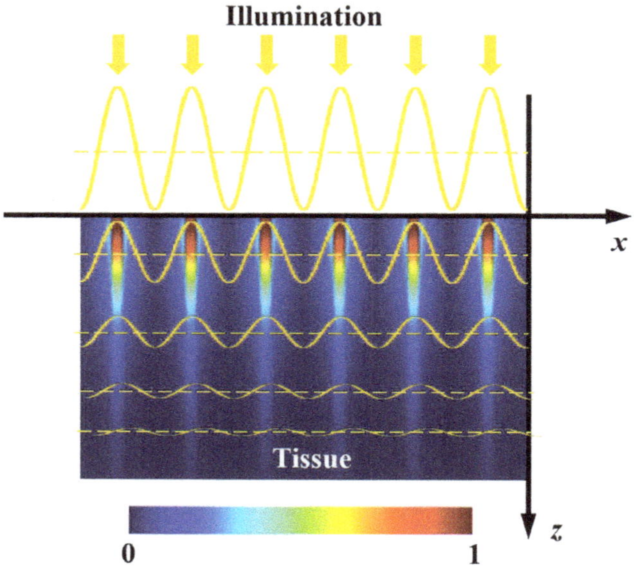

Figure 1. Schematic of light attenuation within a semi-infinite turbid medium under spatially modulated illumination.

Knowledge of light penetration depth sampled by SFDI is of high significance for clinical and preclinical applications in the field of biomedicine [16–18]. For instance, the thickness of burned skin dictates the treatment protocol, highlighting the importance of understanding the detection depth in skin tissue [19]. SFDI has also been explored in deep-tissue applications, where it is essential to understand the penetration of collected photons in order to evaluate the maximum depth of measurable tumor contrast [20]. However, measuring light penetration depths in the field of biomedicine is not without limitations. The experiments generally require patients to remain completely still during a potentially long acquisition time in order to acquire full area scans. The exploration of the light penetration depth often requires prior knowledge about tissue optical properties that may not be valid in damaged tissue. Similar to that in the field of agricultural and food engineering, knowledge of light penetration depth is critical for enhancing the detection performance of early-stage bruises in apples by SFDI. Despite great progress being made for the bruise detection of apples, there still is a lack of exploration of quantifying light penetration depth in apple tissue. As mentioned above, SFDI has a remarkable advantage in subsurface (early-stage) bruise detection of apples, but the light penetration depth is reported to be limited in mm, implying that some internal bruises located in the region of deep tissue cannot be detected. It is thus desirable and also necessary to quantify light penetration depths under varying-frequency spatially modulated illumination, so as to better explore the potential of SFDI for early-stage bruise detection of apples in different depths, as well as to assess the severity of apple bruising. Up to now, the maximum detection depth in apple tissue using SFDI has been uncertain and there are few research studies focusing on studying light penetration depth. Lu and Lu [21] reported that the maximum light penetration depth was confirmed to be no more than three sheets of blank printing paper (or less than 400 μm). Their study investigated the light penetration depth from the aspect of demodulated images, which is different from our research in optical property estimation through inverse computation. Apple is taken as the experimental material in this study, which has different properties from the blank printing paper. The estimated optical property mappings could provide quantitative information in bruise detection (specific values of optical properties for non-bruised and bruised tissues), and thus further exploration of the light penetration depth could be implemented through these quantitative information.

Therefore, in this study, a set of well-designed experiments from theoretical simulation to practical implementation was performed to quantify the light penetration depth (especially for the maximum value) in apple tissue under spatially modulated illumination. The objectives were to (1) explore the light penetrating capacity of demodulated direct component (DC) and amplitude component (AC) images to prove our SFDI system performance, (2) conduct the simulation and practical experiments to investigate the light penetration depths in 'Golden Cream Delicious' apples with and without peels; and (3) validate the conclusion of the maximum light penetration depth in apple tissues by evaluating the performance of bruise detection.

2. Theoretical Formulation for SFDI

2.1. Image Formation and Image Processing

In SFDI, image formation involves two steps [22]: (1) the incident light interacts with the sample through absorption and multiple scattering, and (2) the light reemitted from the sample travels through a series of optical devices (e.g., lens, camera) of the imaging system, and eventually forms a digital image. As a general rule, SFDI is regarded as a linear, space-invariant technique which applies transfer function theory to an optical imaging system [23]. There are several factors negatively affecting the resolution and contrast of resulting images during image formation and processing, such as convolution operation and environmental noise. A mathematical method is used to analyze the image acquired by SFDI, which is generally composed of two parts. The first part is DC, i.e., I_{DC} with the Fourier spectra centered at the origin; the other part is AC termed as I_{AC}, which is composed of an oscillatory or harmonic component, with the Fourier spectra shifted by positive or negative frequency (f_x or $-f_x$) [24].

The process of optical property estimation from the remitted image in the SFDI can be roughly divided into two steps: acquisition of diffuse reflectance image through demodulation and estimation of optical property mapping through inverse computation. Due to the characteristics of high accuracy and easy implementation, phase shifting techniques are widely used for demodulation from sinusoidal fringe patterns. Three-phase demodulation (TPD) is a commonly used and effective method that uses three images with the phase offsets of $-2\pi/3$, 0, and $2\pi/3$. Under the illumination of three phase-shifted sinusoidal patterns, the corresponding intensity images, i.e., $I_1(x,y)$, $I_2(x,y)$, and $I_3(x,y)$, can be expressed as follows [25]:

$$I_1(x,y) = I_{DC} + I_{AC}\cos(2\pi f_x x - 2\pi/3) \tag{1}$$

$$I_2(x,y) = I_{DC} + I_{AC}\cos(2\pi f_x x) \tag{2}$$

$$I_3(x,y) = I_{DC} + I_{AC}\cos(2\pi f_x x + 2\pi/3) \quad (3)$$

where (x,y) represents the spatial coordinates, f_x is the spatial frequency along the x-axis direction, and I_{DC} and I_{AC} are the direct and amplitude components, respectively. For the purpose of simplicity, we will drop off the coordinate notation. From Equations (1)–(3), the DC and AC images can be obtained by the following equations [15]:

$$I_{DC} = \frac{1}{3}(I_1 + I_2 + I_3) \quad (4)$$

$$I_{AC} = \frac{\sqrt{2}}{3}\sqrt{(I_1 - I_2)^2 + (I_1 - I_3)^2 + (I_2 - I_3)^2} \quad (5)$$

2.2. Image Contrast

Light penetration features are of primary concern for the demodulated images, which are also critical for fruit bruise detection. To our knowledge, it is challenging to assess the light penetration capability, because it is largely dependent on tissue physicochemical properties and illumination conditions. For the bruised apples illuminated under spatially modulated illumination with varying frequencies, light penetration capability could essentially determine the thickness of the tissue that light passes through. In this study, we introduce image contrast and the ratio of peak to valley's intensity (PVR) to evaluate the light penetration capability, which will be introduced in Section 3.2. Examination of the composition of photons backscattered from a turbid medium will provide qualitative insights into the relationship between light penetration depth and image contrast. The ballistic photons experience one or more backward and forward scattering events before exiting from the tissue. Due to the shortest traveling path, they suffer from minimal scattering and thus can deliver image information with superior resolution and contrast. However, the information generated by ballistic photons is more about the superficial layer of the medium in one mean free path (MFP) [26], around 100 μm for fruit tissue such as apple (assuming the value of μ'_s is equal to or larger than 1.00 mm^{-1}). The weakly scattered photons provide information on deeper, subsurface tissues, and they are still capable of forming well-resolved images due to limited scattering events. In summary, the tradeoff should be carefully considered between the light penetration depth and image contrast, while selecting spatial frequency in SFDI [26,27].

The I_{DC}, which contains a larger contribution of diffusive photons, probes a deeper region of sample tissues than I_{AC}, while I_{AC} contains more ballistic and weakly scattered photons, resulting in better image contrast [15]. High-frequency illumination is more likely to enhance image contrast. Presented in the following sections are well-designed experiments to quantitatively determine the relationship between image contrast and the depth-resolved imaging feature of SFDI.

2.3. Light Penetration

In diffuse optics, the light penetration depth δ in biological tissues can be attained from the response to an infinitely narrow photon beam normally incident on a semi-infinite medium. For the case where that photon's propagation depth z is larger than the light penetration depth, internal fluence distribution predicted from diffusion theory should be [28]:

$$\varnothing(z) = \varnothing_0\, k\, \exp(-z/\delta) \quad (6)$$

where k is a scalar that depends on the amount of backscattered reflectance, $\varnothing_0$ is the incident irradiance, and $\varnothing(z)$ represents a function of photon fluence. The light penetration depth is defined as [29]:

$$\delta = \frac{1}{\sqrt{3\mu_a(\mu_a + \mu_s(1-g))}} = \frac{1}{\mu_{eff}} \quad (7)$$

where μ_a is the absorption coefficient, μ_s is the scattering coefficient, g is anisotropy factor, and μ_{eff} is the effective attenuation coefficient.

According to this, the light penetration depth is estimated to be 1.50–6.00 mm for apple tissues with typical μ_a and μ_s coefficients of 0.01–0.05 mm^{-1} and 9.00–28.00 mm^{-1}, respectively. However, the estimated depth does not always stand for the actually detectable depth for a general imaging system under spatially extended wide-field or broad-beam illumination. The fluence rate and reflectance properties of spatially modulated photon density plane waves in the SFDI are described in the study of Cuccia, Bevilacqua, Durkin, Ayers, and Tromberg [22], in which the effective penetration depth δ'_{eff} is concisely defined as:

$$\mu'_{eff} = \left(\mu_{eff}^2 + k_x^2 + k_y^2\right)^{1/2} = \frac{1}{\delta'_{eff}} \tag{8}$$

where μ'_{eff} is a scalar attenuation coefficient, k_x and k_y are variable coefficients related to spatial frequencies f_x and f_y ($k_x = 2\pi f_x$, $k_y = 2\pi f_y$), and δ'_{eff} is the effective penetration depth, which is inversely proportional to spatial frequency. The above mathematical formula just provides a simple conceptual framework to understand the transmission of modulated scalar photons in a turbid medium. In practice, the detected signal is mostly due to the photons backscattered close to the illumination source, which corresponds to a far more superficial depth of tissue interrogation than that derived from diffuse light attenuation [30]. This is equivalent to calculating the relative probability that a photon will visit a certain location in tissue before its detection. In the reflectance measurement geometry with spatially modulated illumination, the reemitted light intensity decays by many orders of magnitude within millimeters. Therefore, in using the formula to calculate the light penetration depth, there directly exists some unreasonable aspects. In this study, well-designed experiments were conducted, coupled with two evaluation parameters (image contrast and PVR), to explore the light penetration depth in apple tissues using SFDI.

3. Materials and Methods

3.1. SFDI System

An in-house assembled SFDI system, as illustrated in Figure 2, mainly consisted of a 150 W DC-regulated halogen fiber optic light source (Fiber-Lite DC950, Dolan-Jenner, Boxborough, MA, USA), a light-guide fiber (MSG4-2200S, MORITEX Corporation, Saitama, Japan), an 8-bit camera (MER-131-210U3M NIR, Daheng imaging vision Corporation, Shanghai, China) coupled with a C-mount zoom lens (HN-0816-5M-C2/3X, Daheng imaging vision Corporation, Shanghai, China) for vertically shining over a field of view (FOV, 11.5 × 11.5 cm^2), a filter wheel (BOCIC Co., Ltd., Beijing, China) comprising six bandpass filters (550, 600, 630, 675, 710 and 730 nm), a microcontroller unit (MCU) (STM32F103ZET6, Opendv, Guangzhou, China), a three-axis manual displacement platform (THZ210, Runjia Pneumatic, Shenzhen, China) for holding samples, and an optical projector (DLi6500 1080p Optics Bundles, TI, Austin, TX, USA) for generating sinusoidal patterns. The projector was slightly angled at 12 degrees relative to the vertical axis to mitigate the image distortion, based on our tests and preliminary experiments, which was also confirmed by Lu et al. (2017) for constructing a multiple structured-illumination reflectance imaging system [31]. The MCU could take control of the projector for synchronous pattern projection and image acquisition with the camera. A pair of cross linear polarizers was mounted in front of the lens of the projector and camera to suppress specular reflectance from samples. The aforementioned components were mounted on an optical platform (SPL-R-0910, SPL-Tech, Hangzhou, China) and enclosed in a dark chamber for reducing the influence of external stray light.

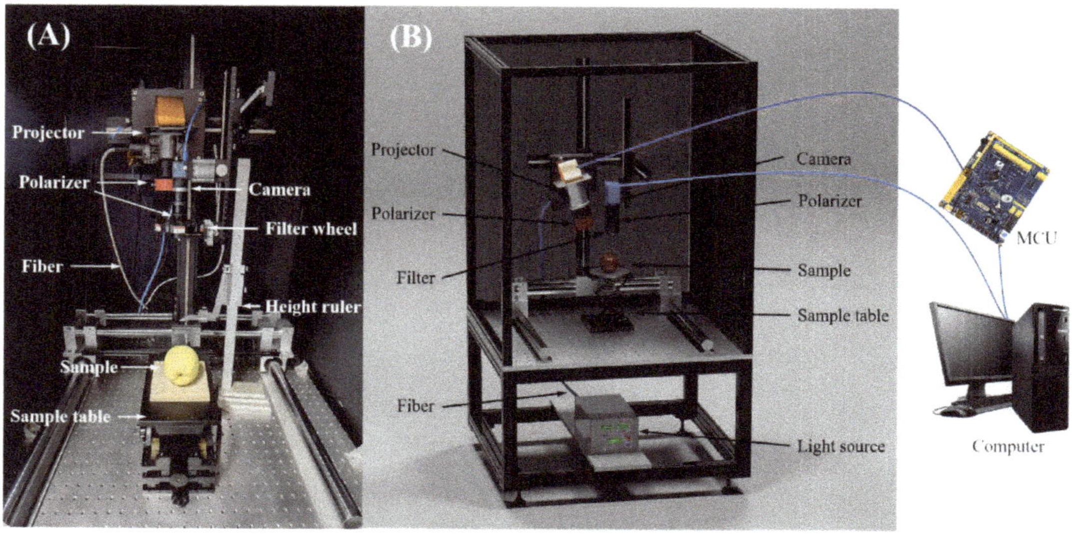

Figure 2. (**A**) Physical and (**B**) schematic maps of spatial-frequency domain imaging system.

3.2. Experiments

3.2.1. Experiment 1: Penetrating Capacity of DC and AC Images

The first experiment was to verify the light-penetrating capacity of DC and AC images and thus prove our SFDI system performance. The experiment was conducted using a nylon slab with high scattering characteristics, which was drilled with five cylindrical holes running parallel with the surface. As shown in Figure 3, the five holes with the diameter of 4 mm or 8 mm were set up at the depths of 1 mm, 11 mm, 6 mm, 4 mm, and 2 mm, from left to right. Number the five holes in sequence as 1, 2, 3, 4 and 5. All the holes were filled with 100-times diluted India ink as absorbers for absorption property comparison with the bull material and then sealed with adhesive black tape. A sequence of sinusoidal patterns, covering 18 frequencies [0.01:0.01:0.15, 0.20, 0.25, 0.30] mm^{-1}, was generated in Matlab R2020a (The Mathworks, Inc., Natick, MA, USA) for sample illumination. Three phase-shifted pattern images were acquired at each spatial frequency with the phase offsets of $-2\pi/3$, 0, and $2\pi/3$, respectively. A standard whiteboard with the calibrated reflectance rate of 99% was imaged first under planar illumination to correct the non-uniformity of the source illumination [32]. Image demodulation was then used to generate the DC and AC images, according to Equations (4) and (5).

3.2.2. Experiment 2: Investigating Light Penetration Depth

The second experiment was to quantify the light penetration depth in apple tissue under sinusoidal illumination. 'Golden Cream Delicious' apples, which were free of visual blemishes or defects and grown in Shandong, China, were purchased from a fruit market. Both simulation and practical experiments were carried out. In the first, the simulation methodology proposed by Hayakawa et al. [33] was adopted to roughly determine the optical sampling depth in apple tissues. In this method, Monte Carlo (MC), which has been widely applied to simulate light propagation in single- and multiple-layered biological tissues, was employed for the simulation experiment [8,29,34]. Optical property parameters (μ_a, μ'_s) of apple tissues at the six wavelengths were measured by our integrating sphere system [35], which were taken as the inputs in the simulation.

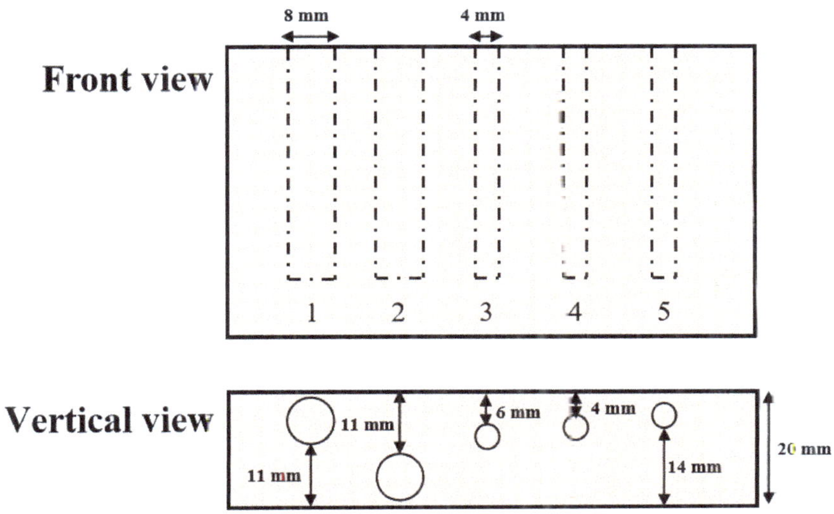

Figure 3. Schematic of relative positions and sizes of holes in the nylon sample filled with absorbing ink solution.

For the practical experiment for validating the results of the simulations, there were two types of samples, apple slices and a USAF-1951 target. A slicer was used to produce apple slices with different thicknesses with peel [0.9, 1.0–1.1, 1.3, 1.5–1.6, 1.6–1.7, 1.8, 2.1–2.2, 2.4–2.5, 2.6, 3.0, 3.4, 3.8, 4.0] mm and without peel [0.8, 1.0, 1.2–1.3, 1.5, 1.6, 1.8, 2.0, 2.3, 2.5–2.6, 2.7–2.8, 3.0, 3.2, 3.6, 4.0] mm. These varying-thickness apple slices were used to cover the USAF-1951 target, which was made of fiber material with high scattering properties, and their combination was imaged by the SFDI system, illuminated with a sequence of frequencies of 0.05, 0.10, 0.15, 0.20, 0.25, and 0.30 mm^{-1}. It is apparent that the light has completely penetrated though the apple slice when the black horizontal bars of the USAF-1951 are recognized, as shown in the region of interest (ROI) in Figure 4. Two parameters in the ROI, i.e., image contrast and PVR, as mentioned above, were calculated to resolve the image details. Image contrast was evaluated based on the Michelson contrast metric (C_M) [36]:

$$C_M = (I_{max} - I_{min})/(I_{max} + I_{min}) \tag{9}$$

where I_{max} and I_{min} denote the maximum and minimum intensities of the image in ROI, respectively. Another evaluating parameter, PVR, defined as the ratio of peak and valley's intensity [37], was also determined from the captured images:

$$PVR = I_{peak}/I_{valley} \tag{10}$$

where I_{peak} and I_{valley} denote the intensities of the peak and valley in the ROI, respectively.

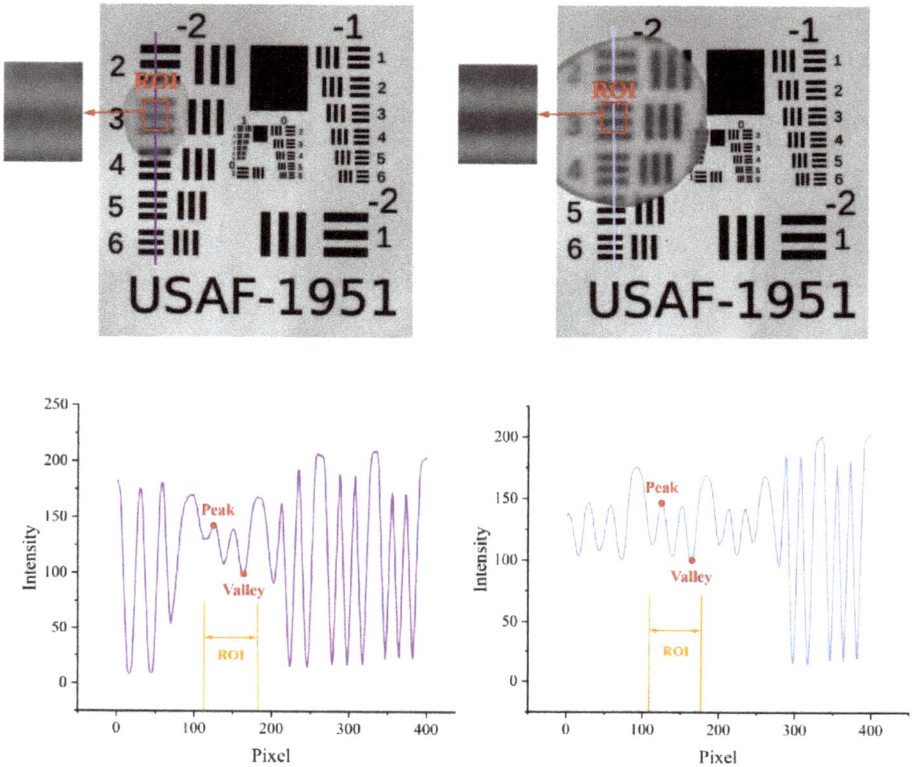

Figure 4. Example images of the USAF-1951 target covered with apple slices with peel (**left**) and without peel (**right**) for calculating image contrast (**up**) and PVR (**down**) in selected region of interest (ROI).

3.2.3. Experiment 3: Detecting Early Bruises in Apples

The third experiment was to test and verify the results of light penetration depths in apple tissues achieved from experiment 2. Impact tests were conducted to induce bruises in the apples with a wooden ball attached to one end. The wooden ball (6 cm in diameter and 105 g in weight) fell freely from the rest position at a certain height to impact the apple at its equatorial area. The peel on the surface of bruised tissue was cut off to eliminate its effect on detection performance. The used frequencies were the same as experiment 2. As shown in Figure 5, apple slices with and without peels in different thicknesses were used to cover the bruised tissues. Image demodulation and inverse parameter estimation were carried out to obtain the optical property (μ_a and μ'_s) mappings of the apples covered with slices. It is supposed that if the bruised tissue could be recognized in mappings, the light can penetrate through the apple slice, and the light penetration depth under this illumination should be equal to or larger than the thickness of the apple slice.

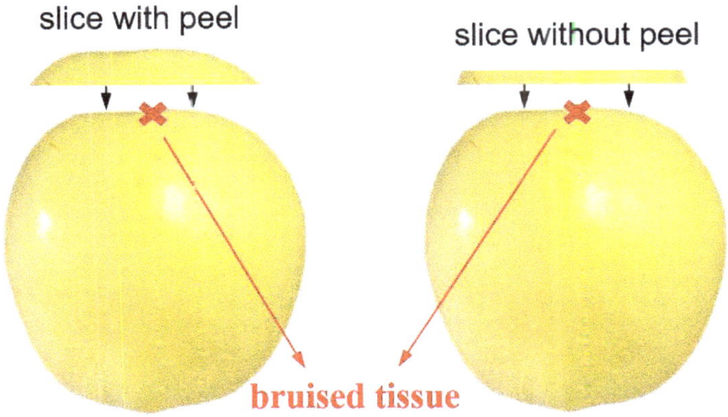

Figure 5. The diagram of peeled bruised apples covering with varying-thickness slices with (**left**) and without peels (**right**).

4. Results and Discussion

4.1. Penetrating Capacity of DC and AC Images

Figure 6 shows the demodulated DC and AC images of the nylon sample. The filled India ink made the hidden holes relatively dark because of its strong light-absorbing capacity. In the DC image (equivalent to uniform illumination), the first hole was easily recognized but with invisibility on the top of the liquid column caused by an unwanted air bubble. The second hole was almost invisible because it was too deep (11 mm) from the tissue surface. For the three hidden holes on the right, the black liquid column became visually fuzzier along with larger distance from the surface. The AC image at the frequency of 0.01 mm^{-1} showed similar details to the DC image. As the frequency increased from 0.02 to 0.12 mm^{-1}, the color of the leftmost column became more and more light, and the other columns revealed a reduction in the image invisibility. It was noticed that as the frequency increased, the surface texture of the nylon sample was revealed to a certain degree, with the images becoming less smooth. These observations implied that higher-frequency illumination brought about less light interrogation with deep tissues, resulting in a shallower light penetration depth. A similar finding was also reported in the previous study [15]. When the frequency reached 0.12 mm^{-1}, all the holes became blurry due to insufficient light interrogation. Given all of that, the DC (0 mm^{-1}) image had more light interrogation with deep tissues, while the AC images showed varying light penetration depths and image resolutions with different frequencies. It is suggested that the low-frequency component could penetrate deeper into the tissues than the high-frequency part, which is called the depth-resolved characteristic of SFDI in this study.

4.2. Investigating Light Penetration Depth

Table 1 showed the measured values of optical property parameters (μ_a, μ'_s) of apple slices without peel using the integrating sphere technique. By inputting the optical property values manually into the developed program, the light penetration depth for the apple tissue could be simulated by consulting a scaled lookup table derived from MC simulations to the radiative transport equation in the spatial-frequency domain. Figure 7 displays the simulated results for the apple tissue at six wavelengths (550, 600, 630, 675, 710, and 730 nm), in which the median sampling depth with a [25–75]% fraction of the total measured diffuse reflectance was recognized as the critical metric for light penetration depth [33]. It was observed in the simulation results that light penetration depths increased slowly with the wavelengths, which is similar to the findings reported by Zhao et al. [17]. At 550 nm, the median sampling depth with [25–75]% was slightly smaller than that of the other five wavelengths, which was approximately in the range of 0.6–2.2 mm. It demonstrated that the 25–75% measured reflectance had the opportunity to interact with the tissue in the depth of 0.6–2.2 mm. Similarly, Lammertyn et al. [38] reported that the maximum light penetration depth in Jonagold apples at 692 nm was about 2 mm, which agrees well with the finding in this study. In the report of Binzoni et al. [39], the light propagation behavior is that the photons reaching the detectors do not go very deep and thus the information contained in the spectral images comes from a depth that does not exceed 2–3 mm. The above experimental results were all consistent with our simulation results. However, the light penetration depth (less than 400 μm) reported by Lu and Lu [21] was much smaller than the 2.2 mm. One potential reason is the evaluation level. Lu and Lu investigated the detection depth through demodulated images, while we studied the penetration depth from the aspect of optical property estimation. Furthermore, the custom-defined acceptable resolution and contrast would also affect the detection depth. Hence, it was concluded that the light penetration depths in apple tissues were close to each other at the six wavelengths, with values of no more than 2.2 mm.

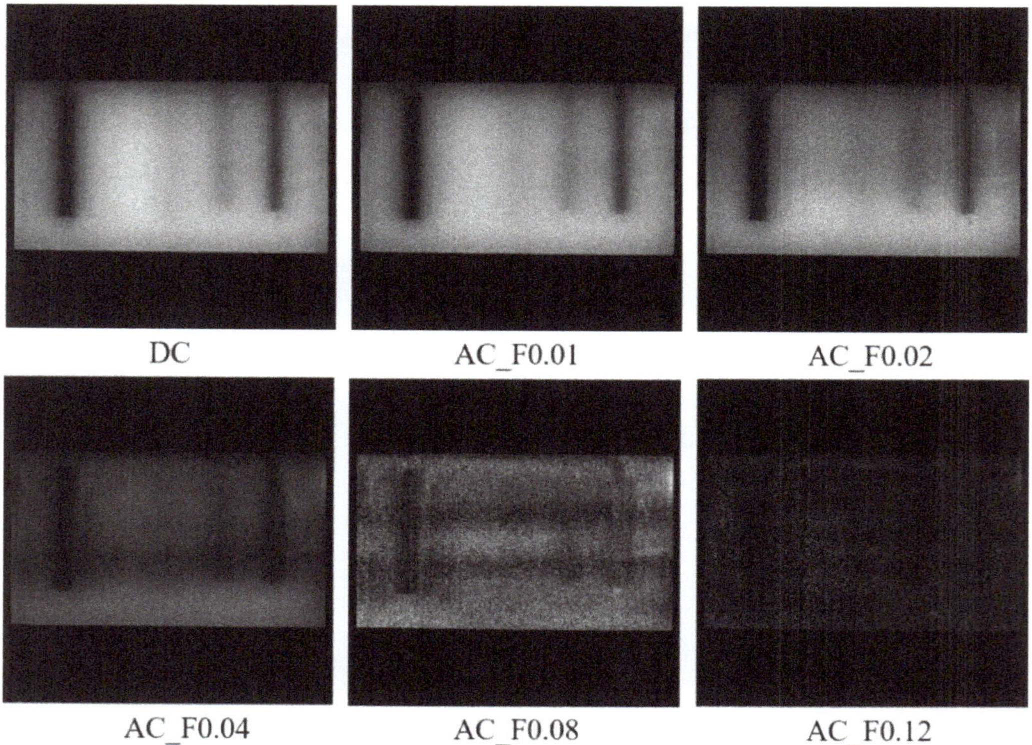

Figure 6. Demodulated DC (direct component) and AC (amplitude component) images of the nylon sample at a sequence of spatial frequencies of 0, 0.01, 0.02, 0.04, 0.08, and 0.12 mm^{-1}, from (**top left**) to (**bottom right**).

Table 1. Optical property parameters (μ_a, μ'_s) of 'Golden Cream Delicious' apple tissues measured by integrating sphere system at six different wavelengths.

Wavelength (nm)	μ_a (mm^{-1})	μ'_s (mm^{-1})
550	0.0302	1.286
600	0.0232	1.273
630	0.0218	1.259
675	0.0224	1.251
710	0.0222	1.256
730	0.0223	1.255

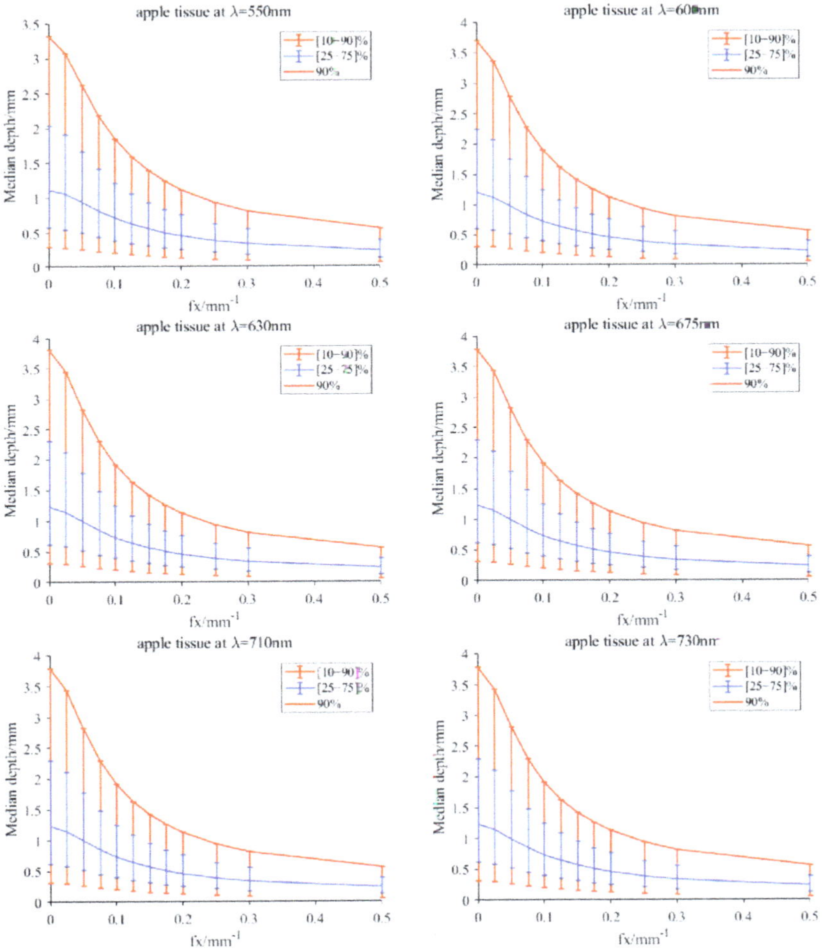

Figure 7. Simulated median optical detection depth for the apple tissue at six wavelengths, which is estimated using a scaled lookup table derived from MC simulations to the radiative transport equation in the spatial-frequency domain [33]. The median detection depth is the depth that encloses the photon trajectories responsible for 50% of the detected light, and accordingly, the vertical-capped lines in the figure correspond to detection depths responsible for 25% (**lower**) and 75% (**upper**) of the detected light.

Figure 8A shows the contrast variation with frequencies for the DC and AC images of the USAF-1951 target covered with apple slices at 630 nm. For both the slices with and without peel, DC images gave almost a constant value of image contrast since they were independent of spatial frequency, while AC images showed much higher contrast values, which rose steadily with the spatial frequency. These findings indicated that AC images, which are unique to SFDI, enhanced image contrast compared to DC images, demonstrating that SFDI is superior to conventional uniform light imaging techniques in image contrast. Figure 8B shows the histogram results of image contrast for the DC images of the USAF-1951 target covered with apple slices at the wavelengths of 600, 630, 675, and 710 nm. It was noticed that the contrast values decreased with the thickness of apple tissue, as well as the wavelength. A special case occurs from 675 nm to 710 nm, in which there is a small rise of the contrast for some thicknesses. This is because the reflected signal intensity was generally poor in 630–690 nm due to strong absorption of chlorophyll. When removing the influence of the peel (right panel in Figure 8B), there was a steady decreasing trend for the image contrast with the wavelength, because the pigment had little effect on apple flesh tissue.

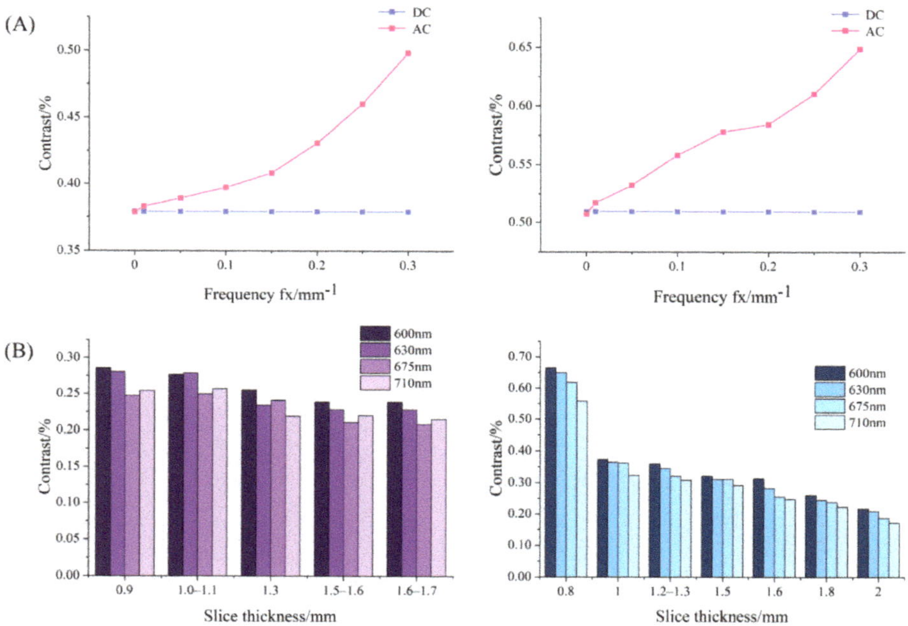

Figure 8. (**A**) Contrast variation with spatial frequencies for the direct component (DC) and amplitude component (AC) images of USAF-1951 target covered with apple slices with (left) and without peel (right); (**B**) histogram of contrast values for the DC images of USAF-1951 target covered with apple slice with (left) and without peel (right) in different thicknesses.

Figure 9 shows statistical results of contrast and PVR at 630 nm for the DC images of the USAF-1951 target covered with different-thickness apple slices. It was observed that the contrast values in the left chart decreased with the thickness of the apple slice. There was an approximate linear relation as the slice thickness ranged from 0.9 mm to 2.5 mm. A similar result was found when analyzing the image contrast in the right chart, with a narrow thickness range of 1.0–2.5 mm. It was noticed that the distribution of image contrast in the left chart (with peel) was more disperse and irregular than that in the right chart. This could be attributed to the influence of pores on the surface tissue of apple peel. On the other hand, PVR showed a gradual decreasing trend with the slice thicknesses. The black horizontal bars of the USAF-1951 target were hardly recognized while PVR was reduced to a certain value. In order to determine the light penetration depth in apple tissue, the threshold value of PVR in the slice thickness range of 0.9–2.5 mm, in which a linear relationship between the contrast and slice thickness was observed, was tested by large-scale experiments and finally set as 1.2. From this aspect, the light penetration depths in apple tissues with and without peels were determined as 0–1.8 mm and 0–2.3 mm, respectively. These results were quite similar to those obtained in the simulation experiments (0–2.2 mm). The differences between the simulation and practical experiments could be caused by many factors. For example, the apple optical properties, which were taken as the inputs in the simulations, are prone to measurement errors of the integrating sphere system, and thus lead to deviations for the simulation results. In addition, it is challenging to consider all the experiment details completely, such as the minute space between the covered apple slice and USAF-1951 target, and they may cause potential effects on our data analysis and final results.

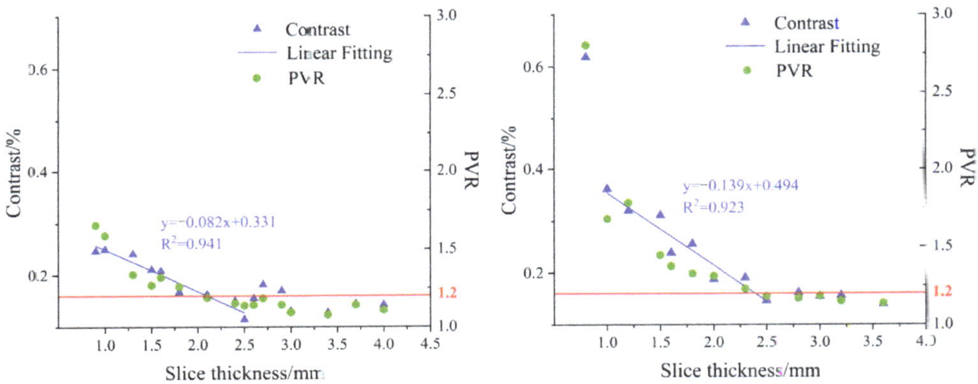

Figure 9. Contrast (left axis) and PVR (right axis) variation at 630 nm for direct component (DC) images of USAF-1951 target covered with different-thickness apple slices ((**left**) with peel; (**right**) without peel).

4.3. Validation in Detecting Early-Stage Bruise of Apple

Experiments on early-stage bruise detection of apples were conducted to validate the results of light penetration depth in apple tissues. The apple was peeled first to make the bruised tissue visible, and then it was covered with different-thickness apple slices with or without peels. Spatial-frequency domain images were acquired, followed with image demodulation and inverse estimation for generating optical property mappings. It was supposed that the bruised apple tissue could be detected if the light penetration depth was equal to or larger than the thickness of the covered apple slice. Figure 10 shows the demodulated images of the bruised apple covered with a 0.8 mm thick apple slice without peel. The AC images at certain spatial frequencies enhanced the bruised feature compared with the DC image. Strong contrast and surface texture variation were observed with the increased frequency. These findings indicated the enhanced capability of SFDI for detecting early-stage bruises of the apples, in comparison with the imaging techniques under uniform or diffuse illumination. It was believed that inverse parameter estimation for optical property mappings would provide more useful information for qualitative and quantitative analyses of bruise detection [32]. Therefore, the absorption and reduced scattering coefficient mappings were produced in our next step. The previous studies reported that bruising changed the optical properties of apples, particularly the reduced scattering coefficient, resulting in the difference between bruised and non-bruised tissues. The bruising detection based on the reduced scattering coefficient mapping was further analyzed to validate the results of light penetration depth in apples.

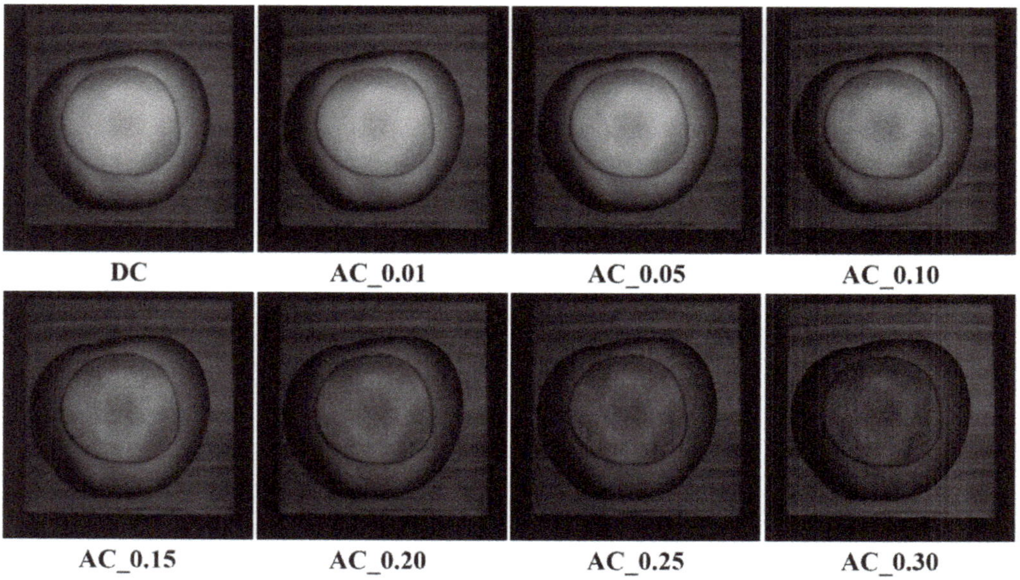

Figure 10. Demodulated images of the bruised apple covered with an 0.8 mm thick slice without peel at a sequence of spatial frequencies of 0, 0.01, 0.05, 0.10, 0.15, 0.20, 0.25, and 0.30 mm^{-1}, from top left to bottom right.

Figure 11 shows the reduced scattering coefficient mappings of the bruised apple covered with pre-prepared apple slices, with the bruised tissue marked in red circles. It was noticed that the bruised apple tissue without being covered by any slice, as shown in Figure 11A, was clearly observed from the reduced scattering coefficient mapping. Figure 11B shows that the bruised apple tissue covered with the 1.2 mm thick slice with peel could still be easily recognized, revealing that the spatially modulated light could completely penetrate through the apple slices with peel with the thickness of 0–1.2 mm. The 1.5 mm thick slice cover without peel (Figure 11D) provided more difficulty for apple bruise detection than the 0.8 mm thick slice (Figure 11C), but both of them could still be penetrated by the light. Through data analysis of all the mapping results with different-thickness slices, it was concluded that the apple slice without peel could be completely penetrated with the thickness range of 0–1.5 mm.

According to the report of Binzoni et al. [39], the number of photons that visit a given tissue voxel situated at a depth larger than 2 mm represents less than the 1% of the total number of photons reaching the corresponding detection pixel. They made the conclusion that the light penetration depth was no more than 2 mm, which confirmed our findings. In addition, Lu and Lu [21] reported that the maximum light penetration depth was no more than three sheets of white paper (or less than 400 μm). In our study, the apple slice, instead of white paper, was used as the cover for experiments, which was more scientific and reasonable for investigating the light penetration depth in apples. It should be pointed out that the results of light penetration depth obtained in this study are based on the apple sample, SFDI system configuration, and data processing method. Zhao et al. [16] reported that shortwave-infrared illumination could penetrate thicker biological tissue than visible light, because there is decreased optical scattering in the shortwave-infrared region compared to visible wavelengths. Hence, the hardware of the SFDI system, including the camera, the light source, and wavelength range, as well as the image processing algorithm, could be improved to increase light penetration depth in the future.

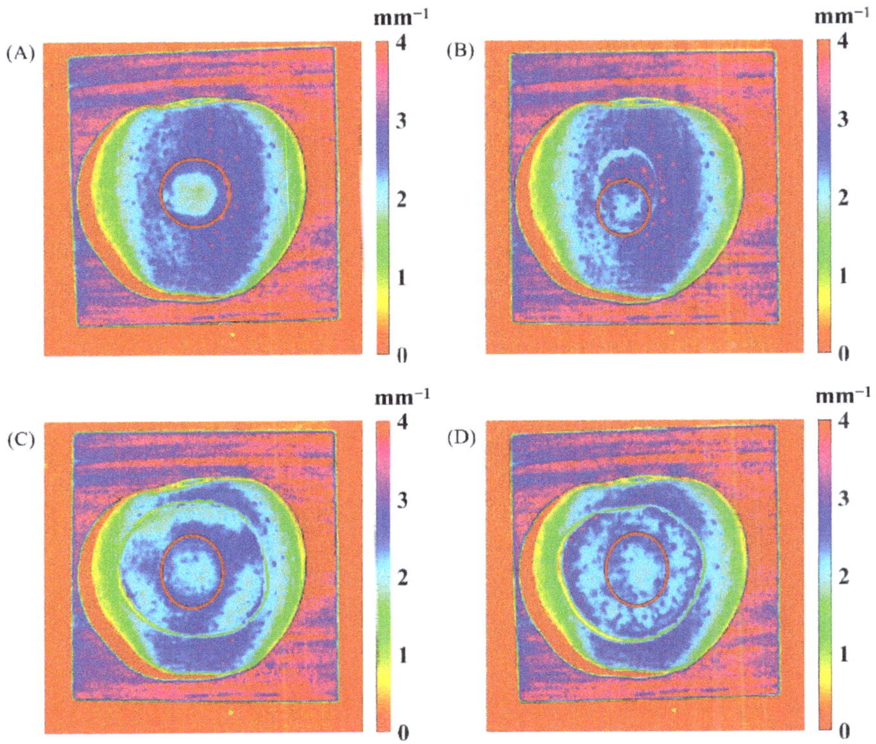

Figure 11. The reduced scattering coefficient mappings of the bruised apple covered without any slice (**A**), with a 1.2 mm thick slice with peel (**B**), and with a 0.8 mm thick (**C**) and 1.5 mm thick (**D**) slices without peel.

5. Conclusions

This paper presents theoretical and experimental analyses of light penetration depth in apple tissues using SFDI technology. The MC simulation coupled with the developed program indicated that the light penetration depth in apples was no more than 2.2 mm. The practical experiment on the USAF-1951 target covered with different-thickness apple slices demonstrated that the maximum light penetration depths were 1.8 mm with peel and 2.3 mm without peel, which were quite similar to the simulation results (0–2.2 mm). An example experiment on early-stage bruise detection was carried out to validate the light penetration depths in apples. The results showed that the bruised apple tissues covered with the slices with the thicknesses of 0–1.5 mm or 0–1.2 mm could be detected, depending on the presence of the apple peel or not. The maximum depth of an apple bruise should be about 1.2 mm beneath the surface peel, otherwise the SFDI technique cannot achieve accurate detection. In summary, SFDI can serve as a subsurface imaging technique for detecting early-stage bruises of thin-skinned fruits (<1.2 mm), such as apples. Improvement of system hardware and the data processing algorithm may provide potential for increasing the light penetration depth. Further work can be directed to explore the full potential of SFDI in bruise detection of other thin-skinned fruits and vegetables, such as pears, peaches, cucumbers, and potatoes.

Author Contributions: Conceptualization: D.H.; Methodology: T.Z.; Data analysis: T.Z. and D.H.; Formal analysis and investigation: D.Q., S.Y., Z.S. and X.S.; Data curation: D.Q. and S.Y.; Writing—original draft preparation: T.Z. and D.H.; Writing—review and editing: D.H. and H.P.; Funding acquisition: D.H. and Y.H.; Supervision: D.H., G.Z., T.S. and H.P. All authors have read and agreed to the published version of the manuscript.

Funding: This research was supported by the National Natural Science Foundation of China (32001414, 32101615) and Natural Science Foundation of Zhejiang Province (LQ20C130002).

Data Availability Statement: The data underlying this article will be shared on reasonable request to the corresponding author.

Conflicts of Interest: The authors declare no conflict of interest.

References

1. Askoura, M.L.; Vaudelle, F.; L'Huillier, J.-P. Use of steady-state imaging setup for assessing the internal optical properties of non-spherical apple samples. *Comput. Electron. Agric.* **2019**, *157*, 181–188. [CrossRef]
2. Luo, W.; Fan, G.Z.; Tian, P.; Dong, W.T.; Zhang, H.L.; Zhan, B.S. Spectrum classification of citrus tissues infected by fungi and multispectral image identification of early rotten oranges. *Spectrochim. Acta Part A Mol. Biomol. Spectrosc.* **2022**, *279*, 121412. [CrossRef] [PubMed]
3. Yin, H.; Li, B.; Liu, Y.D.; Zhang, F.; Su, C.T.; Ou-yang, A. Detection of early bruises on loquat using hyperspectral imaging technology coupled with band ratio and improved Otsu method. *Spectrochim. Acta Part A Mol. Biomol. Spectrosc.* **2022**, *283*, 121775. [CrossRef] [PubMed]
4. Lu, Y.Z.; Lu, R.F. Siritool: A Matlab Graphical User Interface for Image Analysis in Structured-Illumination Reflectance Imaging for Fruit Defect Detection. *Trans. ASABE* **2020**, *63*, 1037–1047. [CrossRef]
5. Luo, Y.F.; Jiang, X.; Fu, X.P. Spatial Frequency Domain Imaging System Calibration, Correction and Application for Pear Surface Damage Detection. *Foods* **2021**, *10*, 2151. [CrossRef]
6. Sun, Z.Z.; Hu, D.; Xie, L.J.; Ying, Y.B. Detection of early stage bruise in apples using optical property mapping. *Comput. Electron. Agric.* **2022**, *194*, 106725. [CrossRef]
7. He, X.M.; Fu, X.P.; Li, T.W.; Rao, X.Q. Spatial frequency domain imaging for detecting bruises of pears. *J. Food Meas. Charact.* **2018**, *12*, 1266–1273. [CrossRef]
8. Zhang, M.Y.; Li, C.Y.; Yang, F.Z. Optical properties of blueberry flesh and skin and Monte Carlo multi-layered simulation of light interaction with fruit tissues. *Postharvest Biol. Technol.* **2019**, *150*, 28–41. [CrossRef]
9. Liu, D.Y.; Xie, D.D.; Guo, W.C. Effect of CPPU on bulk optical properties of kiwifruit during storage in near-infrared range. *J. Sci. Food Agric.* **2020**, *100*, 3111–3119. [CrossRef]
10. Sun, Y.; Lu, R.F.; Pan, L.Q.; Wang, X.C.; Tu, K. Assessment of the optical properties of peaches with fungal infection using spatially-resolved diffuse reflectance technique and their relationships with tissue structural and biochemical properties. *Food Chem.* **2020**, *321*, 126704. [CrossRef]
11. Hu, D.; Lu, R.F.; Ying, Y.B. Spatial-frequency domain imaging coupled with frequency optimization for estimating optical properties of two-layered food and agricultural products. *J. Food Eng.* **2020**, *277*, 109909. [CrossRef]
12. Vanoli, M.; Van Beers, R.; Sadar, N.; Rizzolo, A.; Buccheri, M.; Grassi, M.; Lovati, F.; Nicolaï, B.; Aernouts, B.; Watté, R.; et al. Time- and spatially-resolved spectroscopy to determine the bulk optical properties of 'Braeburn' apples after ripening in shelf life. *Postharvest Biol. Technol.* **2020**, *168*, 111233. [CrossRef]
13. Lu, Y.Z.; Saeys, W.; Kim, M.; Peng, Y.K.; Lu, R.F. Hyperspectral imaging technology for quality and safety evaluation of horticultural products: A review and celebration of the past 20-year progress. *Postharvest Biol. Technol.* **2020**, *170*, 111318. [CrossRef]
14. Lu, Y.Z.; Li, R.; Lu, R.F. Fast demodulation of pattern images by spiral phase transform in structured-illumination reflectance imaging for detection of bruises in apples. *Comput. Electron. Agric.* **2016**, *127*, 652–658. [CrossRef]
15. Lu, Y.Z.; Li, R.; Lu, R.F. Structured-illumination reflectance imaging (SIRI) for enhanced detection of fresh bruises in apples. *Postharvest Biol. Technol.* **2016**, *117*, 89–93. [CrossRef]
16. Zhao, Y.Y.; Pilvar, A.; Tank, A.; Peterson, H.; Jiang, J.; Aster, J.C.; Dumas, J.P.; Pierce, M.C.; Roblyer, D. Shortwave-infrared meso-patterned imaging enables label-free mapping of tissue water and lipid content. *Nat Commun* **2020**, *11*, 5355. [CrossRef]
17. Pilvar, A.; Plutzky, J.; Pierce, M.C.; Roblyer, D. Shortwave infrared spatial frequency domain imaging for non-invasive measurement of tissue and blood optical properties. *J. Biomed. Opt.* **2022**, *27*, 066003. [CrossRef]
18. Lee, S.Y.; Zheng, C.; Brothers, R.; Buckley, E.M. Small separation frequency-domain near-infrared spectroscopy for the recovery of tissue optical properties at millimeter depths. *Biomed. Opt. Express* **2019**, *10*, 5362–5377. [CrossRef]
19. Nguyen, J.Q.; Crouzet, C.; Mai, T.; Riola, K.; Uchitel, D.; Liaw, L.H.; Bernal, N.; Ponticorvo, A.; Choi, B.; Durkin, A.J. Spatial frequency domain imaging of burn wounds in a preclinical model of graded burn severity. *J. Biomed. Opt.* **2013**, *18*, 66010. [CrossRef]

20. Robbins, C.M.; Raghavan, G.; Antaki, J.F.; Kainerstorfer, J.M. Feasibility of spatial frequency-domain imaging for monitoring palpable breast lesions. *J. Biomed. Opt.* **2017**, *22*, 121605. [CrossRef]
21. Lu, Y.Z.; Lu, R.F. Structured-illumination reflectance imaging for the detection of defects in fruit: Analysis of resolution, contrast and depth-resolving features. *Biosys. Eng.* **2019**, *180*, 1–15. [CrossRef]
22. Cuccia, D.J.; Bevilacqua, F.P.; Durkin, A.J.; Ayers, F.R.; Tromberg, B.J. Quantitation and mapping of tissue optical properties using modulated imaging. *J. Biomed. Opt.* **2009**, *14*, 024012. [CrossRef]
23. Goodman, J.W. *Introduction to Fourier Optics*; McGraw-Hill: New York, NY, USA, 1998.
24. Gustafsson, M.G.L. Nonlinear structured-illumination microscopy: Wide-field fluorescence imaging with theoretically unlimited resolution. *Proc. Natl. Acad. Sci. USA* **2005**, *102*, 13081–13086. [CrossRef] [PubMed]
25. Neil, M.A.A.; Juskaitis, R.; Wilson, T. Method of obtaining optical sectioning by using structured light in a conventional microscope. *Opt. Lett.* **1997**, *22*, 1905–1907. [CrossRef]
26. Ntziachristos, V. Going deeper than microscopy: The optical imaging frontier in biology. *Nat. Methods* **2010**, *7*, 603–614. [CrossRef] [PubMed]
27. Gigan, S. Optical microscopy aims deep. *Nat. Photon* **2017**, *11*, 16. [CrossRef]
28. Wilson, B.C.; Jacques, S.L. Optical reflectance and transmittance of tissues: Principles and applications. *IJQE* **1990**, *77*, 2186–2199. [CrossRef]
29. Wang, L.H.; Jacques, S.L. *Monte Carlo Modeling of Light Transport in Multi-Layered Tissues in Standard C*; University of Texas M.D. Anderson Cancer Center: Houston, TX, USA, 1992; Volume 47, pp. 131–146.
30. Bigio, I.J.; Fantini, S. *Quantitative Biomedical Optics: Theory, Methods, and Applications*; Cambridge University Press: Cambridge, UK, 2016.
31. Lu, Y.Z.; Lu, R.F. Development of a Multispectral Structured Illumination Reflectance Imaging (SIRI) System and Its Application to Bruise Detection of Apples. *Trans. ASABE* **2017**, *60*, 1379–1389. [CrossRef]
32. Sun, Z.Z.; Xie, L.J.; Hu, D.; Ying, Y.B. An artificial neural network model for accurate and efficient optical property mapping from spatial-frequency domain images. *Comput. Electron. Agric.* **2021**, *188*, 106340. [CrossRef]
33. Hayakawa, C.K.; Karrobi, K.; Pera, V.; Roblyer, D.; Venugopalan, V. Optical sampling depth in the spatial frequency domain. *J. Biomed. Opt.* **2019**, *24*, 071603. [CrossRef]
34. Hu, D.; Sun, T.; Yao, L.J.; Yang, Z.D.; Wang, A.C.; Ying, Y.B. Monte Carlo: A flexible and accurate technique for modeling light transport in food and agricultural products. *Trends Food Sci. Technol.* **2020**, *102*, 280–290. [CrossRef]
35. Sun, X.L.; Sun, Z.Z.; Zhou, T.T.; Hu, D.; Sun, T.; Yang, Z.D.; Zhou, G.Q. Optical characterization of liquid phantoms in 500–1000 nm using an improved single integrating sphere system. *Infrared Phys. Technol.* **2022**, *127*, 104386. [CrossRef]
36. Makous, P. Spatial frequency, phase, and the contrast of natural images. *J. Opt. Soc. Am. A* **2002**, *19*, 1096–1106.
37. Wang, W.; Li, C.; Tollner, E.W.; Rains, G.C.; Gitaitis, R.D. A liquid crystal tunable filter based shortwave infrared spectral imaging system: Calibration and characterization. *Comput. Electron. Agric.* **2012**, *80*, 135–144. [CrossRef]
38. Lammertyn, J.; Peirs, A.; Baerdemaeker, J.D.; Nicola, B. Light penetration properties of NIR radiation in fruit with respect to non-destructive quality assessment. *Postharvest. Biol. Technol.* **2000**, *18*, 121–132. [CrossRef]
39. Binzoni, T.; Vogel, A.; Gandjbakhche, A.H.; Marchesini, R. Detection limits of multi-spectral optical imaging under the skin surface. *Phys. Med. Biol.* **2008**, *53*, 617–636. [CrossRef]

foods MDPI

Article

High-Stability Bi-Layer Films Incorporated with Liposomes @Anthocyanin/Carrageenan/Agar for Shrimp Freshness Monitoring

Junjun Zhang [1,2,3], Yan Yang [1,2], Jianing Zhang [1,3,4], Jiyong Shi [1,5,*], Li Liu [2], Xiaowei Huang [1,2,5], Wenjun Song [1,3], Zhihua Li [1,3,5], Xiaobo Zou [1,2,3,5] and Megan Povey [6,*]

[1] Agricultural Product Processing and Storage Laboratory, School of Food and Biological Engineering, Jiangsu University, Zhenjiang 212013, China
[2] China Light Industry Key Laboratory of Food Intelligent Detection & Processing, School of Food and Biological Engineering, Jiangsu University, Zhenjiang 212013, China
[3] China Light Industry Engineering Technology Research Center of Central Kitchen Intelligent Equipment, School of Food and Biological Engineering, Jiangsu University, Zhenjiang 212013, China
[4] International Joint Research Laboratory of Intelligent Agriculture and Agri-Products Processing, Jiangsu Education Department, Jiangsu University, Zhenjiang 212013, China
[5] Collaborative Innovation Center for Modern Grain Circulation and Safety, College of Food Science and Engineering, Nanjing University of Finance and Economics, Nanjing 210023, China
[6] School of Food Science and Nutrition, University of Leeds, Leeds LS2 9JT, UK
* Correspondence: shi_jiyong@ujs.edu.cn (J.S.); m.j.w.povey@food.leeds.ac.uk (M.P.); Fax: +86-511-88780201 (J.S.)

Abstract: High-stability bi-layer films were prepared by incorporating anthocyanin-loaded liposomes into carrageenan and agar (A-CBAL) for non-destructive shrimp freshness monitoring. The encapsulation efficiency of the anthocyanin-loaded liposomes increased from 36.06% to 46.99% with an increasing ratio of lecithin. The water vapor transmission (WVP) of the A-CBAL films, with a value of 2.32×10^{-7} g $\cdot$ m^{-1} $\cdot$ h^{-1} $\cdot$ pa^{-1}, was lower than that of the film with free anthocyanins (A-CBA). The exudation rate of the A-CBA film reached 100% at pH 7 and pH 9 after 50 min, while the A-CBAL films slowed down to a value lower than 45%. The encapsulation of anthocyanins slightly decreased the ammonia sensitivity. Finally, the bi-layer films with liposomes successfully monitored shrimp freshness with visible color changes to the naked eye. These results indicated that films with anthocyanin-loaded liposomes have potential applications in high-humidity environments.

Keywords: liposomes; high stability; freshness; bi-layer indicator

Citation: Zhang, J.; Yang, Y.; Zhang, J.; Shi, J.; Liu, L.; Huang, X.; Song, W.; Li, Z.; Zou, X.; Povey, M. High-Stability Bi-Layer Films Incorporated with Liposomes @Anthocyanin/Carrageenan/Agar for Shrimp Freshness Monitoring. *Foods* **2023**, *12*, 732. https://doi.org/10.3390/foods12040732

Academic Editor: M. Angela A. Meireles

Received: 7 December 2022
Revised: 9 January 2023
Accepted: 20 January 2023
Published: 8 February 2023

1. Introduction

The spoilage of meat products, which is extremely harmful and destructive, significantly increases the risk to human health [1]. Therefore, it is necessary to detect meat freshness. In the past, total volatile basic nitrogen (TVB-N) was widely regarded as a useful method for meat freshness monitoring using the Kjeldahl method [2]. However, it is destructive to samples and time-consuming. In recent years, more studies have been interested in intelligent packaging systems for "on-package" tracing in real-time. Intelligent food packaging is an effective tool for monitoring food conditions for consumers through intuitive changes. Meat corruption produces volatile amines, which results in an alkaline packaging environment. Therefore, pH indicator films, as a kind of intelligent sensor, have garnered wide attention because they can reflect freshness information through visual color changes. As a natural extract, anthocyanin presents visible color changes at different pH values and has been used in intelligent pH indicator films in recent years [3]. For instance, Zhang et al. successfully developed a novel film based on a mulberry anthocyanin extract for fish freshness monitoring [4]. However, most reported indicator films are based only on individual anthocyanins in the film-forming matrices, which makes them easily degraded in harsh environments (such as light and temperature). In addition, water-soluble free anthocyanins easily leak out from the film matrix in a high-humidity environment, thereby affecting the indicator stability. This instability creates a barrier that limits the use of intelligent film in practical food packaging. Therefore, it is essential to use an effective method to improve the stability of anthocyanin indicators.

Nano/microencapsulation technology in particular has been verified as a useful method to enhance the stability of anthocyanins. In terms of encapsulation technology, liposomes are popularly prepared with phospholipids, oils, and different solvents [5]. The central aqueous cavity of liposomes can be used to improve the stability of hydrophilic active ingredients and increase bioavailability [6]. Liposomes are attractive because they can encapsulate anthocyanins without changing their structure. The prepared anthocyanin elderberry extract-loaded liposome has the highest encapsulation efficiency of 69% and storage stability [7]. The retention rate of anthocyanins in milk was effectively improved with liposomes prepared by Chi et al. [8]. Up to now, few researchers have reported the use of liposomes to encapsulate free anthocyanins in intelligent packaging films.

Compared with single-layer biopolymer films, bi-layer films have shown excellent mechanical properties and stability [9,10]. One layer serves as an indicator layer containing anthocyanins and the other as a protective layer. Agar, one of the most widely promising agents, is applied in food packaging films due to its good gelling properties and excellent film-forming materials [11]. Carrageenan, a natural biopolymer, is widely studied as a packaging film matrix due to its high gelling capacity [12]. Because of the stronger hydrogen bond interactions between their highly polar hydroxyl groups, agar and carrageenan are used together to improve their mechanical properties.

Therefore, in this study, a liposome was formed by encapsulating a butterfly bean flower anthocyanin extract (BA) in soybean lecithin, which was then added to carrageenan to develop as an indicator layer. In addition, agar was used as the protective layer of a bi-layer indicator to monitor shrimp freshness. The particle size, zeta potential, morphology, and encapsulation efficiency of the liposome were initially analyzed. The bi-layer indicator film was determined with Fourier transform infrared spectroscopy (FTIR), scanning electron microscopy (SEM), and its mechanical properties. Moreover, the temperature stability and the response sensitivity to pH solutions and ammonia of the bi-layer films were investigated before evaluating their application for shrimp freshness monitoring.

2. Materials and Methods

2.1. Materials

The dried butterfly bean flower calyxes and fresh shrimp were obtained from the Zhenjiang Darunfa supermarket. Agar, ammonia, ethanol, potassium chloride, and hydrochloric acid were purchased from Sinopharm Chemical Reagent Co., Ltd. Citric acid, sodium acetate, and sodium dihydrogen citrate were bought from Jiangsu Thorpe Group Co., Ltd. (Zhenjiang, China) Carrageenan, glycerol, and disodium hydrogen phosphate were obtained from Jiangsu Chentong Chemical Co., Ltd. (Zhenjiang, China). Soybean lecithin, cholesterol, TritonX-100, and Tween80 were acquired from Zhenjiang Huadong Chemical Glass Co., Ltd. (Zhenjiang, China).

2.2. Preparation and Characterization of Anthocyanin-Loaded Liposomes

2.2.1. Extraction of Butterfly Bean Flower Anthocyanin

The butterfly bean flower anthocyanin (BA) was obtained according to a previous study [13]. The dried butterfly bean flower calyxes were crushed into a powder. Then, approximately 100 g of the powder was macerated with 1 L of 75% ethanol for 3 h at 60 °C. The solvent extraction solution was obtained using a centrifuge at 3000 r/min for 6 min. After that, the anthocyanin concentrated solution was obtained to remove the ethanol solvent using a rotary evaporator (RE-200A, SHANGHAI YARONG biochemistry instrument factory, China) at 50 °C for 2 h. Finally, the concentrated solution was dried in a vacuum freeze-dryer to obtain a BA powder.

2.2.2. Preparation of BA-Loaded Liposomes

The BA-loaded liposomes (BALs) were prepared using the ethanol injection high-pressure homogenization method with some modifications [14]. Lecithin (0.2%, 0.5%, and 1%), cholesterol (0.2%), and Tween 80 (0.24%) were dissolved in an ethanol solution. The BA solution (0.2%) was prepared in an acetate buffer solution (pH 3.5, 0.05 mol/L). To obtain crude milky liposomes, the BA solution was quickly injected into lecithin mixtures of different concentrations and stirred vigorously for 30 min. Next, the crude liposomes were homogenized using a high-pressure homogenizer (AH-BASIC, Antos Nano Technology Co., Ltd., Suzhou, China) at 20,000 psi for 5 cycles. Then, the cooling solution was passed through a 0.25 μm extruder, and the solvent was removed via rotary evaporation (SY-4000, Shanghai Yarong Co., Ltd., Shanghai, China) to obtain concentrated BALs with different concentrations of lecithin at 0.2%, 0.5%, and 1% (BAL1, BAL2, and BAL3).

2.2.3. Characterization of the Liposomes

The average particle size, zeta potential, and polydispersity index (PDI) of the liposomes were evaluated with the dynamic light scattering technique using a Zeta-sizer Nano ZS (Malvern, Worcestershire, UK). The microstructure of the liposomes was observed using an optical microscope (4XC-W, Jinanchenda, Jinan, China).

The encapsulation efficiency (EE) was measured according to the literature with some modifications [15]. Solutions of various concentrations (0.5, 1.0, 1., 2.0, and 2.5 μL/mL) of the anthocyanin were dissolved in a buffer solution (pH 6.86), and the absorbance was measured at 620 nm. The standard curve of the anthocyanin was analyzed as the equation Y = 0.221x + 0.010 (R^2 = 0.9981). Then, a certain amount of BALs was immersed in a buffer solution and centrifuged at 8000 rpm for 20 min. Free anthocyanins were isolated from the supernatant, and their absorbance was measured at 620 nm. Then, the concentration of free anthocyanins was calculated with the standard curve. Finally, the EE of the anthocyanin was obtained as follows:

$$EE(\%) = \frac{\text{Total anthocyananins} - \text{Free anthocyanins}}{\text{Total anthocyanins}} \quad (1)$$

2.2.4. The Color of BAL in Different pH Solutions

The absorbance of the BA and BAL solutions at different pH values was measured in the range of 450 nm to 700 nm using a UV-visible spectrophotometer (TU10CS, Beijing General Analytical Instrument, Beijing, China).

2.3. Preparation of the Bi-Layer Indicator Films

The bi-layer films were prepared with two individual solvent casting methods. Firstly, 2 g of agar was stirred in 100 mL of distilled water for 2 h at 100 °C. Then, an agar hydrogel was formed as the outer layer by cooling the plastic Petri dish at room temperature. Secondly, 2 g of carrageenan was stirred in 100 mL of water with 2% glycerin for 1 h at 85 °C. After cooling at 65 °C, free anthocyanins and different groups of BALs (BAL1, BAL2, and BAL3, each containing 20 mg of the anthocyanin) were added to the above carrageenan solution. The solutions were thoroughly stirred at 65 °C for 30 min. Finally, the carrageenan solutions containing free anthocyanins and liposomes were dispersed onto the agar protective layer and dried in an oven for 24 h at 35 °C, and the bi-layer films were obtained and termed as the A-CBA, A-CBAL1, A-CBAL2, and A-CBAL3 films, respectively.

2.4. Structural Properties of the Bi-Layer Films

2.4.1. Microstructure

The cross-sections of the bi-layer films were performed with a JSM-3400 (JEOL Ltd., Tokyo, Japan) at an accelerating voltage of 10 keV. Prior to observation, the samples were divided into small pieces and vertically adhered to an aluminum stub with a thin layer of gold.

2.4.2. FTIR Analysis

The FTIR spectra of the films and film-forming materials were measured using a Nicolet 50 spectrometer in the attenuated total reflection mode at 4000–525 cm^{-1} with a resolution of 4 cm^{-1} (Thermo Scientific, Waltham, MA, USA).

2.5. Determination of Physical Properties of Bi-Layer Films

2.5.1. Mechanical Properties

The thickness of the films was determined using a Mitutoyo digital micrometer (Tester Sangyo Co., Ltd., Saitama, Japan). The mechanical properties were defined using a TA-XT Plus texture analyzer (Stable Micro Systems, Godalming, UK). The films were cut into 20 × 60 mm pieces with an initial distance of 40 mm and a proper tensile speed of 0.6 mm/s [16].

2.5.2. Water Vapor Transmission (WVP) Results

WVP was determined using the standard gravimetric method of ASTM E96-05. The films were covered on top of a 50 mL centrifuge tube with 20 mL of water and stored in a desiccator. *WVP* was analyzed according to the centrifuge tube weight every 12 h for 5 days and calculated with the following formula:

$$WVP = \frac{\Delta m \times d}{S \times \Delta P \times t} \tag{2}$$

where d is the average thickness (mm); S is the effective permeation area of the film (m^2); Δm is the mass of water permeation (g); t is the interval time(s); and ΔP is the pressure difference between the 2 sides of the film (3179 Pa).

2.5.3. Color Appearance and Opacity

The colors of the bi-layer films were measured using a portable scanner (G4050, HP, USA) and then expressed as L*, a*, and b* values. The opacity of the films was recorded with a UV-vis spectrophotometer at 200 to 800 nm. The opacity formula was as follows [11]:

$$Opacity = \frac{Abs_{600}}{d} \tag{3}$$

where Abs_{600} is the absorbance at 600 nm, and d is the average thickness (mm).

2.6. Color Stability of Bi-Layer Films

In order to measure the stability, the films were kept at 4 °C or 25 °C at 2-day intervals within 14 days using a portable scanner. The calculation of color changes (ΔE) was as follows:

$$\Delta E = \sqrt{(L - L_0)^2 + (a - a_0)^2 + (b - b_0)^2} \tag{4}$$

where L, a, and b are the color values of the films at storage time; L_0, a_0, and b_0 are the initial color values.

2.7. Color Response and the Leaching Rate under Different pH Buffers of Films

The film samples were immersed in plastic Petri dishes containing 15 mL of buffer solutions (2–10). During the different time intervals, the exudation rate was determined by calculating the concentration of the anthocyanin leaching solution, and the color response was captured using a camera at the beginning time.

2.8. Color Response to Ammonia of Bi-Layer Films

Each of the bi-layer films was placed into the middle-upper layer of a sealed, home-made acrylic box (500 mL). An aqueous ammonia solution was injected into the bottom of each box with 0.1 mL of different concentrations at 0–200 μM, and the color changes were determined using a CM2300 spectrophotometer [17]. The digital values were also expressed as color changes (ΔE).

2.9. Application in Shrimp Freshness Detection of Bi-Layer Films

According to the determinations of the film results, the A-CBA and A-CBAL2 films were used as shrimp freshness indicators. An amount of 50 g of fresh shrimp was placed inside a sealed packing box (700 mL), whose inner surface was attached to a film at 4 °C for 96 h. The color of each film was obtained using a CM2300 spectrophotometer every 12 h. The total volatile basic nitrogen (TVB-N) value was determined according to the method of Zhang et al. [17].

2.10. Data Analysis

All tests were repeated three times with mean ± standard deviation results. Duncan's test was used to analyze the data in SPSS software (Version 21, IBM SPSS Inc, New York, NY, USA), and the differences were considered significant if $p < 0.05$.

3. Results and Discussion

3.1. Characterization of the BAL Liposomes

The results of different liposomes with average particle sizes, Zeta potentials, PDI values, microstructures, and EE values are shown in Table 1. With the addition of lecithin, the average particle sizes of the liposomes obviously increased from 131.39 nm to 311.42 nm, which was attributed to the amount of hydrogen and van der Waals force between the anthocyanins and lecithin [18]. Zeta potential is an important parameter to characterize the stability of liposomes. The higher value of Zeta potential, the greater repulsion strength required to settle and coagulate liposomes [19]. The Zeta potentials of BAL1 and BAL2 were −48.23 mV and −40.16 mV, respectively, indicating the stable dispersion of liposome particles in the solution PDI is an index that reflects the particle size distribution [20]. The smaller the PDI, the better the regularity of dispersion of the particles. A PDI < 0.4 indicates a homogenous particle size distribution in the system [21]. With the addition of lecithin, the PDI increased from 23.96% to 29.51%, indicating the heterogeneous size distribution. This was consistent with the Zeta potential results. These structure formations can also be observed in the microstructures of the multi-compartmental but obvious core–shell structures. Thus, the EE increased from 36.06% to 46.99% with the increasing ratio of lecithin. The above results indicated that the ratio of lecithin was one of the key factors in the characterization of anthocyanin-loaded liposomes.

Table 1. The sizes, Zeta potentials, PDI values, EE values, and microscope pictures of the liposomes.

	Average Particle Size/nm	Zeta Potential/mV	PDI/%	Microscope	EE/%
BAL1	131.39 ± 5.12^{a}	-49.01 ± 1.52^{c}	23.96 ± 2.03^{a}		36.06 ± 1.87^{a}
BAL2	153.49 ± 10.53^{b}	-43.16 ± 0.88^{b}	27.57 ± 3.53^{bc}		44.28 ± 4.25^{b}
BAL3	311.42 ± 15.68^{c}	-19.96 ± 2.31^{a}	29.51 ± 4.25^{c}		46.99 ± 6.17^{bc}

Note: the superscripted characters a, b, c represent significant differences ($p < 0.05$).

3.2. The pH Response of Anthocyanin-Loaded Liposomes

As shown in Figure 1A, both anthocyanins and liposomes showed obvious color changes in different pH values. The color of BA changed from pink to purple, then blue, and finally blue-green. The color of BAL1 changed from pink to purple-green, then cyan, and finally green. The different color changes of anthocyanin were caused by structural transformations, which were found in a previous study [13]. In fact, the different color changes between BA and BAL were attributed to the cavity structure of liposomes, which decreases the structural transformation rate in anthocyanins [22]. As shown in Figure 1B, 2 characteristic absorption peaks can be observed around 574 nm and 620 nm for BA and BAL. At pH 2, the absorption peak of BA was at 552 nm and gradually red-shifted to 574 nm at pH 3–8. With the pH increasing to 9–10, the absorption peak disappeared due to the destroyed structure of the anthocyanin molecular center ring under strong alkaline conditions [23]. The response mechanism of BAL to pH was consistent with that of the anthocyanin solution. However, the peak at 574 nm disappeared at pH 8 for the BAL1 spectrum while occurring at pH 7 for the BAL2 and BAL3 spectrums, respectively. This is mainly because of the encapsulation difference. The ratio of A_{620} to A_{574} reflected the shift changes of the absorption peaks in Figure 1C. This was clearly observed in the variation of the maximum values of the BA and BAL spectra. The above results showed that the coloration degree of the solution obviously decreased after being encapsulated by liposomes, while the color sensor function of the anthocyanins was not hindered.

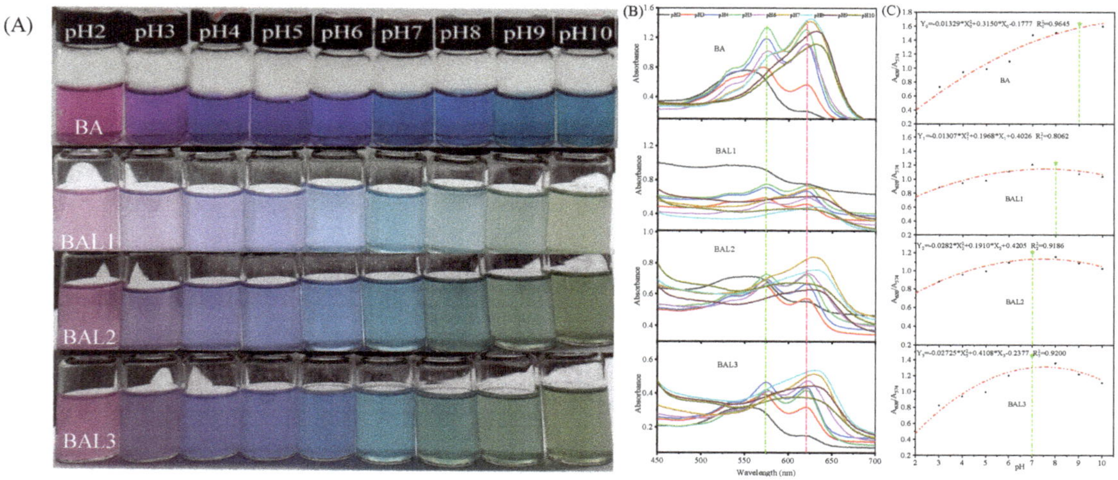

Figure 1. Color (**A**) and ultraviolet-visible spectra (**B**), and the ratio of A_{620} and A_{574} (**C**) of BA and BAL at different pH values.

3.3. The Structural Analysis of the Bi-Layer Films

3.3.1. SEM Analysis of Indicator Films

The film compatibility can be observed in the cross-section of a bi-layer film. As can be seen in Figure 2, all the films presented an obvious two-layer structure, which was attributed to the hydrogel thermal irreversibility processes between agar and carrageenan. Meanwhile, hydrogen bonding, cross-linked agar, and carrageenan prevented the bi-layer films from separating. The agar outer layers appeared relatively uniform except for the parts that were contaminated by the inner anthocyanin layers. In Figure 2A, the A-CBA film with free anthocyanins displays a homogeneous and compact structure. Compared with the free anthocyanins, the liposomes with hydrophobic structures of the A-CBAL film caused a reduction in the cross-linking between the film-forming solution and water molecules. Therefore, the liposomes in the film-forming matrix presented a lower homogeneous dispersion. However, there were no obvious differences between the A-CBAL films, indicating that the anthocyanin encapsulation of liposomes hardly presented a negative effect on the film morphologies. Importantly, the above results indicated that bi-layer films were satisfactorily prepared.

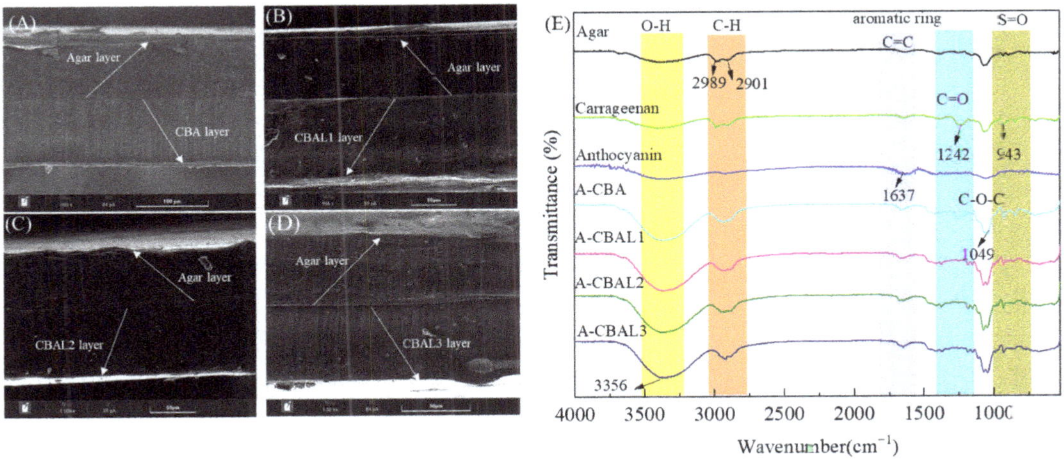

Figure 2. Cross-section SEM images of A-CBA (**A**), A-CBAL1 (**B**), A-CBAL2 (**C**), and A-CBAL3 films (**D**), and FTIR spectra (**E**) of colorimetric films.

3.3.2. FTIR Analysis of Indicator Films

The absorption peaks of the FIIR spectra of the film-forming materials and the bi-layer films are shown in Figure 2E. The band at 3356 cm^{-1} corresponds to the OH stretching vibration of the hydroxyl structure. It occurred in all the spectra but with lower peak intensity changes [24]. The peaks at 2289 and 2901 cm^{-1} were due to the C-H and $-CH_2$ stretching vibrations of alkane groups [25]. The absorption band that appears at 1637 cm^{-1} of the anthocyanins was ascribed to the C=C stretching from the aromatic ring frame of the butterfly bean flower anthocyanin, which is related to the flavonoid fingerprint spectra and was also found in all the bi-layer films [26]. The absorption peaks of carrageenan at 1242 and 943 cm^{-1} were associated with the C=O of the glycoside bond and the S=O of the sulfate ester group, respectively [27]. The other major band at 1049 cm^{-1} was attributed to the C-O-C stretching vibration of the conjugated carbonyl group [28]. In the case of the A-CBAL films, all the peaks presented similar positions with minor intensity changes to the control film (A-CBA). The results indicated that there was no chemical interaction between the liposomes and anthocyanins.

3.4. Physical Performance Analysis

3.4.1. Appearance and Opacity Analysis Results

The intuitive packaging color appearance can easily affect the application efficacy of an indicator film. As presented in Table 2, there were no obvious differences in the L* values of A-CBAL films with different lecithin ratios but they exhibited slightly higher values than the A-CBA film, indicating that liposome films have higher brightness. The decreased negative a* and b* values reflect the lower greenness and blueness strengths of the films with liposomes, which could be attributed to the yellowish color of the liposomes. Therefore, the blue of anthocyanin was covered after being encapsulated by liposomes, increasing the opacity.

3.4.2. Thickness and WVP Analysis Results

As illustrated in Figure 3A, the thicknesses of all the indicator films were not significantly different. The considerable index evaluates whether the packaging quality is WVP, which can represent the ability to block external water vapor of a film. As summarized in Figure 3B, the WVP values of all the A-CBAL films were significantly lower than that of the A-CBA film. It may be that lecithin had hydrophobic tails, which reduced the hydrophilicity of the indicator film. Therefore, the liposome films reflected higher water vapor resistance.

However, the films with different lecithin ratios had little difference between them, and the maximum values did not exceed 2.32×10^{-7} g $\cdot$ m^{-1} $\cdot$ h^{-1} $\cdot$ pa^{-1}.

Table 2. The colors and opacity values of the films.

Film	L*	a*	b*	Opacity	Appearance
A-CBA	84.82 ± 0.42 ^{a}	−10.52 ± 1.89 ^{c}	−7.37 ± 1.01 ^{d}	15.08 ± 0.86 ^{a}	
A-CBAL1	92.65 ± 1.31 ^{b}	−1.54 ± 0.35 ^{a}	−1.08 ± 0.11 ^{a}	32.39 ± 1.18 ^{c}	
A-CBAL2	89.61 ± 1.05 ^{b}	−2.45 ± 0.56 ^{b}	−1.59 ± 0.40 ^{ab}	27.72 ± 2.45 ^{b}	
A-CBAL3	90.17 ± 0.56 ^{b}	−2.25 ± 0.31 ^{b}	−3.11 ± 0.79 ^{c}	26.02 ± 1.82 ^{b}	

Note: the superscripted characters of a, b, c, d represent significant differences ($p < 0.05$).

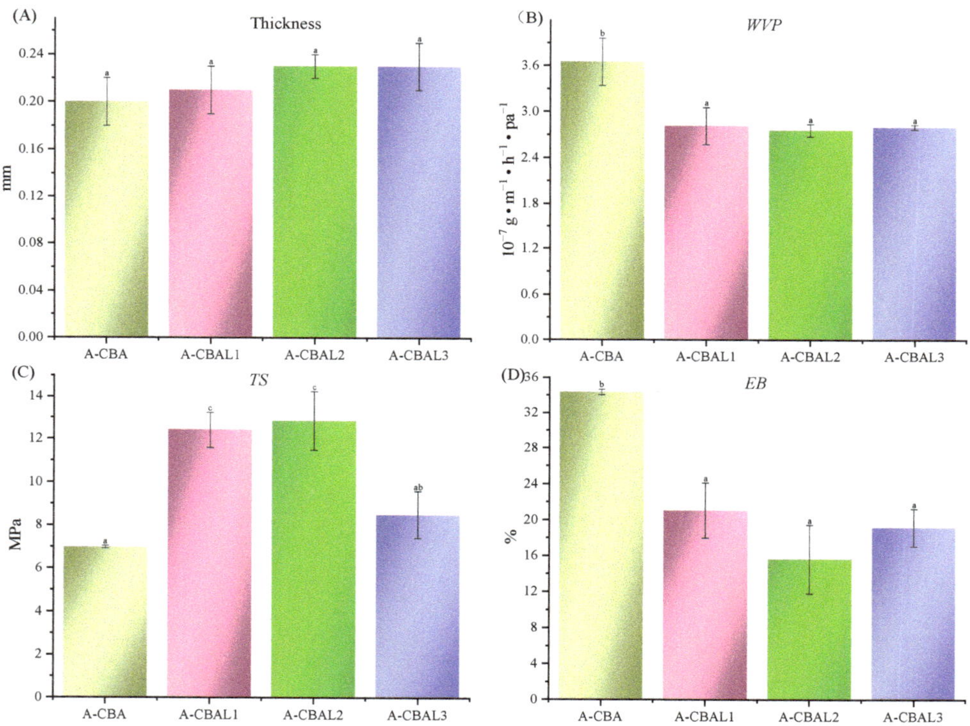

Figure 3. Thicknesses (**A**), *WVP* values (**B**), and mechanical properties with *TS* (**C**) and *EB* values (**D**) of the bi-layer films. Characters represent a significant difference ($p < 0.05$).

3.4.3. Mechanical Properties

Excellent TS and EB values can improve the protection performance of food packaging materials during food transportation periods. Each of the A-CBAL films had a significantly higher TS value and a lower EB value than the A-CBA film. This was probably attributed to the stronger intramolecular chemical bonding force and intermolecular force (van der Waals force and hydrogen bond) between liposomes and the film-forming materials than free anthocyanins [29]. The A-CBAL2 film simulated the maximum TS value with a value of 12.42 MPa, and that of the A-CBAL3 film gradually decreased to 5.81 MPa due to liposome instability. However, there were no significant differences in the EB values of the three A-CBAL films, that is, the change in the lecithin ratio did not destroy the crystal structure of the film-forming matrix.

3.5. Exudation Rate and Color Response of Indicator Films to pH Solution

The issue of anthocyanin leaking out from films causes the failure of the indicator function. The pH behavior of the films differed noticeably, as seen in Figure 4. The exudation rates of bi-layer film with free anthocyanins (A-CBA) exuded rapidly, reaching 80% at pH 2 after 70 min. In addition, the exudation rate of the A-CBA film reached 100% at pH 7 and pH 9 after 50 min, due to the higher degradation of anthocyanins under the alkaline environment [30]. Thus, the films with anthocyanin-loaded liposomes (A-CBAL) slowed down the exudation rate by no more than 45%. However, there was no correlation between the anthocyanin exudation rate and the ratio of lecithin in liposomes. In conclusion, the liposomes enhanced anthocyanin encapsulation, which can improve the stability of an indicator film in a high-humidity environment.

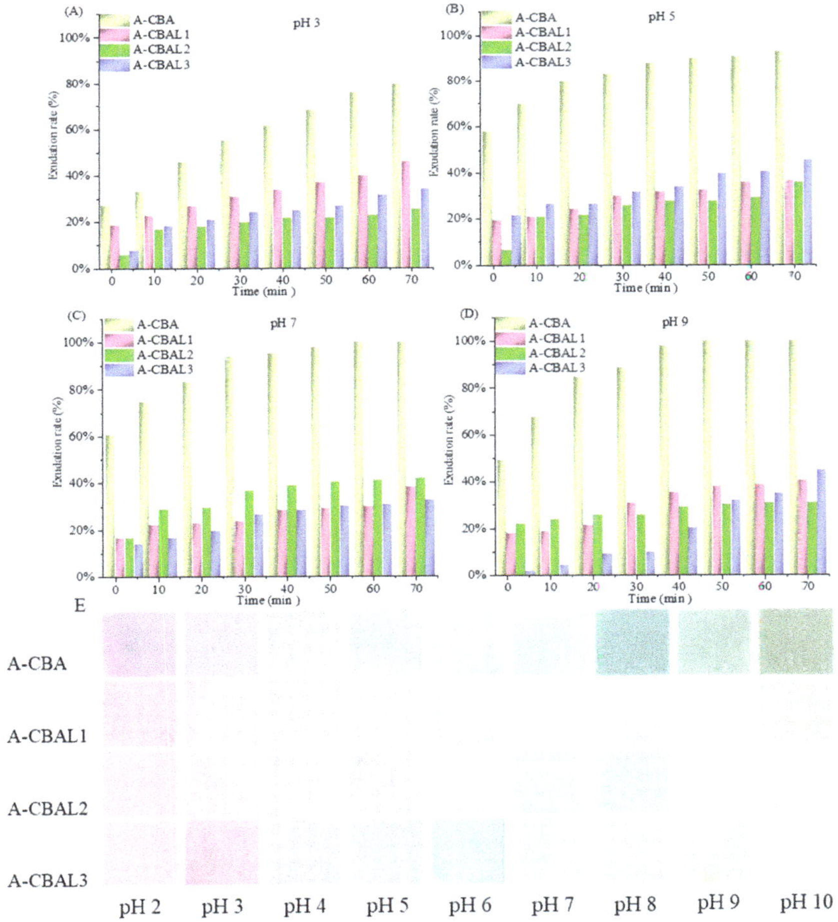

Figure 4. The exudation rates of bi-layer films at pH 3 (**A**), pH 5 (**B**), pH 7 (**C**), and pH 9 (**D**), and the color of the films at pH 2-10 (**E**).

Figure 4E shows the color responses of the indicator films at different pH values. With pH increases, the color of the A-CBAL films changed from pink to purple and then gradually tended toward yellowish green. It can be verified that the color changes of the indicator films were consistent with anthocyanin-loaded liposome solutions, but with different degrees of coloration. However, compared with the A-CBA film, the response chrominance of the A-CBAL films decreased, which corresponded to the color appearance results in the encapsulation of liposomes. The encapsulation hindered the coloration of the butterfly bean flower anthocyanin. With the addition of lecithin, the coloration of the indicator films decreased, but they still presented visible color changes. As a result, in high-humidity food packaging, our bi-layer film with anthocyanin-load liposomes can be used as a pH indicator.

3.6. Color Stability of the Bi-Layer Films

The storage stability of the indicator films was determined under 4 °C and 25 °C, respectively. Generally, when the ΔE value of an indicator is no more than five, it will be difficult to notice with the naked eye [31]. As can be seen from Figure 5A, each of the bi-layer films presented higher stability with a lower ΔE value at 4 °C within 14 days. Thus, the film with free anthocyanins was not stable on the 4th day at 25 °C with an ΔE value of 5.35. The values of the A-CBAL1 and A-CBAL2 films were greater than 5 on the 10th day. At 25 °C, the films were more easily able to form a ring-opened chalcone structure with color changes [32]. The ΔE value of the A-CBAL3 film was 4.48 on the 14th day. This was because more radio lecithin with high encapsulation could protect free anthocyanins from external intrusion.

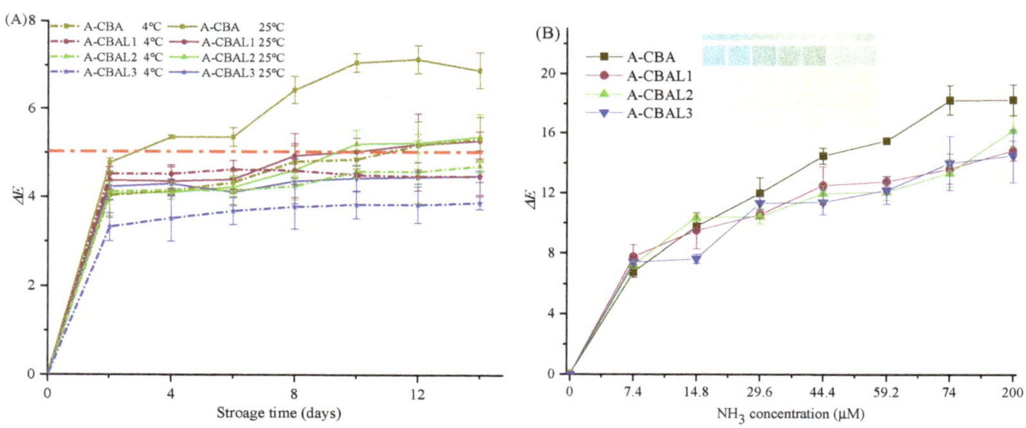

Figure 5. Color changes of the films stored at 4 °C (**A**) and 20 °C (**B**) for 20 d.

3.7. Response Analysis of Indicator Films to Ammonia

The indicator films exposed to ammonia with different ΔE values can be seen in Figure 5B. The colors changed from baby blue to light green and then to yellow green with the increase in ammonia concentration. Moreover, the ΔE values were consistent with the visible colors of the films. All of the indicator films had the same variation trend but with some differences. Compared with A-CBAL films, the film with free anthocyanins presented the highest color changes, with an ΔE value of 18.29. The encapsulation of anthocyanins decreased their ammonia reactive ability. Even so, the A-CBAL2 film also presented visible color changes with ΔE values above 14.28. However, there were slight differences between the films with different liposomes. As a result, a film containing anthocyanin-loaded liposomes also has the potential to serve as a food freshness indicator.

The *Lab* values (A–E), *b* values (F), ΔE (G), and color changes of the films (H) with the concentration of ammonia.

3.8. Application on Monitoring Shrimp Freshness of Bi-Layer Indicator Film

In this study, the A-CBA and A-CBAL2 films were used to monitor shrimp freshness at 4 °C. As shown in Figure 6, the ΔE of the indicator films and the TVB-N of the shrimp exhibited a similar increasing trend during the storage time. The TVB-N increased to 11.20 mg/100 g and the corresponding ΔE_2 was 3.56 for the A-CBAL2 film, with little color change in the first 24 h. Then, the films changed from blue to dark green-yellow with an ΔE_2 value of 6.58, and the ΔE_1 value of the A-CBA film was 7.63 after 48 h. The *TVB-N* was 27.08 mg/100 g at 48 h. The freshness of shrimp was still approved because the legislation limiting level of *TVB-N* is 30 mg/100 g in Seawater shrimp (GB2733-2015). After 60 h, the *TVB-N* increased to 36.93 mg/100 g, which was spoiled, and the ΔE increased dramatically due to the shrimp's deeper putrefaction. The ΔE_1 value was 8.96 for the A-CBA film with a deepened yellow color, and the ΔE_2 value was 7.83 with a light yellowish color. The color of the film with liposomes was lower than that of the film with free anthocyanin, which may be attributed to the encapsulation of the anthocyanin by liposomes, which reduced the color response sensitivity of the anthocyanin. Meanwhile, the correlation analysis between the ΔE of the bi-layer film and the *TVB-N* followed a linear model. For the A-CBA film, the coefficient was 0.8956, and for the A-CBAL2 film, it was 0.9158 (Figure 6B). Therefore, the film with anthocyanin-loaded liposomes can also be used as a good indicator for the detection of the putrefaction period of shrimp.

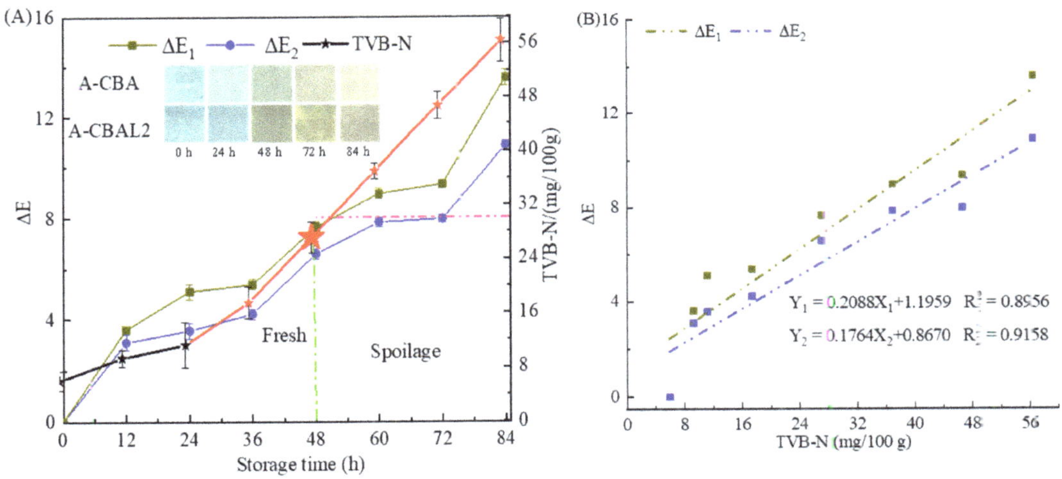

Figure 6. The TVB-N of shrimp and ΔE of the films (**A**). ΔE_1 is A-CBA and ΔE_2 is A-CBAL2 at the storage time. Correlation relation between TVB-N and ΔE (**B**). The illustration is the color change of films during storage.

4. Conclusions

In this study, free anthocyanins and anthocyanin-loaded liposomes were added to carrageenan as the sensor layer of the bi-layer films, respectively, and agar was the outer protective layer. Different ratios of lecithin were used to design the butterfly bean flower anthocyanin extraction into liposomes, and their characterization was investigated. Then the structure, mechanical physical properties (such as TS, EB, and WVP), stability, pH, and ammonia sensitivity of the bi-layer films were individually analyzed with different ratios of lecithin in liposomes. The SEM and FT-IR results indicated that the bi-layer films were satisfactorily prepared via hydrogen bonding interactions. The films with anthocyanin-loaded liposomes had significantly higher TS values and lower EB values than that with free anthocyanins. Importantly, the films with liposomes had a positive effect on the stability of the indicator films in high-humidity environments but slightly decreased the pH and ammonia sensitivity. Finally, the application on the shrimp verified that the bi-layer film can be used as an indicator of meat freshness. However, the encapsulation of anthocyanins by liposomes delayed the sensitivity of the film. Therefore, future exploration could focus on a higher sensitivity method based on liposomes.

Author Contributions: Writing—original draft and methodology: J.Z. (Junjun Zhang) and Y.Y. Formal analysis and software: J.Z. (Jianing Zhang). Validation: X.H. Investigation, visualization, and funding acquisition: X.Z. and J.S. Data Curation: L.L. and W.S. Formal analysis: Z.L. Writing—review and editing draft: M.P. All authors have read and agreed to the published version of the manuscript.

Funding: The authors were funded by the National Key Research and Development Program of China (2022YFD2100603), the Natural Science Foundation of Jiangsu Province (BK20220058), the project supported by the Jiangsu Specially-Appointed Professor Project (202074), the Priority Academic Program Development of Jiangsu Higher Education Institutions (PAPD), the National Natural Science Foundation of China (32272407), and the Natural Science Foundation of Jiangsu Province (BK20200103, BK20220111).

Data Availability Statement: The original contributions presented in the study are included in the article; further inquiries can be directed to the corresponding author.

Acknowledgments: The authors gratefully acknowledge the Project of the Faculty of Agricultural Equipment of Jiangsu University. The authors also appreciate the support provided by Chairman Fen XIAO (alumnus of Jiangsu University) of Shenzhen Fenda Technology Co., Ltd.

Conflicts of Interest: The author Jiyong Shi is employed by the Yixing Institute of Food and Biotechnology Co., Ltd. (researcher). The Yixing Institute of Food and Biotechnology Co., Ltd. is a joint laboratory of universities. All authors declare no conflicts of interest.

References

1. Roy, S.; Rhim, J.-W. Anthocyanin food colorant and its application in pH-responsive color change indicator films. *Crit. Rev. Food Sci. Nutr.* **2021**, *61*, 2297–2325. [CrossRef] [PubMed]
2. Tahir, H.E.; Hashim, S.B.H.; Komla Mahunu, G.; Arslan, M.; Jiyong, S.; Adam Mariod, A.; Zhang, J.; El-Seedi, H.R.; Zhai, X.; Musa, T.H.; et al. Smart films fabricated from natural pigments for measurement of total volatile basic nitrogen (TVB-N) content of meat for freshness evaluation: A systematic review. *Food Chem.* **2022**, *396*, 133674. [CrossRef] [PubMed]
3. Sani, M.A.; Tavassoli, M.; Hamishehkar, H.; McClements, D.J. Carbohydrate-based films containing pH-sensitive red barberry anthocyanins: Application as biodegradable smart food packaging materials. *Carbohydr Polym* **2021**, *255*, 117488. [CrossRef] [PubMed]
4. Zhang, C.; Sun, G.; Cao, L.; Wang, L. Accurately intelligent film made from sodium carboxymethyl starch/κ-carrageenan reinforced by mulberry anthocyanins as an indicator. *Food Hydrocoll.* **2020**, *108*, 106012. [CrossRef]
5. Rosales, T.K.O.; Fabi, J.P. Nanoencapsulated anthocyanin as a functional ingredient: Technological application and future perspectives. *Colloids Surf. B Biointerfaces* **2022**, *218*, 112707. [CrossRef]
6. Sharma, S.; Mulrey, L.; Byrne, M.; Jaiswal, A.K.; Jaiswal, S. Encapsulation of Essential Oils in Nanocarriers for Active Food Packaging. *Foods* **2022**, *11*, 2337. [CrossRef]
7. Bryła, A.; Lewandowicz, G.; Juzwa, W. Encapsulation of elderberry extract into phospholipid nanoparticles. *J. Food Eng.* **2015**, *167*, 189–195. [CrossRef]
8. Chi, J.; Ge, J.; Yue, X.; Liang, J.; Sun, Y.; Gao, X.; Yue, P. Preparation of nanoliposomal carriers to improve the stability of anthocyanins. *LWT* **2019**, *109*, 101–107. [CrossRef]
9. Zhou, X.; Yu, X.; Xie, F.; Fan, Y.; Xu, X.; Qi, J.; Xiong, G.; Gao, X.; Zhang, F. pH-responsive double-layer indicator films based on konjac glucomannan/camellia oil and carrageenan/anthocyanin/curcumin for monitoring meat freshness. *Food Hydrocoll.* **2021**, *118*, 106695. [CrossRef]
10. Ebrahimi, V.; Mohammadi Nafchi, A.; Bolandi, M.; Baghaei, H. Fabrication and characterization of a pH-sensitive indicator film by purple basil leaves extract to monitor the freshness of chicken fillets. *Food Packag. Shelf Life* **2022**, *34*, 100946. [CrossRef]
11. Puscaselu, R.G.; Anchidin-Norocel, L.; Petraru, A.; Ursachi, F. Strategies and Challenges for Successful Implementation of Green Economy Concept: Edible Materials for Meat Products Packaging. *Foods* **2021**, *10*, 3035. [CrossRef] [PubMed]
12. He, F.; Kong, Q.; Jin, Z.; Mou, H. Developing a unidirectionally permeable edible film based on k-carrageenan and gelatin for visually detecting the freshness of grass carp fillets. *Carbohydr. Polym.* **2020**, *241*, 116336. [CrossRef] [PubMed]

13. Liu, L.; Zhang, J.; Shi, J.; Huang, X.; Zou, X.; Zhang, D.; Zhai, X.; Yang, Z.; Li, Z.; Li, Y. Preparation and comparison of two functional nanoparticle-based bilayers reinforced with a κ-carrageenan–anthocyanin complex. *Int. J. Biol. Macromol.* **2020**, *165*, 758–766. [CrossRef] [PubMed]
14. Charcosset, C.; Juban, A.; Valour, J.-P.; Urbaniak, S.; Fessi, H. Preparation of liposomes at large scale using the ethanol injection method: Effect of scale-up and injection devices. *Chem. Eng. Res. Des.* **2015**, *94*, 508–515. [CrossRef]
15. He, B.; Ge, J.; Yue, P.; Yue, X.; Fu, R.; Liang, J.; Gao, X. Loading of anthocyanins on chitosan nanoparticles influences anthocyanin degradation in gastrointestinal fluids and stability in a beverage. *Food Chem.* **2017**, *221*, 1671–1677. [CrossRef]
16. Liu, L.; Zhang, J.; Zou, X.; Arslan, M.; Shi, J.; Zhai, X.; Xiao, J.; Wang, X.; Huang, X.; Li, Z.; et al. A high-stable and sensitive colorimetric nanofiber sensor based on PCL incorporating anthocyanins for shrimp freshness. *Food Chem.* **2022**, *377*, 131909. [CrossRef]
17. Zhang, J.; Huang, X.; Zou, X.; Shi, J.; Zhai, X.; Liu, L.; Li, Z.; Holmes, M.; Gong, Y.; Povey, M. A visual indicator based on curcumin with high stability for monitoring the freshness of freshwater shrimp, Macrobrachium rosenbergii. *J. Food Eng.* **2021**, *292*, 110290. [CrossRef]
18. Yin, Z.; Wu, Y.; Chen, Y.; Qie, X.; Zeng, M.; Wang, Z.; Qin, F.; Chen, J.; He, Z. Analysis of the interaction between cyanidin-3-O-glucoside and casein hydrolysates and its effect on the antioxidant ability of the complexes. *Food Chem.* **2021**, *340*, 127915. [CrossRef]
19. Krivorotova, T.; Cirkovas, A.; Maciulyte, S.; Staneviciene, R.; Budriene, S.; Serviene, E.; Sereikaite, J. Nisin-loaded pectin nanoparticles for food preservation. *Food Hydrocoll.* **2016**, *54*, 49–56. [CrossRef]
20. Hu, Y.; Wu, T.; Wu, C.; Fu, S.; Yuan, C.; Chen, S. Formation and optimization of chitosan-nisin microcapsules and its characterization for antibacterial activity. *Food Control* **2017**, *72*, 43–52. [CrossRef]
21. Ahmad, M.; Mudgil, P.; Gani, A.; Hamed, F.; Masoodi, F.A.; Maqsood, S. Nano-encapsulation of catechin in starch nanoparticles: Characterization, release behavior and bioactivity retention during simulated in-vitro digestion. *Food Chem* **2019**, *270*, 95–104. [CrossRef] [PubMed]
22. Zang, Z.; Chou, S.; Si, X.; Cui, H.; Tan, H.; Ding, Y.; Liu, Z.; Wang, H.; Lang, Y.; Tang, S.; et al. Effect of bovine serum albumin on the stability and antioxidant activity of blueberry anthocyanins during processing and in vitro simulated digestion. *Food Chem* **2022**, *373*, 131496. [CrossRef]
23. Zhai, X.; Zou, X.; Shi, J.; Huang, X.; Sun, Z.; Li, Z.; Sun, Y.; Li, Y.; Wang, X.; Holmes, M.; et al. Amine-responsive bilayer films with improved illumination stability and electrochemical writing property for visual monitoring of meat spoilage. *Sens. Actuators B Chem.* **2020**, *302*, 127130. [CrossRef]
24. Agarwal, S.; Hoque, M.; Bandara, N.; Pal, K.; Sarkar, P. Synthesis and characterization of tamarind kernel powder-based antimicrobial edible films loaded with geraniol. *Food Packag. Shelf Life* **2020**, *26*, 100562. [CrossRef]
25. Li, Z.; Deng, S.; Chen, J. Surface Modification via Dielectric Barrier Discharge Atmospheric Cold Plasma (DBD-ACP): Improved Functional Properties of Soy Protein Film. *Foods* **2022**, *11*, 1196. [CrossRef]
26. Kim, H.-J.; Roy, S.; Rhim, J.-W. Gelatin/agar-based color-indicator film integrated with Clitoria ternatea flower anthocyanin and zinc oxide nanoparticles for monitoring freshness of shrimp. *Food Hydrocoll.* **2022**, *124*, 107294. [CrossRef]
27. Gao, L.; Liu, P.; Liu, L.; Li, S.; Zhao, Y.; Xie, J.; Xu, H. κ-carrageenan-based pH-sensing films incorporated with anthocyanins or/and betacyanins extracted from purple sweet potatoes and peels of dragon fruits. *Process Biochem.* **2022**, *121*, 463–480. [CrossRef]
28. Zhang, Z.; Wells, C.J.R.; Davies, G.-L.; Williams, G.R. The effect of formulation morphology on stimuli-triggered co-delivery of chemotherapeutic and MRI contrast agents. *Int. J. Pharm.* **2021**, *609*, 121155. [CrossRef]
29. Wang, S.; Xia, P.; Wang, S.; Liang, J.; Sun, Y.; Yue, P.; Gao, X. Packaging films formulated with gelatin and anthocyanins nanocomplexes: Physical properties, antioxidant activity and its application for olive oil protection. *Food Hydrocoll.* **2019**, *96*, 617–624. [CrossRef]
30. Kamer, D.D.A.; Kaynarca, G.B.; Yücel, E.; Gümüş, T. Development of gelatin/PVA based colorimetric films with a wide pH sensing range winery solid by-product (Vinasse) for monitor shrimp freshness. *Int. J. Biol. Macromol.* **2022**, *220*, 627–637. [CrossRef]
31. Zhang, J.; Huang, X.; Shi, J.; Liu, L.; Zhang, X.; Zou, X.; Xiao, J.; Zhai, X.; Zhang, D.; Li, Y.; et al. A visual bi-layer indicator based on roselle anthocyanins with high hydrophobic property for monitoring griskin freshness. *Food Chem.* **2021**, *355*, 129573. [CrossRef] [PubMed]
32. Zhang, J.; Huang, X.; Zhang, J.; Liu, L.; Shi, J.; Muhammad, A.; Zhai, X.; Zou, X.; Xiao, J.; Li, Z.; et al. Development of nanofiber indicator with high sensitivity for pork preservation and freshness monitoring. *Food Chem.* **2022**, *381*, 132224. [CrossRef] [PubMed]

foods

MDPI

Article

Detection Method for Tomato Leaf Mildew Based on Hyperspectral Fusion Terahertz Technology

Xiaodong Zhang [1], Yafei Wang [1], Zhankun Zhou [1], Yixue Zhang [2] and Xinzhong Wang [1,*]

[1] College of Agricultural Engineering, Jiangsu University, Zhenjiang 212013, China
[2] Basic Engineering Training Center, Jiangsu University, Zhenjiang 212013, China
* Correspondence: xzwang@ujs.edu.cn; Tel.: +86-138-5298-9966

Abstract: Leaf mildew is a common disease of tomato leaves. Its detection is an important means to reduce yield loss from the disease and improve tomato quality. In this study, a new method was developed for the multi-source detection of tomato leaf mildew by THz hyperspectral imaging through combining internal and external leaf features. First, multi-source information obtained from tomato leaves of different disease grades was extracted by near-infrared hyperspectral imaging and THz time-domain spectroscopy, while the influence of low-frequency noise was removed by the Savitzky Golay (SG) smoothing algorithm. A genetic algorithm (GA) was used to optimize the selection of the characteristic near-infrared hyperspectral band. Principal component analysis (PCA) was employed to optimize the THz characteristic absorption spectra and power spectrum dimensions. Recognition models were developed for different grades of tomato leaf mildew infestation by incorporating near-infrared hyperspectral imaging, THz absorbance, and power spectra using the backpropagation neural network (BPNN), and the models had recognition rates of 95%, 96.67%, and 95%, respectively. Based on the near-infrared hyperspectral features, THz time-domain spectrum features, and classification model, the probability density of the posterior distribution of tomato leaf health parameter variables was recalculated by a Bayesian network model. Finally, a fusion diagnosis and health evaluation model of tomato leaf mildew with hyperspectral fusion THz was established, and the recognition rate of tomato leaf mildew samples reached 97.12%, which improved the recognition accuracy by 0.45% when compared with the single detection method, thereby achieving the accurate detection of facility diseases.

Keywords: tomato; leaf mildew; terahertz time-domain spectroscopy; near infrared hyperspectral technology; multi-source information fusion

Citation: Zhang, X.; Wang, Y.; Zhou, Z.; Zhang, Y.; Wang, X. Detection Method for Tomato Leaf Mildew Based on Hyperspectral Fusion Terahertz Technology. *Foods* **2023**, *12*, 535. https://doi.org/10.3390/foods12030535

Academic Editor: Ana Teresa Sanches-Silva

Received: 1 December 2022
Revised: 14 January 2023
Accepted: 19 January 2023
Published: 25 January 2023

1. Introduction

Crop diseases greatly impact the yield and quality of agricultural products, as they can easily cause stem and leaf death, thereby leading to plant decay [1]. In this way, such diseases affect human food security and food safety. Therefore, research on technologies for crop disease diagnosis is of great significance for the early warning and control of these diseases. The traditional diagnosis method used for crop diseases mainly relies on manual diagnosis, which is based on the experience of the examiner. Although this method is simple and convenient, it consumes a great deal of manpower and allows for a high degree of subjectivity, which can lead to misdiagnosis. Currently, the most objective and accurate disease detection methods available are based on laboratory biochemical tests (e.g., the polymerase chain reaction (PCR), nucleic acid hybridization, and DNA microarray techniques) [2–4]. Although laboratory-based biochemical detection methods feature the advantage of high identification accuracy, their involved sampling and detection steps require professional operation, are associated with high costs, are lengthy to conduct, and are difficult to conduct on a large-scale [5,6]. In recent years, the rapid development of machine vision and spectral imaging technologies has enabled the quick detection of crop diseases. Such technologies include visible/near-infrared imaging, multi-spectral/hyperspectral imaging, and chlorophyll fluorescence imaging, which have all been applied to crop disease detection [7,8]. Although this represents progress, most existing studies only discriminate the grade of crop disease by the reflective properties and apparent outer characteristics of the diseased leaves. Because the internal damage of diseased leaves cannot be detected, it remains difficult to achieve the combined analysis of internal and external damage caused by fungal diseases.

In recent years, hyperspectral technology has attracted increasing research interest in the context of disease detection, owing to its merits of featuring high-resolution and integrated mapping. Spectral imaging technology can obtain the spectral image data cubes of the tested sample, thereby accurately obtaining the image information and spectral reflection intensity distribution characteristics of each test sample in each waveband. Fazari et al. [9] established a three-dimensional CNN model using hyperspectral imaging to classify olive anthrax, which performed with a prediction accuracy of 95.73%. Zhang et al. [10] used visible light imaging on downy mildew in combination with machine learning methods to quickly and accurately estimate the severity of cucumber downy mildew in a greenhouse. Image features that had a high correlation with the actual value of greenhouse cucumber downy mildew severity were then used to construct a shallow machine-learning estimation model. The results showed that there was a good linear relationship between the severity of greenhouse cucumber downy mildew estimated by the model and the actual value. Qin et al. [11] proposed a feature band extraction method combining an improved competitive adaptive reweighting algorithm (CARS) and a successive projections algorithm (SPA) with disease information to establish an early detection model of cucumber downy mildew. With this model, the difficult problem of conducting the early detection of cucumber downy mildew was solved.

Terahertz (THz) radiation refers to long wavelength electromagnetic waves with a frequency range of 0.1–10 THz (corresponding to wavelengths of 30 μm–3 mm). THz waves penetrate deeply into the medium and their high correlation helps to determine the exact refractive index and absorption coefficient of a given sample. THz spectroscopy can be utilized to analyze macromolecules and components inside of crops due to the transmission properties of the radiation, which gives it unique advantages in the application of biological information detection. Some researchers have carried out a preliminary attempt at the THz-based detection of crops and agricultural products [12,13]. Di Girolamo et al. [14] imaged 50 chestnuts that were partially infected with Pygmy fungus in the low THz frequency range by means of a homemade 0–0.1 THz small portable imaging system. By assuming different moisture densities and different physical structures of healthy and unhealthy chestnuts, the relationship between the physical parameters (mass or volume) of chestnuts and the light attenuation of healthy and infected chestnuts was tentatively resolved. The results showed that the index of light attenuation combined with the measurement of chestnut weight or volume could successfully identify whether a given chestnut was healthy or diseased. Li et al. [15] employed a recognition model based on a THz spectroscopy technique to analyze data for apple ring rot and cucumber powdery mildew. The researchers established recognition models for common crop diseases based on K-nearest neighbor, SVM, and BP neural network algorithms, respectively, with a correlation coefficient Rp of 0.9649. Their findings demonstrated that hyperspectral and THz technology could be used to detect crop diseases. However, it remains difficult to obtain the internal and external indicators of crop diseases from either external characterization or by using only a single method, and the prediction accuracy also needs to be further improved.

Tomato leaf mold, also known as black mold and black hair, is a tomato disease caused by *Fulvia fulva* (Cooke) Cif. Tomato leaf mildew mainly affects the leaves of infected plants, and in severe cases, also affects the stems, flowers, and fruits. In the early stages of the disease, yellow-green spots with obscure edges appear on the front of affected leaves, while a grayish-white mildew layer appears on the back of the leaves. When the humidity is high, the leaf surface lesions can also grow a mildew layer. After the conidia of tomato leaf mold invade the tomato leaves, they cause changes in the sugars, lipids, proteins, and nucleic acids inside of the leaves. Existing crop disease detection models employ only a single detection method, and such existing methods are unable to fully reflect the condition of the diseased crops. Therefore, this study acquired the near-infrared hyperspectral data, THz power spectrum, and absorbance time-domain spectral data of tomato leaf mildew samples from different infection grades, and carried out a study on a detection model combining both internal and external features of tomato leaf mildew. Through the spectral analysis of tomato leaves under different characteristic frequency bands, a high-precision prediction model of tomato leaf mildew was established.

2. Materials and Methods

2.1. Experimental Method

The experiment took tomato leaf mildew samples as the research object and collected test samples with different percentages of diseased spot areas. Using a hyperspectral imaging system and THz time-domain spectral measurement system, the near-infrared spectrum, power spectrum, and absorbance time-domain spectral information of samples with different grades of disease were collected. Algorithm optimization was used to remove interference, remove redundancy, and perform feature extraction. Finally, based on the extracted spectral feature data of different grades of disease, single-dimensional and multi-dimensional fusion tomato leaf mildew recognition models were established, respectively. The specific process is shown in Figure 1.

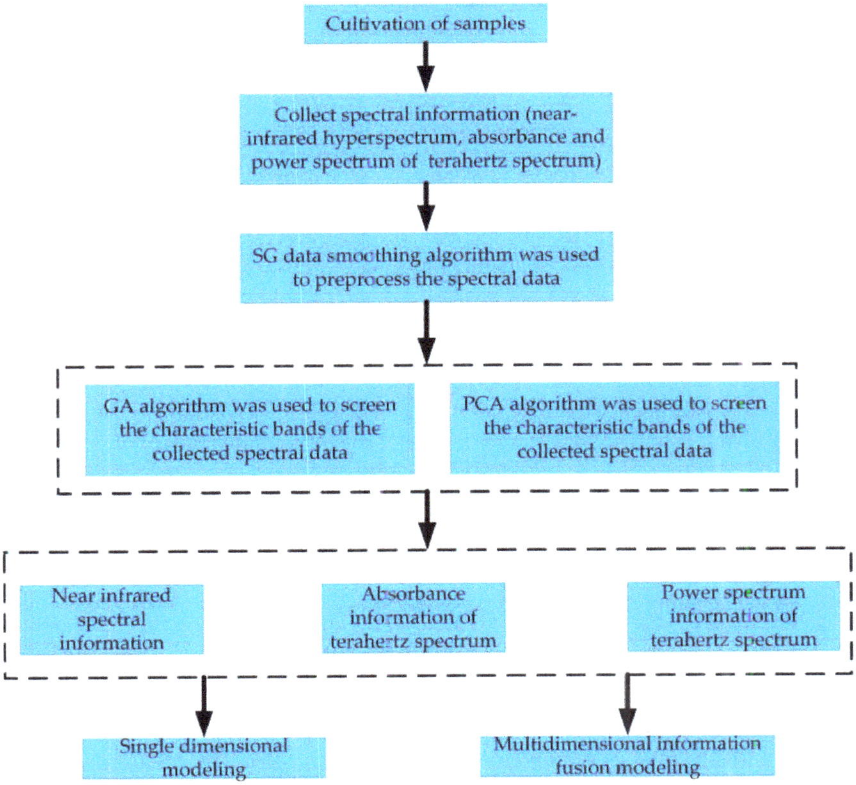

Figure 1. Flow chart of the experiment.

2.2. Cultivation of Samples

Experimental samples of the tomato variety "Cooperative 906" were cultivated in the Venlo greenhouse of Jiangsu University. A 32 × 56 cm rectangular black plastic plate was used for raising the seedlings. Peat, perlite, and vermiculite were mixed to comprise the cultivation substrate, and the seeds were sown in the seedling tray. After budding, the seeds were transplanted into a flowerpot with a diameter of 23.8 cm and a height of 35 cm, and then cultivated by soilless potting using perlite as the substrate nutrient solution.

To eliminate interference, a standard concentration of Yamazaki nutrient solution was used to provide the same nutrients for the samples. High temperature and humidity conditions characterized the greenhouse to allow for the development of tomato leaf mildew. After 15 days of infection with tomato leaf mold, the mold was collected from lesioned areas. After collection, the mold was immediately placed into fresh-keeping sealed bags and placed in a portable refrigerated incubator in order to prevent evaporation and minimize the impact of external conditions. Finally, 240 effective samples were obtained, including those obtained from 40 healthy leaves and those obtained from 200 infected leaves. All tomato leaf samples were divided into four disease grades according to GB/T 17980.26-2000. Pictures of these four tomato leaf mildew grades are shown in Figure 2.

(a) (b) (c) (d)

Figure 2. Tomato leaf mildew grades. (**a**) Level 0, (**b**) level 1, (**c**) level 3, (**d**) level 5.

Statistics of the effective sample sizes are shown in Table 1. We randomly arranged the samples of the four different tomato leaf mildew grades and randomly divided the training set and prediction set in a proportion of 2:1.

Table 1. Statistics of effective sample sizes.

Disease Level	Number	Training Set	Prediction Set
Level 0 (healthy samples)	42	28	14
Level 1 (disease spot area < 5%)	76	51	25
Level 3 (6% < disease spot area < 10%)	65	43	22
Level 5 (11% < disease spot area < 25%)	57	38	19
Total samples	240	160	80

2.3. Equipment Used for Experiments

The HIS-VSNIR scanning hyperspectral measurement system (Shanghai Wuling Opto-electronic Technology Co., Ltd.) was used in the experiment. The system is composed of a near-infrared camera (NIR, 871.6–1766.3 nm), ImpectorN17E spectrometer, OLES30 lens, DC adjustable light source, glass fiber symmetrical line light source, stage, self-propelled displacement stage, stepping motor controller, computer, and display. The structure of this hyperspectral imaging system is illustrated in Figure 3.

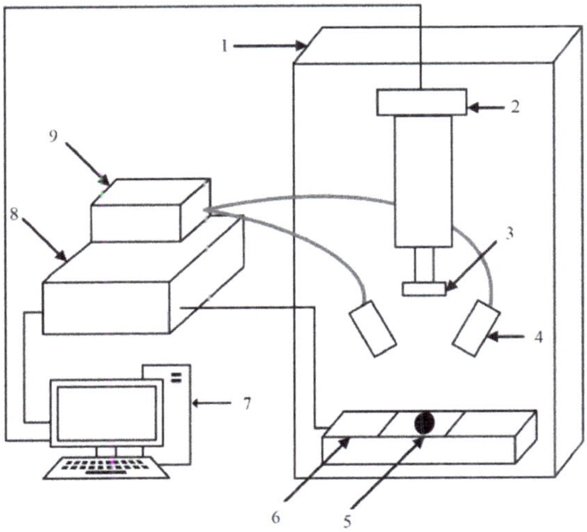

Figure 3. Structure of hyperspectral imaging system. (1) Light box, (2) near-infrared camera, (3) lens, (4) light conduction device, (5) sample, (6) load bearing platform, (7) industrial control machine, (8) displacement control box, (9) light source.

Conducting pre-sampling tests on tomato leaves is required prior to NIR hyperspectral data acquisition. In order to ensure good clarity and no distortion of the imaging data, the initial exposure scanning time of the hyperspectral imaging system was set to 15 ms, the scanning speed was set to 1.32 mm/s, and the maximum peak reflection imaging intensity of the leaf pre-sampled image data was set to 3000. The dark current generated in the measured sample was required to be calibrated in a black-and-white field in advance, and the reflection intensity range was set to 0–4096. In the sample test, the sample was placed on a full black background separately, and the whole image acquisition and test process was completed in a dark room. The original hyperspectral imaging of the sample was corrected in black and white. The correction formula is as follows:

$$R = \frac{R_r - R_d}{R_w - R_d} \quad (1)$$

where R is the corrected sample image; R_r is the original image of the sample; R_d is the dark field fixed image; and R_w is the standard whiteboard calibration image.

In this experiment, the TS7400 THz time-domain spectral measurement system (Advantest Corporation of Japan) was used to collect the THz information of samples, which was specially customized for the detection of agricultural biological information. A structure introduction diagram of the THz time-domain spectrum measurement system is shown in Figure 4.

The measurement range of the TS7400 THz time-domain spectral measurement system was 0.1–4.0 THz and the frequency sampling interval selected for testing was 0.0038 THz, which can be used to detect 225 cm^2 samples. This meets the detection requirements of tomato leaves. In order to improve the accuracy of the acquired data and reduce the effect of moisture on the THz time-domain spectrum, before scanning the tomato samples, the tomato leaves were first freeze-dried using a vacuum freeze-dryer set to −65 °C and then left for 36 h to reduce their moisture content to less than 3%. Additionally, the THz time-domain spectral scanning cabinet was filled with nitrogen to keep the maximum relative humidity below 5%.

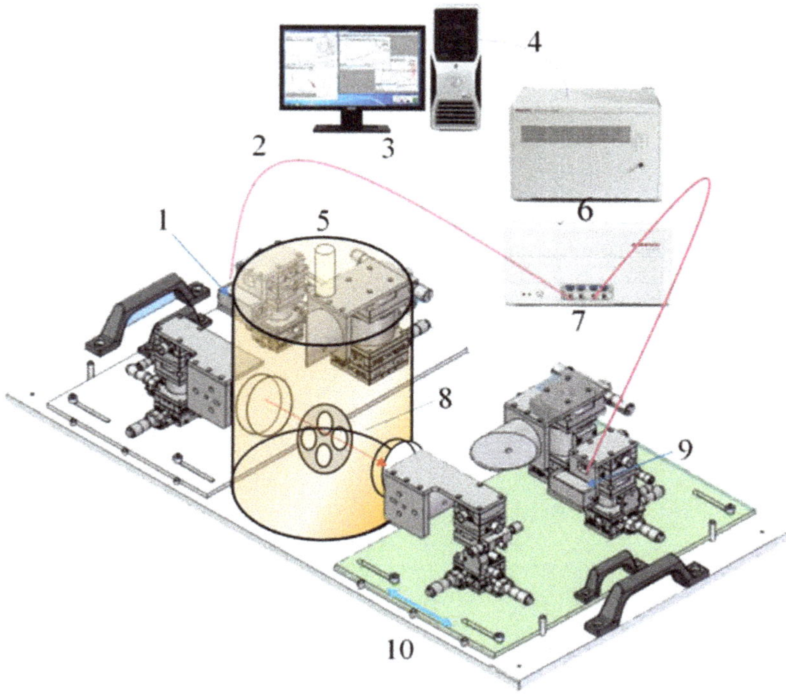

Figure 4. Composition of the measuring system. (1) THz transmitter, (2) optical fiber, (3) operation/analysis computer, (4) ethernet, (5) low-temperature thermostat transmission module, (6) analysis unit, (7) measuring unit, (8) sample stage, (9) THz detector, (10) movable support.

In order to obtain the best response information for tomato leaf mildew samples, this study used the power spectrum and absorbance information for sample analysis. 'Power spectrum' is an abbreviation for the power spectrum density function, which is defined as the signal power within the unit frequency band. It represents the variation of the signal power with frequency, i.e., the distribution of the signal power in the frequency domain. Absorbance is used to express the degree of light absorption by substances. Samples of different grades of tomato leaf mildew have different absorbances.

2.4. Data Processing

2.4.1. Data Smoothing

The SG smoothing algorithm is commonly used in data pre-processing, which features the advantages of being simple, convenient, fast, and efficient [16]. The principle of the algorithm is to first take a window with an odd number of points in width, use the least squares method to fit through the translation of the window, and then replace the original value with the fitting value of the point in the window to achieve the effect of smoothing the data. In this study, the SG smoothing algorithm was used to preprocess the data, and the window width was 7 points/time. This algorithm can be used to effectively reduce interference signals and improve both modeling efficiency and accuracy. After the above preprocessing, the before-and-after data comparison of the spectral data of tomato leaf mildew samples was obtained, as shown in Figure 5.

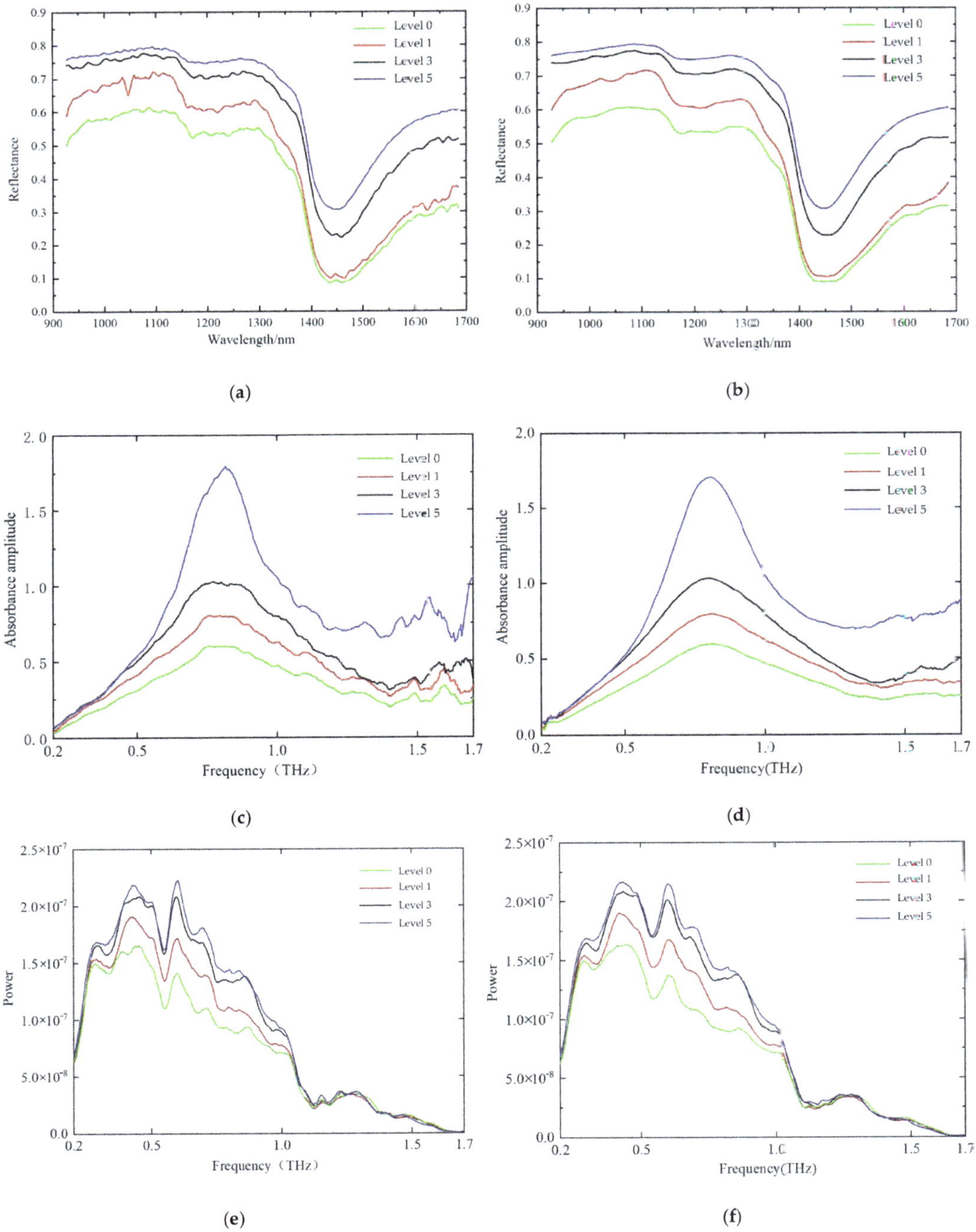

Figure 5. Data of tomato leaf mildew samples before and after SG smoothing preprocessing. (**a**) Near-infrared primary spectrum, (**b**) near-infrared spectra after SG smoothing, (**c**) THz absorbance spectrum, (**d**) THz absorbance spectrum after SG smoothing, (**e**) THz power spectrum, (**f**) THz power spectrum after SG smoothing.

2.4.2. Characteristic Band screening

Because the collected spectral data contains many redundant and collinear information characteristics, this interferes with the extraction of effective spectral information, consequently leading to the effective spectral information extraction model being too complex and hence difficult to calculate. In this paper, a genetic algorithm (GA) and principal component analysis (PCA) were used to select the characteristic wavelength in order to reduce the influence of information redundancy and collinearity, simplify the model, and reduce the amount of calculation. The use of a GA algorithm represents an intelligent optimization method that simulates the evolutionary process that occurs by the natural selection of organisms [17]. When running the GA to screen the near-infrared hyperspectral characteristic bands in the current study, the crossover probability was set to 0.5, the population size was set to 30, and the mutation probability was 0.01. The characteristic wavelength was determined as the wavelength with the highest frequency of 100 GA iterations.

PCA is a multivariate statistical method used for analyzing correlations among multiple variables. The method converts a group of variables that may correlate with a group of linearly unrelated variables through orthogonal transformation [18]. The new variables obtained through PCA can reduce the number of variables while preserving the original feature information as much as possible. Therefore, PCA is a suitable method for the dimension reduction and feature extraction of THz time-domain spectral data.

2.4.3. Establishment of the Model

The backpropagation neural network (BPNN) is a powerful learning system that can realize highly nonlinear mapping between the input and output [19]. The number of units in the input layer of the BPNN model is the number of principal component feature variables, while its output layer is the disease spot area percentage; that is, the grade of tomato leaf mildew in this study. The non-linear Sigmoid type function was selected as the action function of the model, the learning rate was set to 0.6, the number of iterations was set to 300, the target deviation was set to 10–5, and other settings were kept as the default settings of the MATLAB self-contained toolbox. The activation function of the hidden layer was tansig and the activation function of the output layer was purelin.

Bayesian reasoning is a commonly used method of statistical reasoning. The main way to obtain information and evidence is by the updating of probability assumptions by the Bayesian theorem [20]. The steps for the classification and recognition of tomato leaf mildew samples by Bayesian reasoning are as follows.

(1) Calculate the prior probability; that is, the proportion of each level in the tomato leaf mildew sample. The prior probability formula is as shown below:

$$P(Y = c_k) = \frac{\sum_{i=1}^{N}(y_i = c_k)}{N}, k = 1, 2, \cdots, K \tag{2}$$

(2) Calculate the conditional probability; that is, the conditional probability of each attribute in the training data set:

$$P\left(X^{(j)} = a_{jl} \middle| Y = C_k\right) = \frac{\sum_{i=1}^{N} I\left(X_i^{(j)} = a_{jl}, y_i = c_k\right)}{\sum_{i=1}^{N} I(y_i = c_k)}$$
$$j = 1, 2, \cdots, n, l = 1, 2, \cdots, s_j, k = 1, 2, \cdots, K \tag{3}$$

(3) For a given example $x_i = \left(x^{(1)}, x^{(2)}, \cdots, x^{(n)}\right)^T$, *a posteriori* probability is calculated.

(4) Calculate the maximum *a posteriori* probability and determine the class of instance x according to the value of the maximum *a posteriori* probability:

$$y = \arg\max_{c_k} P(Y = c_k) \prod_{j=1}^{n} P\left(X^{(j)} = x^{(j)} \mid Y = c_k\right) \tag{4}$$

There are three types of node variables in the Bayesian network model: hyperspectral characteristic band nodes representing the health status of tomato leaves

$f_a = \{f_{a1}, f_{a2}, \cdots, f_{aN}\}$, THz characteristic band nodes representing the health status of tomato leaves $f_b = \{f_{b1}, f_{b2}, \cdots, f_{bN}\}$, and parameter nodes representing the health status of tomato leaves $Y = \{Y_1, Y_2, \cdots, Y_M\}$. The functional relationship between hyperspectral, TH, and parameter characteristic band nodes representing the health status of tomato leaves is as shown below:

$$Y = F(u, f_a, f_b) \tag{5}$$

After introducing the new node λ, the health status analysis of tomato leaves based on the Bayesian network model is obtained, as shown in Figure 6. Bayesian networks can be introduced by virtue of the prior distribution of health parameters. In the Bayesian network model, λ is the percentage of the diseased spot area; that is, the threshold value, which is set to 0.5.

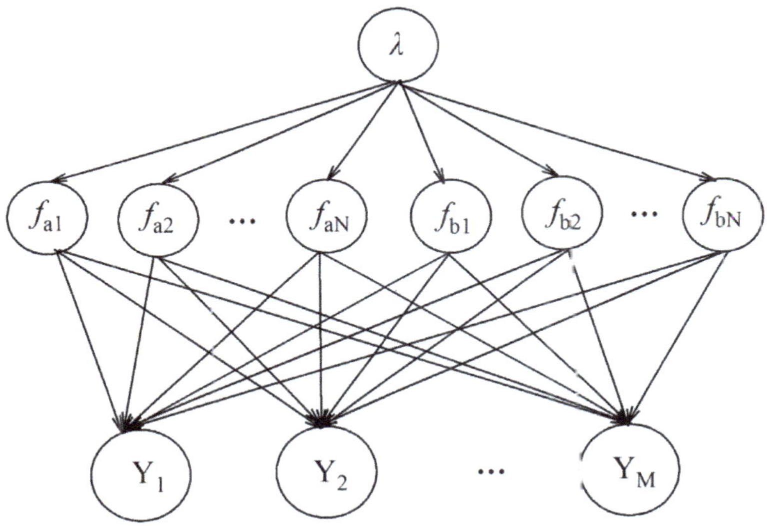

Figure 6. Improved Bayesian network model for the health state analysis of tomato leaves.

3. Results and Discussion

3.1. Screening of Near-Infrared Hyperspectral Characteristic Bands

Figure 7 shows the selected frequency of each variable of the tomato leaf mildew samples. The variables that were selected more than 35 times became the final selected variables, and GA greatly reduced the number of variables from hundreds to only several. The GA operation screened eight near-infrared hyper-spectral characteristic wavebands of tomato leaf mildew samples, which corresponded to 1016 nm, 1019.9 nm, 1157.1 nm, 1160.5 nm, 1163.9 nm, 1338.7 nm, 1553.3 nm, and 1556.7 nm, respectively.

3.2. Terahertz Time-Domain Spectral Data Processing Results

3.2.1. Terahertz Time-Domain Spectral Analysis

The average values of the sample power spectrum and absorbance spectrum can be obtained by THz time-domain spectroscopy. Figure 5e shows the average value curve of the power spectrum of the four tomato leaf mildew grades at 0.1–2.0 THz, with clear absorption peaks observed at approximately 0.43 THz and 1.27 THz, as well as a faint absorption peak at approximately 0.53 THz. Figure 5c shows the mean absorbance curves for the four tomato leaf mildew classes at 0.1–2.0 THz, with a clear absorption peak observed at approximately 0.79 THz. For level 3 mold leaves, a relatively clear absorption peak was observed at approximately 1.89 THz. However, the other three grades of leaf mildew in leaves did not have this absorption peak, indicating that this absorption may be an error caused by the equipment itself, and hence should not be directly judged as the peak of the absorbance sample. The identification of each sample should be achieved by mathematical modeling.

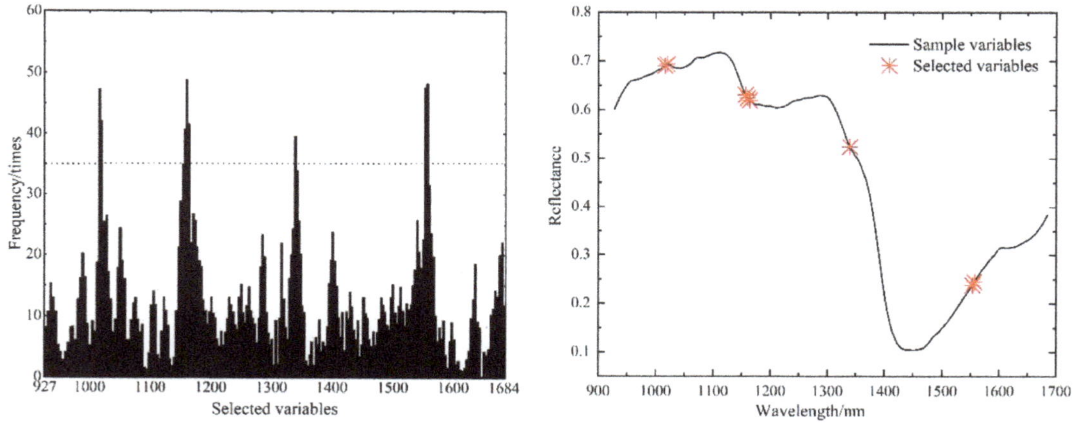

(a) (b)

Figure 7. Running process of the genetic algorithm. (**a**) Selected times of each wavelength point during genetic iteration, (**b**) schematic diagram of characteristic bands screened by the genetic algorithm.

Figure 8 shows the THz frequency domain image at 0.4 THz derived from the data distribution. It can be seen that the difference between the diseased and healthy areas of the leaves is reflected by the color information corresponding to the strength of the frequency domain values, which indicates that the processed THz feature image can reflect the changes in crops from a visual perspective.

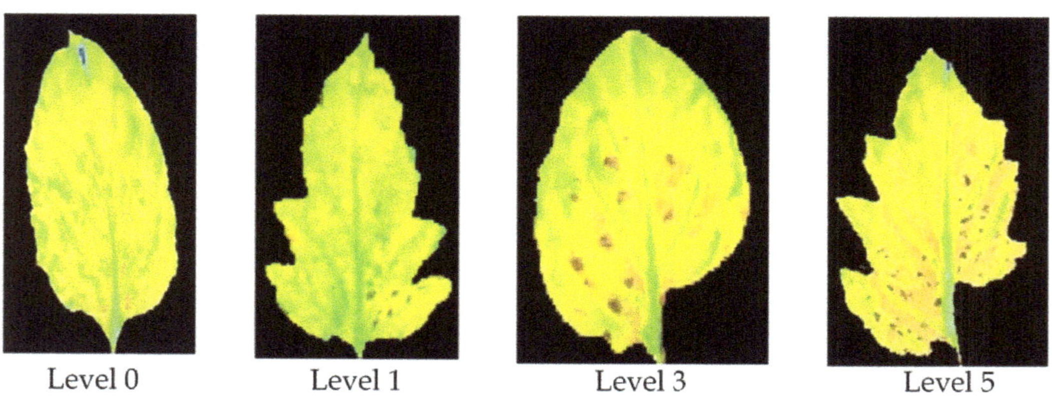

Figure 8. Terahertz images of tomato leaves with different disease grades.

3.2.2. Screening of the Terahertz Time-Domain Spectrum Characteristic Frequency Band

PCA enables the original spectral bands to obtain principal components through linear combination, and also determines the characteristic wavelength according to the absolute value of the loadings of the principal components. The loading refers to the correlation coefficient between the principal component and the original wavelength variable, which is used to reflect the closeness degree between the principal component and each wavelength variable [21]. Loading curves of the first three principal components of tomato leaf mildew samples are shown below in Figure 9. The absolute value of loadings at the peak and trough of each principal component curve was large and the corresponding wavelength was the characteristic wavelength. Therefore, after smoothing the power spectrum, five characteristic wavelengths were obtained: 0.413 THz, 0.752 THz, 1.394 THz, 1.457 THz, and 1.622 THz, respectively. Using the same method, the smoothed absorbance spectrum obtained six characteristic wavelengths: 0.249 THz, 0.567 THz, 0.813 THz, 1.243 THz, 1.771 THz, and 1.892 THz, respectively.

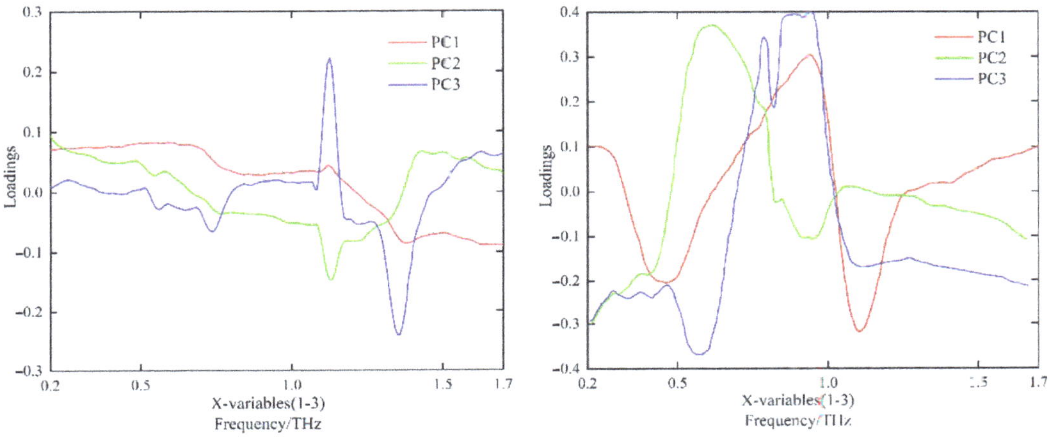

(a) (b)

Figure 9. Load curves of the first three principal components of tomato leaf mildew samples. (**a**) absorbance dimension, (**b**) power dimension.

To further compare the visualized images in different frequency domains, THz frequency domain imaging was performed for five characteristic spectra, as shown in Figure 10. The images of the samples were relatively distinct at the 0.413 THz, 0.752 THz, and 1.394 THz frequencies. At the frequency of 0.413 THz, the image of the sample was the clearest and the recognition effect was the best. However, at the 1.457 THz and 1.622 THz frequencies, the sample images became blurred.

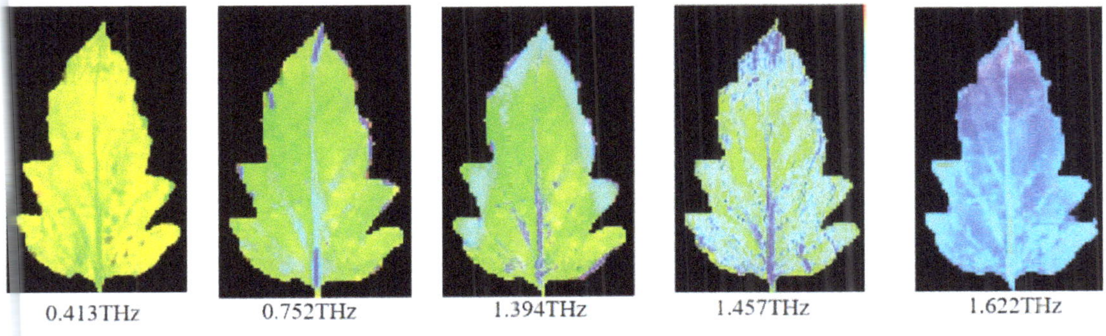

Figure 10. Terahertz time-domain spectral characteristic image.

The PCA method was used to establish the identification model of different tomato leaf mildew grades on the power spectrum dimension and the absorbance dimension of the THz time-domain spectrum. Table 2 shows the PCA results of the spectral data in both dimensions combined with the preprocessing of the SG smoothing algorithm. As shown in

Table 2, the cumulative variance contribution of the first two principal components (PC1 and PC2) to the level variable of tomato leaf mildew was above 85% [22]. Hence, PC1 and PC2 were selected for the analysis.

Table 2. Prediction accuracy under each model.

Principal Component/Cumulative Contribution Rate (%)	PC1	PC2	PC3
absorbance	72.345	92.368	94.522
power spectrum	69.657	89.672	93.914

According to Figure 11, it can be seen that the confidence ellipse of the absorbance data of different grades of tomato leaf mildew exhibited an intertwined state with a discrimination rate of 19.8%. This is because the recognition rate of level 1 grade tomato leaves was 84.9%, while the recognition rates of tomato leaves classed as grades 0, 3, and 5 were lower. The confidence ellipse of the power spectrum data of different grades of tomato leaf mildew also exhibited an intertwined state, with a discrimination rate of 24.7%. The above results show that the recognition rate of tomato leaf mildew using the SG smoothing preprocessing algorithm combined with the PCA model was low, and that the PCA method could not fully mine the spectral information of tomato leaves with different disease grades. Hence, it is necessary that other algorithms are used to build models to improve the prediction accuracy.

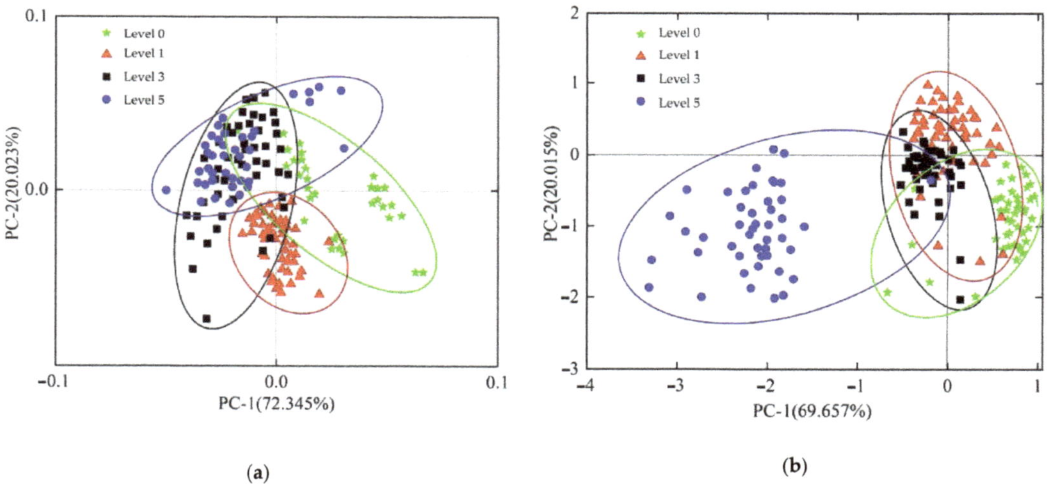

Figure 11. Scatter diagram of tomato leaf mildew sample distribution. (**a**) absorbance scatter, (**b**) power scatter.

3.3. *Single-Model Analysis*

After using the GA and PCA algorithms to reduce the dimension of the data and screen the characteristic variables, a prediction model of tomato leaf mildew disease was developed based on the screened feature variables by the BPNN method. Before the model was established, it was necessary to carry out PCA and extract the sub-vectors of the principal components to form the input of pattern recognition. During the training process of the model, the number of principal component variables affects both the accuracy and stability of the model. Too few principal component factors will lead to excessive loss of information and reduce the accuracy of the model. However, if the number of principal component factors is too great, an excessive amount of redundant information will be introduced, which both influences the robustness of the model and lengthens the data processing time [22]. Therefore, it is important to select the appropriate number of principal component factors for the establishment of the model.

Figure 12 shows the recognition results of the BPNN model training and prediction under different numbers of principal component factors. It can be seen that, initially, with the increasing number of principal component factors, the recognition rates in the training and prediction sets generally exhibited an increasing trend, while after the number of principal component factors reached 7, the recognition rates of the models stabilized, and then even exhibited a moderately decreasing trend.

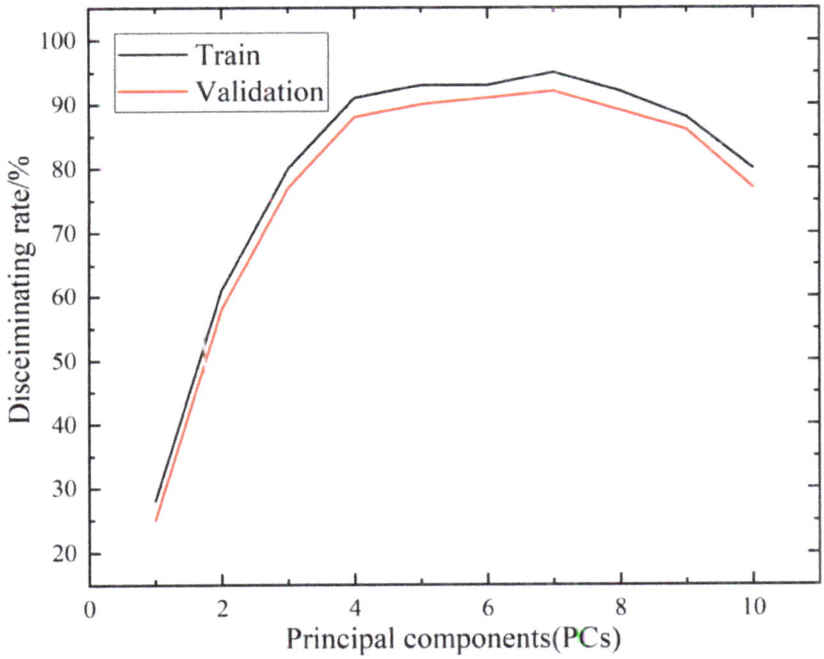

Figure 12. Recognition results of training and prediction under different principal component factors.

Figure 13a shows the BPNN performance graph, which shows that the minimum MSE was 0.6792. Figure 13b shows the BPNN training status graph, which shows that the actual training times were 189. Figure 13c–e shows the BPNN regression analysis graph. When the test set classification index falls within the threshold of the training set classification index, the recognition result is correct. The converse indicates that the classification recognition is incorrect. The precision of the proposed model under the near-infrared hyperspectrum was determined to be R = 0.9367, while under the THz absorbance dimension it was R = 0.9573, and under the THz power spectrum dimension it was R = 0.9431. Based on the actual classification diagram and prediction classification diagram of all the test sets, it was found that the BPNN model was able to identify almost all tomato leaves with leaf mildew.

To evaluate the detection accuracy of the model, this study comprehensively evaluated the recognition results with the recognition accuracy variable P, which is an indicator used to measure the detection signal-to-noise ratio; that is, the percentage of the 'correct' detection results among all detection results. The calculation formula is shown below [23]:

$$P = \frac{T_P}{T_P + F_P} \quad (6)$$

where T_P represents the correctly identified tomato leaf mildew samples, and F_P represents the incorrectly identified tomato leaf mildew samples.

In this study, tomato leaf mildew was divided into four grades, so the prediction accuracy of each level was taken as the evaluation index used for statistics. The results are shown in Table 3.

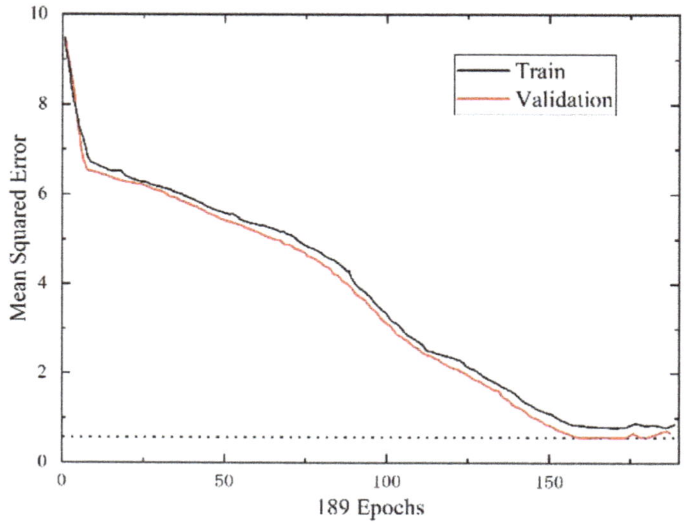

(a)

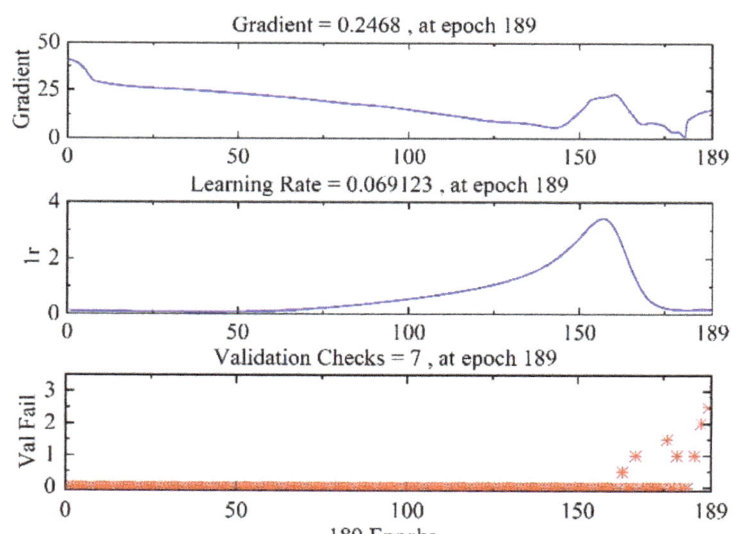

(b)

Figure 13. *Cont.*

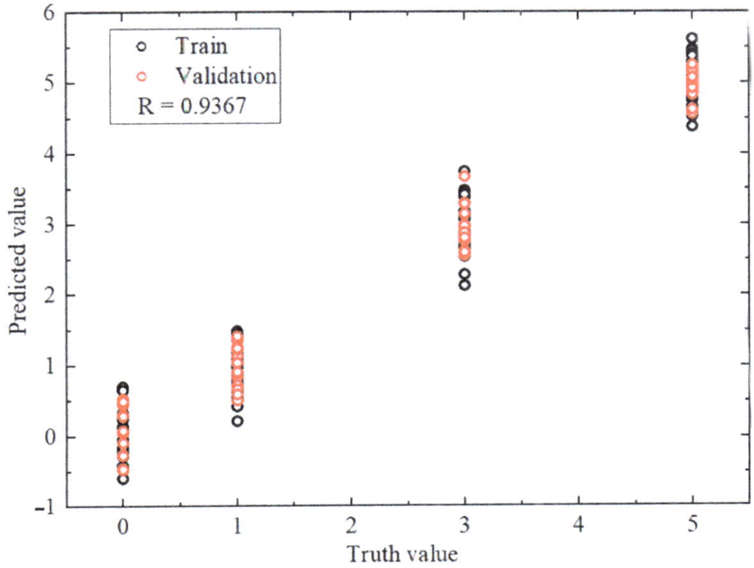

(c)

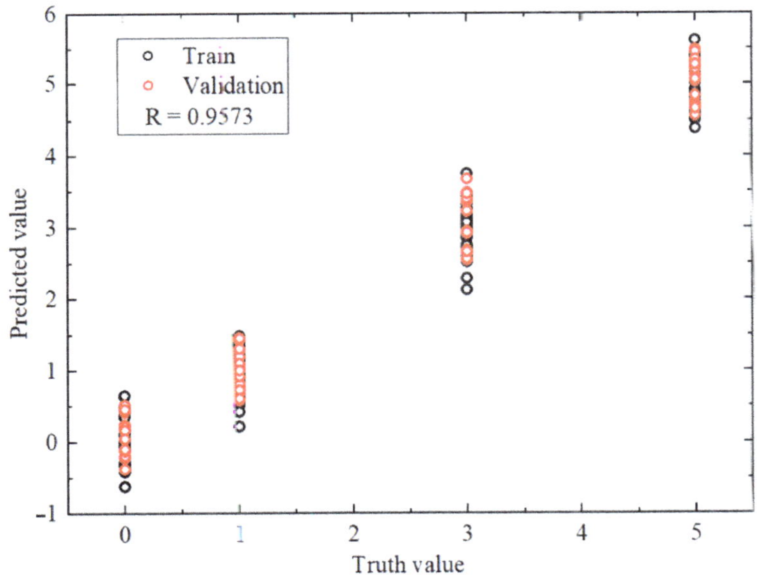

(d)

Figure 13. *Cont.*

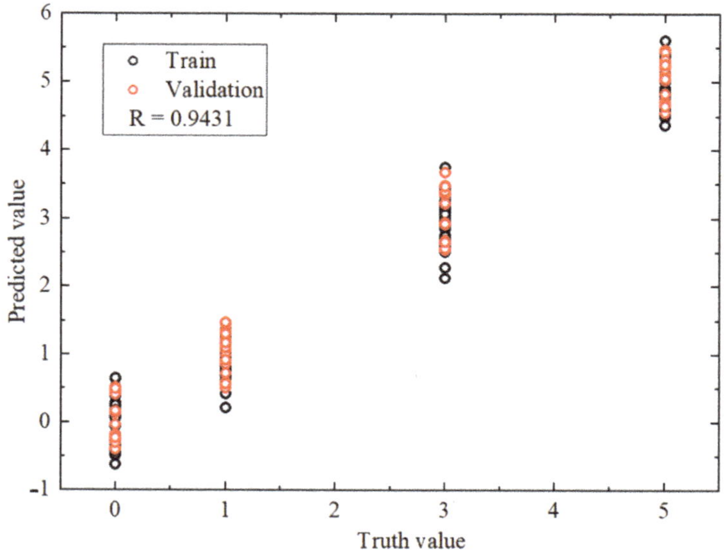

(e)

Figure 13. (**a**) Performance diagram of the backpropagation neural network, (**b**) training status of the backpropagation neural network, (**c**) regression analysis of the backpropagation neural network using the near-infrared hyperspectrum, (**d**) regression analysis of the backpropagation neural network using the THz absorbance, (**e**) regression analysis of the backpropagation neural network using the THz power spectrum.

Table 3. Prediction accuracy of each model.

Dimensions	Model	Number of Characteristic Variables	Prediction Accuracy (%)				
			Level 0	Level 1	Level 3	Level 5	Total
Near-infrared hyperspectrum	GA-BPNN	8	100	96	90.90	94.74	95
THz power spectrum	PCA-BPNN	5	100	96	95.45	94.74	96.67
THz absorbance	PCA-BPNN	6	100	92	95.45	94.74	95

The results show that in the model established by the characteristic variables, the overall detection accuracy of the samples was more than 90%, featuring high accuracy. The highest and lowest detection accuracy rates for the Level 1 samples were 96% and 92%, respectively. The average accuracy rate was 94.67%. Compared to Level 3, the recognition effect in Level 1 was better. Compared to Level 5, the recognition rate was slightly lower. Each model had the highest detection accuracy rate for the Level 0 samples. Hence, the PCA-BPNN model of the power spectrum dimension is the optimal model for comprehensive evaluation. Its prediction accuracy for grades 0, 1, 2, 3, and 4 was 100%, 96%, 95.45%, and 94.74%, respectively, with an overall prediction accuracy of 96.67%.

3.4. Fusion Model Analysis

Figure 14 shows the Gibbs sampling dynamics of the health parameters under the condition of tomato leaves infected with leaf mildew. Figure 14a represents the frequency of tomato leaves infected with leaf mildew, while Figure 14b,c each represent a health parameter map of a hyperspectral THz characteristic band.

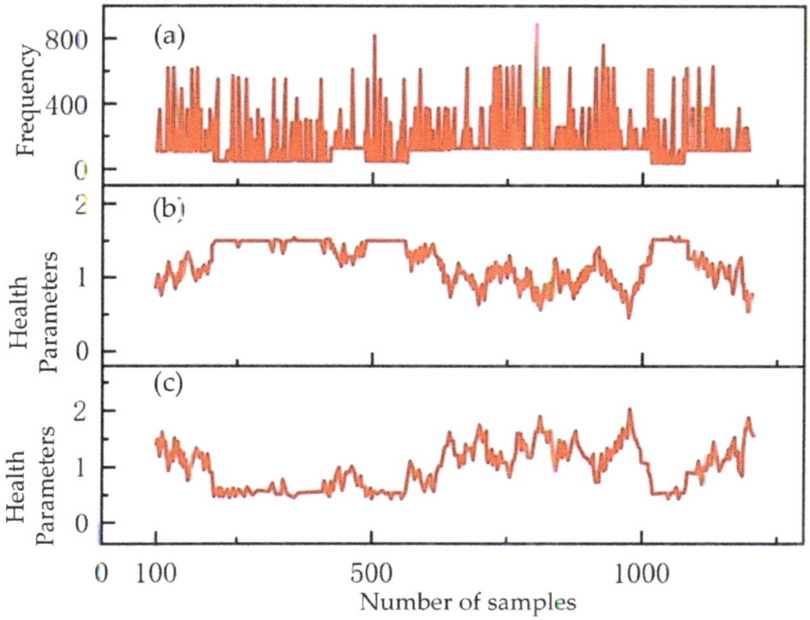

Figure 14. Gibbs sampling diagram. (**a**) represents the frequency of tomato leaves infected with leaf mildew, while (**b**,**c**) each represent a health parameter map of a hyperspectral THz characteristic band.

In Figure 15, a probability density diagram was used to characterize the leaf health parameters of tomato leaf mold. Type I information fusion refers to THz spectral absorbance feature band fusion, while type II information fusion refers to THz spectral power spectrum feature band fusion, and type III information fusion refers to hyperspectral feature band fusion. These three types of information are fused to re-evaluate the health parameter indicators and calculate the recognition rate. After fusing the three types of prior information, it can be seen from the figure that the estimation results were significantly improved after fusing type I information. The posterior distribution of tomato pests and diseased leaves illustrates this point more clearly. The health parameters of tomato leaf mildew posterior samples were also all distributed around 1.75, indicating that the modified Bayesian network model is effective in identifying tomato leaf mildew samples. After the fusion of the prior information, the variables and the actual values increased in agreement, and the final obtained health parameters and posterior distribution of tomato leaves in the state of infection with pests and disease were very close to the actual values.

As shown in Table 4, the overall recognition rate of the improved Bayesian inference for tomato leaf mildew was finally obtained as 97.12%. Therefore, the hyperspectral fusion THz-based technique is feasible for application in tomato leaf mildew recognition.

Table 4. Prediction accuracy of each model.

Number of Characteristic Variables	Prediction Accuracy (%)				
	Level 0	Level 1	Level 2	Level 3	Total
19	99.36	95.57	96.20	97.35	97.12

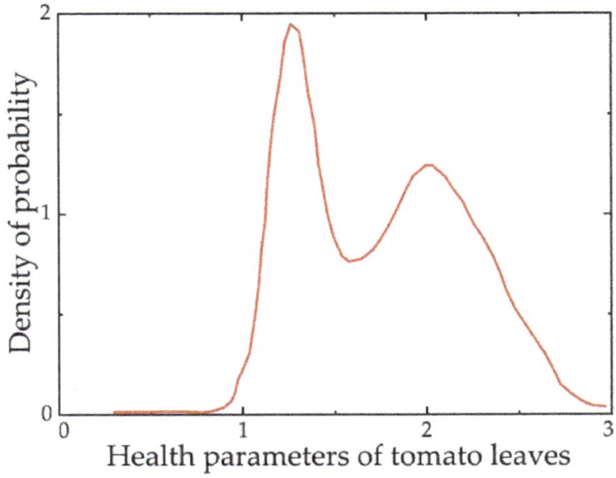

(a)

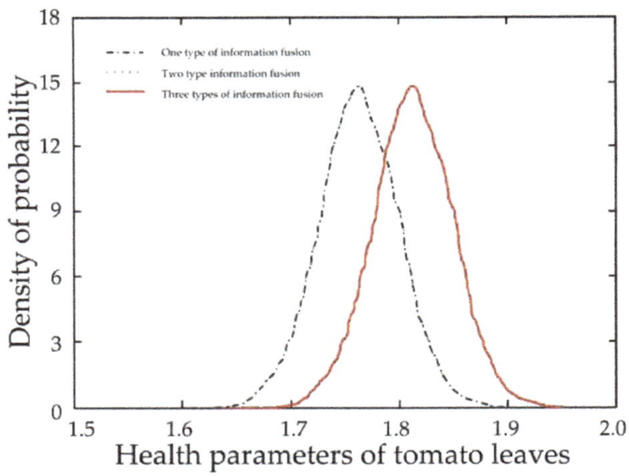

(b)

Figure 15. Schematic diagram of the probability density of health parameters. (**a**) Posterior distribution of changes in health parameters, (**b**) posterior distribution of changes in health parameters after information fusion.

4. Conclusions

In this study, a new method was proposed for the multi-source detection of tomato leaf mildew by THz hyperspectral imaging through the fusion of internal and external features. First, multi-source information obtained from diseased tomato leaves of different grades was extracted by near-infrared hyperspectral imaging and THz time-domain spectroscopy, while the influence of low-frequency noise was removed by the Savitzky Golay (SG) smoothing algorithm. A genetic algorithm (GA) was used to optimize the characteristic near-infrared hyperspectral band. Principal component analysis (PCA) was employed to optimize the THz characteristic absorption spectra and power spectrum dimensions. Based on the near-infrared hyperspectral features, THz time-domain spectrum features, and classification model, the probability density of the posterior distribution of tomato leaf health parameter variables was recalculated by the use of the Bayesian network. Finally, a fusion diagnosis and health evaluation model of tomato leaf mildew using hyperspectral THz was established, and the recognition rate of tomato leaf mildew samples reached 97.12%. This study has therefore successfully developed a method to realize the detection of tomato leaf mildew which can provide a scientific basis for the subsequent monitoring of the disease and provide theoretical support for the development of disease detection instruments.

Author Contributions: Conceptualization, X.Z. and X.W.; methodology, Z.Z., software, Y.W. and X.Z.; validation, Y.W. and Z.Z.; data curation, X.Z.; writing—original draft preparation, Y.W., Z.Z. and Y.Z.; writing—review and editing, Y.W., Y.Z. and X.Z.; project administration, X.Z.; funding acquisition, X.W. All authors have read and agreed to the published version of the manuscript

Funding: This research was funded by Project of Agricultural Equipment Department of Jiangsu University (NZXB20210106),the Key Laboratory of Modern Agricultural Equipment and Technology (Jiangsu University), Ministry of Education (Grant No. MAET202111), the National Key Research and Development Program (2022YFD2002302), the National Natural Science Foundation of China (32071905 and 61771224), the Scientific and Technological Project of Henan Province (Grant No. 212102110029), the Key Laboratory of Modern Agricultural Equipment and Technology (Ministry of Education), and the High-tech Key Laboratory of Agricultural Equipment and Intelligence of Jiangsu Province (Grant No. JNZ201901).

Data Availability Statement: Data is contained within the article.

Conflicts of Interest: The authors declare no conflict of interest.

References

1. Mao, H.P.; Wang, Y.F.; Yang, N.; Liu, Y.; Zhang, X.D. Effects of nutrient solution irrigation quantity and powdery mildew infection on the growth and physiological parameters of greenhouse cucumbers. *Int. J. Agric. Biol. Eng.* **2022** *15*, 68–74. [CrossRef]
2. Bandamaravuri, K.B.; Nayak, A.K.; Bandamaravuri, A.S.; Samad, A. Simultaneous detection of downy mildew and powdery mildew pathogens on Cucumis sativus and other cucurbits using duplex-qPCR and HRM analysis. *AMB Express* **2020**, *10*, 135. [CrossRef] [PubMed]
3. Saberi, Z.; Rezaei, B.; Khayamian, T. A fluorescent aptasensor for analysis of adenosine triphosphate based on aptamer–magnetic nanoparticles and its single-stranded complementary DNA labeled carbon dots. *Luminescence* **2018**, *33*, 640–646. [CrossRef] [PubMed]
4. Chen, J.J.; Li, K.T. Analysis of PCR Kinetics inside a Microfluidic DNA Amplification System. *Micromachines* **2018**, *9*, 48. [CrossRef]
5. Sireesha, Y.; Velazhahan, R. Rapid and specific detection of Peronosclerospora sorghi in maize seeds by conventional and real-time PCR. *Eur. J. Plant Pathol.* **2018**, *150*, 521–526. [CrossRef]
6. Wang, Y.F.; Mao, H.P.; Zhang, X.D.; Liu, Y.; Du, X.X. A Rapid Detection Method for Tomato Gray Mold Spores in Greenhouse Based on Microfluidic Chip Enrichment and Lens-Less Diffraction Image Processing. *Foods* **2022**, *10*, 3011. [CrossRef]
7. Marin-Ortiz, J.C.; Gutierrez-Toro, N.; Botero-Fernandez, V.; Hoyos-Carvajal, L.M. Linking physiological parameters with visible/near-infrared leaf reflectance in the incubation period of vascular wilt disease. *Saudi J. Biol. Sci.* **2022**, *27*, 88–99. [CrossRef]
8. Feng, X.P.; Yu, C.L.; Chen, Y.; Oeng, J.Y.; Ye, L.H.; Shen, T.T.; Wen, H.Y.; He, Y. Non-destructive Determination of Shikimic Acid Concentration in Transgenic Maize Exhibiting Glyphosate Tolerance Using Chlorophyll Fluorescence and Hyperspectral Imaging. *Front. Plant Sci.* **2018**, *9*, 468. [CrossRef] [PubMed]
9. Fazari, A.; Pellicer-Valero, O.J.; Gomez-Sanchis, J.; Bernardi, B.; Cubero, S.; Benalia, S.; Zimbalatti, G.; Blasco, J. Application of deep convolutional neural networks for the detection of anthracnose in olives using VIS/NIR hyperspectral images. *Comput. Electron. Agric.* **2021**, *187*, 106252. [CrossRef]
10. Zhang, L.X.; Tian, X.; Li, Y.X.; Chen, Y.Q.; Chen, Y.Y.; Ma, J.C. Estimation of Disease Severity for Downy Mildew of Greenhouse Cucumber Based on Visible Spectral and Machine Learning. *Spectrosc. Spectr. Anal.* **2020**, *40*, 227–232. [CrossRef]
11. Qin, L.F.; Zhang, X.; Zhang, X.Q. Early Detection of Cucumber Downy Mildew in Greenhouse by Hyperspectral Disease Differential Feature Extraction. *Trans. Chin. Soc. Agric. Mach.* **2022**, *51*, 212–220.
12. Cao, Q.H.; Lin, H.M.; Zhou, W.; Li, Z.X.; Zhang, T.J.; Huang, H.Q.; Li, X.M.; Li, D.H. Water Quality Analysis Based on Terahertz Attenuated Total Reflection Technology. *Spectrosc. Spectr. Anal.* **2022**, *42*, 31–37. [CrossRef]
13. Zhang, X.D.; Wang, P.; Wang, Y.F.; Hu, L.; Luo, X.W.; Mao, H.P. Cucumber powdery mildew detection method based on hyperspectra-terahertz. *Front. Plant Sci.* **2022**, *13*, 1035731. [CrossRef] [PubMed]
14. Di Girolamo, F.V.; Pagano, M.; Tredicucci, A.; Bitossi, M.; Paoletti, R.; Barzanti, G.P.; Benvenuti, C.; Roversi, P.F.; Toncelli, A. Detection of fungal infections in chestnuts: A terahertz imaging-based approach. *Food Control* **2021**, *123*, 107700. [CrossRef]
15. Li, B.; Zhang, D.P.; Shen, Y. Study on Terahertz Spectrum Analysis and Recognition Modeling of Common Agricultural Diseases. *Spectrochim. Acta Part A Mol. Biomol. Spectrosc.* **2022**, *243*, 118820. [CrossRef] [PubMed]

16. Zhang, X.D.; Duan, C.H.; Gao, H.Y.; Shi, Q.; Wang, Y.F.; Shen, B.G.; Zhang, X. Tomato water stress state detection model by using terahertz spectroscopy technology. *Trans. Chin. Soc. Agric. Eng.* **2021**, *37*, 121–128.
17. Jin, C.; Mao, H.P.; Chen, Y.; Shi, Q.; Wang, Q.R.; Ma, G.X.; Liu, Y. Engineering-oriented dynamic optimal control of a greenhouse environment using an improved genetic algorithm with engineering constraint rules. *Comput. Electron. Agric.* **2020**, *177*, 105698. [CrossRef]
18. Zhang, F.; Cui, X.H.; Zhang, Y.K.; Wang, Y.X. Relationship Between Visible/Near Infrared Spectral Data and Fertilization Information at Different Positions of Hatching Eggs. *Spectrosc. Spectr. Anal.* **2021**, *41*, 3064–3068. [CrossRef]
19. Wang, Y.F.; Zhang, X.D.; Ma, G.X.; Du, X.X.; Shaheen, N.; Mao, H.P. Recognition of weeds at asparagus fields using multi-feature fusion and backpropagation neural network. *Int. J. Agric. Biol. Eng.* **2021**, *14*, 190–198. [CrossRef]
20. Wang, Z.F.; He, X.Q.; Shen, H.L.; Fan, S.J.; Zeng, Y.L. Multi-source information fusion to identify water supply pipe leakage based on SVM and VMD. *Inf. Process. Manag.* **2022**, *59*, 102819. [CrossRef]
21. Li, L.; Zhu, F.; Sun, H.; Hu, Y.Y.; Yang, Y.Y.; Jin, D.W. Multi-source information fusion and deep-learning-based characteristics measurement for exploring the effects of peer engagement on stock price synchronicity. *Inf. Fusion* **2021**, *69*, 1–21. [CrossRef]
22. Chen, Y.Y.; Cheng, Q.Q.; Fang, X.M.; Yu, H.H.; Li, D.L. Principal component analysis and long short-term memory neural network for predicting dissolved oxygen in water for aquaculture. *Trans. Chin. Soc. Agric. Eng.* **2018**, *34*, 183–191. [CrossRef]
23. Wang, Y.F.; Mao, H.P.; Xu, G.L.; Zhang, X.D.; Zhang, Y.K. A Rapid Detection Method for Fungal Spores from Greenhouse Crops Based on CMOS Image Sensors and Diffraction Fingerprint Feature Processing. *J. Fungi* **2022**, *8*, 374. [CrossRef] [PubMed]

foods

MDPI

Article

Identification of Defective Maize Seeds Using Hyperspectral Imaging Combined with Deep Learning

Peng Xu [1], Wenbin Sun [1], Kang Xu [1], Yunpeng Zhang [2], Qian Tan [2], Yiren Qing [2] and Ranbing Yang [2,*]

[1] College of Information and Communication Engineering, Hainan University, Haikou 570228, China
[2] College of Mechanical and Electrical Engineering, Hainan University, Haikou 570228, China
* Correspondence: yangranbing@hainanu.edu.cn; Tel.: +86-0898-66267576

Abstract: Seed quality affects crop yield and the quality of agricultural products, and traditional identification methods are time-consuming, complex, and irreversibly destructive. This study aims to establish a fast, non-destructive, and effective approach for defect detection in maize seeds based on hyperspectral imaging (HSI) technology combined with deep learning. Raw spectra collected from maize seeds (200 each healthy and worm-eaten) were pre-processed using detrending (DE) and multiple scattering correction (MSC) to highlight the spectral differences between samples. A convolutional neural network architecture (CNN-FES) based on a feature selection mechanism was proposed according to the importance of wavelength in the target classification task. The results show that the subset of 24 feature wavelengths selected by the proposed CNN-FES can capture important feature information in the spectral data more effectively than the conventional successive projections algorithm (SPA) and competitive adaptive reweighted sampling (CARS) algorithms. In addition, a convolutional neural network architecture (CNN-ATM) based on an attentional classification mechanism was designed for one-dimensional spectral data classification and compared with three commonly used machine learning methods, linear discriminant analysis (LDA), random forest (RF), and support vector machine (SVM). The results show that the classification performance of the designed CNN-ATM on the full wavelength does not differ much from the above three methods, and the classification accuracy is above 90% on both the training and test sets. Meanwhile, the accuracy, sensitivity, and specificity of CNN-ATM based on feature wavelength modeling can reach up to 97.50%, 98.28%, and 96.77% at the highest, respectively. The study shows that hyperspectral imaging-based defect detection of maize seed is feasible and effective, and the proposed method has great potential for the processing and analysis of complex hyperspectral data.

Keywords: hyperspectral imaging; maize seeds; defect detection; feature selection; convolutional neural network

Citation: Xu, P.; Sun, W.; Xu, K.; Zhang, Y.; Tan, Q.; Qing, Y.; Yang, R. Identification of Defective Maize Seeds Using Hyperspectral Imaging Combined with Deep Learning. *Foods* **2023**, *12*, 144. https://doi.org/10.3390/foods12010144

Academic Editors: Kamruzzaman Mohammed

Received: 14 November 2022
Revised: 22 December 2022
Accepted: 23 December 2022
Published: 27 December 2022

1. Introduction

Maize (*Zea mays* L.) is one of the three major food crops in China, and it is also an essential feed and industrial raw material [1]. The quality of maize seed directly affects maize yield, food security, and the agricultural economy, making the issue of seed quality particularly important. In addition, with the development of modern agricultural machinery technology, especially the application of precision sowing machinery, the quality of seeds has been increasingly demanding [2]. Maize seeds are prone to damage, defects, and mildew during storage and transportation, where defects are significant indicators for seed quality evaluation [3]. Maize seeds of the same variety have many similar characteristics (such as color and shape) that make it difficult for human vision to distinguish them. Traditional detection methods (such as morphology, molecular biology, and genetic markers) are time-consuming, inefficient, and professionally demanding [4]. It can also cause irreversible damage to the sample, which is not conducive to a non-destructive and rapid evaluation of maize seed quality in industrial production [5]. In recent years, many scholars have applied machine vision, deep learning, and spectral sensing technology to seed quality detection, and it has achieved some research achievements [6]. Conventional machine vision techniques can detect severe external defects in seeds. However, when pest symptoms are not obvious, internal information about the sample cannot be obtained for identification, resulting in poorer detection accuracy. Hyperspectral imaging (HSI) is an effective non-destructive inspection technique that combines traditional spectral information (reflecting chemical composition) with image information (reflecting physical properties) [7]. It can also be used to solve the problem that the defective areas of the seeds are obscured during the detection process. The subdivided wavelength data in hyperspectral has a unique role in reflecting subtle changes in the intrinsic physiological properties of seeds, such as moisture, starch, protein, and fat [8]. Therefore, the most sensitive hyperspectral characteristic parameters of different quality seeds can be obtained by analyzing the spectral response of seed wavelengths.

Sellami proposed a new hyperspectral image classification method based on a multi-view deep neural network that fuses spectral and spatial features using only a small number of labeled samples [9]. Alimohammadi successfully identified and classified three varieties of maize seeds using hyperspectral nondestructive imaging in the wavelength range of 400–1000 nm [10]. Zhang developed a new method for identifying different varieties of commercial maize seeds using visible and hyperspectral imaging techniques in the near-infrared band to develop a new method for identifying different varieties of commercial maize seeds [11]. Liu proposed a method for identifying the purity of hyperspectral rice seeds based on the LASSO logistic regression model by combining the advantages of the sparsity property of the least absolute shrinkage selection operator (LASSO) algorithm and the classification property of the logistic regression model (LRM) [12]. Due to the complexity of hyperspectral data, not all spectral variables are related to the target components, and the full spectrum inevitably contains a large amount of noise, invalid information, and even interfering variables [13]. This situation increases the complexity of the model but also seriously affects its accuracy and reliability. Therefore, feature selection has become a critical step in the hyperspectral modeling process, which can quickly obtain the optimal subset of feature wavelengths [12].

Feature selection has been a hot research topic in recent years in hyperspectral analysis techniques and chemometrics, where the more common method is the traditional spectral feature selection algorithm [14]. Nagasubramanian used a genetic algorithm (GA) as an optimizer for optimal band selection, and the accuracy of identifying charcoal rot infection in soybean with the selected combination of six bands reached 97% [15]. Pang used successive projections algorithm (SPA) to select 25 feature variables, which could achieve 99.77% accuracy in predicting Quercus variabilis seed vigor [16]. Song used competitive adaptive reweighted sampling (CARS) to screen 60 sensitive wavelengths, in which the root mean square error (RMSE) and calibration set accuracy of the diagnostic model were 1.97 and 0.87, respectively, which could accurately predict chlorophyll content in maize canopy and provide a reference for rational use of fertilizer [17]. Zhang showed that the CARS algorithm (29 wavelengths) outperformed the SPA algorithm (24 wavelengths) in feature variable selection and achieved an accuracy of 89% in the validation set based on the deep forest (DF) model, which can effectively identify the sound or slightly sprouted wheat kernels (slightly sprouted wheat kernels) [18]. However, the above traditional feature selection methods utilize only spectral data without considering the labels in the dataset, which may lead to a low relevance of the selected feature subset to the final target. With the rapid development of deep learning techniques, convolutional neural network (CNN)-based methods have been successfully applied to hyperspectral band selection [19]. Yuan proposed a point-centered convolutional neural network incorporating embedded feature selection for feature selection, extraction, and classification, and the accuracy of the selected five critical bands reached 97.98% for non-destructive and rapid identification of moldy peanuts [20]. Sharma proposed a CNN named DeepFeature applied to non-image data for feature selection with 98% classification accuracy on an independent test set, which can provide a powerful method for identifying biologically relevant gene sets [21].

With the development and research of artificial intelligence, the application of deep learning combined with HSI technology to agricultural tasks has gradually become a research hotspot where more and more researchers are using deep learning models represented by CNN to solve the problem of seed quality detection [6,22]. Yu achieved the recognition of hybrid okra seeds using HSI (948.17–1649.20 nm) technology, and its recognition rate reached 97.68%, demonstrating that the CNN model has reliable advantages in achieving high accuracy and stability [23]. Pang showed that building a CNN model based on spectral images (497.72–998.16 nm) has excellent results for identifying the seed viability of Sophora japonica, which is significant for the study of seed vigor and spectral change mechanisms [24]. Gao used a one-dimensional convolutional neural network (1D-CNN) to classify aflatoxins in maize and peanuts based on hyperspectral data (292–865 nm), and the classification accuracy was 96.35% and 92.11%, respectively, which remarkably improved the detection efficiency [25]. Li proposed a method based on a deep convolutional generative adversarial network (DCGAN) and near-infrared hyperspectral imaging technique (866.4–1701.0 nm) for identifying unsound wheat seeds, and the accuracy could reach up to 96.67% [26]. The above shows that the joint model based on DCGAN and CNN has good performance for identifying small samples and can learn more feature information from spectral data. In order to further optimize the deep learning model, some scholars have introduced the attention mechanism to focus on critical information selectively [27,28]. Of these, Wang developed an attention-based CNN approach (Geo-CBAM-CNN) for crop classification using time-series Sentinel-2 images. The results show that the convolutional block attention module (CBAM) can help mitigate the effect of geographical heterogeneity and suppress unnecessary information [28]. The proposed model achieves an accuracy of 97.82%, demonstrating its superior performance in large-scale applications. In summary, CNN is an accurate identification model, and combining it with HSI can improve the overall model performance by using the internal and external feature information of samples. Therefore, it is important to establish an identification model for maize seed defects using CNN combined with HSI and to explore the effects of different feature selection methods and classification models on the identification results.

This study investigates the feasibility of HSI combined with CNN for the identification of maize seed defects. The objectives of this study are as follows: (1) to collect image information of defective maize seeds, extract spectral data of the region of interest, and preprocess them to analyze the differences between defective and healthy maize seeds; (2) to propose a CNN model (CNN-FES) based on feature selection mechanism for selecting key feature wavelength variables that are beneficial to the target task, and to evaluate the impact of the variables selected by different feature selection methods on the model; (3) to design a CNN model (CNN-ATM) based on attentional classification mechanism in order to discriminate the samples, and the best model for the identification of maize seed defects was selected by comparing the classification results of different machine learning methods (LDA, RF, and SVM) and CNN-ATM.

2. Materials and Methods

2.1. Sample Preparation and Hyperspectral Image Acquisition

Maize seeds of "Zhengdan 958" collected in 2022 from the local farmers' market in Haikou City (20.06°N, 110.33°E), Hainan Province, were used as material for this study; this variety is widely grown in China. All samples were identified and selected by agricultural cultivation experts, and they were analyzed in three aspects: morphology, structure (with endosperm), and end-use type (cultivation). From the samples, 200 worm-eaten seeds (Figure 1A–C) and another 200 healthy seeds (Figure 1D) were chosen as the control group, numbered in order, and placed in sealed bags. In order to eliminate the interference of moisture during spectral data collection, all samples were stored at room temperature (26 °C) for 12 h. These maize seeds did not perform other physical or chemical treatments before acquiring spectral information, and the shapes and colors between them were roughly the same. In this study, the division of the training set and test set was carried out randomly in the ratio of 7:3, with 280 samples in the training set and 120 samples in the test set.

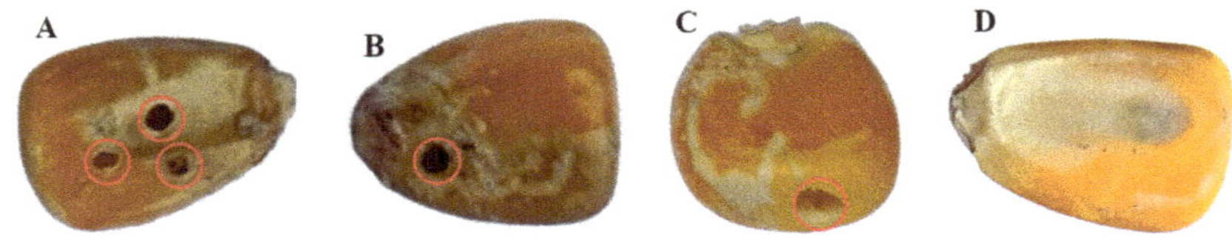

Figure 1. Images of maize seed samples: (**A**) embryonic surface worm-eaten; (**B**) non-embryonic surface worm-eaten; (**C**) lateral surface worm-eaten; (**D**) healthy.

As shown in Figure 2, the hyperspectral image acquisition system is the "GaiaSorter" hyperspectral sorting equipment manufactured by Zolix Co., Ltd. (Beijing, China). The whole system consists of five parts: a uniform light source (2900-ER+9596-E, Illumination, East Syracuse, NY, USA), a "Image − λ" series of spectral cameras (Imspector N17E; TE-cooled InGaAs CCD, Spectral Imaging Ltd., Oulu, Finland), an electronically controlled mobile platform (MTS 120, Beijing optical instrument factory, Beijing, China), a computer (OptiPlex 7090, Dell, Round Rock, TX, USA), and control software (SpecView Ltd., Uckfield, UK). The system has a spectral acquisition range of 866.4–1701 nm and contains 254 bands, and a three-dimensional data cube containing image information and spectral information of all maize seed samples can be obtained by one scan at a time. In addition, the system includes a dark box to isolate the interference of external light or noise and an object stage to place the seed samples.

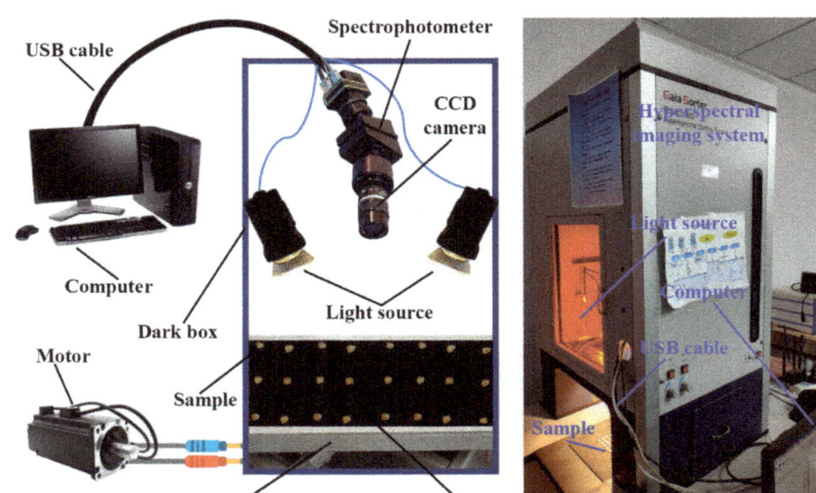

Figure 2. Hyperspectral image acquisition system.

The equipment was allowed to warm up for 0.5 h before capturing the spectral images, and then the camera exposure time was tuned and determined to be 30 ms. The distance between the camera and the sample was adjusted to 36 cm, the image resolution at full frame was 1392 × 1040, and the movement speed of the mobile platform was 2 mm/s. The maize seeds were placed single on the blackboard of the object stage in 7 rows and 6 columns. The computer-controlled mobile platform was moved from left to right, and the camera gathered spectral images of the samples, which were then transferred to the computer for storage. Due to the inhomogeneous intensity distribution of the light source and the presence of dark current noise, black and white correction of the original hyperspectral images is required to improve the signal-to-noise ratio of the acquired images [1]. First, we placed a whiteboard for white reference to obtain a corrected image with a reflectivity of about 100%. Then, we placed a blackboard for black reference to obtain a corrected image with a reflectivity of 0. Finally, we calculated the relative reflectivity of the calibrated image according to Formula (1) [29].

$$R_e = \frac{I_r - I_d}{I_w - I_d} \tag{1}$$

where R_e is the relative reflectivity of the calibrated image, I_r is the spectral reflectivity of the original image, I_w is the spectral reflectivity of the corrected whiteboard, and I_d is the spectral reflectivity of the corrected blackboard.

2.2. Spectral Data Extraction and Preprocessing

In order to obtain the representative spectral information of the samples, the data processing of the original hyperspectral images is required, and the implementation process is shown in Figure 3. First, the background and noise of the original hyperspectral image were removed using Matlab R2020a (MathWorks, Natick, Massachusetts, MA, USA) combined with the morphological filtering method, which was converted to a grayscale image by threshold segmentation, and the single seed region was retained as the region of interest (ROI) of the image. Next, the grayscale image was converted into a mask image based on the difference between the spectral reflectance values of the sample and the background, and the hyperspectral image of the ROI was separated from the background by multiplying the original hyperspectral image with the mask pixel points to obtain the spectral reflectance image of each seed. Last, the average value of all pixel points (per band) within the ROI was extracted as the average spectrum of the sample using ENVI 5.3.

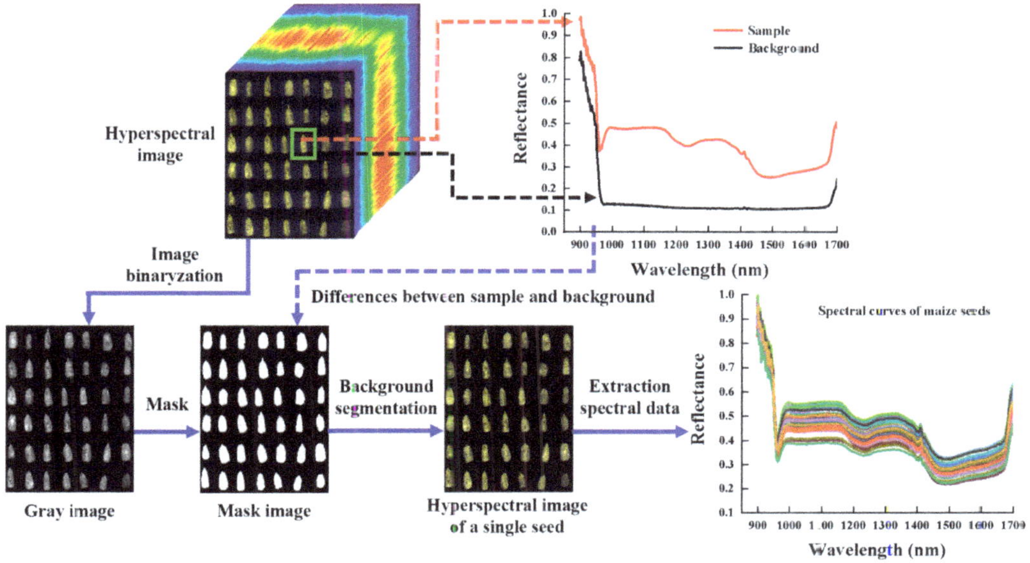

Figure 3. The process of hyperspectral image segmentation and spectral extraction.

The experiments are disturbed by external factors such as the measurement state of maize seeds, the working state of the equipment, and the environment, which lead to random noise and baseline drift in the obtained spectral data, which can have a significant impact on the results of spectral data analysis. It is necessary to pre-process the spectral data to highlight the valid information of the spectral to reduce the influence of these conditions on the robustness of the model and improve its prediction accuracy. This study uses Matlab R2020a to denoise the spectral data, and after the Savitzky-Golay (SG) processing [30], the DE and MSC methods were used to pre-process the data and compare these two methods, respectively [31]. Among them, the SG can improve the smoothness of the spectra and reduce noise interference, and a smoothing window of 11 and a polynomial number of 1 are selected to process it in this paper. DE can improve the baseline offset phenomenon and focus the analysis on the fluctuation of the data trend itself. MSC can improve the signal-to-noise ratio of the spectral, effectively eliminate the effect of scattering, and enhance information on the spectral absorption related to the component content.

2.3. Traditional Feature Selection and Machine Learning Methods

The wavelength range of the original spectra was 866.4–1701.0 nm (254 bands). In order to remove noise interference caused by factors such as equipment and environment, the wavelength range of 897.4–1701 nm (245 bands) was selected for analysis in this study. In practical applications, the amount of spectral data is usually enormous. Using spectral data at the full wavelength will increase the complexity of model calculations, and its modeling effects will be affected by redundant information and collinearity problems. Therefore, to further investigate the feasibility of applying hyperspectral techniques to identify defects in maize seeds, first, conventional SPA and CARS were used for feature selection to reduce the amount of modeling data and ensure its identification on the model, both of which were implemented using Matlab R2020a. Second, traditional machine learning algorithms, including LDA, RF, and SVM, were used to establish a hyperspectral classification model to identify healthy and defective maize seeds. These three methods are all implemented using Python's sklearn toolkit (https://scikit-learn.org/stable/ (accessed on 13 November 2022).

2.3.1. Feature Selection Method

SPA is a method that minimizes the overlap of spectral information, using vector projection to select effective feature wavelengths that have low redundancy yet reflect the critical information of the sample spectrum [32]. The algorithm can eliminate collinearity between the combinations of feature variables through recursive computation to select new wavelengths, thus improving the conditions of classification and regression tasks [33]. It can also remove spectral regions with high noise and irrelevant information to achieve screening of sensitive variables and improve crop identification [34]. In addition, CARS is an alternative approach to feature selection that allows assessing the importance of each variable based on the absolute value of the regression coefficients of the partial least squares model [35]. Monte Carlo resampling was performed iteratively and competitively to make the distribution of the selected band positions more uniform [36]. Different subsets were evaluated using cross-validation, and the selected best variables combination is more suitable for subsequent modeling [37].

2.3.2. Machine Learning Method

LDA is a powerful supervised learning technique that can significantly increase the discrimination ability between classes based on the distance between projections and effectively classify data [38]. LDA pays more attention to the inter-class distance and intra-class distance of the projected samples in the new dimension space, ensuring that the model has the best separability in the subspace [39]. The method used in this study to solve the LDA hyperplane eigenmatrix is singular value decomposition (SVD), and the threshold used for rank estimation in the SVD solver is 1×10^{-4}.

SVM is designed based on statistical principles and follows the structural risk minimization principle to obtain stable classification results by maximizing the decision boundary [40]. The radial basis function (RBF) is used to transform nonlinear problems into linear ones [41]. The optimal values of its penalty coefficient C and kernel function gamma are determined by the grid search method, and their search range is set to 10^{-5}–10^{5}. This study uses a 10-fold cross-validation strategy to set C and gamma to be 10 and 0.004, respectively.

RF is an integrated learning method that can be used for classification, combining the results of decision tree-based modeling and obtaining the final estimation results by voting [42]. The grid search method seeks the appropriate number of decision trees (n_estimators), the number of randomly selected variables at nodes (random_state), and the maximum number of features (max_features) to correct the overfitting due to the inductive preference of decision trees [41]. This study uses the Gini coefficient as a measure to determine n_estimators, random_state, and max_features to be 15, 2, and 6, respectively.

2.4. *Convolutional Neural Network Architecture for Feature Selection and Classification*

CNN is a deep learning method with structural diversity and nonlinear transformation that has achieved significant achievements in many fields such as image processing, speech recognition, and text data [20,24]. In recent years, CNN has been intensively researched and explored in spectral analysis and extended to applications for one-dimensional (1D) data (such as pixel-level spectra) and three-dimensional (3D) data [43]. This study proposed a novel 1D-CNN model for hyperspectral analysis, in which the feature shape of the spectral data of the sample is 1×245. The model consists mainly of two parts. One is a CNN architecture based on a feature selection function for optimal wavelength selection; the other is a CNN architecture based on an attention mechanism for identifying defective seeds. We define them as CNN-FES and CNN-ATM. These two custom modules were annotated through purple dashed boxes, and the processing process of the whole network is shown in Figure 4. First, the input data of shape $N \times L$ (N is the number of samples, and L is the number of features) was processed by the feature selection module, which evaluates the weight of the coefficients in the network iteration through the loss function, according to the output importance score (IS) selects the input. Then, the selected combination of feature variables ($N \times L_1$) was input to the convolutional classification module for processing, and healthy and defective seeds were identified based on category output.

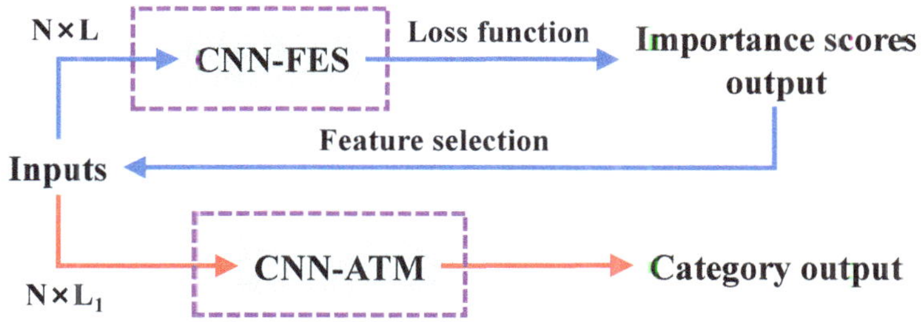

Figure 4. CNN network processing flow.

2.4.1. CNN Architecture Based on the Feature Selection Mechanism

For the CNN-FES module, the architecture is shown in Figure 5. The feature selection mechanism allows the network to select certain vital variables of the input data while ignoring the selection of unimportant variables. Referring to the Vaswani algorithm, we calculate the value of the loss function in the CNN network to indicate the importance of each feature in the target classification task and select the feature wavelengths based on IS. The feature weight (FW) block was annotated with a red dotted box, and the weight of FW is output in the same shape as the input.

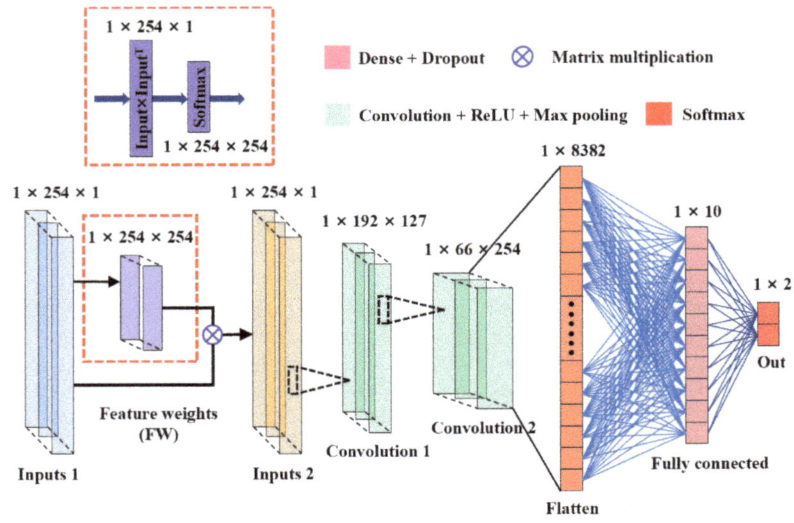

Figure 5. Architecture of CNN-FES module.

Figure 5 shows the processing of the CNN-FES module. First, the input data was processed by the nonlinear activation function Softmax after the FW operation, and then the output weight coefficients were multiplied by the input. The weighted vector of the FW block is defined as the output and calculated as shown in Equation (2) for the subsequent convolution operation and the calculation of IS.

$$Y_{FW} = f_S(W_{FW}) * X \tag{2}$$

where $X = [X_1, X_2, \ldots, X_L]$, $W_{FW} = X * X^T$, X^T denotes the transpose matrix of X and $*$ denotes the multiplication of the corresponding elements in the two matrices. W_{FW} is the inner product of the vector, Y_{FW} is the weighted vector of the FW block, f_S is the activation function, and $f_S(W_{FW})$ is the weight coefficient. The geometric meaning of W_{FW} is the angle between two vectors, which is the projection of one vector onto another vector. The larger the value of W_{FW}, the higher the correlation between the two vectors, thus giving them more attention in the feature extraction of convolution operation, and the IS will be higher. In addition, alternatives to f_S are Sigmoid, Tanh, and ReLU. Sigmoid makes the output not centered on 0, which will reduce the efficiency of weight update. Tanh will make part of the input negative, which is not conducive to model training. ReLU sets part of the output to 0 and deletes related feature information. The denominator of Softmax combines all the factors of the original output value, making the outputs associated with each other. Therefore, Softmax was selected in this study to output the weight coefficients.

Next, the Y_{FW} output from the FW block was reconstructed into N × L × 1 for convolutional operation, and two 1D convolutions (Convolution 1/Convolution 2) blocks (number of kernels, kernel size, and strides were set to 127/245, 64/32, and 1/1, respectively), one flatten layer, and one dense layer (number of neurons is 10) were processed, and the last is the output layer (Softmax). Of them, the convolutional layers use the ReLU activation function with L2 regularization added. The MaxPooling layer (pool size and strides of 2, and 1, respectively) was downsampled after each convolutional layer, and the dropout layer (rate is 0.25) was performed after the dense layer.

In the feature selection process, features unrelated to labels or random noise will be related through model training, resulting in the neglect of a lot of potentially valuable information. Referring to the idea of the Altmann algorithm [44], "permutation importance" was used for processing in this study. We disordered the labels five times, then obtained the feature importance (IF) under the false labels, selected the features based on the difference between the IF under the true and false labels, and defined the IS calculation as shown in Formula (3).

$$IS = log[(1e - 10 + S1_k)/(S2_k)] \quad (3)$$

where k denotes the kth characteristic variable (k = 1, 2, ..., 245), $S1_k$ denotes the IF of variable k before no shuffle, and $S2_k$ denotes the 75% quantile of the IF of all variables k after multiple shuffles. IF is the loss value calculated by the model at each iteration based on the loss function ($Loss_{yp}$); $1e - 10$ and 1 are smoothing methods used to avoid numerator and denominator zeros during computation, and they have no practical significance. If the value of $IS < 0$, indicates that the feature variable k is not a vital feature of the classification task.

Finally, the Adam optimization algorithm was used to train the feature selection model to minimize the classification error while retaining a minimum $f_S(W_{FW})$ value, and the defined loss function is $Loss_{yp}$ shown in Formula (4).

$$Loss_{yp} = -\frac{1}{N}\sum_{i=1}^{N}\sum_{j=1}^{M} y_{ij}log(p_{ij}) + \lambda\sum_{j=1}^{M}\left(softmax\left(w_j^2\right)\right) \quad (4)$$

N indicates the number of samples, M indicates the number of categories, y_{ij} indicates the label value, p_{ij} indicates the predicted value, w_j indicates the weight vector, and λ indicates the regularization coefficient. $Loss_{yp}$ consists of two parts: one is the cross-entropy loss, which controls the classification accuracy of the target task; the other is the sum of the weight coefficients in the FW block, which makes the absolute value of the weights tend to decrease overall and the IS of the features can take full advantage of all the inputs from the upper layer.

2.4.2. CNN Architecture Based on Attention Classification Mechanism

For the CNN-ATM module, the architecture is shown in Figure 6. The attention classification mechanism makes the network focus more on the information in the input data that is more critical to the current task and reduces the attention to other non-critical information. By referring to the SeNet algorithm [45], the importance degree (ID) of the feature channel was obtained through network learning, and then enhances the beneficial features and suppresses the features that are useless for the current according to the ID, and uses it to achieve the identification of defective maize seeds. The attention score (AS) block was annotated with a blue dashed box, and the weights of AS are output in the same shape as the input.

Figure 6 illustrates the processing of the CNN-ATM module. First, the input data was reshaped after global average pooling, and the weights of each feature channel were obtained using the nonlinear activation function Sigmoid after the two fully connected, and the weights of the output were reshaped again and multiplied with the input. The weighted vector of AS block was defined as the output, and the computation was performed according to the following Formula (5) for subsequent convolution operation and ID calculation.

$$Y_{AS} = f_S(W_2 * f_R(W_{1Z})) * X \quad (5)$$

where, $X = [X_1, X_2, \ldots, X_L]$, $*$ represents the multiplication of corresponding elements in two matrices, the dimension of W_{1Z} is $1 \times 1 \times L/r$, r is the scaling parameter, which was taken as 8 in this study, the dimension of W_2 is $L \times L/r$, Y_{AS} is the weighting vector of the AS block, f_S and f_R are the activation functions, and $f_S(W_2 * f_R(W_{1Z}))$ was defined as the weight coefficient.

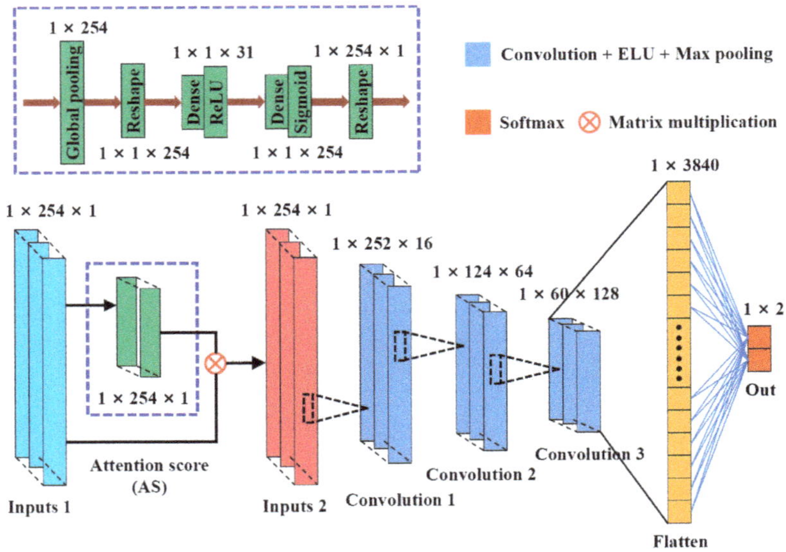

Figure 6. Architecture of CNN-ATM module.

Second, the output (Y_{AS}) from the AS block continued to complete the convolution operation and performed three 1D convolutions (Convolution 1/Convolution 2/Convolution 3) blocks (the number of kernels, kernel size, and strides were set to 16/64/128, 3/3/3, and 1/1/1, respectively), and one flatten layer, and the last part is output layer (Softmax). Where the ELU activation function was used for the convolutional layers, and L2 regularization was added. The downsampling operation of the MaxPooling layer (pool size and strides of 2, and 1, respectively) was performed after each convolutional layer. Finally, the Adam optimization algorithm was used to train the classification model to obtain the best detection effect of defective seeds.

2.5. Model Training Process and Evaluation Metric

In this study, the training set (280) was used for model training and the test set (120) was used for the performance evaluation of the model. The sample features were normalized (z-score) to compensate for scale differences between the data before they were fed into these classification models. In addition, the obtained spectral data were statistically analyzed using IBM SPSS 26 using paired sample t-tests to assess statistical differences in the spectral characteristics of the samples. The mean, standard deviation, and standard error mean of the samples were calculated by mean comparison analysis to analyze if significant differences were presented between healthy and defective maize seeds. The remaining parameters of the models for LDA, SVM, and RF were set as default, and then their classification results were compared with the CNN models. We optimize the hyperparameters of both CNN-FES and CNN-ATM models by analyzing the accuracy of the training and test sets on epoch to obtain the output value of the minimized loss function and the model with good robustness. The number of epochs on CNN-FES was set at 100, the batch size was 4, λ was 0.15, and the weight decay value was 0.1. The number of epochs on CNN-ATM was 200, the batch size was 32, and the weight decay value was 0.008. The initial value of the learning rate for both was 0.001, the momentum value of the gradient descent optimizer was set to 0.9, and the learning rate decays automatically according to the number of iterations. The above CNN model was built and implemented using the programming language Python 3.8 (https://www.python.org/ (accessed on 13 November 2022)) in the machine learning platform TensorFlow 2.3 (https://devdocs.io/tensorflow~2.3/ (accessed on 13 November 2022).) and the deep learning framework Keras 2.4.3 (https://keras.io/ (accessed on 13 November 2022)), using an AMD Ryzen 7 3800 × 8-core processor, 3.90 GHz 16 GB, and NVIDIA GeForce RT × 3050 graphics processing unit for training and testing.

For the evaluation strategy, this study uses the standard metrics of the classification task to calculate the accuracy (A_C), sensitivity (S_E), and specificity (S_P) of the model based on the confusion matrix using Formulas (6)–(8), respectively, to measure the ability of the model to discriminate the samples.

$$A_C = \frac{TP + TN}{TP + FP + TN + FN} \tag{6}$$

$$S_E = \frac{TP}{TP + FN} \tag{7}$$

$$S_P = \frac{TN}{TN + FP} \tag{8}$$

In this case, TP, TN, FP, and FN represent true positive, true negative, false positive, and false negative of the confusion matrix, respectively.

3. Results and Discussion

3.1. Spectral Data Analysis

This study analyzed the differences in spectral data using paired-sample t-tests. Among them, the mean spectral value of healthy maize seeds (0.448) was significantly higher than that of defective ones (0.426), and they showed a significant difference ($t = 21.259$, $p = 0.001 < 0.05$). The large difference in the spectral characteristics data between the two implies that the presence of defects in maize seeds leads to a change in their internal structure. This change affects the magnitude of its spectral reflectance and can better reflect the actual state (healthy or defective) of the maize seeds. The average spectral information of the maize seed samples in the wavelength range of 897.4–1701 nm is shown in Figure 7. It can be seen that the general trend and characteristic peaks of the spectral curves of healthy and worm-eaten seeds are the same, with the spectral curves showing a significant decrease in the wavelength range of 897.4–959.3 nm and a leveling off in the wavelength range of 993.4–1666.4 nm. Although some regions overlap, the reflectance of different wavelengths is slightly different, and the spectral data can be further studied and analyzed. The spectral reflectance of the defective seeds at 952.4–1484.5 nm was lower than that of healthy maize seeds due to changes in their protein, starch, and moisture composition. Among them, the average spectral curves show absorption peaks or valleys at 959.3 nm, 996.9 nm, 1159.1 nm, 1239.1 nm, and 1331.5 nm, which are more affected by the damage to the structure and tissues inside the maize seeds. Traditional visual detection methods are based on surface defects of seeds in visible light, but most of the spectral signals are invisible to the human eye, which can provide a great deal of information about the internal defects of seeds.

The troughs around 980 nm were caused by the joint action of stretching vibrations of N-H in protein and C-O and C-N in soluble sugar, and the absorption peak near 1000 nm is related to the third overtone of N-H stretching in proteins [46]. There are two noticeable troughs around 1235 and 1500 nm, where the former represents the absorption wavelength of the second overtone of C-H stretching in carbon hydrate, and the latter represents the absorption wavelength of the first overtone of O-H stretching in water and N-H stretching in protein, respectively [23]. The absorption peak near 1350 nm was correlated with the second overtone of C-H stretching in starch, and it is also the absorption region of C-O stretching vibration, and the valley of absorption around 1660 nm is related to the fat content [47]. These characteristic peaks reflect the degree of absorption of different wavelengths of near-infrared light by the high molecular compound in maize seeds. In addition, worm-eaten seeds lost some of their original higher structures, resulting in broken chemical bonds and disruption of the molecular structure of substances such as proteins, starch, and water, which led to differences in reflectance on the spectral curves [2]. These differences demonstrate the distinction between healthy and worm-eaten maize seeds for one thing and the feasibility of HSI for defect detection of maize seeds for another.

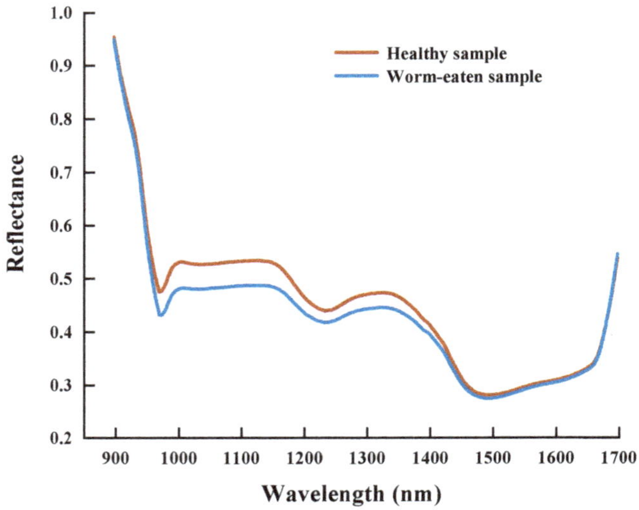

Figure 7. The average spectral curve of maize seeds.

Figure 8B shows the spectral curves of the samples after SG pretreatment, from which it can be shown that the SG method retains almost all the information of the original spectral curves, and the modeling results after the preprocessing may not differ much from the modeling results using the original spectra directly. Figure 8C,D show the spectral curves of the samples pretreated by detrending (DE) and multiple scattering correction (MSC), respectively, from which it can be seen that the DE method changes the overall trend of the original spectral, while the MSC method retains the overall trend of the original spectral. However, it can be found that both of them enhanced the characteristics of the spectral absorption curves and decreased the discreteness of the curve after pretreatment, and the use of these two preprocessing methods can effectively reduce noise interference in the spectral data.

3.2. Results of Feature Wavelength Selection

Set the selection range of the characteristic wavelength of SPA to 1–50, and the result is shown in Figure 9, taking the spectral data processed by MSC as an example. In this study, the RMSE value was used as the main evaluation index to screen the best combination of variables, and a total of 24 characteristic variables were obtained at the minimum RMSE value of 0.14194 (the coefficient of determination R^2 of the model is 0.9864), accounting for 9.80% of the total wavelengths. From Figure 9A, it can be seen that the RMSE values showed an overall decreasing trend when the number of variables was less than 24, and the changing trends tended to be gentle when the number of variables was greater than 24. Figure 9B indicates the selection of specific variables, and the red "□" represents the selected variables with the smallest RMSE value corresponding to the optimal number of characteristic wavelengths. The above indicates that the selected characteristic wavelengths contained information about worm-eaten maize seeds and had higher discrimination than healthy ones. Therefore, the reflectance values corresponding to the 28 wavelengths were selected as the data for subsequent modeling.

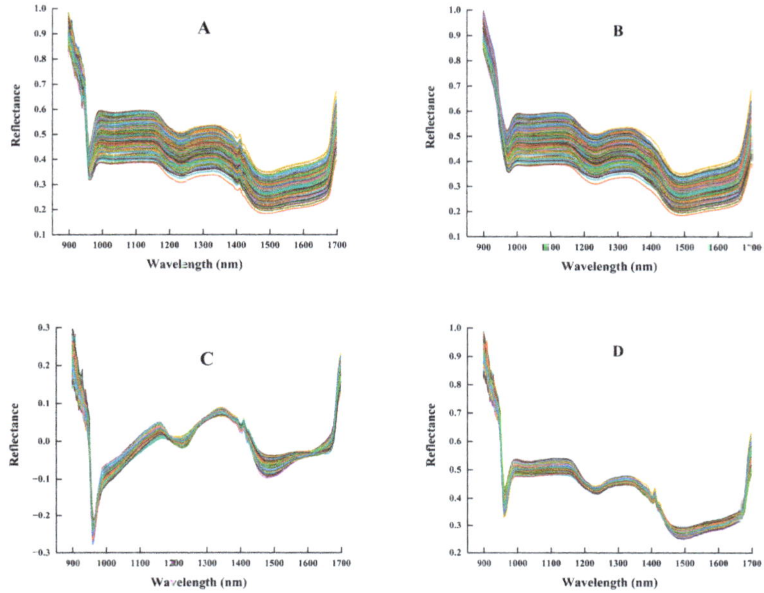

Figure 8. Spectral curves of maize seed samples under preprocessing conditions: (**A**) Raw; (**B**) SG; (**C**) DE; (**D**) MSC.

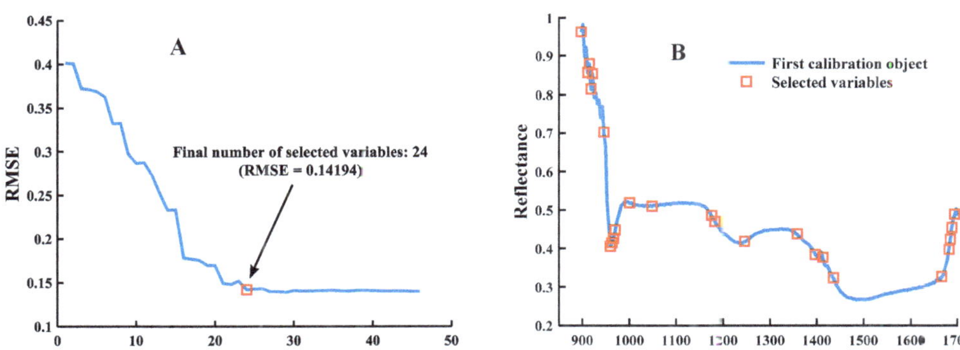

Figure 9. SPA selected feature wavelengths: (**A**) the value of RMSE varies with the number of variables; (**B**) the selected wavelength variables.

The results selected by CARS (set the number of Monte Carlo sampling N to 50 and 5-fold cross-validation) are shown in Figure 10, taking the spectral data processed by DE as an example. As can be seen in Figure 10A, the subset of spectral data changed at each sampling, resulting in a trend of decreasing and then increasing root mean square error of cross-validation (RMSECV) values. The above indicates that at the initial stage of sampling (0–21), a large amount of information irrelevant to identifying defective maize seeds or partially co-linear information was removed, and the RMSECV achieved a minimum value (0.2104) when the sampling number was 21. With the sampling number increased, and the removal of critical information from the spectral data, the model performance becomes progressively worse. Figure 10B shows the trend of the regression coefficient of each wavelength variable in the process of variable optimization, and the position marked by the red dashed vertical line is the minimum value of RMSECV, at which the subset of spectral data obtained by sampling is the optimal wavelength combination, containing 34 wavelength variables, accounting for 13.39% of the total wavelengths.

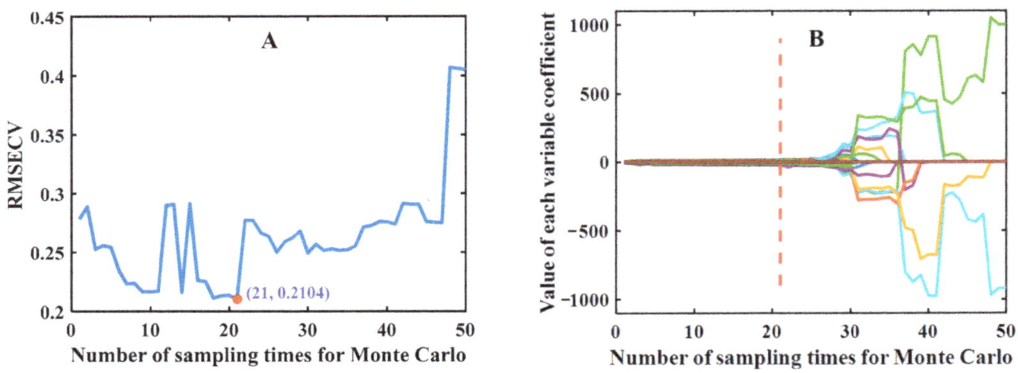

Figure 10. CARS selected feature wavelengths: (**A**) the value of RMSECV varies with the number of sampling; (**B**) the value of the variable's coefficient.

The result of the feature selection of the CNN-FES network proposed in this study is shown in Figure 11 (taking the spectral data processed by MSC as an example), and its score is of great significance for the object classification task. Figure 11A shows the IS at each wavelength obtained by inputting the raw hyperspectral data into the CNN-FES network. Since this method shows good selectivity, the IS value of some unimportant wavelengths will be less than 0, so 24 wavelength variables with higher values (the first 10%) can be selected (which can be adjusted according to the needs of practical applications), and the detailed distribution of its selection results on the spectral curve is shown in Figure 11B. It can be seen that there are more continuous wavelengths in the selected subset of effective wavelengths (the red "□" represents selected variables). This situation indicates that the correlation between adjacent feature wavelengths is high, and the selected subset has less redundant information, which may be more favorable for the model to achieve better classification results. In addition, the selected feature wavelengths are all at and near the wave peaks on the average spectral curve, indicating that the wavelength selection method based on the CNN-FES model proposed in this study can effectively capture important feature information in the spectral data. Moreover, this phenomenon can be explained by the difference caused by the stretching vibration caused by the absorption of chemical functional groups in the biochemical components of maize seeds, indicating that our proposed selection method is representative and interpretable, which can be used to solve practical agricultural problems.

Feature selection is one of the important methods for data dimensionality reduction, and it has been a hot topic of research in the field of spectral data analysis. Therefore, in this study, the proposed CNN-FES method was compared with two traditional SPA and CARS feature selection methods, and the results of feature wavelengths selected from the preprocessed spectral data (245 variables) are shown in Table 1. As can be seen, the feature wavelengths selected by these algorithms are very different. This case is because they use different principles, with SPA focusing on the comparison of projection vector sizes, CARS tending to use thresholds to control the number of variables selected, and CNN-FES focusing on the interactions and differences between variables.

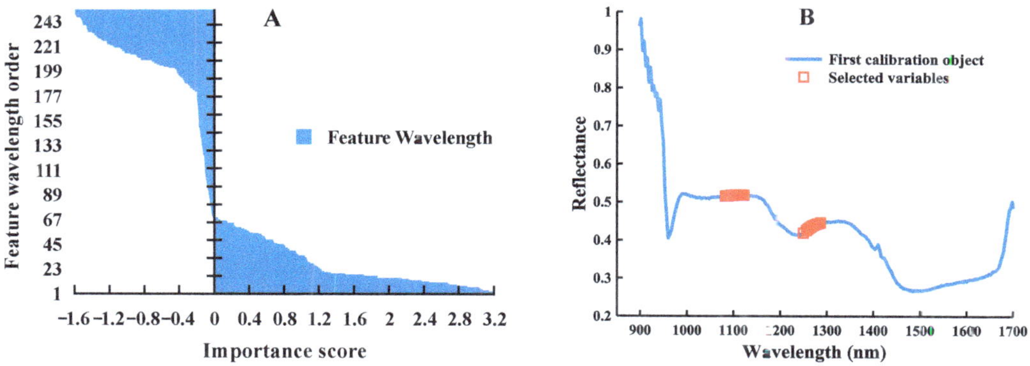

Figure 11. CNN-FES network selected feature wavelengths: (**A**) the value of IS of each feature wavelength; (**B**) the selected wavelength variables.

Table 1. Feature wavelengths selected by different algorithms.

Method	Algorithm (Number)	Feature Wavelengths (nm)
DE	SPA (23)	911, 922, 925, 935, 959, 970, 980, 987, 1000, 1142, 1183, 1345, 1397, 1413, 1436, 1657, 1682, 1685, 1689, 1692, 1695, 1698, 1701
	CARS (34)	956, 1256, 1259, 1262, 1269, 1272, 1276, 1282, 1286, 1292, 1295, 1299, 1302, 1305, 1309, 1312, 1315, 1318, 1322, 1328, 1355, 1358, 1361, 1364, 1368, 1371, 1374, 1378, 1381, 1423, 1619, 1644, 239, 1666
	CNN-FES (24)	1239, 1243, 1249, 1252, 1256, 1259, 1262, 1266, 1269, 1272, 1276, 1279, 1282, 1286, 1289, 1292, 1295, 1299, 1302, 1305, 1312, 1368, 1381, 1407
MSC	SPA (24)	897, 911, 915, 918, 922, 946, 959, 963, 966, 970, 1000, 1048, 1176, 1183, 1246, 1358, 1397, 1413, 1436, 1666, 1682, 1685, 1689, 1695
	CARS (29)	980, 990, 1017, 1024, 1028, 1031, 1045, 1243, 1266, 1269, 1276, 1279, 1282, 1289 1305, 1315, 1361, 1364, 1371, 1374, 1378, 1384, 1420, 1436, 1439, 1443, 1578, 1619, 1632
	CNN-FES (24)	897, 911, 915, 918, 922, 946, 959, 963, 966, 970, 1000, 1048, 1176, 1183, 1246, 1358, 1397, 1413, 1436, 1666, 1682, 1685, 1689, 1695

3.3. Analysis of Modeling Results

3.3.1. Detection Results Based on Full Wavelength

In order to compare the discrimination results of different models, this study inputs the spectral data preprocessed by DE and MSC into the LDA, RF, and SVM models and the proposed CNN-ATM network for modeling, respectively. Figure 12A,B show the accuracy curve and loss function curve of CNN-ATM training, respectively, which can observe the training process and classification effect in real-time. The proposed model achieved an accuracy of 85% at 20 epochs, and it improved rapidly as the number of epochs increased. The loss dropped to 0.3 in the 100th epoch, and it continued to decrease around this low loss. This result indicates that the model has a good convergence speed and generalization. After 200 iterations, the model achieved its best performance and showed higher stability in classification tasks.

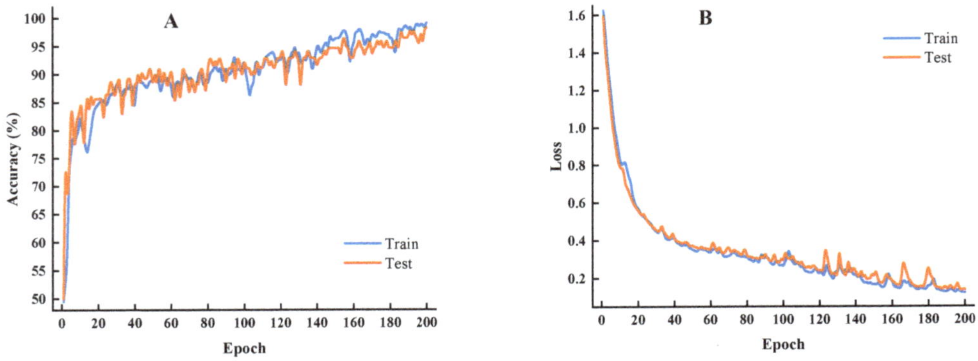

Figure 12. The training results of the CNN-ATM network: (**A**) classification accuracy of the model; (**B**) loss of the model.

The detection results of the four models on the full wavelength were compared in this study, and their accuracy in the training set and test set is shown in Table 2. The classification accuracy of all models was above 91%. The RF model showed the worst results, and its A_C (91.67%), S_E (87.50%), and S_P (96.43%) on the test set were lower than those of the other models. It can be seen that the spectral data preprocessed by DE has a better classification effect than those preprocessed by MSC. The A_C of the CNN-ATM model is higher than 98% for both training and test sets, which is significantly better than the modeling results of the other methods. The results show that the method can achieve accurate classification for the objective mission compared with traditional machine learning methods because the CNN-ATM network has a powerful feature learning capability to achieve sufficient mining of feature information. However, the high redundancy of hyperspectral data and the drawback of large data volume lead to the difficulty of processing hyperspectral data directly, reducing the computational speed and robustness of the model. Therefore, it is necessary to select feature wavelengths to remove redundant information to obtain a subset of wavelengths with rich information and low correlation.

Table 2. Detection results based on full wavelength.

Method	Model	A_C(%)		S_E(%)		S_P(%)		Time (s)
		Train	Test	Train	Test	Train	Test	
DE	LDA	100	97.50	100	96.55	100	98.39	1.36
	RF	98.93	93.33	97.93	89.06	100	98.21	1.34
	SVM	98.21	96.67	97.24	93.55	99.26	100	1.38
	CNN-ATM	98.21	98.33	98.59	100	97.83	96.67	11.73
MSC	LDA	100	95.00	100	93.33	100	96.67	1.37
	RF	99.29	91.67	98.61	87.50	100	96.43	1.36
	SVM	95.00	95.83	92.11	93.44	98.44	98.31	1.35
	CNN-ATM	97.86	98.21	98.59	100	96.38	98.39	12.18

3.3.2. Detection Results Based on Feature Wavelength

To better evaluate the effectiveness and importance of the proposed CNN-FES network for feature wavelength selection, we compare the performance of hyperspectral data input models after feature selection, and Table 3 shows the classification results for LDA, RF, SVM, and CNN-ATM models. The results show that CNN-FES does not increase the training time, but improves the accuracy and efficiency of the classification task to some extent. Overall, the MSC preprocessing method has slightly inferior modeling results than the DE, and the feature wavelength selection algorithm CNN-FES differs less from SPA and CARS. After analyzing the experimental results, it can be seen that when the values of A_C, S_E, and S_P are close or equal, the number of sample subsets selected by CNN-FES is less or equal to that of SPA and CARS, and are concentrated at the peak of the spectral curve. The high accuracy obtained from the model indicates that the wavelength variables selected by the method are chemically significant to some extent.

Table 3. Detection results based on feature wavelength.

Method	Feature Select	Model	A_C(%)		S_E(%)		S_P(%)		Time (s)
			Train	Test	Train	Test	Train	Test	
DE	SPA	LDA	99.28	100	100	100	98.55	100	1.38
		RF	97.14	92.50	99.30	100	94.93	85.48	1.37
		SVM	98.57	98.33	98.59	100	98.55	96.77	1.38
		CNN-ATM	98.57	96.67	97.18	100	97.10	93.55	8.38
	CARS	LDA	98.93	100	100	100	97.83	100	1.36
		RF	100	92.50	100	100	98.28	87.10	1.37
		SVM	98.57	99.17	98.59	100	98.55	98.39	1.35
		CNN-ATM	92.14	95.00	89.44	94.83	94.93	95.16	9.55
	CNN-FES	LDA	99.64	100	100	100	99.28	100	1.37
		RF	100	91.67	100	91.38	100	89.06	1.35
		SVM	95.36	97.50	95.07	100	95.65	95.16	1.34
		CNN-ATM	93.57	97.48	98.55	100	91.30	93.55	8.36
MSC	SPA	LDA	97.50	99.17	100	100	94.93	98.39	1.34
		RF	98.93	89.17	99.30	93.10	98.55	85.48	1.38
		SVM	96.43	97.50	99.30	100	93.48	95.16	1.38
		CNN-ATM	94.29	92.50	98.59	96.55	89.86	88.71	9.17
	CARS	LDA	98.21	99.17	100	100	96.38	98.39	1.37
		RF	98.93	92.50	100	96.55	97.83	88.71	1.36
		SVM	97.50	98.33	99.30	95.65	100	96.77	1.34
		CNN-ATM	97.50	97.50	96.48	98.28	98.55	96.77	8.86
	CNN-FES	LDA	98.93	99.17	100	100	97.83	98.39	1.36
		RF	100	90.00	100	93.10	100	87.10	1.36
		SVM	91.79	94.17	94.37	94.83	89.13	93.55	1.37
		CNN-ATM	98.21	97.50	100	98.28	96.38	96.77	8.50

Compared to the results based on full-wavelength modeling, the results of RF and SVM decreased on average by 3.33% and 2.50%, respectively. This may be due to the relatively small differences in spectral characteristics of some samples, resulting in a slight decrease in the detection accuracy. However, the accuracy of all models is above 90% on both the training and test sets, and their discriminatory ability is LDA > SVM > CNN-ATM > RF in descending order, with the highest accuracy of 97.50% for CNN-ATM. It indicates that the performance of the spectral data does not degrade significantly after feature selection, which proves the effectiveness and rationality of the CNN-FES feature selection method in this study.

CNN-ATM reduces overfitting and improves generalization by exploiting attention mechanisms to improve the expressiveness of spectral features with the learning ability of the model. It also utilizes the joint action of multiple convolutional layers to reduce the complexity of the model while retaining key feature information. Moreover, when the model performance was evaluated, the excellent S_E (98.28%) and S_P (96.77%) values proved the high accuracy of CNN-ATM on the classification task. Although CNN-ATM has an average of 7 s more running time than the machine learning model, it can reach a steady state in a shorter time. This indicates that the proposed model is more robust and has more potential for achieving end-to-end problem-solving.

Although researchers have combined CNN with hyperspectral techniques to detect maize seed quality, fewer studies have addressed the identification of maize seed defects. Detailed comparisons between related literature are difficult due to the differences in crops, methods, and disciplines. However, when considering the use of the same detection method (spectroscopic techniques), experimental object (agricultural products), and agricultural task (defect detection). This study presents a rough comparison of the proposed method with related studies on similar problems. As shown in Table 4, this comparison considered five main aspects: agricultural product type, device type, sample size, spectral range, and accuracy. It can be seen that the accuracy rates for all agricultural tasks are above 83.00%. Among them, the experimental results of this paper (97.50%) are better than the related studies in Table 4. This indicates that the method proposed in this study performs better and enables accurate identification of defective maize seeds.

Table 4. Comparison of the proposed method with related studies.

Agricultural Product Type	Device Type	Sample Size	Spectral Range	Accuracy	References
Sugar beet seed	Terahertz time-domain spectroscopy	100	0.25–0.35 THz	87.00%	[48]
Maize kernel	Multispectral imaging	910	375–970 nm	83.00%	[49]
Wheat kernel	Terahertz time-domain spectroscopy	240	0.1–3.5 THz	96.00%	[50]
Cowpea seed	Raman spectroscopy	105	400–1800 nm	93.70%	[51]
Maize kernel	Hyperspectral imaging	240	953–2517 nm	93.30%	[52]
Maize seed	Hyperspectral imaging	400	900–1700 nm	97.50%	This study

In summary, CNN-FES and CNN-ATM proposed in this study can simultaneously perform spectral feature wavelength selection and sample classification and achieve good performance and results. These results indicate that 1D-CNN is a feasible method for hyperspectral data analysis, which is of great help to the detection and application of defective maize seeds. Therefore, the CNN model of feature selection and classification proposed by us is of great significance in practical production, which can be applied to hyperspectral data sets for other agricultural tasks, such as disease detection and abiotic stress response. In addition, as the number of wavelengths decreases, portable spectral imaging devices can be developed to improve data acquisition efficiency for practical applications. At present, the detection technology of hyperspectral is developing toward online applications and industrial production. In future work, it is necessary to deeply optimize the network architecture to maintain the accuracy and robustness of the model. At the same time, the variety and number of maize seed samples should be increased to provide accurate quantitative analysis. Based on this, image feature information and spectral feature information can be fused to improve the performance of the wavelength selection algorithm and the efficiency of defect detection.

4. Conclusions

This study successfully detected defective maize seeds using HSI combined with deep learning. DE and MSC algorithms were used to pre-process the raw spectra and highlight the spectral differences between different classes. We propose a novel CNN-FES-based feature selection method to select effective feature wavelengths from spectral data. The results show that the number of selected features is fewer than SPA and CARS, and it has better representativeness, interpretability, and classification performance. In addition, a CNN-ATM model based on the attention classification mechanism was proposed for detecting the presence of defects in maize seeds and compared with LDA, RF, and SVM models. The results show that the classification accuracy based on full-wavelength modeling is 98.21%, which is not significantly different from the classification performance of machine learning methods, and the attention mechanism has an excellent ability to emphasize the valid information of the spectral. It demonstrates the feasibility of using hyperspectral data combined with CNN for defect detection in maize seeds, even though these samples have similar shapes and colors. In addition, the modeling accuracy based on the characteristic wavelength variables also reached 97.50%. This indicates that the choice of feature wavelength in the spectral analysis is an effective way to achieve a simplified model. Therefore, the optimal wavelength can be set in advance when using a portable spectral camera to collect sample spectral data, which can effectively improve the processing speed of the model and its applicability to industrial production. In addition, the data obtained were statistically analyzed and compared with related research works in this field. In conclusion, the 1D-CNN proposed in this study is a promising method for hyperspectral wavelength selection and analysis with the potential for online detection, which can provide a reference for the seed industry to improve seed quality.

Author Contributions: Conceptualization, P.X., W.S., K.X., and R.Y.; methodology, P.X., Y.Z., Q.T., Y.Q., and R.Y.; software, P.X.; validation, P.X. and R.Y.; formal analysis, P.X., W.S., K.X., and R.Y.; investigation, Y.Z. and Q.T.; resources, R.Y.; data curation, P.X.; writing—original draft preparation, P.X., W.S., and K.X.; writing—review and editing, P.X., Y.Q., and R.Y.; visualization, P.X.; supervision, Y.Q. and R.Y.; project administration, P.X.; funding acquisition, R.Y. All authors have read and agreed to the published version of the manuscript.

Funding: This research was funded by National Talent Foundation Project of China, grant number T2019136.

Institutional Review Board Statement: Not applicable.

Informed Consent Statement: Not applicable.

Data Availability Statement: The data that support the findings of this study are available upon request from the authors.

Conflicts of Interest: The authors declare no conflict of interest.

References

1. Wang, Z.; Fan, S.; Wu, J.; Zhang, C.; Xu, F.; Yang, X.; Li, J. Application of Long-Wave near Infrared Hyperspectral Imaging for Determination of Moisture Content of Single Maize Seed. *Spectrochim. Acta—Part A Mol. Biomol. Spectrosc.* **2021**, *254*, 119666. [CrossRef] [PubMed]
2. Wang, Z.; Tian, X.; Fan, S.; Zhang, C.; Li, J. Maturity Determination of Single Maize Seed by Using Near-Infrared Hyperspectral Imaging Coupled with Comparative Analysis of Multiple Classification Models. *Infrared Phys. Technol.* **2021**, *112*, 103596. [CrossRef]
3. Wang, L.; Liu, J.; Zhang, J.; Wang, J.; Fan, X. Corn Seed Defect Detection Based on Watershed Algorithm and Two-Pathway Convolutional Neural Networks. *Front. Plant Sci.* **2022**, *13*, 730190. [CrossRef] [PubMed]
4. Yu, L.; Liu, W.; Li, W.; Qin, H.; Xu, J.; Zuo, M. Non-Destructive Identification of Maize Haploid Seeds Using Nonlinear Analysis Method Based on Their near-Infrared Spectra. *Biosyst. Eng.* **2018**, *172*, 144–153. [CrossRef]
5. Yang, G.; Wang, Q.; Liu, C.; Wang, X.; Fan, S.; Huang, W. Rapid and Visual Detection of the Main Chemical Compositions in Maize Seeds Based on Raman Hyperspectral Imaging. *Spectrochim. Acta—Part A Mol. Biomol. Spectrosc.* **2018**, *200*, 186–194. [CrossRef]
6. Elmasry, G.; Mandour, N.; Al-Rejaie, S.; Belin, E.; Rousseau, D. Recent Applications of Multispectral Imaging in Seed Phenotyping and Quality Monitoring—An Overview. *Sensors* **2019**, *19*, 1090. [CrossRef]
7. Zhang, C.; Zhao, Y.; Yan, T.; Bai, X.; Xiao, Q.; Gao, P.; Li, M.; Huang, W.; Bao, Y.; He, Y.; et al. Application of Near-Infrared Hyperspectral Imaging for Variety Identification of Coated Maize Kernels with Deep Learning. *Infrared Phys. Technol.* **2020**, *111*, 103550. [CrossRef]
8. Feng, L.; Zhu, S.; Liu, F.; He, Y.; Bao, Y.; Zhang, C. Hyperspectral Imaging for Seed Quality and Safety Inspection: A Review. *Plant Methods* **2019**, *15*, 91. [CrossRef]
9. Sellami, A.; Tabbone, S. Deep Neural Networks-Based Relevant Latent Representation Learning for Hyperspectral Image Classification. *Pattern Recognit.* **2022**, *121*, 108224. [CrossRef]
10. Alimohammadi, F.; Rasekh, M.; Afkari Sayyah, A.H.; Abbaspour-Gilandeh, Y.; Karami, H.; Rasooli Sharabiani, V.; Fioravanti, A.; Gancarz, M.; Findura, P.; Kwaśniewski, D. Hyperspectral Imaging Coupled with Multivariate Analysis and Artificial Intelligenceto the Classification of Maize Kernels. *Int. Agrophys.* **2022**, *36*, 83–91. [CrossRef]
11. Zhang, X.; Liu, F.; He, Y.; Li, X. Application of Hyperspectral Imaging and Chemometric Calibrations for Variety Discrimination of Maize Seeds. *Sensors* **2012**, *12*, 17234–17246. [CrossRef] [PubMed]
12. Liu, W.; Zeng, S.; Wu, G.; Li, H.; Chen, F. Rice Seed Purity Identification Technology Using Hyperspectral Image with Lasso Logistic Regression Model. *Sensors* **2021**, *21*, 4384. [CrossRef] [PubMed]
13. Cui, H.; Cheng, Z.; Li, P.; Miao, A. Prediction of Sweet Corn Seed Germination Based on Hyperspectral Image Technology and Multivariate Data Regression. *Sensors* **2020**, *20*, 4744. [CrossRef] [PubMed]
14. Huang, L.; Li, T.; Ding, C.; Zhao, J.; Zhang, D.; Yang, G. Diagnosis of the Severity of Fusarium Head Blight of Wheat Ears on the Basis of Image and Spectral Feature Fusion. *Sensors* **2020**, *20*, 2887. [CrossRef]
15. Nagasubramanian, K.; Jones, S.; Sarkar, S.; Singh, A.K.; Singh, A.; Ganapathysubramanian, B. Hyperspectral Band Selection Using Genetic Algorithm and Support Vector Machines for Early Identification of Charcoal Rot Disease in Soybean Stems. *Plant Methods* **2018**, *14*, 86. [CrossRef]

16. Pang, L.; Wang, J.; Men, S.; Yan, L.; Xiao, J. Hyperspectral Imaging Coupled with Multivariate Methods for Seed Vitality Estimation and Forecast for Quercus Variabilis. *Spectrochim. Acta—Part A Mol. Biomol. Spectrosc.* **2021**, *245*, 118888. [CrossRef]
17. Song, D.; Gao, D.; Sun, H.; Qiao, L.; Zhao, R.; Tang, W.; Li, M. Chlorophyll Content Estimation Based on Cascade Spectral Optimizations of Interval and Wavelength Characteristics. *Comput. Electron. Agric.* **2021**, *189*, 106413. [CrossRef]
18. Zhang, L.; Sun, H.; Rao, Z.; Ji, H. Non-Destructive Identification of Slightly Sprouted Wheat Kernels Using Hyperspectral Data on Both Sides of Wheat Kernels. *Biosyst. Eng.* **2020**, *200*, 188–199. [CrossRef]
19. Zheng, Z.; Liu, Y.; He, M.; Chen, D.; Sun, L.; Zhu, F. Effective Band Selection of Hyperspectral Image by an Attention Mechanism-Based Convolutional Network. *RSC Adv.* **2022**, *12*, 8750–8759. [CrossRef]
20. Yuan, D.; Jiang, J.; Gong, Z.; Nie, C.; Sun, Y. Moldy Peanuts Identification Based on Hyperspectral Images and Point-Centered Convolutional Neural Network Combined with Embedded Feature Selection. *Comput. Electron. Agric.* **2022**, *197*, 106963. [CrossRef]
21. Sharma, A.; Lysenko, A.; Boroevich, K.A.; Vans, E.; Tsunoda, T. DeepFeature: Feature Selection in Nonimage Data Using Convolutional Neural Network. *Brief. Bioinform.* **2021**, *22*, bbab297. [CrossRef] [PubMed]
22. Gao, T.; Chandran, A.K.N.; Paul, P.; Walia, H.; Yu, H. Hyperseed: An End-to-End Method to Process Hyperspectral Images of Seeds. *Sensors* **2021**, *21*, 8184. [CrossRef] [PubMed]
23. Yu, Z.; Fang, H.; Zhangjin, Q.; Mi, C.; Feng, X.; He, Y. Hyperspectral Imaging Technology Combined with Deep Learning for Hybrid Okra Seed Identification. *Biosyst. Eng.* **2021**, *212*, 46–61. [CrossRef]
24. Pang, L.; Wang, L.; Yuan, P.; Yan, L.; Yang, Q.; Xiao, J. Feasibility Study on Identifying Seed Viability of Sophora Japonica with Optimized Deep Neural Network and Hyperspectral Imaging. *Comput. Electron. Agric.* **2021**, *190*, 106426. [CrossRef]
25. Gao, J.; Zhao, L.; Li, J.; Deng, L.; Ni, J.; Han, Z. Aflatoxin Rapid Detection Based on Hyperspectral with 1D-Convolution Neural Network in the Pixel Level. *Food Chem.* **2021**, *360*, 129968. [CrossRef] [PubMed]
26. Li, H.; Zhang, L.; Sun, H.; Rao, Z.; Ji, H. Discrimination of Unsound Wheat Kernels Based on Deep Convolutional Generative Adversarial Network and Near-Infrared Hyperspectral Imaging Technology. *Spectrochim. Acta—Part A Mol. Biomol. Spectrosc.* **2022**, *268*, 120722. [CrossRef]
27. Li, Z.; Zhou, G.; Song, Q. A Temporal Group Attention Approach for Multitemporal Multisensor Crop Classification. *Infrared Phys. Technol.* **2020**, *105*, 103152. [CrossRef]
28. Wang, Y.; Zhang, Z.; Feng, L.; Ma, Y.; Du, Q. A New Attention-Based CNN Approach for Crop Mapping Using Time Series Sentinel-2 Images. *Comput. Electron. Agric.* **2021**, *184*, 106090. [CrossRef]
29. Eshkabilov, S.; Lee, A.; Sun, X.; Lee, C.W.; Simsek, H. Hyperspectral Imaging Techniques for Rapid Detection of Nutrient Content of Hydroponically Grown Lettuce Cultivars. *Comput. Electron. Agric.* **2021**, *181*, 105968. [CrossRef]
30. Yang, J.; Sun, L.; Xing, W.; Feng, G.; Bai, H.; Wang, J. Hyperspectral Prediction of Sugarbeet Seed Germination Based on Gauss Kernel SVM. *Spectrochim. Acta—Part A Mol. Biomol. Spectrosc.* **2021**, *253*, 119585. [CrossRef]
31. Zhang, L.; Sun, H.; Rao, Z.; Ji, H. Hyperspectral Imaging Technology Combined with Deep Forest Model to Identify Frost-Damaged Rice Seeds. *Spectrochim. Acta—Part A Mol. Biomol. Spectrosc.* **2020**, *229*, 117973. [CrossRef] [PubMed]
32. Araújo, M.C.U.; Saldanha, T.C.B.; Galvão, R.K.H.; Yoneyama, T.; Chame, H.C.; Visani, V. The Successive Projections Algorithm for Variable Selection in Spectroscopic Multicomponent Analysis. *Chemom. Intell. Lab. Syst.* **2001**, *57*, 65–73. [CrossRef]
33. Sun, J.; Zhou, X.; Hu, Y.; Wu, X.; Zhang, X.; Wang, P. Visualizing Distribution of Moisture Content in Tea Leaves Using Optimization Algorithms and NIR Hyperspectral Imaging. *Comput. Electron. Agric.* **2019**, *160*, 153–159. [CrossRef]
34. Zhang, Y.; Gao, J.; Cen, H.; Lu, Y.; Yu, X.; He, Y.; Pieters, J.G. Automated Spectral Feature Extraction from Hyperspectral Images to Differentiate Weedy Rice and Barnyard Grass from a Rice Crop. *Comput. Electron. Agric.* **2019**, *159*, 42–49. [CrossRef]
35. Li, H.; Liang, Y.; Xu, Q.; Cao, D. Key Wavelengths Screening Using Competitive Adaptive Reweighted Sampling Method for Multivariate Calibration. *Anal. Chim. Acta* **2009**, *648*, 77–84. [CrossRef] [PubMed]
36. Feng, Z.-H.; Wang, L.-Y.; Yang, Z.-Q.; Zhang, Y.-Y.; Li, X.; Song, L.; He, L.; Duan, J.-Z.; Feng, W. Hyperspectral Monitoring of Powdery Mildew Disease Severity in Wheat Based on Machine Learning. *Front. Plant Sci.* **2022**, *13*, 828454. [CrossRef] [PubMed]
37. Long, Y.; Huang, W.; Wang, Q.; Fan, S.; Tian, X. Integration of Textural and Spectral Features of Raman Hyperspectral Imaging for Quantitative Determination of a Single Maize Kernel Mildew Coupled with Chemometrics. *Food Chem.* **2022**, *372*, 131246. [CrossRef] [PubMed]
38. Saeidan, A.; Khojastehpour, M.; Golzarian, M.R.; Mooenfard, M.; Khan, H.A. Detection of Foreign Materials in Cocoa Beans by Hyperspectral Imaging Technology. *Food Control* **2021**, *129*, 108242. [CrossRef]
39. Zhang, L.; Sun, H.; Li, H.; Rao, Z.; Ji, H. Identification of Rice-Weevil (*Sitophilus oryzae* L.) Damaged Wheat Kernels Using Multi-Angle NIR Hyperspectral Data. *J. Cereal Sci.* **2021**, *101*, 103313. [CrossRef]
40. Arun Kumar, M.; Gopal, M. Least Squares Twin Support Vector Machines for Pattern Classification. *Expert Syst. Appl.* **2009**, *36*, 7535–7543. [CrossRef]
41. Zhang, J.; Yang, Y.; Feng, X.; Xu, H.; Chen, J.; He, Y. Identification of Bacterial Blight Resistant Rice Seeds Using Terahertz Imaging and Hyperspectral Imaging Combined with Convolutional Neural Network. *Front. Plant Sci.* **2020**, *11*, 821. [CrossRef] [PubMed]
42. Svetnik, V.; Liaw, A.; Tong, C.; Christopher Culberson, J.; Sheridan, R.P.; Feuston, B.P. Random Forest: A Classification and Regression Tool for Compound Classification and QSAR Modeling. *J. Chem. Inf. Comput. Sci.* **2003**, *43*, 1947–1958. [CrossRef] [PubMed]

43. Zhou, L.; Zhang, C.; Taha, M.F.; Wei, X.; He, Y.; Qiu, Z.; Liu, Y. Wheat Kernel Variety Identification Based on a Large Near-Infrared Spectral Dataset and a Novel Deep Learning-Based Feature Selection Method. *Front. Plant Sci.* **2020**, *11*, 575810. [CrossRef] [PubMed]
44. Altmann, A.; Toloşi, L.; Sander, O.; Lengauer, T. Permutation Importance: A Corrected Feature Importance Measure. *Bioinformatics* **2010**, *26*, 1340–1347. [CrossRef] [PubMed]
45. Hu, J.; Shen, L.; Albanie, S.; Sun, G.; Wu, E. Squeeze-and-Excitation Networks. *IEEE Trans. Pattern Anal. Mach. Intell.* **2020**, *42*, 2011–2023. [CrossRef]
46. Wang, Y.; Peng, Y.; Qiao, X.; Zhuang, Q. Discriminant Analysis and Comparison of Corn Seed Vigor Based on Multiband Spectrum. *Comput. Electron. Agric.* **2021**, *190*, 106444. [CrossRef]
47. Ma, T.; Tsuchikawa, S.; Inagaki, T. Rapid and Non-Destructive Seed Viability Prediction Using near-Infrared Hyperspectral Imaging Coupled with a Deep Learning Approach. *Comput. Electron. Agric.* **2020**, *177*, 105683. [CrossRef]
48. Gente, R.; Busch, S.F.; Stubling, E.-M.; Schneider, L.M.; Hirschmann, C.B.; Balzer, J.C.; Koch, M. Quality Control of Sugar Beet Seeds with THz Time-Domain Spectroscopy. *IEEE Trans. THz Sci. Technol.* **2016**, *6*, 754–756. [CrossRef]
49. Sendin, K.; Manley, M.; Williams, P.J. Classification of White Maize Defects with Multispectral Imaging. *Food Chem.* **2018**, *243*, 311–318. [CrossRef]
50. Ge, H.; Jiang, Y.; Zhang, Y. THz Spectroscopic Investigation of Wheat-Quality by Using Multi-Source Data Fusion. *Sensors* **2018**, *18*, 3945. [CrossRef]
51. Sanchez, L.; Farber, C.; Lei, J.; Zhu-Salzman, K.; Kurouski, D. Noninvasive and Nondestructive Detection of Cowpea Bruchid within Cowpea Seeds with a Hand-Held Raman Spectrometer. *Anal. Chem.* **2019**, *91*, 1733–1737. [CrossRef] [PubMed]
52. Sendin, K.; Manley, M.; Marini, F.; Williams, P.J. Hierarchical Classification Pathway for White Maize, Defect and Foreign Material Classification Using Spectral Imaging. *Microchem. J.* **2021**, *162*, 105824. [CrossRef]

foods

Article

An Improved YOLOv5 Model: Application to Mixed Impurities Detection for Walnut Kernels

Lang Yu, Mengbo Qian *, Qiang Chen, Fuxing Sun and Jiaxuan Pan

College of Optical Mechanical and Electrical Engineering, Zhejiang A & F University, Hangzhou 311300, China
* Correspondence: qianmengbo@zafu.edu.cn; Tel.:+86-189-6802-0808

Abstract: Impurity detection is an important link in the chain of food processing. Taking walnut kernels as an example, it is difficult to accurately detect impurities mixed in walnut kernels before the packaging process. In order to accurately identify the small impurities mixed in walnut kernels, this paper established an improved impurities detection model based on the original YOLOv5 network model. Initially, a small target detection layer was added in the neck part, to improve the detection ability for small impurities, such as broken shells. Secondly, the Tansformer-Encoder (Trans-E) module is proposed to replace some convolution blocks in the original network, which can better capture the global information of the image. Then, the Convolutional Block Attention Module (CBAM) was added to improve the sensitivity of the model to channel features, which make it easy to find the prediction region in dense objects. Finally, the GhostNet module is introduced to make the model lighter and improve the model detection rate. During the test stage, sample photos were randomly chosen to test the model's efficacy using the training and test set, derived from the walnut database that was previously created. The mean average precision can measure the multi-category recognition accuracy of the model. The test results demonstrate that the mean average precision (*mAP*) of the improved YOLOv5 model reaches 88.9%, which is 6.7% higher than the average accuracy of the original YOLOv5 network, and is also higher than other detection networks. Moreover, the improved YOLOv5 model is significantly better than the original YOLOv5 network in identifying small impurities, and the detection rate is only reduced by 3.9%, which meets the demand of real-time detection of food impurities and provides a technical reference for the detection of small impurities in food.

Keywords: YOLOv5; walnut kernels; impurities detection; small object detection

Citation: Yu, L.; Qian, M.; Chen, Q.; Sun, F.; Pan, J. An Improved YOLOv5 Model: Application to Mixed Impurities Detection for Walnut Kernels. *Foods* **2023**, *12*, 624. https://doi.org/10.3390/foods12030624

Academic Editor: Antonello Santini

Received: 26 October 2022
Revised: 17 January 2023
Accepted: 22 January 2023
Published: 1 February 2023

1. Introduction

Food safety has always been a social health issue of great concern to people. Impurity pollution accounts for a large proportion of food pollution and is difficult to avoid [1]. Impurity pollution refers to the presence of other substances or foreign substances in food other than the food itself [2], which will cause physical and psychological harm to consumers. Taking walnut kernels as an example, impurities of walnut kernels can be divided into exogenous impurities (stones, metal parts, and plastic fragments) and endogenous impurities (walnut shells; spoiled walnut kernels) [3], which will seriously affect consumers satisfaction. Therefore, impurity detection is one of the important links to ensure the high quality of nut food [4]. In recent years, many researchers have tried to use imaging detection technology to detect impurities in food, including X imaging technology, terahertz detection technology, spectral detection analysis technology, machine vision detection technology, etc.

In the field of machine vision detection, object detection methods based on deep learning are developing rapidly. According to the nature of the algorithm stages, the current mainstream algorithms can be divided into two categories: one of them is the two-stage algorithm R-CNN series, and the representative algorithms include R-CNN, SPP-Net, and Faster R-CNN [5]. The series of algorithms first generate regions and then classify samples through convolutional neural networks. Xie et al. used the Fsater-RCNN + VGG16 neural network model to detect bone impurities in salmon meat [6]. Wang et al. used the Faster RCNN ResNet101 for detecting the potato surface defects and verified the high recognition accuracy of the model [7]. The point of the R-CNN series of algorithms is that the detection accuracy is high, but the detection speed is defective. The other type is a single-stage algorithm, and the representative algorithm is the YOLO series. The YOLO (You Only Look Once) algorithm directly inputs the entire image into the model network and returns the classification category and location of the bounding box at the output, so it can extract all features in the image and predict all objects [8,9]. Based on the results of various studies, the YOLOv5 algorithm in the YOLO series has better comprehensive detection ability than other YOLO models due to its accuracy and detection accuracy [10–12]. Many researchers have applied the YOLOv5 model or the improved YOLOv5 model to the food safety field for object detection [13]. Jubayer et al. used the YOLOv5 model to detect molds on food surfaces and successfully identified the types of molds on food surfaces [14]. On the basis of the original YOLOv5 network model, Chen et al. added a new involution bottleneck module, which reduced the parameters and calculation amount, and introduced the SE module to improve the sensitivity of the model to channel features, establishing a plant pest identification model [15]. Qi et al. borrowed the human visual attention mechanism and added the squeeze-and-excitation module to the YOLOv5 model to achieve a key feature extraction [16]; the trained network model was evaluated on the tomato virus disease test set, and the accuracy rate reached 91.07%. Han et al. adopted the YOLOv5 model based on the flood filling method to achieve cherry quality detection [17].

This paper takes walnut impurities as the detection target. There is a high requirement for the real-time detection rate of impurities for the fast running speed of the walnut processing line [18]. The YOLOv5 model can maintain a higher detection accuracy, while maintaining a higher detection rate [19], so this paper chooses YOLOv5 as the detection model. However, the original YOLOv5 model is challenging to extract image features of impurities in walnut kernels under complex backgrounds. It is hard to detect small impurities such as broken shells, resulting in a low impurity recognition rate. In order to solve the above problems, we take the pursuit of a balance between detection performance and detection rate as the goal and improve the original YOLOv5 network, so that it can more accurately detect the impurities in the image without losing the detection rate. Firstly, a small target detection layer is added to the neck part to improve the model's ability to detect small impurities. Secondly, the Tans-E module is proposed to replace some of the convolution blocks in the original network. Thirdly, the CBAM module is added to improve the sensitivity of the model to channel features, which is convenient for finding prediction regions in dense objects. Finally, the GhostNet module is introduced to make the model lighter and improve the model detection rate.

2. Materials and Methods

2.1. Samples Used in the Experiments

There is uncertainty in walnut processing; thus, we selected random sampling without repetition. From March 2022 to April 2022, we randomly collected about 20 kg of walnut kernels with impurities before the manual sorting process from the walnut processing line of the Nut Fried Goods Base in Longgang Town, Lin'an District. All samples were collected three times. The walnut kernels were mixed with broken walnut shells, unqualified walnut kernels and other impurities. The mixture was divided into 40 groups of samples evenly, according to quality.

2.2. Images Acquisition System and Dataset Creation

2.2.1. Images Acquisition System

The image acquisition system has two functions: simulating the walnut processing line and take sample pictures. The system consists of a conveyor belt, aluminum profile, the computer, a camera (D435i from Intel, Santa Clara, CA, USA), a black inspection chamber, and an LED light belt [3]. The conveyor belt is divided into two stages: high speed and low speed. The density of the walnut kernel is changeable by controlling the speed difference of the conveyor belt. The camera is used to capture images with resolutions of 1920 × 1080 pixels. The color model is RGB. The camera is located 400 mm above the second conveyor belt [20]. The black inspection chamber, which was made by diffuse reflection plates, is set to cover the camera. Four equal power light belts (10 W each) are set in the black inspection chamber to provide light, as shown in Figure 1.

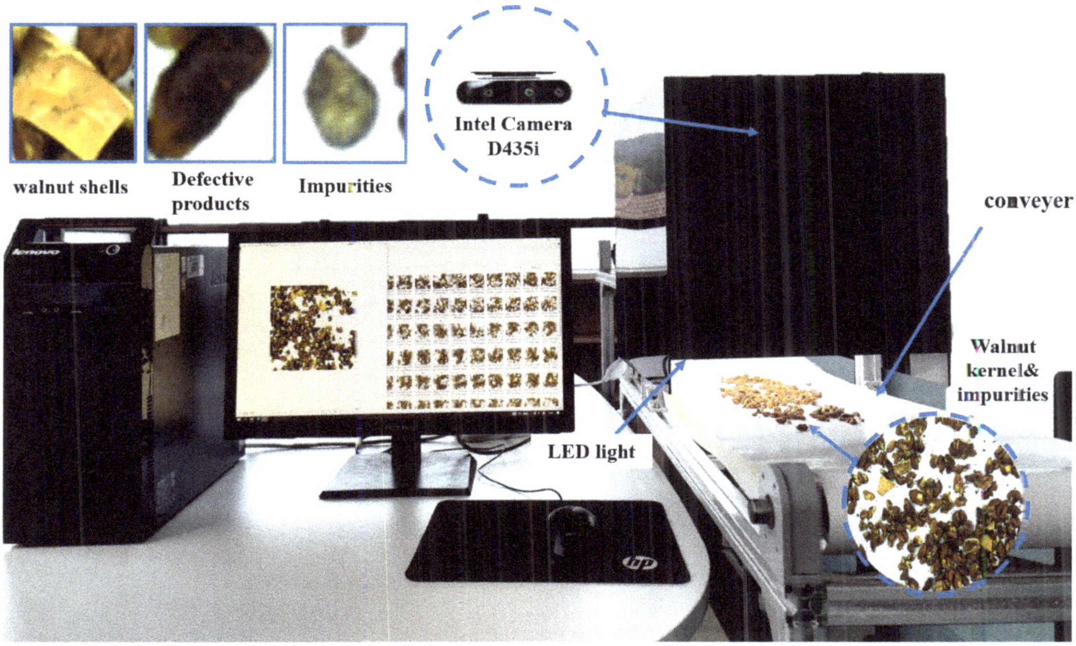

Figure 1. The hardware of the image acquisition system in the lab.

Before the images acquisition system began, one group of the walnut kernels mixed with impurities was manually placed on the first-step conveyor belt, and the driving motor was started. After entering the second stage of the conveyor belt, the sample is paved. The walnut image was captured by the camera and stored in the computer. The image acquisition frequency was 2/s. About 130 images of walnut mixed with different impurities can be obtained by each group of the sample.

2.2.2. Dataset Production

In order to improve the effectiveness of training and increase the diversity of samples, the collected image data were screened before training, and the images with low definition were removed. Finally, 1320 walnut kernel images were obtained and stored in JPG format. After processing by Matlab, the image resolution was set to 512 pixels × 512 pixels. In this paper, the dataset is enhanced by changing the adaptive contrast, rotation, translation, cropping and other methods, and the dataset is expanded to 5732 images [21]. The dataset contains four categories of labels: walnut shell, small impurities (diameter less than 5 mm), foreign impurities and metamorphic walnut kernels, as shown in Figure 2. The gray value range of the walnut kernel is the basis for identifying the deterioration degree of walnut kernels. All the images of walnut kernels are gray processed, and the gray value range of the metamorphic walnut kernel is from 20 to 35 after testing and statistics. The image labeling software is Labelimg, which is used to label the real bounding box and categories [22]. Then, according to the ratio of 3:1:1, all the enhanced images are divided into the training set, validation set and test set. There are 3439 images in the training set, 1146 images in the validation set and 1146 images in the test set.

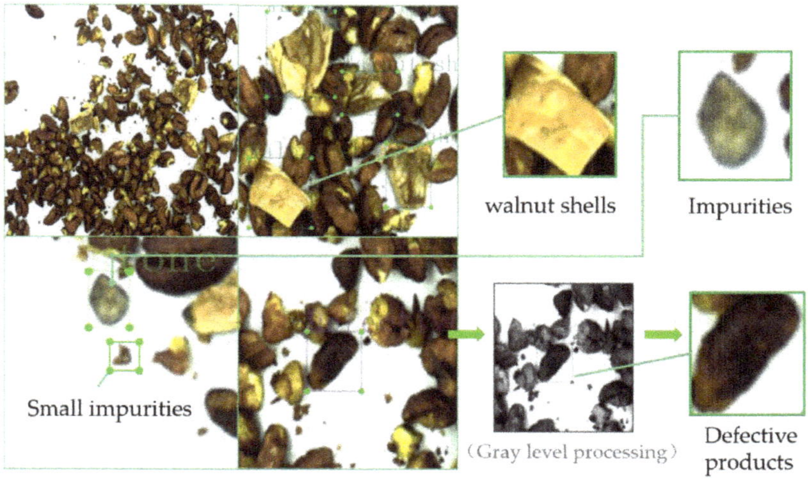

Figure 2. Walnut kernel impurity type labeling.

2.2.3. Experimental Equipment

The training of this model is conducted based on the Windows 10 operating system and the Pytorch framework. The CPU model of the test equipment is Intel®Core™ i7\11800H CPU@3.70 GHz, the GPU model is GeForce RTX 3080 10 G, and the software environment is CUDA 11.3, CUDNN 7.6 and Python3.8. The original YOLOv5 and the im-proved YOLOv5 are trained separately. The specific parameters are presented in Table 1.

Table 1. Test environment setting and parameters.

Parameter	Configuration
Operating system	Windows 10
Deep Learning Framework	Pytorch2.6
Programming language	Python3.8
GPU accelerated environment	CUDA 11.3
GPU	GeForce RTX 3080 10 G
CPU	Intel®Core™ i7\11800H CPU@3.70 GHz

2.3. *Walnut Kernel Impurity Detection Based on YOLOv5*

Currently, the target detection algorithms applied in food detection have high recognition accuracy, but the detection models often have too many parameters and large volumes, and are too complex and challenging to meet the needs of real-time detection [23]. Since the actual application site of walnut impurity detection is located in the food assembly line, the detection model should not only meet the requirements of recognition accuracy but also meet the real-time requirements of detection. YOLOv5 has a higher detection accuracy and a lighter model volume, so it has a faster response speed. Therefore, this paper adopts the YOLOv5 model for the detection of walnut impurities; its frame is shown in Figure 3.

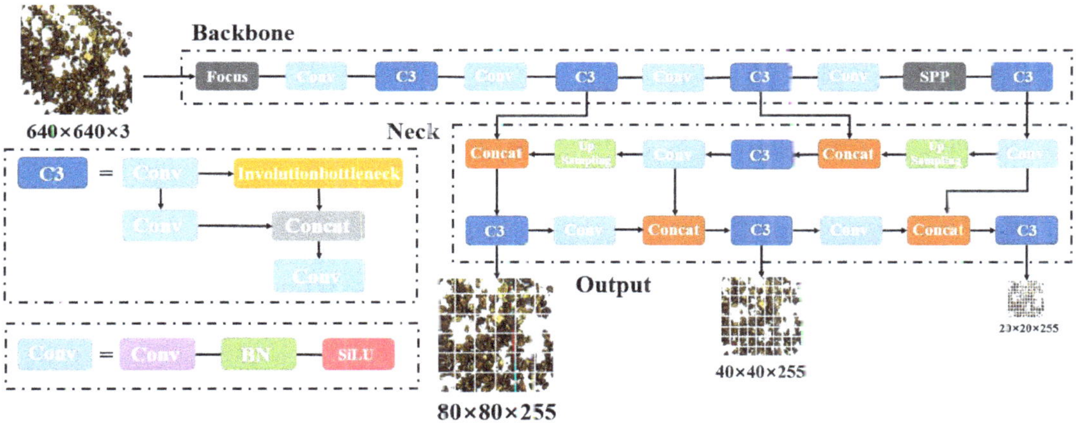

Figure 3. The impurity detection model of walnut kernel based on YOLOv5.

2.4. Walnut Kernel Impurity Detection Based on YOLOv5

2.4.1. Small Object Recognition Layer

There are small impurities, such as broken shells in the walnut images, and the detecting model used must be able to detect small objects. In the process of using the original YOLOv5 model, impurities such as the small broken shells of walnut kernels are small. The feature map in the YOLOv5 network structure is too small, while the multiple of the downsampling is large; thus, it is difficult for the deeper feature map to learn the features of small targets' information, which lead to omissions of small impurities. To solve this problem, this paper tries to add a small object detection layer to the original YOLOv5 head, which will continue to process the feature map for expansion. After the 17th layer of the head part, it performs upsampling and other processing on the feature map so that the feature map continues to expand. At the 20th layer, the acquired feature map with a size of 160 × 160 is concated with the feature map of the second layer in the backbone to obtain a larger feature map for small target detection.

As shown in Figure 4, the function of upsampling is to enlarge the feature map so that the displayed image has a higher resolution, which is more conducive to detecting and recognising small targets. The upsampling process in this paper is implemented by the method of transposed convolution. Unlike the ordinary convolution, transposed convolution is adding a unit-step null pixel between each two pixels of the input image, so that the obtained Feature Map size becomes larger.

2.4.2. Trans-E Block

The Transformer was first used in the field of natural language machine translation, and its most significant feature is the self-Attention mechanism. The main working modules in the Transformer structure are the encoder and decoder. During machine translation, the encoder part models the input sequence. It extracts the output value of the last time step at the structural output as a representation of the input sequence. The decoder then takes the input sequence representation as its input value and generates the translation with maximum probability. This paper simulates the encoder function in the Transformer structure, and proposes a Transformer-Encoder (Trans-E) block and tries to apply it to the image impurity detection. The structure of the Trans-E block is shown in the Figure 5. The Trans-E block consists of two sub-layers, the multi-head attention layer and the fully-connected layer. Among them, the multi-head attention layer is to perform multiple linear mappings of different sub-region representation spaces through multiple heads under the consideration of parallel computing; thus, it can obtain more comprehensive information under different sub-spaces at different locations. The main function of theconnected layer is to map the feature space calculated by the previous layer to the sample label space. A residual structure connects the two sub-layers. This article replaces the bottleneck blocks and some Cnov blocks of CSPDarknet53 in the original YOLOv5 with Trans-E blocks. Compared with CSP bottleneck blocks, Trans-E blocks have more advantages in capturing global information.

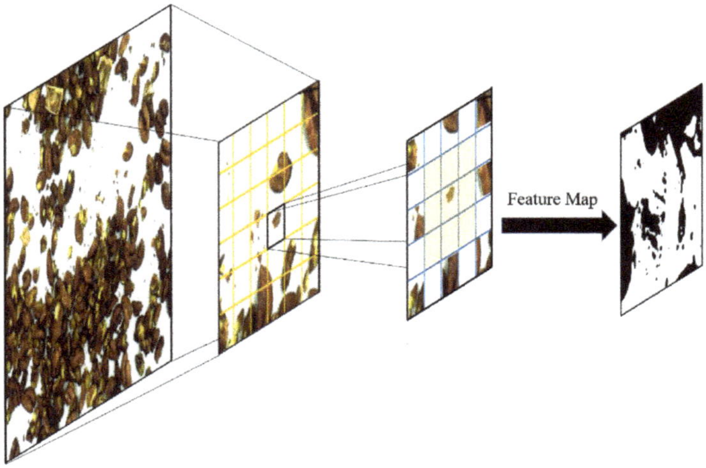

Figure 4. Schematic diagram of upsampling.

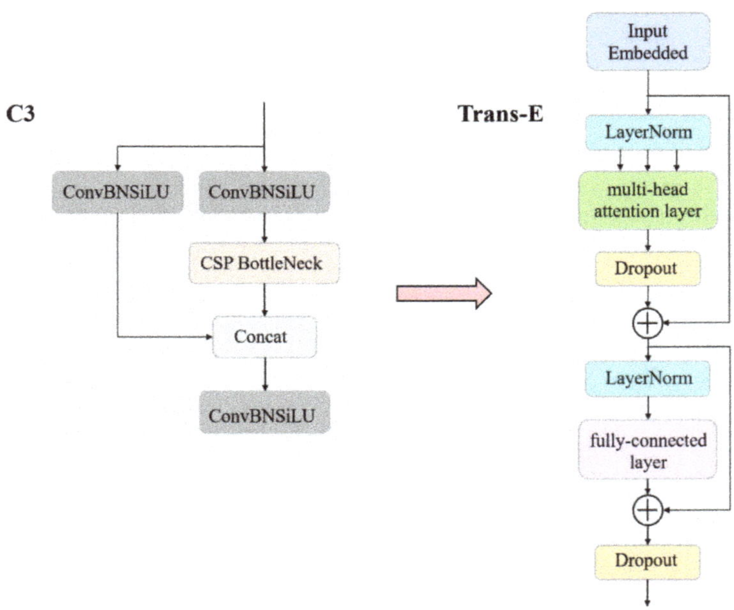

Figure 5. The architecture of Tran-E block.

2.4.3. CBAM Attention Mechanism

Since there is much useless information in the walnut kernel image, such as the walnut kernel itself, in order to suppress other useless image information, we increase the effective image feature weight, reduce the invalid weight, and make the training network model produce the best results. This paper introduces the based YOLOv5 Convolutional Block Attention Module (CBAM). The working principle of this module is as follows: take the global max pooling and global average pooling operations based on width and height, respectively for the input feature map F ($H \times W \times C$), and the output result is two $1 \times 1 \times C$ feature maps. Then, the obtained feature maps are sent to the neural network (MLP), respectively The number of layers in the neural network is two layers. The number of neurons in the first layer is C/r (r is the reduction rate), the activation function is Relu, and the second layer is the number of neurons. The number of neurons in the layer is C. Then, an element-wise-based sum operation is performed on the output features, and the final channel attention feature, namely M_c, is generated after the sigmoid activation operation. Finally, the element-wise multiplication operation is performed on M_c and the input feature map F to generate the input features required by the Spatial attention module. The specific calculation is as follows:

$$M_C(F) = \sigma(MLP(AvgPool(F)) + MLP(MaxPool(F))) \quad (1)$$

$$= \sigma\left(W_1\left(W_0\left(F^c_{avg}\right)\right) + W_1(W_0(F^c_{max}))\right) \quad (2)$$

The output of the channel attention module is taken as an input into the spatial attention module, which is also subjected to maximum pooling and average pooling. Then the two are stacked through the Concat operation, which only compresses the channel dimension but not the spatial dimension to focus on the target's location information. The mechanism is shown in Figure 6.

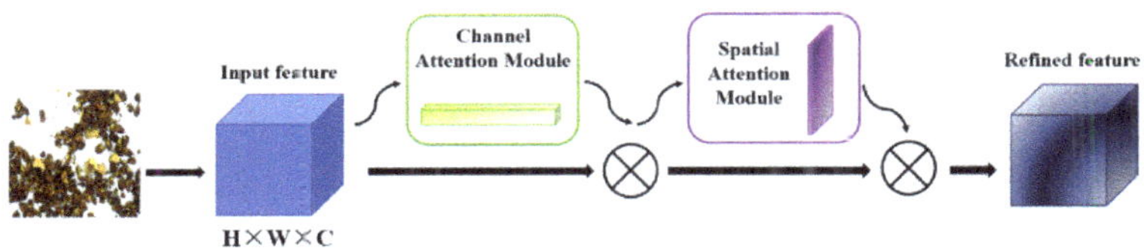

Figure 6. CBAM attention mechanism.

In this paper, the CBAM module is added after the C3 module and the Trans-E module in the neck part so that the image features of walnut shells and foreign objects are weighted and combined, which increases the network at the cost of a small amount of computation, so that the network pays more attention to the key information of foreign objects such as walnut shells, which helps to train a better network.

2.4.4. Ghostconv Makes Models Lightweight

Since the main part of the original YOLOv5 adopts the C3 structure for feature extraction, after adding the small target detection layer, the Trans-E block and the CBAM module based on the original network, the overall network has a large number of parameters. When the detection rate is low, it will be difficult to meet the real-time detection requirements. The actual scene of the walnut kernel impurity detection is a moving conveyor belt, so the detection model must have a relatively lightweight model and low detection delay. This paper applies the GhostConv block in GhostNet and replaces some ordinary convolution block in the current network model to make the detection model more lightweight.

Different from traditional convolution blocks, GhostConv performs feature map extraction on images in two steps [24]. The first step is still using the normal convolution calculation, and the feature map channel obtained at this time is less. The second step uses cheap operation (depthwise conv) to perform feature extraction again to obtain more feature maps, and then concat the feature maps obtained twice to form a new output.

As can be observed from Figure 7, the cheap operation will perform cheap computations on each channel to enhance feature acquisition and increase the number of channels. This mode requires significantly less computation than traditional convolution computations.

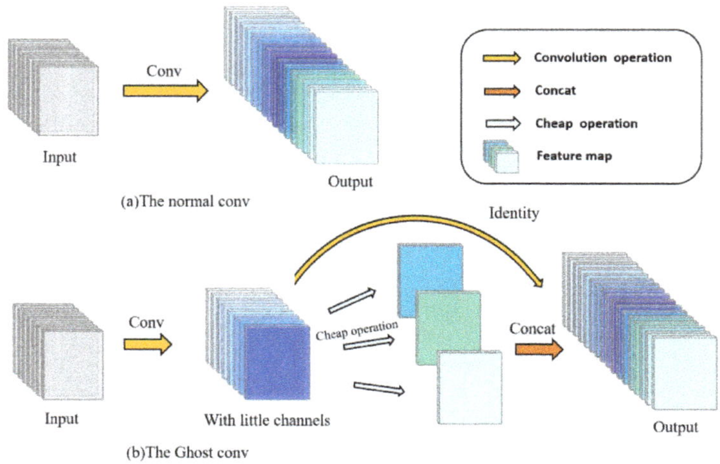

Figure 7. (**a**)The ordinary convolution. (**b**)The Ghost convolution.

In order to solve the problem that the original YOLOv5 network cannot detect small impurities well and the detecting accuracy of individual near-color foreign objects is low, this paper combines the small object detection layer, Trans-E block, CBAM module and GhostConv to construct the entire improved YOLOv5 network model framework, as shown in Figure 8.

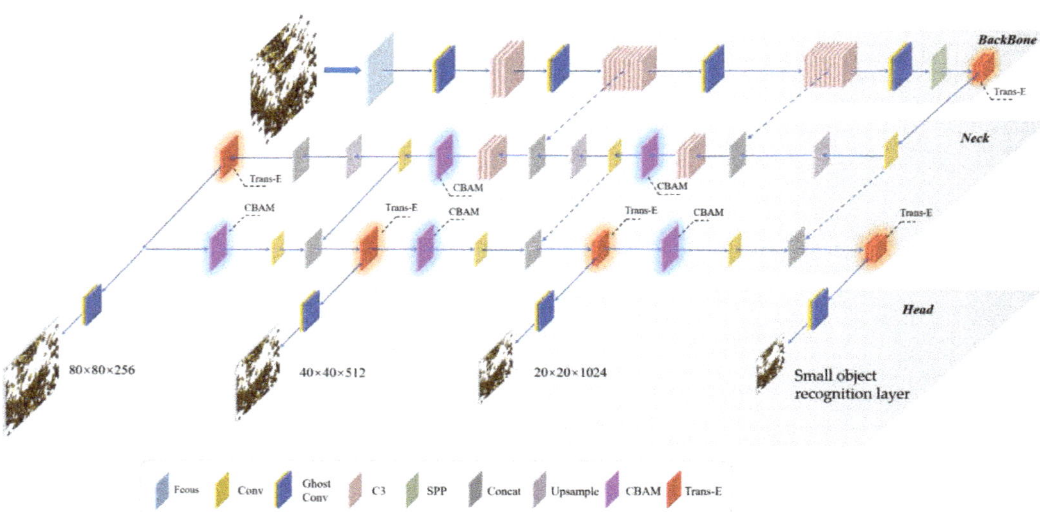

Figure 8. Walnut kernel impurity detection model based on improved YOLOv5 network.

2.5. Experiment Process

First, the manual labeling method is used to mark each walnut image to obtain the training label image, and then the walnut image set is divided into training set, validation set and test set according to the ratio of 3:1:1. The training set is input into the improved YOLOv5 network for training. During the training process, the stochastic gradient descent algorithm is used to optimize the network model, and the optimal network weights are obtained when the training is completed. Subsequently, the images in the validation set of weight values are used to test the performance of the network model and compare with the test results of the original YOLOv5 model and other prediction models. The feasibility of the walnut kernel impurity detection model based on the improved YOLOv5 was verified. The test process is shown in Figure 9.

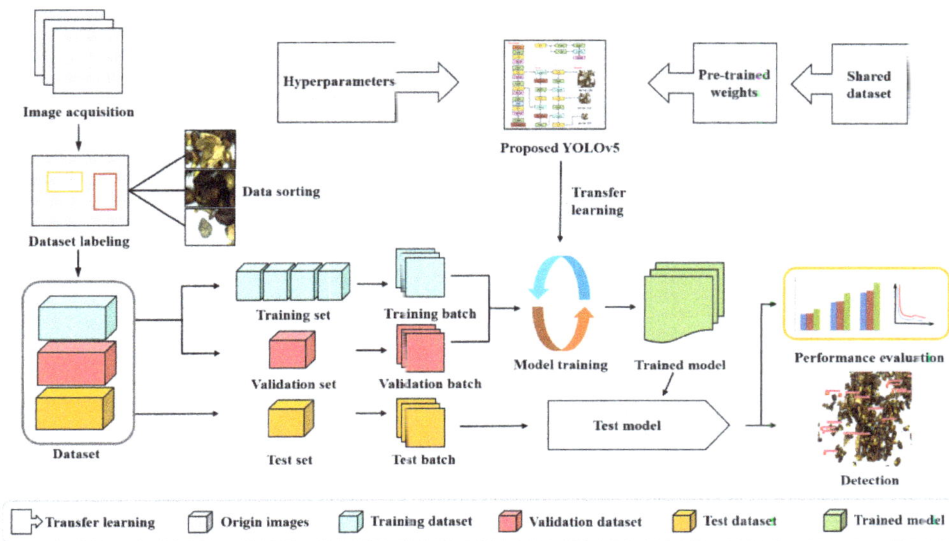

Figure 9. Flowchart of the overall workflow methodology for the proposed detection model.

2.6. Model Evaluation Index

The model loss of YOLOv5 consists of bounding box loss, object loss and classification loss, which can be used to test the target prediction performance of the model. Precision (Pre) and recall (Rec) can intuitively reflect the accuracy of target prediction, which are calculated by the ratio of the number of *TP*, *FP*, *TN*, and *FN* [25], where *TP* represents the number of correctly detected positive samples, and FP represents the error Number of negative samples detected, *FN* indicates the number of positive samples not detected. The *F*1 score is the weighted average of precision and recall. The AP value of each class is the area composed of the label P-R map of that class. The mean average precision (*mAP*) is the average of the *AP* values of various labels; thus, it can represent the global detection performance of the model.

$$loss = l_{bbox} + l_{object} + l_{classification} \tag{3}$$

$$Pre = \frac{TP}{(TP + FP)} \tag{4}$$

$$Rec = \frac{TP}{(TP + FN)} \tag{5}$$

$$F_1 = \frac{2 \times Pre \times Rec}{Pre + Rec} \tag{6}$$

$$AP = \int_0^1 Pre(Rec)dRec \tag{7}$$

$$mAP = \frac{1}{|Q_R|} \sum_{q=Q_R} AP(q) \tag{8}$$

3. Results and Discussion

3.1. Model Training Results

According to the data set type, the loss function of the prediction model can be divided into training loss and validation loss, and the curve is shown in the Figure 10a. It can be observed from the figure that in the process of model training, when the number of iterations is between 0 and 150, the training loss and validation loss decrease rapidly, and when the number of iterations reaches more than 250, the loss value of the prediction model begins to stabilize gradually. In this paper, the training model with 300 iterations is selected as the final walnut kernel impurity detection model. In addition, it can be observed from the *mAP* curves of the training set and the validation set in the Figure 10b that the trained prediction model does not appear to be overfitting.

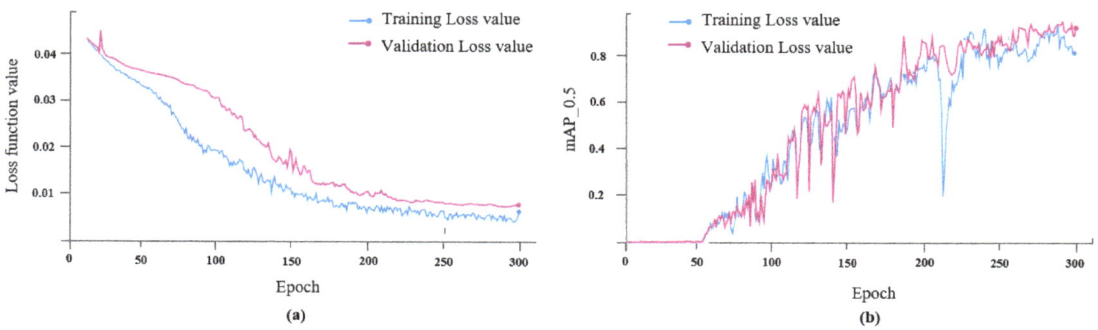

Figure 10. Training results of the improved YOLOv5 model. (**a**) Training and validation loss. (**b**) *mAP*_0.5 of training and validation sets. *mAP*_0.5: mean average precision when the threshold of IoU is 0.5.

3.2. Model Test Results and Analysis

In order to verify the performance of the detection model, the number of impurities for each category in the random 300 images in the validation set was counted and calculated, and then compared with the test results of the model. Among the 300 images in the validation set, the number of walnut shell impurities is 2059, the number of metamorphic walnut kernels is 786, the number of small impurities is 2621, and the number of other impurities is 432. The precision rate, recall rate, *F*1 score and *mAP* were used to evaluate the prediction accuracy of the model for various impurities. The predicted results are shown in Table 2.

Table 2. Recognition results of targets using improved YOLOv5 model.

Class	Num	*Pre* (%)	Rec (%)	*mAP* (%)	*F*1 (%)
Shell	2059	92.21	96.32	94.20	94.56
Small_impurities	2624	83.56	87.84	85.12	86.21
metamorphic_walnut	786	89.24	93.37	90.98	91.26
Other impurities	432	90.25	94.93	92.21	92.87
Total	5901	89.69	93.42	91.25	91.77

The confusion matrix can intuitively reflect the prediction results of classification problems, showing the prediction probability for each category. From the confusion matrix in the Figure 11, it can be observed that among the four types of impurities, the detection accuracy of the walnut shell is the highest, which can reach 92.21%, and the detection accuracy of small impurities is the lowest. Since impurities are located at the boundary of the image, the annotation information is accurate, resulting in a small part of the spoiled walnut kernels being predicted as other impurities.

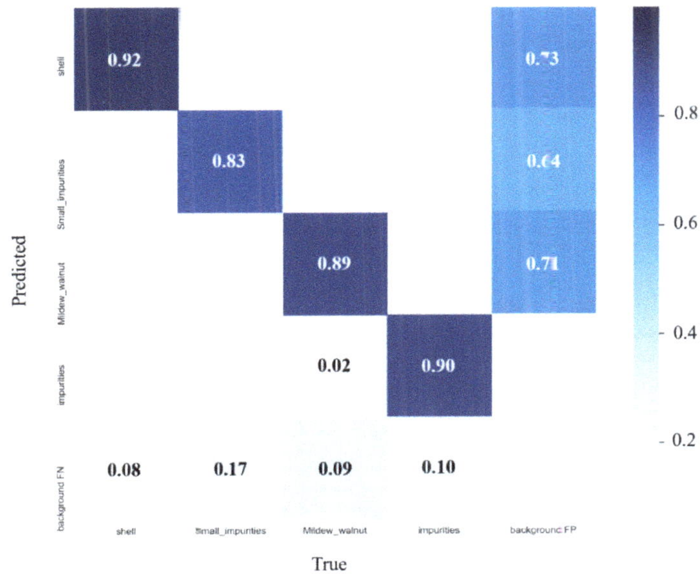

Figure 11. Confusion matrix of four kinds of impurities.

3.3. Performance Comparison of Different Models

In order to better verify the performance of the improved walnut kernel impurity detection model, 300 images in the above validation set were used as the test objects, and the original YOLOv5, YOLOv4, Faster R-CNN, and SSD300 models were used to test and compare the test results [26]. Similarly, the accuracy, F1 score and *mAP* are used as indicators to evaluate the performance of the model. Considering that in actual nut processing, the detection rate of the model is high to meet the needs of real-time detection, so it is also necessary to use the model size and the average GPU detection speed as the evaluation indicators of the model. The test results of each model are shown in Table 3.

Table 3. Comparison of precision, recall, F1-score, mean Average Precision, detection speed and ModelSizes between proposed model and other advanced models.

Model	P (%)	R (%)	F1-Score (%)	*mAP* (%)	Dect. Time (ms)	ModelSizes (M)
Faster-RCNN	87.36	89.25	88.39	81.62	121.86	110.770
SSD300	67.75	75.38	65.43	69.36	89.07	82.781
YOLOv4	82.56	90.14	85.56	85.62	400	245.5
YOLOv5	85.32	88.97	86.43	83.25	43.64	41.489
Proposed YOLOv5	90.25	91.56	90.81	88.9	45.38	43.562

As can be observed from the data in the Figure 12, the detection accuracy of the improved YOLOv5 model is 5.77% higher than that of the original YOLOv5, and both are higher than other detection models, *mAP* has increased by 6.79%, and F_1 has increased by 5.06%. The result is also better than the fire inspection small target detection model based on YOLO algorithm, whose *mAP* is 80.23% and F_1 is 73% [27]. The experiment proves that the introduction of the small target detection layer, the replacement of the Trans-E block, and the introduction of the CBAM module on the basis of the original YOLOv5 model can help improve the accuracy and performance of walnut kernel impurity detection.

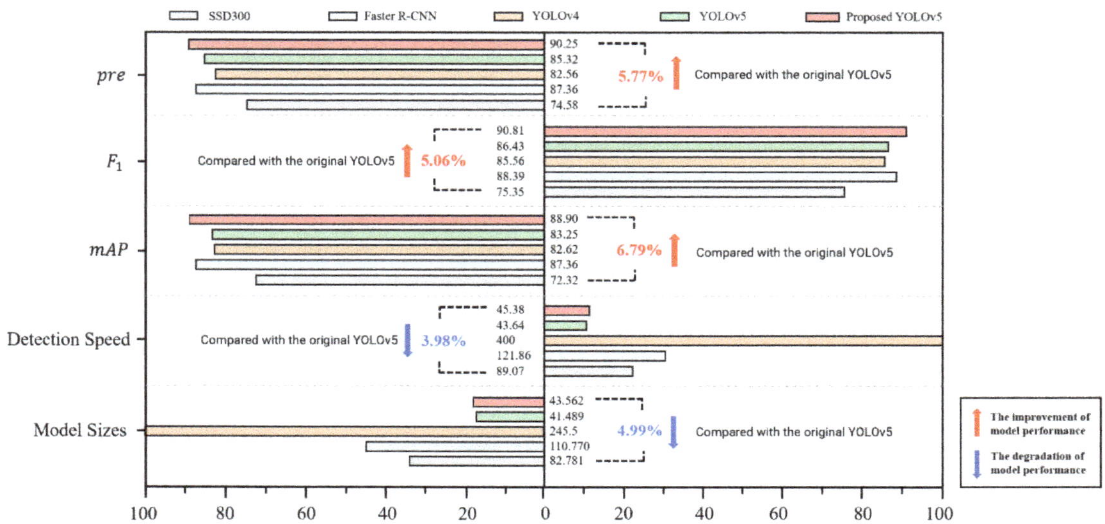

Figure 12. Comparison of the performance from different network models.

Model detection speed is also one of the important performance indicators for real-time detection of food impurities. While improving the accuracy of impurity detection, the YOLOv5 model parameters have increased, and the model size has also increased by 1.74 M. At the same time, the detection time of a single image is increased to 65.25 ms, which is 21.51 ms longer than the original YOLOv5 single image detection time. In order to reduce the detection time of a single image and improve the efficiency of real-time detection of impurities, this paper replaces the conventional Conv of the main part and the detection head part with Ghostconv to make the model more lightweight. After replacing Conv with Ghostconv, the single image impurity detection time is reduced from 65.25 ms to 45.38 ms, which is only 4.99% longer than the original YOLOv5 detection time. Compared with the improved SE-YOLOv5, the detection response time is reduced by 10.4%. [17] This model also leads to other commonly used detection models such as YOLOv4 in the detection rate performance. Therefore, the improved YOLOv5-based walnut kernel impurity modeling model is a suitable detection model.

3.4. Comparison of Recognition Result

Figure 13 compares the results of the original YOLOv5 and the improved YOLOv5 for detecting impurities in walnut kernels. The brown boxes in the figure represent walnut shells, the green boxes represent mildewed walnut kernels, and the red boxes represent small impurities. It can be observed from the figure that the missed detection rate of small impurities in the improved YOLOv5 model is greatly reduced, and the corresponding target confidence is improved. Under the background of high-density walnut kernels and extremely small impurities, the original YOLOv5 has a weak ability to extract features, resulting in the inability to accurately predict the impurity target. The detection performance of the improved YOLOv5 model is significantly better than the former, with a large number of detected small targets and high accuracy, and better performance in detecting small impurity targets.

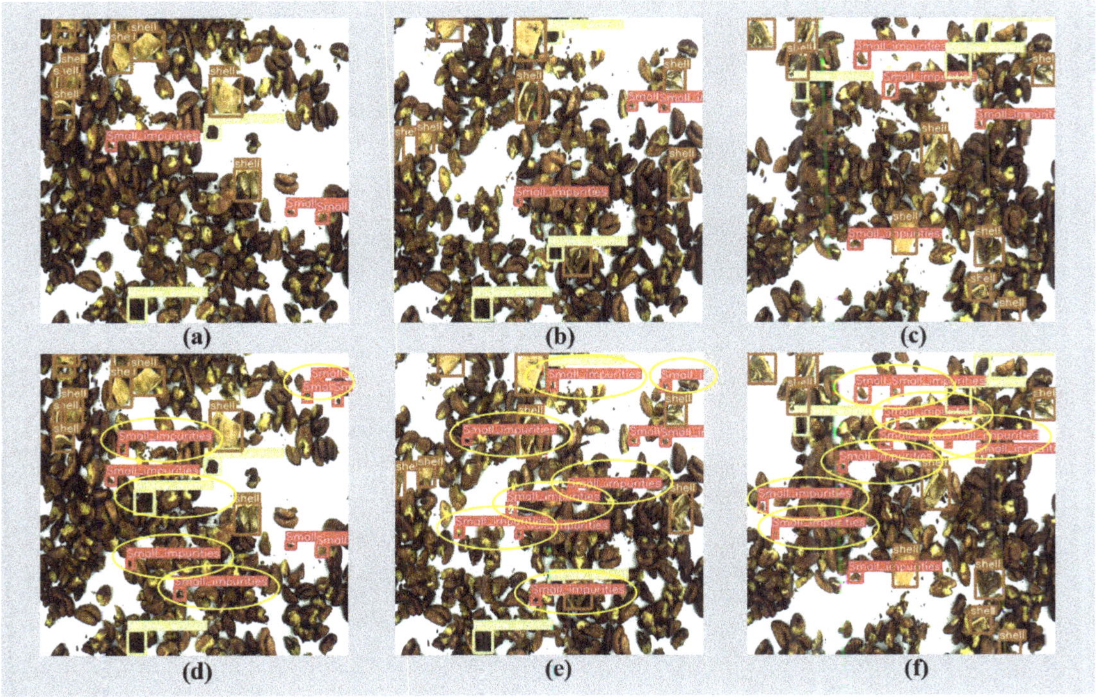

Figure 13. Comparison of the recognition effects of the YOLOv5 models. (**a–c**) recognition effects of the original YOLOv5; (**d–f**) proposed YOLOv5 network.

4. Conclusions and Future Research

The detection of walnut impurities is of great significance to the safety of nut food. In this paper, an impurity detection model of walnut kernels based on the improved YOLOv5 network is established: a small target recognition layer is added to the original prediction head of the model to obtain more small impurities feature information. Then, some convolution blocks in the network are replaced by Trans-E blocks, which can capture more comprehensive information in different subspaces at different locations. The CBAM attention module is added to the neck part of the network model for feature fusion, which improves the network performance at a small cost. Finally, Ghostconv is introduced to replace the original Conv, which reduces the computational burden of the model and improves the detection speed. The improved model detection *mAP* can reach 88.9% and *F*1 can reach 90.81%, which is better than the original YOLOv5 network and other networks. Moreover, the improved network model has not only a high detection rate, but also a significant improvement in the identification rate of small target impurities. The model improvement studied in this paper is to maintain a balance between detection performance and detection speed, so as to meet the demand of the real-time detection of walnut impurities. Near infrared spectroscopy is an important tool in the field of food impurity detection [28]. However, it requires demanding hardware. The detection technology based on YOLOv5 has a higher detection rate, lighter detection equipment and a wider range of application objects when compared to the near infrared spectroscopy. It also has certain advantages in detection accuracy. The research content is also applicable to other nut food impurity detection fields, and provides technical reference for the detection of snack food impurities.

However, the improved YOLOv5 model has limitations, such as a fraction of missing and wrong detection cases for small foreign bodies. Therefore, the detection accuracy of the model still needs to be improved. Improving the resolution of the camera is conducive to improving the detection accuracy. Then, due to the influence of external light source, the illumination of the image is biased. Fan Youchen et al. improved the YOLOv5 combined with dark channel enhancement to solve the problem of insufficient illumination. [29] This method can be applied to solve the illumination problem of the image. In addition, making the detection model lighter is one of the key points of future research. Chu et al. proposed a real-time apple flower detection method based on YOLOv4 and using the channel pruning method. [30] Isa Iza Sazanita et al. used the adaptive moment estimation optimizer and the function reducing-learning-rate-on-plateau to optimize the model's training scheme [31]. In the future, we can try to replace the backbone network with other lightweight networks to reduce the number of model parameters.

Author Contributions: Conceptualization, L.Y.; Data curation, Q.C.; Formal analysis, F.S.; Funding acquisition, M.Q.; Investigation, J.P.; Project administration, M.Q.; Resources, L.Y.; Software, Q.C.; Visualization, F.S. and J.P.; Writing—original draft, L.Y.; Writing—review and editing, M.Q. All authors have read and agreed to the published version of the manuscript.

Funding: This research was funded by National Key R&D Program of China (2017YFC1600800) and "Pioneer" and "Leading Goose" R&D Program of Zhejiang (2022C02042 & 2022C02057).

Informed Consent Statement: Not applicable.

Data Availability Statement: The data are available from the corresponding author.

Acknowledgments: The authors would like to thank the anonymous reviewers and editors whose insightful comments and valuable suggestions are crucial to the improvement of the manuscript. Financial support from the above funds and organizations are gratefully acknowledged.

Conflicts of Interest: The authors declare no conflict of interest.

References

1. Djekic, I.; Jankovic, D.; Rajkovic, A. Analysis of foreign bodies present in European food using data from Rapid Alert System for Food and Feed (RASFF). *Food Control* **2017**, *79*, 143–149. [CrossRef]
2. Aladjadjiyan, A.; Luning, P.A.; Devlieghere, F.; Verhé, J.R. Physical hazards in the agri-food chain. *Saf. Agric.* **2006**, *4*, 1082.
3. Rong, D.; Wang, H.; Xie, L.; Ying, Y.; Zhang, Y.J.C. Impurity detection of juglans using deep learning and machine vision. *Comput. Electron. Agric.* **2020**, *178*, 105764. [CrossRef]
4. Ok, G.; Kim, H.J.; Chun, H.S.; Choi, S.W. Foreign-body detection in dry food using continuous sub-terahertz wave imaging. *Food Control* **2014**, *42*, 284–289. [CrossRef]
5. Shen, Y.; Yin, Y.; Li, B.; Zhao, C.; Li, G. Detection of impurities in wheat using terahertz spectral imaging and convolutional neural networks. *Comput. Electron. Agric.* **2021**, *181*, 6781. [CrossRef]
6. Xie, T.; Li, X.; Zhang, X.; Hu, J.; Fang, Y. Detection of Atlantic salmon bone residues using machine vision technology. *Food Control* **2020**, *123*, 107787. [CrossRef]
7. Wang, C.; Xiao, Z. Potato Surface Defect Detection Based on Deep Transfer Learning. *Agriculture* **2021**, *11*, 863. [CrossRef]
8. Kuznetsova, A.; Maleva, T.; Soloviev, V. *In Detecting Apples in Orchards Using YOLOv3 and YOLOv5 in General and Close-Up Images*; International Symposium on Neural Networks; Springer: Cairo, Egypt, 2020.
9. Cengil, E.; Inar, A. Poisonous Mushroom Detection Using YOLOV5. *Turk. J. Sci. Technol.* **2021**, *16*, 119–127.
10. Thuan, D. Evolution of Yolo Algorithm and Yolov5: The State-of-the-Art Object Detention Algorithm. Bachelor's Thesis, Oulu University of Applied Sciences, Oulu, Finland, 2021.
11. Shao, H.; Pu, J.; Mu, J. Pig-posture recognition based on computer vision: Dataset and exploration. *Animals* **2021**, *11*, 1295. [CrossRef]
12. Krizhevsky, A.; Sutskever, I.; Hinton, G.E.J.C.; Otto, A. Imagenet classification with deep convolutional neural networks. 2017, 60, 84–90. *Commun. ACM* **2017**, *60*, 84–90. [CrossRef]
13. Wang, Y.; Wu, J.; Deng, H.; Zeng, X. Neuroscience, Food Image Recognition and Food Safety Detection Method Based on Deep Learning. *Comput. Intell.* **2021**, *2021*, 297.
14. Jubayer, F.; Soeb, J.A.; Mojumder, A.N.; Paul, M.K.; Barua, P.; Kayshar, S.; Akter, S.S.; Rahman, M.; Islam, A. Mold Detection on Food Surfaces Using YOLOv5. *Curr. Res. Food Sci.* **2021**, *4*, 724–728. [CrossRef] [PubMed]
15. Du, Z.; Fang, S.; Zhe, L.; Zheng, J. *Tomato Leaf Disease Detection Based on Deep Feature Fusion of Convolutional Neural Network*; China Science Paper: Beijing, China, 2020.
16. Qi, J.; Liu, X.; Liu, K.; Xu, F.; Guo, H.; Tian, X.; Li, M.; Bao, Z.; Li, Y. An improved YOLOv5 model based on visual attention mechanism: Application to recognition of tomato virus disease. *Comput. Electron. Agric.* **2022**, *194*, 106780. [CrossRef]

17. Han, W.; Jiang, F.; Zhu, Z. Detection of Cherry Quality Using YOLOV5 Model Based on Flood Filling Algorithm. *Foods* **2022**, *11*, 1127. [CrossRef] [PubMed]
18. Liu, M.; Li, C.; Cao, C.; Wang, L.; Li, X.; Che, J.; Yang, H.; Zhang, X.; Zhao, H.; He, G. Walnut fruit processing equipment: Academic insights and perspectives. *Food Eng. Rev.* **2021**, *13*, 822–857. [CrossRef]
19. Glenn, J. Yolov5. Git Code. 2020. Available online: https://github.com/ultralytics/yolov5 (accessed on 10 June 2021).
20. Chen, Z.; He, L.; Ye, Y.; Chen, J.; Wang, R. Automatic sorting of fresh tea leaves using vision-based recognition method. *J. Food Process Eng.* **2020**, *1*, e13474. [CrossRef]
21. Tzutalin, D. LabelImg. Git Code. 2015. Available online: https://github.com/tzutalin/labelImg (accessed on 10 June 2021).
22. Li, S.; Zhang, S.; Xue, J.; Sun, H. Lightweight target detection for the field flat jujube based on improved YOLOv5. *Comput. Electron. Agric.* **2022**, *202*, 107391. [CrossRef]
23. Yang, F. A Real-Time Apple Targets Detection Method for Picking Robot Based on Improved YOLOv5. *Remote Sens.* **2021**, *13*, 2972.
24. Han, K.; Wang, Y.; Tian, Q.; Guo, J.; Xu, C.; Xu, C. Ghostnet: More Features from Cheap Operations. In Proceedings of the IEEE/CVF Conference on Computer Vision and Pattern Recognition, Seattle, WA, USA, 13–19 June 2020; pp. 1580–1589.
25. Garcia-Garcia, A.; Orts-Escolano, S.; Oprea, S.; Villena-Martinez, V.; Garcia-Rodriguez, J. A Review on Deep Learning Techniques Applied to Semantic Segmentation; 2017. *arXiv* **2017**, arXiv:1704.06857.
26. Fan, Y.; Zhang, S.; Feng, K.; Qian, K.; Wang, Y.; Qin, S. Strawberry maturity recognition algorithm combining dark channel enhancement and YOLOv5. *Sensors* **2022**, *22*, 419. [CrossRef]
27. Zhao, L.; Zhi, L.; Zhao, C.; Zheng, W. Fire-YOLO: A Small Target Object Detection Method for Fire Inspection. *Sustainability* **2022**, *14*, 288. [CrossRef]
28. Ma, L.; Ma, J.; Han, J.; Li, Y. Research on target detection algorithm based on YOLOv5s. *Comput. Knowl. Technol* **2021**, *17*, 100–103.
29. Ma, S.; Lu, H.; Wang, Y.; Xue, H. YOLOX-Mobile: A Target Detection Algorithm More Suitable for Mobile Devices. *J. Phys. Conf. Ser.* **2022**, *2203*, 012030. [CrossRef]
30. Wu, D.; Lv, S.; Jiang, M.; Song, H. Using channel pruning-based YOLO v4 deep learning algorithm for the real-time and accurate detection of apple flowers in natural environments. *Comput. Electron. Agric.* **2020**, *178*, 105742. [CrossRef]
31. Isa, I.S.; Rosli, M.S.A.; Yusof, U.K.; Maruzuki, M.I.F.; Sulaiman, S.N. Optimizing the Hyperparameter Tuning of YOLOv5 For Underwater Detection. *IEEE Access* **2022**, *10*, 52818–52831. [CrossRef]

foods MDPI

Article

Evaluation of Rice Degree of Milling Based on Bayesian Optimization and Multi-Scale Residual Model

Weidong Chen [1,2,*], Wanyu Li [1] and Ying Wang [1]

1 College of Information Science and Engineering, Henan University of Technology, Zhengzhou 450001, China
2 National Engineering Research Center for Grain Storage and Logistics (Wheat), Zhengzhou 450001, China
* Correspondence: chenweid@haut.edu.cn

Abstract: Traditional machine learning-based methods for the detection of rice degree of milling (DOM) that are not comprehensive in feature extraction and have low recognition rates fail to meet the demand for fast, non-destructive, and accurate detection. This paper presents a digital image processing technology combined with deep learning to implement the classification of DOM of rice. An improved multi-scale information fusion model of the InceptionResNet–Bayesian optimization algorithm (IRBOA) was constructed based on the Inception-v3 structure and residual network (ResNet) model. It enables to automatically extract more comprehensive features of rice and determine the DOM of rice. Additionally, the important hyperparameters in the model were tuned by the BOA to optimize the recognition rate of rice DOM. The results show the hyperparameters optimized using the BOA are those that would not be chosen in manual tuning. The classification precision of the IRBOA model reached 99.22%, 94.92%, and 96.55% for well-milled, reasonably well-milled, and substandard rice, respectively, with an average accuracy of no less than 96.90%. This model improved 7.41% over the traditional machine learning model and at least 1.35% over the fashionable CNN model with strong generalization performance. This method effectively completes rapid, non-destructive, and accurate intelligent detection of rice DOM, which can supply a reliable and accurate technical mean for rice processing enterprises to guide the rice processing process.

Keywords: degree of milling; multi-scale information fusion; residual network model; Bayesian optimization algorithm

Citation: Chen, W.; Li, W.; Wang, Y. Evaluation of Rice Degree of Milling Based on Bayesian Optimization and Multi-Scale Residual Model. *Foods* **2022**, *11*, 3720. https://doi.org/10.3390/foods11223720

Academic Editor: Elena Canellas

Received: 10 October 2022
Accepted: 17 November 2022
Published: 19 November 2022

1. Introduction

Paddy is a major grain in the world. As the worldwide population grows, the requirement for rice is expected to rise by 30% in 2050 [1]. Therefore, the processing and production of rice have a vital role. At present, there are prominent problems in the rice market, such as the one-sided pursuit of appearance quality (fine, white, and nice taste), backward control means of the DOM, and nutrient loss caused by over-processing, which threaten food security [2]. Thus, an efficient and rapid method of estimating the DOM of rice can instruct enterprises to adjust the parameters in the rice milling process in real-time. Additionally, enterprises can perform such approaches to moderately process rice and achieve efficient rice loss reduction through technological innovation. It has essential significance for guiding paddy processing, rice storage, distribution, and trade.

According to the regulations of the Chinese National Standard of "Milled rice (GB/T 1354-2018) [3]", rice DOM refers to the degree of germ remaining and the residual bran layer on the surface and back grooves of a rice grain after processing, which is divided into three levels: well-milled, reasonably well-milled, and substandard. Well-milled, reasonably well-milled, and substandard rice represent rice with skin retention less than 2%, between 2% and 7%, and more than 7%, respectively. The skin retention of rice is defined as the sum of the residual skin and rice embryo projection area as a percentage of the projection area of the sample. In rice processing enterprises, detecting the DOM of rice is still at the stage of human eye inspection or staining method to auxiliary implementation. These approaches have the disadvantages of strong professionalism, being time-consuming and labor-intensive, poor repeatability, etc. Foreign researchers found that rice DOM is closely related to its chemical composition content [4]. They extracted the lipid content of the milled rice surface by chemical extraction to quantify the DOM of rice [5]. However, this method cannot meet the requirements of modern rice DOM for rapid, non-destructive, efficient, and objective detection.

Machine vision technology provides the advantages of high efficiency, fast speed, and accurate detection, which is currently a research hot spot in the field of crop detection [6–8]. Xu et al. [9] and Wood et al. [10] detected the DOM of rice by digital image processing technology combined with the staining method, but the staining process was cumbersome and destructive. Zhang et al. [11] obtained the rice DOM by the bran degree of RGB images of rice. Wan and Long [12] and Wan et al. [13] proposed detection methods based on gray-gradient co-occurrence matrix and color features incorporated with machine learning, respectively, and the corresponding discrimination accuracy reached 94% and 92.17%. Fang et al. [14] used grayscale values of rice to measure DOM. Zareiforoush et al. [15] adopted the fuzzy logic reasoning method to realize the recognition of five rice milling grades, and the overall confidence reached 89.80%. Hortinela et al. [16] used the support vector machine to classify milled rice with an adaptive enhancement algorithm, and the average accuracy was 86.67%. Although the above methods achieved positive detection results, they all need to design and extract features manually, and there is the problem that incomplete feature extraction leads to low accuracy.

In recent years, CNN has achieved remarkable achievements in face recognition [17], handwritten digit recognition [18], pedestrian detection [19], and other fields, bringing new opportunities for the development of rice DOM detection technology. In terms of DOM detection of rice, Qi et al. [20] combined the hypercolumn technology, max-relevance and min-redundancy feature selection algorithm, extreme learning machine technique, and improved VGG16 to identify rice DOM with an overall accuracy of 97.32%. For the quality inspection of rice, Patel and Joshi [21] used the transfer learning-based VGG16 model for fine rice, broken rice, and variety determination. A four-layer CNN model to realize head and broken rice classification was adopted by Hong Son and Thai-Nghe [22]. Li and Li [23] improved Inception-v3 by introducing fine-grained classification to learn local features of rice and to identify the integrity of the rice germ. Li et al. [24] refined the Inception-v3 model to detect the integrity of the germ with the addition of mutual channel loss and mlpconv. Li et al. [25] identified rice germ integrity based on the EfficientNet-B3 model with the introduction of the double attention network (DAN).

To summarize, existing research on rice is mostly quality examination, while the determination of rice DOM has essential guidance for maintaining food nutrition and reducing food waste. The current research is unable to acquire the feature details of rice well, and there is still a lack of deep learning-based methods that can effectively and correctly identify the DOM of rice. Therefore, the main contributions of this study are as follows:

(1). Simple image preprocessing and single-grain rice segmentation methods are used to segment single-grain rice images from multiple-grain rice images. Then, they are fed into the improved IRBOA model for rice DOM classification.
(2). The Inception-v3 structure with ResNet34 are combined to fuse rice features at different scales and enrich the feature representation, thereby enabling the detection of rice DOM and enhancing the recognition accuracy of the model.
(3). We used BOA to search for the hyperparameters that lead to the optimal model performance in order to avoid the problem of manual setting of hyperparameters that fail to obtain the peak accuracy. The method can increase the discrimination rate of the model via upgrading the efficiency of manual search.

2. Materials and Methods

2.1. Experimental Materials and Image Acquisition

Standard samples of early indica rice DOM (SAC LS/T 15121-2020), including well-milled, reasonably well-milled and substandard, were selected from the Anhui grain and oil products quality supervision and testing station in Hefei, Anhui Province, China. A total of 50 g of each class of rice was used for sample preparation. Each five grams of rice was packed in a sealed bag as a group, and each type of rice was packed in 10 groups. Finally, there were 30 groups of three types of rice, marked with the corresponding serial numbers, and stored in a refrigerator at 0–5 °C to prevent the influence of sample deterioration on the inspection results.

According to the requirements of rice image acquisition, a Phantom h9 flatbed scanner was used to acquire RGB images of rice in multiple mixed poses with the background of a black frosted Acrylic plate. The contrast ratio, brightness, resolution, and image size of the flatbed scanner were set to 65, 30, 600 dpi, and 5000 pixels × 7000 pixels, respectively. Image acquisition was carried out in units of five grams, and each group of rice was placed on the draft table of the scanner with the help of a separating sieve to avoid the adhesion of rice grains. Then, image scanning was performed. Next, the operation of random placement and scanning was executed again to fully utilize the sample and obtain two different images. Finally, the scanned rice was put into the corresponding sealed bag, and the other group of rice was repositioned on the scanner. The above steps were performed on 30 groups of samples of well-milled, reasonably well-milled, and substandard in turn. Finally, a total of 60 valid images were obtained, some of which are shown in Figure 1.

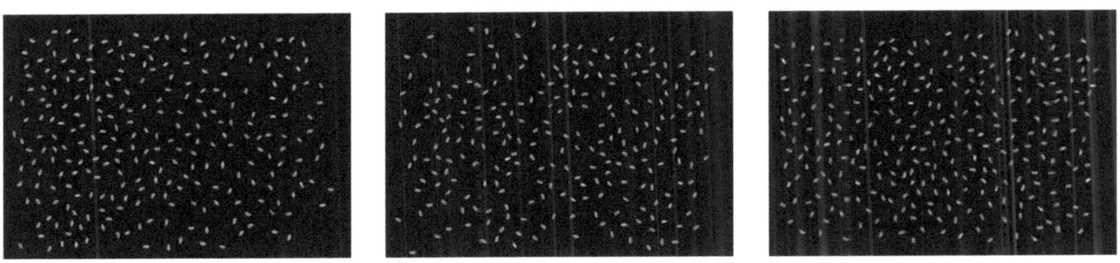

(a) (b) (c)

Figure 1. Images of the original multi-grain rice. (**a**) Well-milled. (**b**) Reasonably well-milled. (**c**) Substandard.

2.2. Image Preprocessing

The image quality of the original images of multi-grain rice is affected by noise due to the limitation of the shooting conditions. So, a series of preprocessing operations were taken for the images to selectively highlight effective features and eliminate irrelevant information in order to improve the image quality and increase the classification and recognition accuracy. Meanwhile, in this research, we performed image smoothing, binarization, and segmentation of single-grain rice on the original rice images before inputting the single-grain rice images into the CNN model.

2.2.1. Image Smoothing and Binarization

We first converted each color image to grayscale using an image grayscale transform. Image smoothing was achieved by median filtering that can eliminate image noise while preserving image edge information before implementing image segmentation [26]. We used a fixed threshold to complete the image binarization operation, which avoided the situation of separating rice endosperm and bran by other methods. Finally, we performed a morphological opening and closing operation on the binarized image to smooth the image and fill the holes inside the target rice.

2.2.2. Segmentation of Single-Grain Rice Images

The Canny algorithm of contour detection was used to detect the edge of each grain of rice. The minimum circumscribed rectangle of each rice was drawn, and its four vertex coordinates and rotation angle were gained. Next, the original rice image was rotated by the derived rotation angle. Finally, image segmentation of single-grain rice in a vertical state was realized by extending the coordinates of the rotated rectangle vertex to the surroundings by 5 pixels as the boundary. Figure 2 shows the sample data of three kinds of DOM rice after single-grain segmentation.

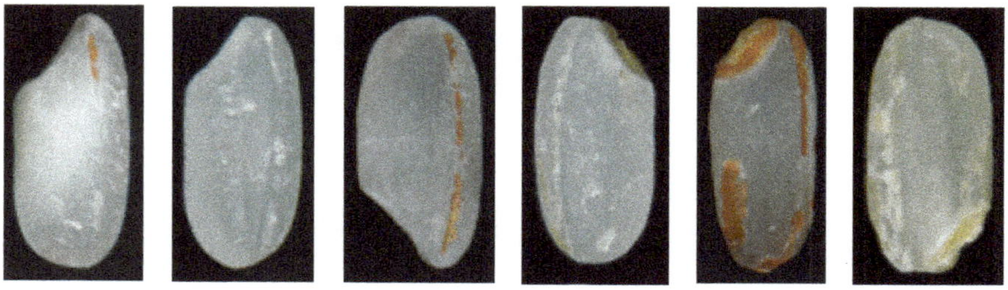

(a) (b) (c) (d) (e) (f)

Figure 2. Single-grain rice images of three kinds of DOM. (**a**,**b**) Well-milled. (**c**,**d**) Reasonably well-milled. (**e**,**f**) Substandard.

2.3. Data Augmentation

A dataset was established based on the segmented single-grain rice images, and 5800 valid images each of well-milled, reasonably well-milled, and substandard rice was obtained, for a total of 17,400 images. Each category of rice dataset was divided into a training set, validation set, and test set with a ratio of 6:2:2 for each category. That means obtaining 3480 images per class of rice for the training set and 1160 images for the validation and test sets, respectively. The training set is used for training the model, while the validation set is employed to optimize the model structure and hyperparameters, and the test set is only designed to test the performance of the model to enhance its generalization ability.

It is essential to enhance the training set data to reduce the incidence of overfitting when the data are limited. Firstly, each rice was cropped to an image of the same size (224 pixels $\times$ 224 pixels) by the center cropping for input into the CNN model. Secondly, 30% of the training data were randomly selected for horizontal and vertical flipping, respectively. Then, a random rotation was executed for each image with rotation angles ranging from 35° to 135°. Finally, the mean and standard deviation of the three color channels of all training set images were calculated and fed into the normalization function to realize the normalization of each image. The training set was expanded according to the above steps to derive sufficient data to train models.

2.4. Proposed Approach

CNN is one of the most popular deep learning models and is widely used in image classification tasks at present. It is not only able to extract features of target objects in images automatically and comprehensively but also possesses the characteristic of weight sharing, which reduces the training parameters of the network and makes the model simpler [27]. We constructed an IRBOA model which can fuse multi-scale information based on the integration of the Inception-v3 structure and ResNet model to classify rice from three kinds of DOM. The model used was as described below.

2.4.1. Inception Structure

Inception structure is a significant breakthrough in the development history of CNN models. Its purpose is to execute multiple convolution operations or pooling operations on the input image in parallel and concatenate all the outputs to attain more comprehensive image features. This structure was first introduced by GoogLeNet and called Inception-v1 [28]. Subsequently, it was improved to the Inception-v2 structure by applying batch normalization (BN) [29] and convolutional decomposition. Then, it evolved into the Inception-v3 network by adding asymmetric convolution, auxiliary classifiers, etc. The architecture not only accelerates the computation but also improves the generalization ability of the model while eliminating the use of dropout in the batch normalization network [30]. Currently, the Inception structure has been developed to the Inception-v4 [31].

2.4.2. ResNet Model

ResNet, which emerged in 2015, marks a milestone in deep learning [32]. It adjusts the structure of the traditional CNN models, in which the most critical residual structure adds an identity mapping to the basic network unit [33]. The residual structures are shown in Figure 3. The original fitting target of the residual structure is $H(x)$, and it becomes extremely difficult to learn $H(x)$ with the gradual deepening of the network level. Thus, transforming the fitting target into the fitted residual function $F(x)$ ($F(x) = H(x) - x$) through the residual structure and turning the output into a superposition of the fit, and the input will make the learning of the network relatively easy. The residual learning is adopted for each stacked layer in ResNet, and the residual learning formula is defined as:

$$y = F(x, \{w_i\}) + x \tag{1}$$

where x and y are the input and output vectors of the residual structure of this layer, and $F(x, \{w_i\})$ represents the residual mapping to be learned. For the example in Figure 3 that has two layers, $F = w_2 ReLU(w_1 x)$ in which ReLU denotes ReLU activation function. In addition, the dimensions of $F(x, \{w_i\})$ and x should be consistent. w_S, a square matrix, can be conducted through identity mapping to match the dimensions when the input or output dimension information needs to be changed, as shown in Figure 3b.

$$y = F(x, w_i) + w_S x \tag{2}$$

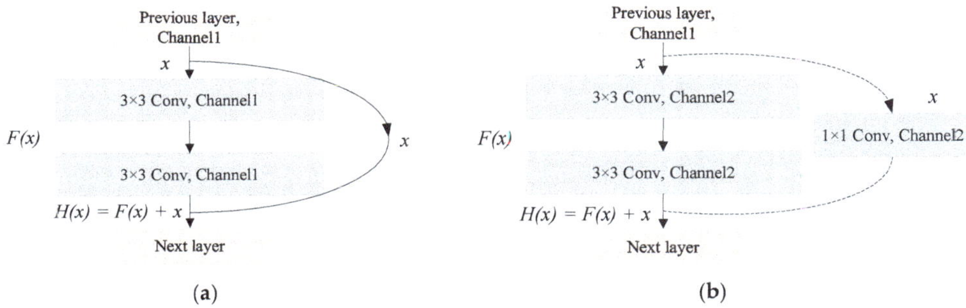

Figure 3. Residual structure. (**a**) Residual-A structure. (**b**) Residual-B structure.

2.4.3. Custom Model

The Inception-v3 structure offers the characteristics of fusing multi-scale features and accelerating network computation, while the residual structure in ResNet prevents gradient explosion, gradient disappearance, and network degradation when the number of network layers is deepened. Consequently, in this study, we integrated the Inception-v3 structure and residual module and established a multi-scale information fusion CNN model based on ResNet34 architecture, named InceptionResNet–BOA model, or IRBOA model for short. The model was adopted to enrich the rice feature information and promote the recognition effect. The structure of the IRBOA model is shown in Figure 4. The input of the model is a 224 × 224 × 3 color image, and the model architecture consists of an Inception-A structure as shown in Figure 5a, a maximum pooling layer, five Residual-A structures, two Residual-B structures, an Inception-B structure as shown in Figure 5b, and an average pooling layer. The input of the fully connected layer is the number of flattened characteristic maps of the average pooled layer. While the count of neurons of this layer is the amount of rice DOM types to classify rice DOM.

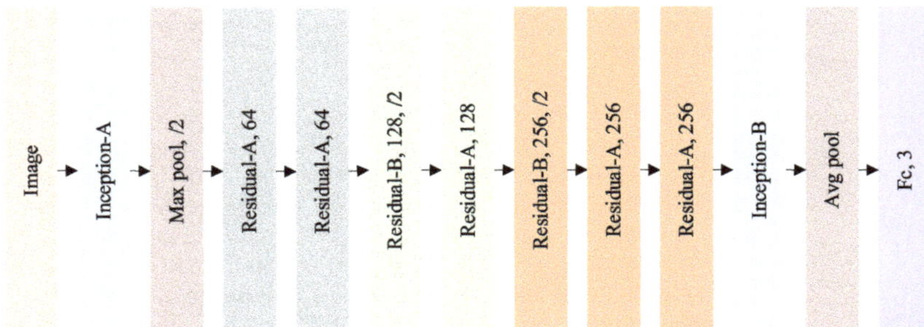

Figure 4. The architecture of the IRBOA model.

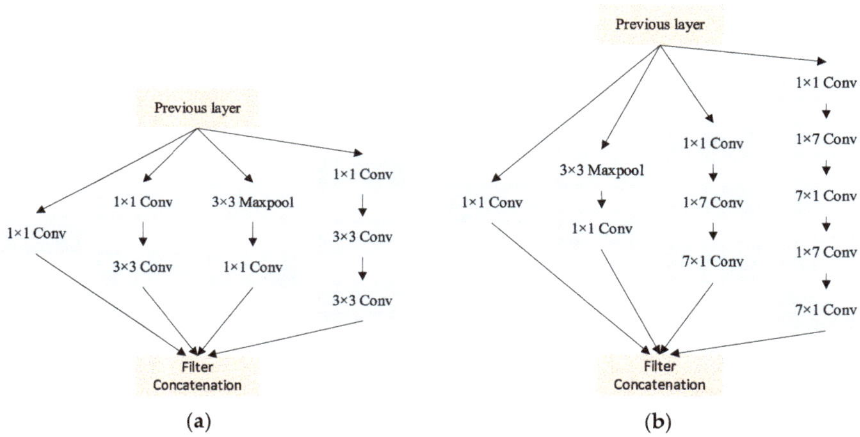

Figure 5. Inception-v3 structure. (**a**) Inception-A structure. (**b**) Inception-B structure.

Table 1 displays the parameter settings for each layer of the IRBOA model. The Inception-A structure is a parallel combination of a series of 1 × 1 convolution layers, 3 × 3 convolution layers, and a 5 × 5 convolution layer replaced by two 3 × 3 convolution layers, with the number of convolution kernels from branch1 to branch4 being 8, 12, 24, 8, 12, 24, 24, respectively. The Residual-A structure contains two convolutional layers with 3 × 3 kernels and an identity mapping, and the number of convolutional kernels in Residual-A1 to A4 are 64, 128, 256, and 256, respectively. Residual-B structure matches the number of channels in the two pathways by 1 × 1 convolution at identity mappings based on the Residual-A structure, with 128 and 256 convolution kernels for Residual-B1 to B2. The Inception-B structure is combined by 1 × 1 convolution layers, asymmetric 1 × 7 convolution layers, and 7 × 1 convolution layers. The number of convolution kernels from branch1 to branch4 are 64, 128, 64, 64, 128, 192, 192, 192, 192, and 128, respectively.

Table 1. Parameters of the IRBOA model structure.

Name of Layer		Parameters (Kernel_Size, Kernel_Num, Padding, Stride)
Image input		224 × 224 × 3
Inception-A	branch1	1 × 1 Conv, 8, 0, 1
	branch2	1 × 1 Conv, 12, 0, 1 3 × 3 Conv, 24, 1, 1
	branch3	3 × 3 Maxpool, –, 1, 1 1 × 1 Conv, 8, 1, 1
	branch4	1 × 1 Conv, 12, 0, 1 3 × 3 Conv, 24, 1, 1 3 × 3 Conv, 24, 1, 1
Filter concatenation		224 × 224 × 64
MaxPool		3 × 3 MaxPool, –, 1, 2
Residual-A1		–, 64, 1, 1
Residual-A2		–, 64, 1, 1
Residual-B1		–, 128, 1, 2
Residual-A3		–, 128, 1, 1
Residual-B2		–, 256, 1, 2
Residual-A4		–, 256, 1, 1
Residual-A5		–, 256, 1, 1
Inception-B	branch1	1 × 1 Conv, 64, 0, 1
	branch2	3 × 3 MaxPool, –, 1, 1 1 × 1 Conv, 128, 0, 1
	branch3	1 × 1 Conv, 64, 0, 1 1 × 7 Conv, 64, [0, 3], 1 7 × 1 Conv, 128, [3, 0], 1
	branch4	1 × 1 Conv, 192, 0, 1 1 × 7 Conv, 192, [0, 3], 1 7 × 1 Conv, 192, [3, 0], 1 1 × 7 Conv, 192, [0, 3], 1 7 × 1 Conv, 128, [3, 0], 1
Filter concatenation		28 × 28 × 512
Avg_pool		1 × 1 × 512
Fc		3

"–" represents that there is no corresponding parameter.

2.5. Optimization Methods of the Model

2.5.1. BOA

Determining how to select appropriate hyperparameters has become a key issue in image classification tasks in the circumstance that the performance of the model largely depends on the selection of hyperparameters. The method of manual optimization is difficult and time-consuming to find the optimal parameters. Recently, the widely used methods of automatic parameter tuning of machines include the grid search algorithm (GSA), the random search algorithm (RSA), and the BOA. The essence of the GSA is the enumeration method, which is costly in terms of time spent when the objective function is more complex [34]. Although the RSA no longer tests all values within a parameter range, randomly selected sample points in the search range may ignore optimal values [35]. The BOA is one of the most popular methods for tuning hyperparameters in deep learning models [36]. Its main idea is that, given an objective function to be optimized, the posterior distribution of the objective function is updated by continuously adding sample points until the posterior distribution approximately corresponds to the true distribution or the function is executed for a predetermined number of iterations. It is a technique for adjusting hyperparameters based on the priori information, which is faster, more effective, and more efficient than the previous two algorithms. The major problem scenarios of the BOA are as follows:

$$X^* = \arg_{x \in S} \max f(x) \tag{3}$$

Here, S is the candidate set of x and $f(x)$ is the objective function. The target of the BOA is to pick an x from S such that the value of $f(x)$ is maximized or minimized.

The BOA was used to optimize the hyperparameters of the back propagation neural network (BPNN), AlexNet, VGG16, ResNet34, and IRBOA models. The activation function adopted for each model was ReLU with each batch_size set to 64, and the training epoch for the BPNN and CNN models were 5000 and 100, respectively. The cross-entropy function was employed for the loss function and the accuracy of the validation set was selected for the objective function of the BOA. The optimized variables are those proposed in 2.5.2, 2.5.3, and 2.5.4, including the number of neurons in the hidden layer of the BPNN (hidden), optimizer, learning_rate, the update interval in the learning rate decay algorithm (step_size), the multiplication factor for updating the learning rate (gamma), and L2 regular term parameters (weight_decay). Table 2 shows the search space of each hyperparameter.

Table 2. Hyperparameters search space based on BOA.

Model	Hyperparameter	Search Space
BPNN	hidden	{10, 12, 14, 16, 18, 20, 22, 24}
	optimizer	{SGD, Adam}
	learning_rate	[0.1, 0.00001]
	step_size	{600, 800, 1000, 1200, 1400}
	gamma	[0.1, 0.00001]
AlexNet, VGG16, ResNet34, IRBOA	optimizer	{SGD, Adam}
	learning_rate	[0.1, 0.00001]
	step_size	{10, 15, 20, 25, 30}
	gamma	[0.1, 0.00001]
	weight_decay	[0.1, 0.00001]

2.5.2. Optimizer

The optimizer is designed to minimize the loss in the training process through gradient descent, thereby enhancing the accuracy of the model. The stochastic gradient descent (SGD) algorithm and the adaptive momentum estimation (Adam) algorithm are two superior optimizers for image classification tasks in deep learning. Each of them has its advantages and disadvantages, hence the optimizer was selected to make the model optimal by employing the BOA in Section 2.5.1.

2.5.3. Learning Rate

Learning rate is a very crucial hyperparameter in CNN classification models and impacts the recognition accuracy of the model. It is difficult and extremely important to choose the appropriate learning rate. In this paper, the model was trained by the equal-interval learning rate decay method, where the values of step_size and gamma were determined by BOA. The equation for the equal-interval learning rate decay is as follows.

$$new\_lr = initial\_lr \times gamma^{\frac{epoch}{step\_size}} \quad (4)$$

where new_lr is the learning rate after decay, $initial\_lr$ is the learning rate before decay, $gamma$ is the decay rate less than 1, $epoch$ is the number of training rounds, and $step\_size$ is the decay step.

2.5.4. Regularization

Regularization is performed by adding penalty terms for the loss function to reduce model complexity and instability to avoid overfitting the model. L2 regularization not only prevents overfitting but also makes the process of optimizing the solution stable and fast through weight decay. Therefore, the L2 regularization method was adopted to solve the problem of model overfitting, and the regular term parameter was calculated by BOA.

2.6. Performance Evaluation Indicators for the Model

Confusion matrix, accuracy, precision, recall, and F1-score are usually used to evaluate the performance of models for single-label image classification issues [37]. The confusion matrix is mainly used to compare the objective results with the predicted results when evaluating the recognition accuracy of the images. Accuracy refers to the probability of predicting correct samples among all samples. Precision indicates the proportion of samples with positive predictions that are correctly predicted. Recall denotes the proportion of correctly predicted outcomes in the actual sample of true examples. In the actual situation, precision and recall are mutually "restricted". Therefore, we need the F1-score, a weighted average of precision and recall, to comprehensively evaluation the performance of models. The higher the F1-score, the better the performance of the model. The calculation formula of each indicator is as follows.

$$\text{Precision (P)} = \frac{TP}{TP + FP} \tag{5}$$

$$\text{Recall (R)} = \frac{TP}{TP + FN} \tag{6}$$

$$\text{Accuracy (Acc)} = \frac{TP + TN}{TP + TN + FP + FN} \tag{7}$$

$$\text{F1-score} = \frac{2 \times \text{P} \times \text{R}}{\text{P} + \text{R}} \tag{8}$$

Here, TP is the number of samples where the actual case is true, and the predicted outcome is positive. TN is the number of samples where the actual case is true and the predicted outcome is negative, and the same for FP and FN. They can be calculated by a confusion matrix.

2.7. Experimental Environment

All models used in this study were trained and tested based on the Windows 10 operating system and the following specifications: Intel ® Core™ i7-11800H CPU @ 2.30 GHz, 16 GB RAM, NVIDIA GeForce RTX 3060 GPU under CUDA v11.1 and cuDNN v8.0.5, PyTorch v1.9.0 (Facebook, America).

3. Results and Discussion

Rice image datasets with different DOMs were trained on BPNN, AlexNet, VGG16, ResNet34, and IRBOA models. In addition, we compared the five models to find the optimal rice DOM inspection model. The training epochs for the BPNN and CNN models were 5000 and 100, respectively. Figure 6 shows the loss and accuracy curves of the four CNN models on the training set. The horizontal axis in the graph is the number of training epochs, and the vertical axes are the loss value (Loss) and accuracy (Acc) of the model, respectively. With the continuous increase of training epochs, the classification error of the training set shows a downward trend, and the accuracy shows an opposite trend. When the training epochs of the IRBOA model reach 69, the training loss is close to a stable value. The stable value of the average loss is 0.087, which is lower than the other three CNN models, and the accuracy is significantly higher than other models. In conclusion, the IRBOA model designed in this paper is reasonable and provides satisfactory training results.

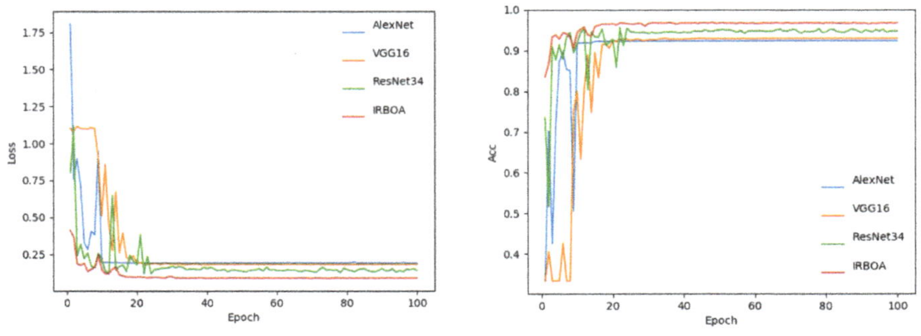

(a) (b)

Figure 6. Comparison of the learning curves of the four CNN models. (**a**) Loss curve. (**b**) Accuracy curve.

The hyperparameter optimization result of the IRBOA model is shown in Figure 7. The horizontal axis (Trial) in Figure 7 represents the number of iterations of the BOA, when it is 98, the objective function value is 0.9690 and the best result is obtained. However, the value of the objective function is still changing as the number of iterations increases. The effect indicates that the BOA is still trying to explore other optimal positions while approaching the optimal value. Table 3 lists hyperparameters obtained by the BOA for the five models, from which we can see that the hyperparameters are those that would normally not be set manually. The algorithm saves time and achieves results that cannot be captured by manual search. The models were trained and tested based on the optimized hyperparameters and the recognition rates were calculated for each model based on the test set. According to the comparative analysis in Table 3, we found that the detection accuracy of the IRBOA model for recognizing rice images was higher than that of the other four models, at 96.90%.

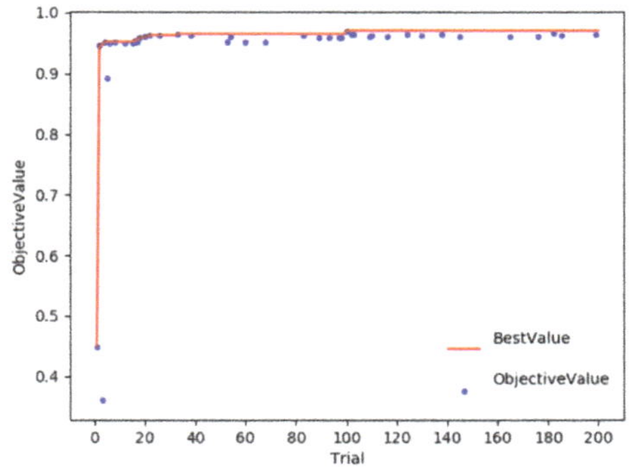

Figure 7. The Bayesian optimization process for the IRBOA model.

Accuracy is not sufficient to describe the practical application performance of the model in the case of significant differences and imbalances in the data samples. Confusion matrices were plotted for several models based on the test set (Figure 8) to accurately assess the classification performance of the above five classification models for rice DOM. The actual categories (horizontal axis) are compared with the predicted category (vertical axis) in Figure 8 to depict the individual classification performance of each category. 'A' in the diagram for well-milled, 'B' for reasonably well-milled, and 'C' for substandard. These results demonstrated that the classification effect of the CNN models was better than that of BPNN, with the IRBOA model offering the best classification efficiency. The recognition precision of this model was 99.22%, 94.92%, and 96.55% for well-milled, reasonably well-milled and substandard rice, respectively, with an average correct detection rate of 96.90%. The accuracy of the IRBOA model is 7.41% higher than that of traditional machine learning and no less than 1.35% higher than that of the classic CNN models.

Table 3. Hyperparameter results for the five models of Bayesian optimization.

Model	Parameter (Hidden, Optimizer, Learning_Rate, Weight_Decay, Step_Size, Gamma)	Accuracy (%)
BPNN	12, Adam, 0.054, –, 1000, 0.00034	89.49
AlexNet	–, SGD, 0.035, 0.0001, 10, 0.0005	92.30
VGG16	–, SGD, 0.016, 0.00027, 20, 0.053	92.93
ResNet34	–, Adam, 0.00011, 0.00012, 25, 0.0001	95.55
IRBOA	**–, Adam, 0.00019, 0.0002, 15, 0.085**	**96.90**

"–" represents that there is no corresponding parameter.

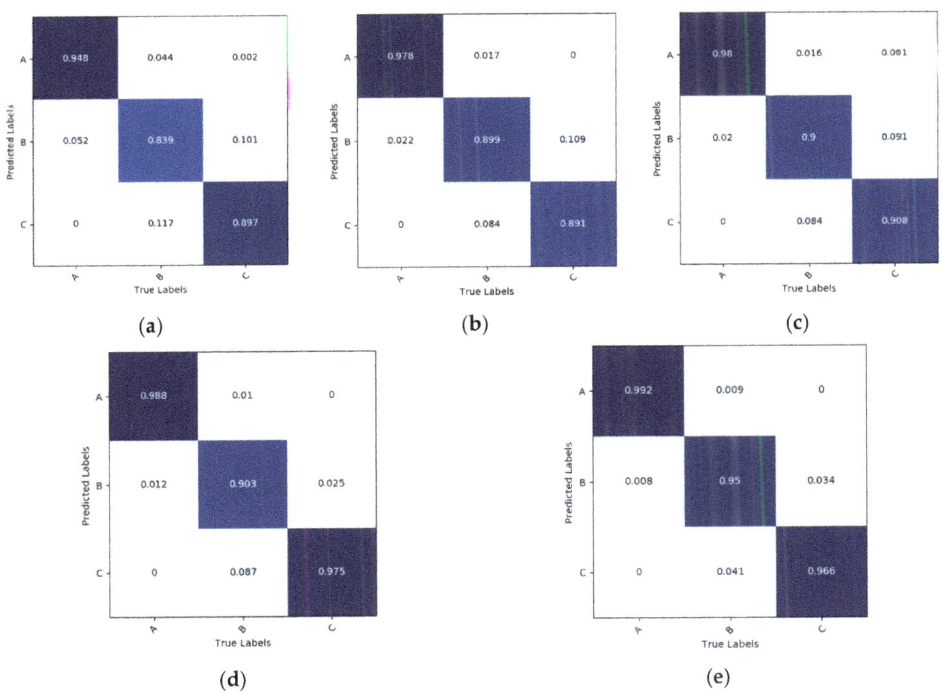

Figure 8. Confusion matrix of five models. (**a**) BPNN. (**b**) AlexNet. (**c**) VGG16. (**d**) ResNet34. (**e**) IRBOA.

According to the prediction value in the confusion matrix, four different statistical indicators were attained, namely, *TP*, *TF*, *FP*, and *FN*. Moreover, the four evaluation indicators of accuracy, precision, recall, and F1-score, as well as the training time and single image test time of each model were calculated to compare the performance of several classification models (Table 4). The precision, recall, and F1-score of the IRBOA model were all 96.90% from Table 4. The corresponding values of BPNN, AlexNet, VGG16, and ResNet34 were all lower than the model proposed. Their F1-scores were 89.43%, 92.32%, 92.94%, and 95.59%. The experiments indicated that the recognition performance of the IRBOA model is better than that of the remaining four models, with higher accuracy and generalization performance. Meanwhile, we found that the BPNN took a longer time when testing the network on a single piece of data although its training time of it was much faster than the CNN model. The reason for this consequence is the BPNN takes a large amount of time in extracting the color and texture feature parameters and in reducing the dimension of the feature parameters using principal component analysis. The IRBOA model for recognizing rice DOM is characterized by its long training time, but high detection accuracy and less than 20 milliseconds for a single image among the four CNN models. The effect of the model proposed can meet the actual needs in terms of temporal and model recognition performance.

Table 4. Detection performance indicators for the five models.

Model	Accuracy (%)	Precision (%)	Recall (%)	F1-Score (%)	Training Time (h)	Single Image Detection Time (s)
BPNN	89.49	89.44	89.43	89.43	0.10	20.69
AlexNet	92.30	92.35	92.30	92.32	0.56	2.87
VGG16	92.93	93.00	92.93	92.94	3.52	6.61
ResNet34	95.55	95.64	95.54	95.59	1.78	3.74
IRBOA	**96.90**	**96.90**	**96.89**	**96.90**	**4.22**	**12.93**

4. Conclusions

The nutritional value of rice decreases with the fineness of the rice DOM, while the processing process causes unnecessary food waste and affects national food security.

The purpose of this study was to solve the problems of the high labor intensity of traditional manual detection of rice DOM with manual feature extraction and a low recognition rate of existing classification methods based on machine learning. This paper presents an IRBOA model capable of extracting multi-scale rice features to identify classified rice DOM to further guide the processing process of rice enterprises.

The classical CNN model was improved by fusing the Inception-v3 structure and the residual structure. IRBOA, a multi-scale information fusion model, was constructed and its identification accuracy was enhanced relative to other classical networks. In addition, we used the BOA to seek the hyperparameters that led to the optimal performance of the model and increased the correct classification rate of the model. The IRBOA model, which performed hyperparameter optimization by BOA, achieved a recognition rate of 96.90% for rice DOM, while the testing time for a single image was less than 20 ms. The accuracy of IRBOA improved by 7.41 and no less than 1.35 percentage points relative to traditional machine learning methods and classic CNN models, respectively. The model enhances the feature representation and has better classification performance and generalization ability.

This study has demonstrated the feasibility of the inspection method proposed, which can provide a certain guidance to the processing work of rice enterprises and provide a reliable and accurate technical means for the classification of rice DOM level. More importantly, real-time rice DOM level evaluation can be achieved in the actual production process. Subsequently, the model can be combined with specific sorting apparatus to sort rice that has reached a certain DOM level in the rice milling section. It avoids the rice being over-milled in the next milling stage, so as to reach the goal of moderate processing and grain saving.

However, there are still some shortcomings in the research of this paper, and we will improve our current work in the following two aspects in the future work: (1) The model is prone to error attributed to the acquisition of single-sided images due to the different bran degrees on two sides of different DOMs rice. In the future, we will adopt the method of double-sided image acquisition [38] to improve the recognition rate of the model. (2) The chalky region of rice will have an impact on the discrimination of DOM level. In future research, we will search for effective image processing means to reduce the influence of the chalky areas of rice. (3) The accuracy of the model proposed only reaches 96.90%, which not only takes a long training time but also requires a large number of training samples. In the future, we can try to use the lightweight model [39,40] with small samples to save training time, or use the transfer learning model [41,42] to improve the recognition accuracy while reducing training time and samples.

Author Contributions: Conceptualization, W.C. and W.L.; methodology, W.L.; software, W.L.; validation, W.L. and Y.W.; formal analysis, W.C.; investigation, W.L.; resources, W.C.; data curation, W.L. and Y.W.; writing—original draft preparation, W.L.; writing—review and editing, W.C. and W.L.; visualization, W.L.; supervision, W.C.; project administration, W.C.; funding acquisition, W.C. All authors have read and agreed to the published version of the manuscript.

Funding: This research received no external funding.

Institutional Review Board Statement: This study does not involve humans or animals.

Informed Consent Statement: This study does not involve humans.

Data Availability Statement: The data presented in this study are available on request from the corresponding author.

Conflicts of Interest: The authors declare that they have no conflict of interest.

References

1. Zhang, G.; Sun, B.; Zhao, H.; Wang, X.; Zheng, C.; Xiong, K.; Ouyang, Z.; Lu, F.; Yuan, Y. Estimation of Greenhouse Gas Mitigation Potential through Optimized Application of Synthetic N, P and K Fertilizer to Major Cereal Crops: A Case Study from China. *J. Clean. Prod.* **2019**, *237*, 117650. [CrossRef]
2. Fan, Q.; Liu, M.Y.; Qi, H.Q. Research on Loss in Rice Processing and Governance Measures. *Sci. Technol. Cereal. Oils Foods* **2015**, *23*, 117–120. [CrossRef]
3. State Administration for Market Regulation; Standardization Administration of the People's Republic of China. *Milled Rice: GB/T 1354-2018*; State Administration for Market Regulation: Beijing, China; Standardization Administration of the People's Republic of China: Beijing, China, 2018.
4. Yoon, S.-H.; Kim, S.-K. Physicochemical Properties of Rice Differing in Milling Degrees. *Food Sci. Biotechnol.* **2004**, *13*, 57–62.
5. Matsler, A.L.; Siebenmorgen, T.J. Evaluation of Operating Conditions for Surface Lipid Extraction from Rice Using a Soxtec System. *Cereal Chem. J.* **2005**, *82*, 282–286. [CrossRef]
6. Nagoda, N.; Ranathunga, L. Rice Sample Segmentation and Classification Using Image Processing and Support Vector Machine. In Proceedings of the 2018 IEEE 13th International Conference on Industrial and Information Systems (ICIIS), Rupnagar, India, 1–2 December 2018; pp. 179–184.
7. Wu, A.; Zhu, J.; Yang, Y.; Liu, X.; Wang, X.; Wang, L.; Zhang, H.; Chen, J. Classification of Corn Kernels Grades Using Image Analysis and Support Vector Machine. *Adv. Mech. Eng.* **2018**, *10*, 168781401881764. [CrossRef]
8. Hu, Y.; Du, Y.; San, L.; Tian, J. Research on Rice Grain Shape Detection Method Based on Machine Vision. In Proceedings of the 2019 5th International Conference on Control, Automation and Robotics (ICCAR), Beijing, China, 19–22 April 2019; pp. 300–304.
9. Xu, L.; Qian, M.; Fang, R.; Luo, Y. Image Process Technique to Cognize the External Qualities and Milling Degree of Rice. *Trans. Chinese Soc. Agric. Eng.* **1996**, *12*, 172–175.
10. Wood, D.F.; Siebenmorgen, T.J.; Williams, T.G.; Orts, W.J.; Glenn, G.M. Use of Microscopy to Assess Bran Removal Patterns in Milled Rice. *J. Agric. Food Chem.* **2012**, *60*, 6960–6965. [CrossRef]
11. Zhang, H.; Meng, Y.; Zhou, Z.; Yang, H. Analyzing Rice Milling Degree Based on Digital Image Technology. *J. Chinese Cereal. Oils Assoc.* **2006**, *21*, 135–137.
12. Wan, P.; Long, C. An Inspection Method of Rice Milling Degree Based on Machine Vision and Gray-Gradient Co-Occurrence Matrix. In *Computer and Computing Technologies in Agriculture IV, Proceedings of the International Conference on Computer and Computing Technologies in Agriculture, Nanchang, China, 22–25 October 2010*; Springer: Berlin/Heidelberg, Germany, 2011; pp. 195–202
13. Wan, P.; Tan, H.; Yang, W.; Pan, H. A Judging Method of Rice Milling Degree Based on the Color Characteristic and BP Neural Network. *J. Chinese Cereal. Oils Assoc.* **2015**, *30*, 103–107.
14. Fang, C.; Hu, X.; Sun, C.; Duan, B.; Xie, L.; Zhou, P. Simultaneous Determination of Multi Rice Quality Parameters Using Image Analysis Method. *Food Anal. Methods* **2015**, *8*, 70–78. [CrossRef]
15. Zareiforoush, H.; Minaei, S.; Alizadeh, M.R.; Banakar, A. A Hybrid Intelligent Approach Based on Computer Vision and Fuzzy Logic for Quality Measurement of Milled Rice. *Measurement* **2015**, *66*, 26–34. [CrossRef]
16. Hortinela, C.C.; Balbin, J.R.; Fausto, J.C.; Catli, A.D.; Cui, K.J.R.; Tan, J.A.F.; Zunega, E.O.S. Milled Rice Grain Grading Using Raspberry Pi with Image Processing and Support Vector Machines with Adaptive Boosting. In Proceedings of the 2020 IEEE 12th International Conference on Humanoid, Nanotechnology, Information Technology, Communication and Control, Environment, and Management (HNICEM), Manila, Philippines, 3–7 December 2020; pp. 1–6.

17. Singh, R.; Om, H. Newborn Face Recognition Using Deep Convolutional Neural Network. *Multimed. Tools Appl.* **2017**, *76*, 19005–19015. [CrossRef]
18. Ali, S.; Li, J.; Pei, Y.; Aslam, M.S.; Shaukat, Z.; Azeem, M. An Effective and Improved CNN-ELM Classifier for Handwritten Digits Recognition and Classification. *Symmetry* **2020**, *12*, 1742. [CrossRef]
19. Song, H.; Wang, W.; Wang, J.; Wang, R. Collaborative Deep Networks for Pedestrian Detection. In Proceedings of the 2017 IEEE Third International Conference on Multimedia Big Data (BigMM), Laguna Hills, CA, USA, 19–21 April 2017; pp. 146–153.
20. Qi, C.; Zuo, Y.; Chen, Z.; Chen, K. Rice Processing Accuracy Classification Method Based on Improved VGG16 Convolution Neural Network VGG16. *Nongye Jixie Xuebao Trans. Chin. Soc. Agric. Mach.* **2021**, *52*, 301–307. [CrossRef]
21. Patel, V.A.; Joshi, M.V. Convolutional Neural Network with Transfer Learning for Rice Type Classification. In Proceedings of the Tenth International Conference on Machine Vision (ICMV 2017), Vienna, Austria, 13–15 November 2017; Zhou, J., Radeva, P., Nikolaev, D., Verikas, A., Eds.; SPIE: Bellingham, DC, USA, 2018; Volume 10696, p. 21. [CrossRef]
22. Hong Son, N.; Thai-Nghe, N. Deep Learning for Rice Quality Classification. In Proceedings of the 2019 International Conference on Advanced Computing and Applications (ACOMP), Nha Trang, Vietnam, 26–28 November 2019; pp. 92–96.
23. Li, B.; Li, S. Recognition Algorithm of Rice Germ Integrity Base on Improved Inception V3. In Proceedings of the 2019 International Conference on Intelligent Computing, Automation and Systems (ICICAS), Chongqing, China, 6–8 December 2019; pp. 497–501.
24. Li, S.; Li, B.; Li, J.; Liu, B. Brown Rice Germ Integrity Identification Based on Deep Learning Network. *J. Food Qual.* **2022**, *2022*, 6709787. [CrossRef]
25. Li, B.; Liu, B.; Li, S.; Liu, H. An Improved EfficientNet for Rice Germ Integrity Classification and Recognition. *Agriculture* **2022**, *12*, 863. [CrossRef]
26. Yildirim, M. Analog Circuit Implementation Based on Median Filter for Salt and Pepper Noise Reduction in Image. *Analog Integr. Circuits Signal Process.* **2021**, *107*, 195–202. [CrossRef]
27. Wang, Y.; Li, Z.; Hao, H.; Yang, H.; Zheng, Y. Research on Visual Perception Technology of Autonomous Driving Based on Improved Convolutional Neural Network. In Proceedings of the 2020 4th International Workshop on Advanced Algorithms and Control Engineering, IWAACE 2020, Shenzhen, China, 21–23 February 2020; Institute of Physics Publishing: Shenzhen, China, 2020; Volume 1550.
28. Szegedy, C.; Liu, W.; Jia, Y.; Sermanet, P.; Reed, S.; Anguelov, D.; Erhan, D.; Vanhoucke, V.; Rabinovich, A. Going Deeper with Convolutions. In Proceedings of the 2015 IEEE Conference on Computer Vision and Pattern Recognition (CVPR), Boston, MA, USA, 7–12 June 2015; pp. 1–9.
29. Ioffe, S.; Szegedy, C. Batch Normalization: Accelerating Deep Network Training by Reducing Internal Covariate Shift. In Proceedings of the 32nd International Conference on Machine Learning, ICML 2015, Lile, France, 6–11 July 2015; International Machine Learning Society (IMLS): Lile, France, 2015; Volume 1, pp. 448–456.
30. Szegedy, C.; Vanhoucke, V.; Ioffe, S.; Shlens, J.; Wojna, Z. Rethinking the Inception Architecture for Computer Vision. In Proceedings of the 29th IEEE Conference on Computer Vision and Pattern Recognition, CVPR 2016, Las Vegas, NV, USA, 26 June–1 July 2016; IEEE Computer Society: Las Vegas, NV, USA, 2016; Volume 2016, pp. 2818–2826.
31. Szegedy, C.; Ioffe, S.; Vanhoucke, V.; Alemi, A.A. Inception-v4, Inception-ResNet and the Impact of Residual Connections on Learning. In Proceedings of the 31st AAAI Conference on Artificial Intelligence, AAAI 2017, San Francisco, CA, USA, 4–10 February 2017; AAAI Press: San Francisco, CA, USA, 2017; pp. 4278–4284.
32. He, K.; Zhang, X.; Ren, S.; Sun, J. Deep Residual Learning for Image Recognition. In Proceedings of the 2016 IEEE Conference on Computer Vision and Pattern Recognition (CVPR), Las Vegas, NV, USA, 27–30 June 2016; pp. 770–778.
33. Wang, T.; Sun, M.; Hu, K. Dilated Deep Residual Network for Image Denoising. In Proceedings of the 2017 IEEE 29th International Conference on Tools with Artificial Intelligence (ICTAI), Boston, MA, USA, 6–8 November 2017; pp. 1272–1279.
34. Fadil, I.; Helmiawan, M.A.; Sofiyan, Y. Optimization Parameters Support Vector Regression Using Grid Search Method. In Proceedings of the 9th International Conference on Cyber and IT Service Management, CITSM 2021, Bengkulu, Indonesia, 22–23 September 2021; Institute of Electrical and Electronics Engineers Inc.: Bengkulu, Indonesia, 2021.
35. Buslim, N.; Rahmatullah, I.L.; Setyawan, B.A.; Alamsyah, A. Comparing Bitcoin's Prediction Model Using GRU, RNN, and LSTM by Hyperparameter Optimization Grid Search and Random Search. In Proceedings of the 2021 9th International Conference on Cyber and IT Service Management (CITSM), Bengkulu, Indonesia, 22–23 September 2021; pp. 1–6.
36. Candelieri, A. A Gentle Introduction to Bayesian Optimization. In Proceedings of the 2021 Winter Simulation Conference, WSC 2021, Phoenix, AZ, USA, 12–15 December 2021; Institute of Electrical and Electronics Engineers Inc.: Phoenix, AZ, USA, 2021; pp. 1–16.
37. Heydarian, M.; Doyle, T.E.; Samavi, R. MLCM: Multi-Label Confusion Matrix. *IEEE Access* **2022**, *10*, 19083–19095. [CrossRef]
38. Xiang, Y.; Lin, J.; Li, Y.; Hu, Z.; Xiong, Y. Mango Double-Sided Maturity Online Detection and Classification System. *Nongye Gongcheng Xuebao Trans. Chin. Soc. Agric. Eng.* **2019**, *35*, 259–266. [CrossRef]
39. Bao, W.; Yang, X.; Liang, D.; Hu, G.; Yang, X. Lightweight Convolutional Neural Network Model for Field Wheat Ear Disease Identification. *Comput. Electron. Agric.* **2021**, *189*, 106367. [CrossRef]
40. Liang, K.; Wang, Y.; Sun, L.; Xin, D.; Chang, Z. A Lightweight-Improved CNN Based on VGG16 for Identification and Classification of Rice Diseases and Pests. In Proceedings of the International Conference on Image, Vision and Intelligent Systems (ICIVIS 2021), Jinan, China, 15–17 August 2022; pp. 195–207.

41. Jiang, Z.; Dong, Z.; Jiang, W.; Yang, Y. Recognition of Rice Leaf Diseases and Wheat Leaf Diseases Based on Multi-Task Deep Transfer Learning. *Comput. Electron. Agric.* **2021**, *186*, 106184. [CrossRef]
42. Jenipher, V.N.; Radhika, S. An Automated System for Detecting Rice Crop Disease Using CNN Inception V3 Transfer Learning Algorithm. In Proceedings of the 2022 Second International Conference on Artificial Intelligence and Smart Energy (ICAIS), Coimbatore, India, 23–25 February 2022; pp. 88–94.

foods

MDPI

Article

A Method for Capture and Detection of Crop Airborne Disease Spores Based on Microfluidic Chips and Micro Raman Spectroscopy

Xiaodong Zhang [1,2], Fei Bian [1,2], Yafei Wang [1,2], Lian Hu [3], Ning Yang [4,*] and Hanping Mao [1,2]

1 School of Agricultural Engineering, Jiangsu University, Zhenjiang 212013, China
2 Key Laboratory of Modern Agricultural Equipment and Technology, Ministry of Education, Jiangsu University, Zhenjiang 212013, China
3 Key Laboratory of Key Technology on Agricultural Machine and Equipment, Ministry of Education, South China Agricultural University, Guangzhou 510640, China
4 School of Electrical and Information Engineering, Jiangsu University, Zhenjiang 212013, China
* Correspondence: yangn@ujs.edu.cn; Tel.: +86-187-9608-7751

Abstract: Airborne crop diseases cause great losses to agricultural production and can affect people's physical health. Timely monitoring of the situation of airborne disease spores and effective prevention and control measures are particularly important. In this study, a two-stage separation and enrichment microfluidic chip with arcuate pretreatment channel was designed for the separation and enrichment of crop disease spores, which was combined with micro Raman for Raman fingerprinting of disease conidia and quasi identification. The chip was mainly composed of arc preprocessing and two separated enriched structures, and the designed chip was numerically simulated using COMSOL multiphysics5.5, with the best enrichment effect at W2/W1 = 1.6 and W4/W3 = 1.1. The spectra were preprocessed with standard normal variables (SNVs) to improve the signal-to-noise ratio, which was baseline corrected using an iterative polynomial fitting method to further improve spectral features. Raman spectra were dimensionally reduced using principal component analysis (PCA) and stability competitive adaptive weighting (SCARS), support vector machine (SVM) and back-propagation artificial neural network (BPANN) were employed to identify fungal spore species, and the best discrimination effect was achieved using the SCARS-SVM model with 94.31% discrimination accuracy. Thus, the microfluidic-chip- and micro-Raman-based methods for spore capture and identification of crop diseases have the potential to be precise, convenient, and low-cost methods for fungal spore detection.

Keywords: micro Raman; microfluidic chip; fungal spores; crop disease; numerical simulation

Citation: Zhang, X.; Bian, F.; Wang, Y.; Hu, L.; Yang, N.; Mao, H. A Method for Capture and Detection of Crop Airborne Disease Spores Based on Microfluidic Chips and Micro Raman Spectroscopy. *Foods* **2022**, *11*, 3462. https://doi.org/10.3390/foods11213462

Academic Editor: Lili He

Received: 7 October 2022
Accepted: 27 October 2022
Published: 1 November 2022

Publisher's Note: MDPI stays neutral with regard to jurisdictional claims in published maps and institutional affiliations.

1. Introduction

Crop diseases cause huge losses to agricultural production and directly affect the economic development and national food security of many countries in the world [1]. Among them, fungal diseases can cause huge losses to the growth and yield of crops. In addition, fruits will still be damaged by fungi after being picked [2]. Fungal diseases mostly exist in the form of spores before infecting crops and fruits. In addition to causing damage to crops and fruits, fungal spores can also enter the lungs through the human respiratory tract and spread to other organs of the human body, causing various fungal diseases [3,4] Therefore, it is necessary to capture and identify megaspores quickly and accurately.

The identification and counting of spores under the traditional microscope mainly depend on naked eye observation. Due to the large number of spores captured, this method is labor-intensive, time-consuming, and inefficient. The accuracy of observation depends on the professional experience of operators, sometimes leading to large errors. Image processing methods are used to automatically detect and count spores, including image segmentation using K-means clustering algorithm, recognition based on shape factor and area, and spore contour segmentation based on concavity and contour segment merging.

The automatic detection of disease spores has good effectiveness and accuracy, but spore recognition based on image processing technology cannot contain too many features [5,6]. Using deep neural network FSNet to detect fungal spores can automatically identify and count fungal spores in microscopic images, but image and deep learning methods are still not accurate enough to identify spores with similar shape and size [7]. PCR is the gold standard method for microbial detection. It is often used for microbial identification with high accuracy. However, this method requires professionals to crack the spores under strict experimental conditions. Only through tedious processing can fungal spores be detected [8,9]. The existing microscopic image method can realize rapid detection and recognition through morphology, but it cannot accurately identify spores with similar shape and size. The PCR method has high accuracy, but the detection conditions are harsh, destructive sampling is required, and the timeliness is poor, and the cost is high. Therefore, it is urgent to develop a rapid and accurate spore detection and identification technology. Raman spectroscopy is a light scattering technology with low cost and high speed. It can reflect various vibration frequencies and related vibration levels of biological components and can be used to identify the molecular composition and structure of biological samples [10,11]. Micro Raman spectroscopy has been applied to the identification and analysis of bacteria, which can be classified and identified [12,13]. However, the content of spores in the air is low, and there are a lot of impurities, so it is difficult to detect them directly using Raman spectroscopy. Therefore, a separation and enrichment method is needed to achieve spore detection and improve detection accuracy.

In recent years, microfluidic technology has provided powerful tools for detection applications due to its portability, miniaturization, automation, multi-channel sample detection, and cost saving. Compared with traditional methods, the greatest advantage of microfluidic control is to create a controllable microenvironment, which can accurately drive and control the microfluidic flow in the microchannel, thus improving the detection sensitivity [14,15]. Among them, the impactor designed according to the principle of aerodynamics is widely used in the separation of atmospheric particles with good results [16], and it is feasible to collect fungal spores using a microfluidic chip with a virtual impactor structure. This method has high collection efficiency and low cost [17,18]. In addition, microfluidic chip composite diffraction is used to detect and identify spores. For spores with large differences in size and shape, the identification accuracy is high. Compared with microscopic observation, its detection field of vision is much expanded, and the detection efficiency and speed are greatly improved [19,20]. However, the recognition accuracy of spores with similar shape and size is not high due to morphological detection. Therefore, it is necessary to design a separation and enrichment microfluidic chip with compound Raman spectroscopy to capture and collect spores and identify them accurately.

In this study, a two-stage separation and enrichment microfluidic chip with a semi arc pretreatment structure is proposed, which can be combined with micro Raman. Using numerical analysis and experiment, the best design parameters are obtained, and the feasibility of microfluidic chip is discussed. At the same time, micro Raman was used to detect the collected spores to obtain the Raman spectrum of fungal spores. The Raman spectrum was preprocessed and modeled for analysis. The fungal spores were identified, and their fingerprints were established to provide evidence for the subsequent prevention and treatment of fungal diseases.

2. Materials and Methods

2.1. Sample Preparation

In this study, *Ustilaginoidea virens* (*U. virens*), *Rice blast* (*R. blast*), *Aspergillus niger* (*A. niger*), and *Aspergillus carbonarius* (*A. carbonarius*) were taken as the research objects, among which the size of *R. blast* spores, *A. niger* spores, and *A. carbonarius* spores was 3–5 μm. The shape is similar to oval or round, and the spores of rice false smut are pear-shaped, with a size of 6–8 μm. The spores of *R. blast* and *U. virens* were provided by China Rice Research Institute in Hangzhou, China, and the spores of *A. niger* and *A. carbonarius* were purchased from Beijing Biological Preservation Center. Four fungi were cultured on PDA medium at 28 °C for 7 days, and mature spores with high activity and consistent activity were obtained. During the experiment, fresh conidia on PDA were scraped off and diluted with sterile distilled water in a sterile environment.

2.2. Composite Micro Raman Detection Method for Microfluidic Chips

The microfluidic chip composite micro Raman detection method proposed in this study has the characteristics of rapidity, accuracy, and simple operation. As shown in Figure 1, this method mainly consists of three parts: spore capture, separation and enrichment, micro Raman collection, and Raman data processing and modeling. In this study, nanospheres (3–8 μm) were mixed with four spore suspensions; the four kinds of spores were *U. virens*, *R. blast*, *A. niger*, and *A. carbonarius*. The spore concentration in the real environment is very low, and there will be disease spores in the air unless the disease is about to occur. Therefore, in order to simulate the real experimental environment, the spores were mixed into an aerosol generator and an aerosol was generated, and the aerosol was released into the air. In order to simulate the real use environment, the spore concentration in the air was consistent with the concentration before the outbreak of a disease, 200 spore /m^3 [21]. An air pump was used to pump air into the microfluidic chip at the exit of the chip for about 3 min per acquisition. After the aerosol entered the microfluidic chip, the enrichment of 3–4 μm and 6–8 μm spores and impurities was completed through the two enrichment areas of the chip. The particles larger than 8 μm remained in the pretreatment channel due to their large inertia. The target particles entered the corresponding two enrichment areas, while the particles smaller than 3 μm flowed out of the chip through the outlet.

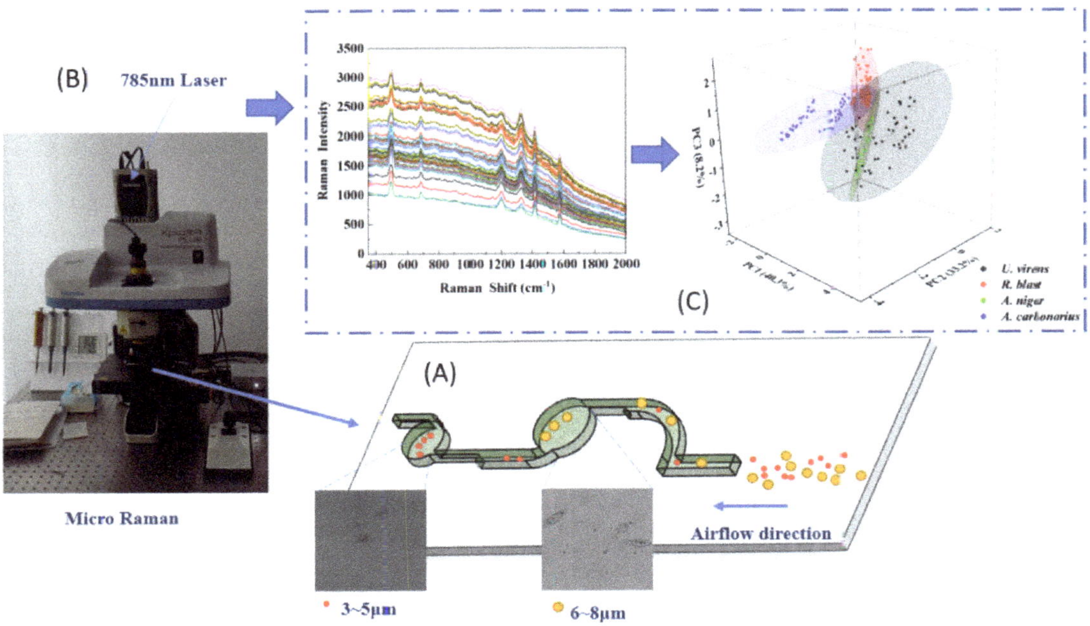

Figure 1. Detection of crop disease spores by microfluidic chip combined with micro Raman spectroscopy: (**A**) spore capture, separation, and enrichment; (**B**) micro Raman collection; (**C**) Raman data processing and modeling.

In this study, a total of 50 sets of Raman spectra of 4 species of spores were collected, and the collected 200 sets of Raman spectra were randomly divided into training set and test set at a ratio of 3:1. The samples were collected with a frequency shift of 200–2000 cm^{-1}

using an XploRA PLUS Raman microscope (HORIBA, France), and the collected samples had 1053 features. The spectral acquisition parameters were set as follows: the excitation power was set to 365 mW, the 50 times objective lens was selected, spot size was 5 µm. Before the Raman spectrum was collected, the Raman spectrometer needed to be calibrated with wave number to eliminate the significant difference between the instrument response and the measured Raman spectrum value on the wave number axis and the true value. Wavenumber calibration requires measurement of the reference Raman spectrum from a standard material with well-defined Raman bands. In this study, the Raman spectrum peak of silicon wafer was selected to calibrate the spectrometer. When the first-order Raman spectrum peak of silicon wafer was located at the 520.7 cm^{-1} frequency shift, the instrument was calibrated. In the laser illumination channel, the narrowband single-mode continuous light with the wavelength of 785 nm was selected as the excitation light of Raman scattering. This is because the 785 nm excitation light can effectively reduce the background spontaneous fluorescence noise and improve the signal-to-noise ratio of the collected Raman signal.

In this study, spectral data were randomly divided into training set (140) and test set (60). Spore identification model was established according to the test set. Before data analysis, the collected Raman spectra were preprocessed by SG smoothing and SNV, and the iterative polynomial fitting method was used for baseline correction to eliminate the interference of baseline drift and spectrum noise of Raman spectra.

The principle of polynomial replacement fitting is to continuously compare and adjust the original spectral data during polynomial replacement fitting, and directly compare the adjusted spectral data with the points on the fitting curve. The advantage of baseline correction with this method is to gradually adjust the coefficients of the polynomial so as to gradually approach the actual baseline shape, and the calculated baseline function form is closer to the actual baseline [22]. Standard normalized variate (SNV) algorithm refers to a deviation method to standardize variable values. Through the numerical standardization and transformation of the original variables, the transformation results will eventually fall within the range of [0, 1]. The premise of SNV algorithm is that all wavelength variables present normal distribution, and then they are standardized. The main way to remove noise is to remove light scattering [23].

Principal component analysis (PCA) is an algorithm for dimension reduction of data features. Spectral data were transformed from high to low dimensions by linear variation. Low-order principal components were retained and high-dimensional and invalid information was removed, reducing the data dimension. Using the most relevant low-dimensional data for classification identification can effectively reduce the difficulty and complexity of data analysis [24,25]. Usually, PCA needs to retain the principal components to make the variance contribution rate reach more than 85%. In this study, all PCAs with cumulative contribution rate greater than 95% were selected. The stability compatible reweighted sampling (SCARS) algorithm measures the magnitude of the stability of a variable, and the larger the stability value, the more likely the variable is to be selected, and the more consistent the bands selected at each iteration. This enables guaranteeing stable and rapid variable selection. The principles of the SCARS algorithm are to take each wavelength as one individual, use the adaptive reweighted sampling and exponentially revealing function to remove regression coefficients, take band points with small weights from the partial least squares model, pick out band points with large stable values, and retain the subset with the lowest RMSECV for interaction validation to find the optimal combination of variables with high efficiency [26].

SVM is a supervised machine learning method based on finite sample statistical learning theory. According to the structural risk minimization (SRM) principle, small samples and nonlinearity can be solved by constructing an optimal classification hyperplane in high-dimensional space. BPANN is a powerful learning algorithm that enables highly nonlinear mapping between inputs and outputs by training sample data, constantly modifying network weights and thresholds to minimize the error function in the direction of a negative gradient, ultimately approaching the expected output [27].

2.3. Chip Design and Simulation

To realize the purification and enrichment of fungal disease spores, a two-stage separation and enrichment microfluidic chip with arcuate pretreatment channel was designed. When a particle enters the inertial impactor with accompanying air, its trajectory is related to the size of the particle. Some small mass particles can cross stream lines and be separated. However, other small mass particles can flow away with the deflection of airflow. This behavior of particles in the curved channel can be characterized by Stokes number [17,28].

$$stk = \frac{\rho_p d_p^2 C_c Q}{9\mu W} \tag{1}$$

where d_p is the particle size (m), ρ_p is the particle density (1000 kg m^{-3}), μ is the air viscosity (1.81×10^{-5} N $\cdot$ s $\cdot$ m^{-2}), Q is the air velocity at the inlet of the microfluidic device (m $\cdot$ s^{-1}), and W is the nozzle width (m). C_c is the Cunningham sliding correction coefficient based on particle size, which can be obtained by Equation (2) [16]:

$$C_C = 1 + \frac{2A\lambda}{d} + \frac{2Q\lambda}{d} e^{-\frac{bd}{2\lambda}} \tag{2}$$

where A = 1.234, Q = 0.413, b = 0.904, and λ is the average free path of an air molecule with a value of 6.95×10^{-8} m. Thus, according to Equation (3) it can be reduced to

$$C_C \approx \begin{cases} 1 + 2.52\frac{\lambda}{d}, d > 2\lambda \\ 1 + 3.29\frac{\lambda}{d}, d < 2\lambda \end{cases} \tag{3}$$

Furthermore, stk_{50} represents a Stokes number corresponding to a particle collection efficiency of 50%, which can be rearranged by Equation (1):

$$d_{50} = \sqrt{\frac{9\mu W_{stk_{50}}}{\rho_p C_c V}} \tag{4}$$

The d_{50} is defined as the cut-off size of the particles producing 50% collection efficiency at each impact stage, and for this study, *U. virens*, *R. blast*, *A. niger*, and *A. carbonarius* collected at d_{50} were set as 3–5 μm and 4–6 μm to obtain the up-to-size W for the two separated enrichment stages of the chip.

The microfluidic chip consists of an arc-shaped preprocessing channel and a two-stage separation and enrichment structure, and the particles enter the chip inlet jointly with air, then enter the preprocessing channel; the particles bonded together will be scattered and then enter the first enrichment area, and the remaining particles enter the second enrichment area due to the constant velocity at the inlet, while the particles that have a wide width channel enter the narrow width channel, so the airflow velocity is elevated, using velocity variation to manipulate particle enrichment versus rounding, which greatly increases collection efficiency. The structure of the microfluidic chip is schematically shown in Figure 2.

The preprocessing channel includes particle inlet, channel 1, and arc channel. The first separation structure includes channel 1, collection area 1, and channel 2. The second separation structure includes channel 3, enrichment zone 2, channel 4, and particle outlet. R1 is the radius of the pretreatment channel, and D1 and D2 are the diameters of the collection area, which are 5000 μm and 3700 μm, respectively. The length of channel 0 is 7000 μm, and the width is set to W0 = 1300 μm. Channel 1 has a length of 3500 μm and a width of 800 μm. Channel 2 has a length of 5650 μm and a width of 1100 μm. The length of channel 3 was set to 4400 μm, and the width was set to 400 μm. The length of channel 4 was set to 3600 μm, and the width was set to 600 μm.

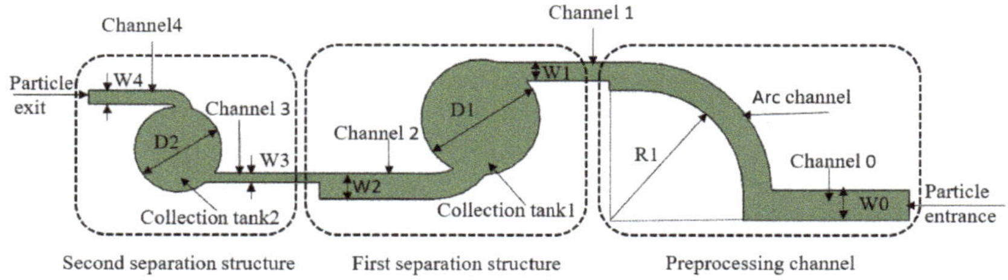

Figure 2. A 2D diagram of the microfluidic chip.

2.4. Numerical Simulation

The numerical analysis software COMSOL multiphysics5.5 was used for simulation analysis in this study. The laminar flow module and particle trajectory tracking module in the software were used to simulate the separation process. The particle trajectory tracking module couples the effect of multiple force fields on the particle trajectory, calculating the trajectory of particles in the channel.

First, a 2D sketch of the microfluidic chip was drawn by AutoCAD2021 and then imported into COMSOL multiphysics5.5. The model needed to be meshed before the analysis, and the meshing was finer to achieve a better simulation. According to the actual situation of the subject, the Reynolds number did not exceed 1000, so the fluid was set as laminar flow. The Reynolds numbers in collection area 1 and 2 were 173 and 346, respectively. The particle density was 1.05 g cm^{-3} and the aerodynamic diameter was 0.5~8 μm. The inlet flow rate was 12.5 mL min^{-1}, and the wall was set to be slip-proof. During the simulation, the collisions between the particles and the walls of the microfluidic device were inelastic, and the particles stuck to the walls to calculate the particle collection rate. Particle tracking uses Newton's law of motion to solve differential equations. The research object of this topic is the movement of spores in the air. The spores were mainly driven by drag force in the microfluidic chip. The particle tracking module was used to simulate the particle trajectory. The drag force satisfies Stokes' law, and the temperature and absolute pressure are standard states. Then, the simulation boundary was, the corresponding entrance and exit were set, and 100 particles were released at the entrance each time to ensure the reliability of particle collection efficiency. Since the simulation of this study needs to study the laminar flow and particle motion state in the microfluidic chip, transient and steady-state solvers were configured. The steady-state solver was used to study laminar flow, and the transient solver was used to study the particle. In the case of motion, the steady-state calculation amount is small, and one can choose to solve it directly. The solver can choose PARDISO to meet the requirements, and then choose the multi-threaded nested analysis pre-sorting algorithm, and the configuration of the steady-state solver is completed. The transient solver is more complicated to solve, and it needs to be solved iteratively. The GMRES solver was selected to solve the trajectory of the particle.

2.5. Chip Making

The microfluidic chip was fabricated using a conventional soft-lithography process. First, the photosensitive film was used to make a mask according to the structure drawn by AutoCAD2021, and the photosensitive film was covered on the copper plate to make the thickness reach 100 μm as the layer height of the channel. Then, ultraviolet lithography was used to expose the photosensitive film and a developing solution was used to develop it, so that only the punch of the microfluidic chip was left on the copper plate. Then, the outer mold was used to fix the channel chip range, and then the polydimethylsiloxane (PDMS) was poured into the mold. It was then put into a 70° oven to cure for 4 h, and finally the mold was taken out of the cured chip, and a plasma bonder was used to bond the chip and the glass slide to form a microfluidic chip. The physical map is shown in Figure 3.

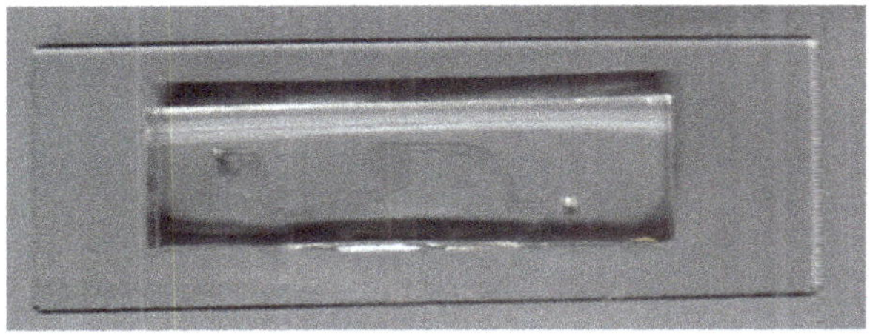

Figure 3. Chip physical map.

3. Results and Discussion

3.1. Numerical Simulation of Microfluidic Chip

According to the set simulation conditions, it is necessary to simulate the parameters of the microfluidic chip to obtain a reasonable structure of the chip to obtain a good enrichment effect. First, the sub-bureau set the conditions to obtain the pressure map of the microfluidic chip in Figure 4A and the gas velocity map of Figure 4B in the microfluidic chip. In this study, the simulation of the channel width changed to obtain the best enrichment effect. The enrichment effect evaluation index is to release 100 particles, and the number of particles obtained in the corresponding enrichment area determines the enrichment efficiency. The number of particles can be counted in a specific area by selecting specific particles from the derived values in the simulation results of COMSOL multiphysics5.5, and the enrichment rate can be expressed as the percentage of the number of enriched particles to the total number of released particles.

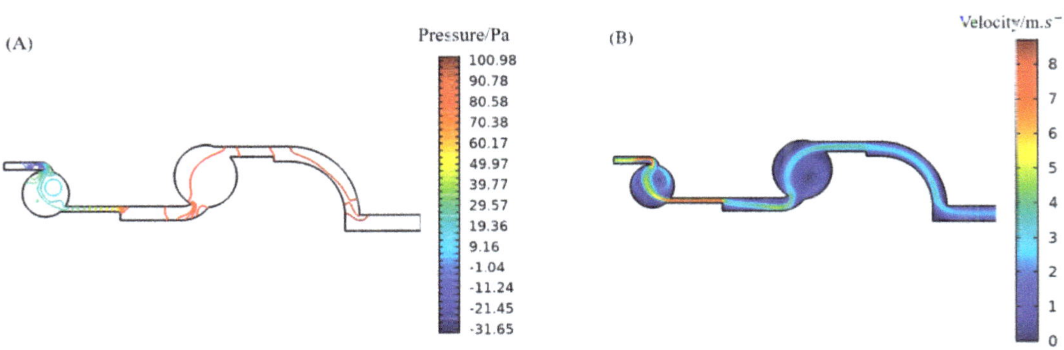

Figure 4. Microfluidic chip simulation diagram: (**A**) microfluidic chip pressure map; (**B**) microfluidic chip velocity map.

According to the simulation condition setting, and after simulation with the width adjustment of the microfluidic chip, the enrichment rate of the microfluidic chip under different ratios of the channel width was obtained. Figure 5A,B show the effect of the W2/W1 ratio on the separation and enrichment effect of particles, taking 6 μm and 8 μm particles as examples. It can be seen from Figure 4A,B that when W1 = 700 μm, the enrichment effect of 6 μm and 8 μm particles is better, and when W2/W1 = 1.6, the enrichment effect of the first enrichment zone is the best, the enrichment rate of 6 μm is 93%, and the enrichment rate of 8 μm is 93%. The enrichment rate was 94%, and the channel widths were fixed at W1 = 700 μm and W2 = 1120 μm. It can be seen from Figure 5C,D that when W3 = 400 μm, the enrichment effect of 3 μm and 5 μm particles is better, while W4/W3 is 1.1, the enrichment efficiency of 3 μm particles is 93%, and the enrichment rate of 5 μm particles is 94%; at this time the fixed channel widths were W3 = 400 μm and W4 = 440 μm.

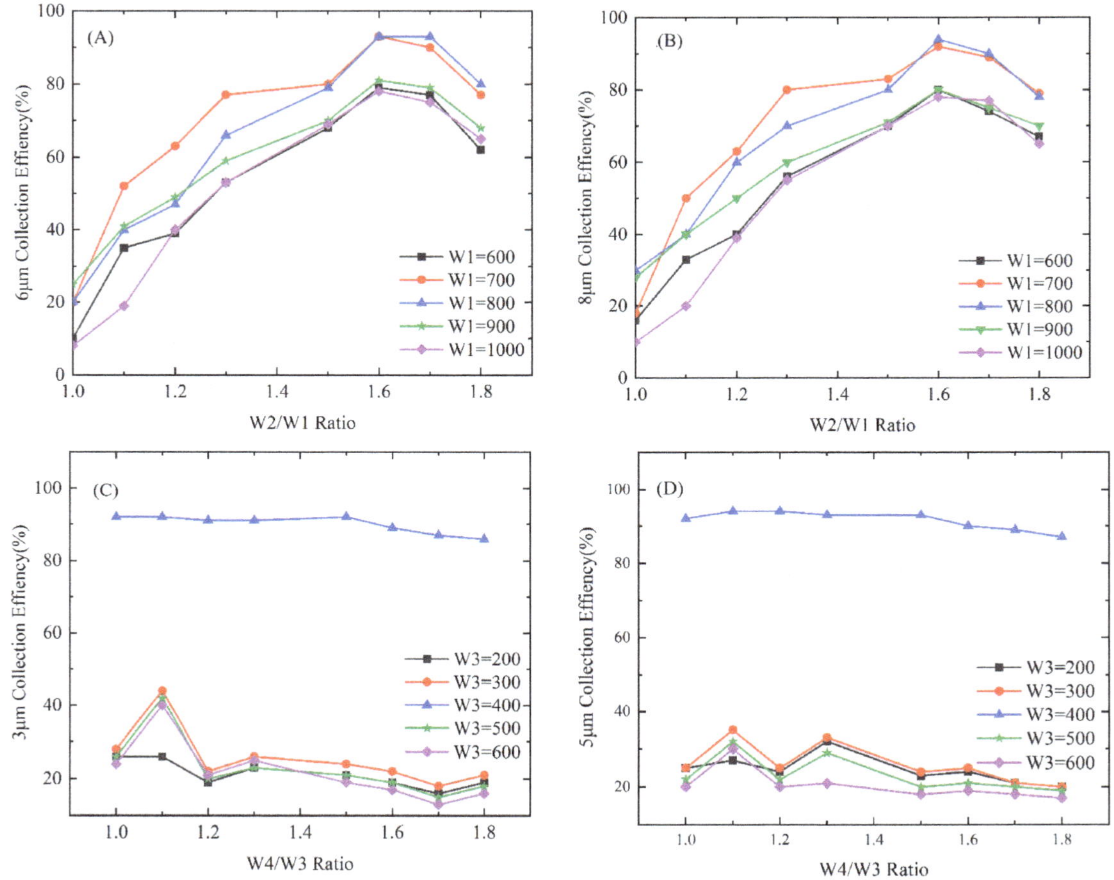

Figure 5. Statistical graph of particle enrichment imitation rate: (**A**) 6 μm collection efficiency; (**B**) 8 μm collection efficiency; (**C**) 3 μm collection efficiency; (**D**) 5 μm collection efficiency.

According to the above simulation and analysis, the optimal enrichment parameters of 6 μm, 8 μm, 3 μm, and 5 μm were obtained, and the best enrichment effect was obtained by simulation according to the optimal parameters, as shown in Figure 6.

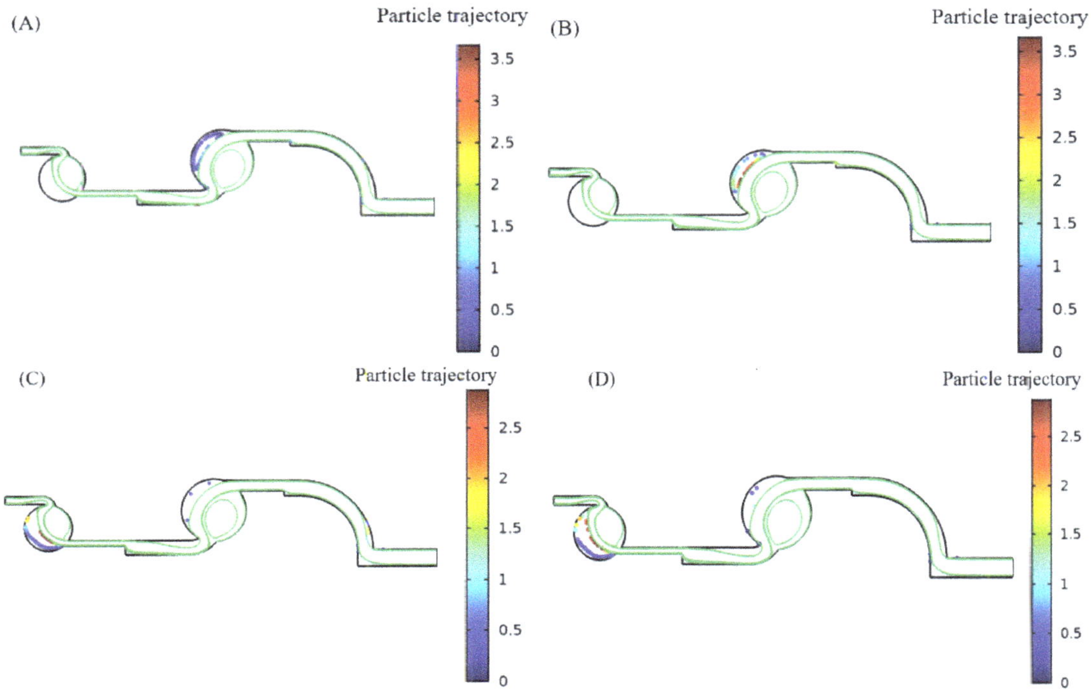

Figure 6. Statistical graph of particle enrichment imitation rate: (**A**) 6 μm collection effect; (**B**) 8 μm collection effect; (**C**) 3 μm collection effect; (**D**) 5 μm collection effect.

3.2. Raman Analysis

A total of 200 spectra of four fungi were collected in this study. Since the original spectral data have more spectral noise and high fluorescence background interference, effective spectral preprocessing is very important. Preprocessing the spectrum by selecting a reasonable spectral preprocessing method can effectively reduce spectral noise, retain useful information, simplify the modeling process, and improve the stability of the model. During the Raman spectrum acquisition process, the detector has a small probability of receiving various interference rays such as cosmic rays in the environment, forming sharp peaks in the spectrum, which affects the stability of the spectrum. Therefore, spectra with cosmic spikes need to be identified and eliminated before data analysis. This paper adopted SG smoothing and SNV to remove the influence of noise on the model. Among them, SG smoothing can eliminate the noise interference and uneven fluorescence intensity of the original Raman spectrum. As shown in Figure 7A, and SNV can construct an ideal spectrum by taking the average value of the spectrum, thereby eliminating the effect of particle scattering, as shown in Figure 7B. Baseline calibration is achieved by an iterative polynomial fitting method for baseline drift phenomena in the spectrum. The preprocessing steps are: (1) removing the cosmic spike Raman curve; (2) SG smoothing; (3) SNV correction; (4) baseline calibration.

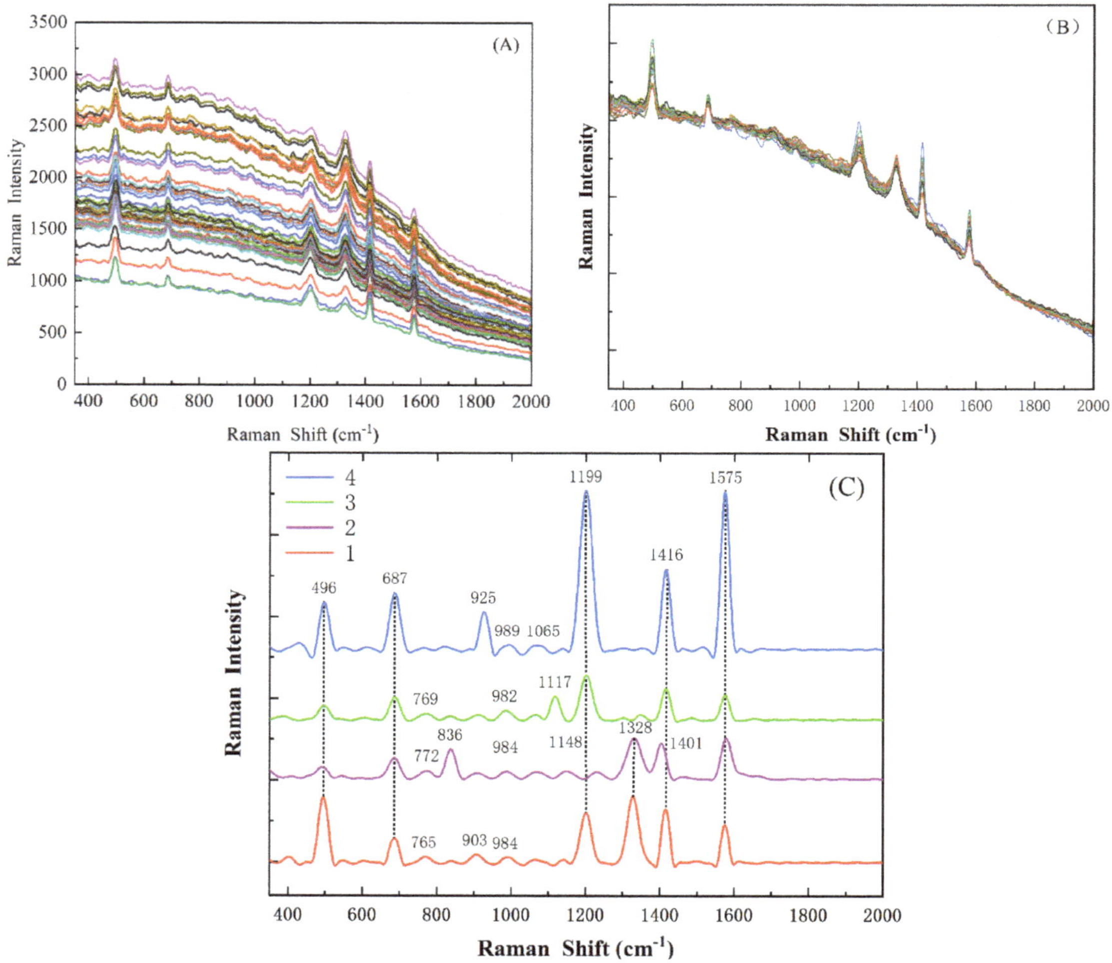

Figure 7. Raman analysis: (**A**) SG-smoothed spectrum after removing cosmic spikes; (**B**) spectra processed by SNV; (**C**) average Raman spectra of four diseased spores.

It can be seen that the SNV greatly reduces the influence of baseline drift and noise on the spectra, while preserving the important spectral information of the fungus. Figure 7C shows the processed average spectra collected from four fungi to provide Raman fingerprints important for the identification of fungal cells. Since Raman scattering depends on the change in molecular polarizability during atomic vibrations, non-polar groups such as S–S, C–C, S–H, and N–N vibrations have strong corresponding signals in Raman, reflecting that various structural information of organic compounds has been obtained [29]. Diseased spores contain cell walls and abundant mRNA. The main components of cell walls are polysaccharides and a small number of proteins and lipids. Different spores contain different types of polysaccharides [30–32].

According to the existing research and experimental data, all characteristic spectral bands and spectral assignments of the four fungi are shown in Table 1. In all Raman spectra, the peaks at 493–497 cm^{-1} and 1416 cm^{-1} are characteristic peaks for galactomannan and chitin, which are important components of fungal cell walls [29,32]. The peak at

686–687 cm^{-1} is attributed to Guanine, Thymine [29,33]. The peaks at 765cm^{-1}–772 cm^{-1} were assigned to (O-P-O) stretching RNA, respectively [33]. The peak at 984 cm^{-1}–989 cm^{-1} is attributed to C=C deformation, C–N stretching [12,29]. The peaks at 1065 cm^{-1} and 1117 cm^{-1} are galactomannan [29]. The peak at 1148 cm^{-1} was attributed to C–O ring aromatic [12]. The peaks at 1200 cm^{-1} and 1202 cm^{-1} are Amide III (random) and Thymine [32,33]. The peak at 1328 cm^{-1} is attributed to C–O Amide III (protein), C–H deformation [12]. The peak at 1570 cm^{-1}–1577 cm^{-1} is Adenine, Guanine (ring stretching) [12,32]. The Raman signals of diseased spores have common components, and there are also differences with their own characteristics. Therefore, the Raman fingerprints of the four fungi measured in this study provide a basis for species identification.

Table 1. Peak assignment of the average spectrogram of the four spores.

Raman Shift (cm^{-1})	*U. virens*	*R. blast*	*A. niger*	*A. carbonarius*	Tentative Assignments	Reference
493–497	496	493	497	497	Galactomannan, chitin	[29]
686–687	686	686	687	687	Guanine, Thymine (ring breathing)	[29,33]
765–798	765	772	769	769	(O-P-O) stretching RNA	[33]
930–990	984	984	982	989	C=C deformation, C–N stretching	[12,29]
1065–1117	-	-	1117	1065	galactomannan	[29]
1150–1185	1148	1148	-	-	C–O ring aromatic amino acid in protein	[12]
1200–1274	1202	-	1202	1200	Amide III (random), Thymine	[32,33]
1315–1325	1328	1328	-	-	Amide III (protein), C–H deformation	[12]
1416	1416	1401	1416	1416	Chitin	[32]
1570–1595	1575	1577	1575	1575	Adenine, Guanine (ring stretching)	[12,32]

Combining Figure 7 and Table 2, it can be concluded that the Raman spectra of the four disease spores have some significant common characteristic peaks, which indicates that they contain many of the same components, and there are also some distinctive characteristic peaks unique to spores, indicating their unique composition. However, there were also some insignificant shared characteristic peaks and unique characteristic peaks, which also indicated some shared and unique compositions of diseased spores. Then, classification modeling of spores by only significant characteristic peaks led to inaccurate classification of spores, and it was necessary to find all characteristic peaks by training the algorithm on all bands of the Raman spectrum.

Table 2. Accuracy statistics of spore classification model.

Serial Number	Algorithm	Calibration Set Accuracy (%)	Prediction Set Accuracy (%)
1	SVM	85.65	86.32
2	BPANN	88.46	87.61
3	PCA-SVM	90.25	91.24
4	PCA-BPANN	88.34	87.55
5	SCARS-SVM	93.41	93.43
6	SCARS-BPANN	94.94	94.31

3.3. Fungal Spore Recognition Model

The large number of Raman spectral features and the existence of a large number of redundant features greatly slow down the speed of modeling and analysis, so dimensionality reduction was required before modeling and analysis. After dimensionality reduction, the Raman spectrum was modeled and analyzed by SVM and BPANN, and a classification model was established.

The SCARS algorithm uses the stability of the variable as a measure. The greater the stability value, the greater the possibility of the variable being selected, and the frequency bands selected for each iteration can be consistent. It can ensure the stability and speed of variable selection. The optimal potential frequency band variable is selected by the Monte Carlo cross-validation method, and an RMSECV value can be obtained in each cycle. Due to the large number of sampling times, in order to obtain a better combination of characteristic frequency bands, it is necessary to compare through repeated trials. The subset combination corresponding to the minimum RMSECV is obtained. When the number of cyclic samplings is set to 25, the running result tends to be stable. The running result of the algorithm is shown in Figure 8. The algorithm filtered out 69 bands, as shown in Figure 8.

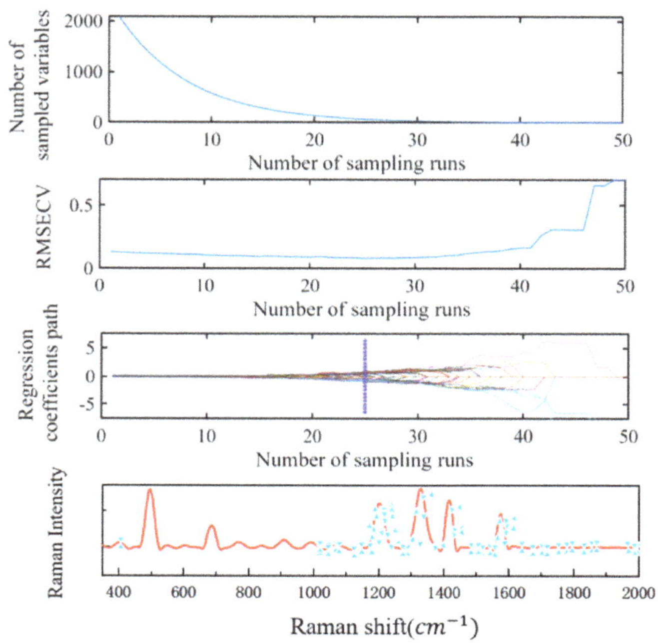

Figure 8. Example of running result of SCARS algorithm.

PCA selects different PCA variables in full and shortened spectral intervals. The top 15 PCAs are the most significant for the raw spectral data, with a cumulative contribution rate of over 95%. Figure 9 shows the top three PCA taxonomic groups of fungi and observed preliminary taxonomic results for four diseased spores.

After the disease spore Raman spectrum was dimensionally reduced, the data were input into the classification model for classification. The data after dimensionality reduction by PCA is a matrix of 200 × 15, and the data after dimensionality reduction by SCARS is a matrix of 200 × 65. In this study, two excellent classification algorithms were selected to classify the dimensionality-reduced data. When the SVM was running, the radial basis function (RBF) kernel function was selected, and then the optimal parameters of the penalty function c of the model and the kernel parameter g of the kernel function were obtained through grid search. The optimal parameters are 0.00094, and the accuracy rates of the test set and prediction set are 94.38% and 86.63%. When using BPANN, the hidden layer transfer function was set to Tansig, the output layer transfer function to purelin, the network training function to trainbfg, the number of training iterations to be 1000, and the target error to be 0.0001 [29]. The discriminative accuracies of the training set and prediction set are 88.46% and 87.61%, respectively. Then, the Raman data reduced by PCA and SCARS were used to classify the classification model. Since the random division of the test set and the prediction set and other factors will affect the classification effect of the pull model, the model was run five times and its average accuracy was used. The results are shown in Table 2. The calibration set and prediction set of SCARS-BPANN classification have the highest accuracy, 94.94% and 94.31%, respectively.

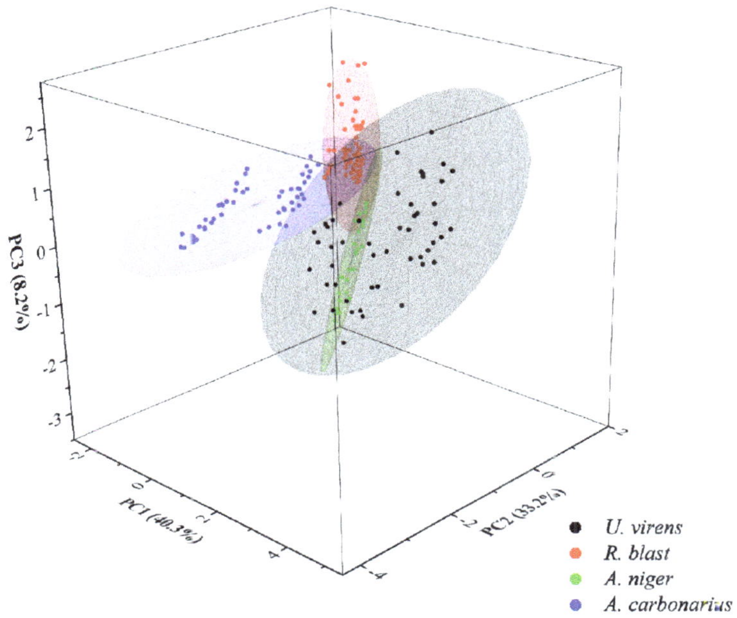

Figure 9. Top 3 PCA distribution results.

3.4. Enrichment Experiment

In order to verify the enrichment effect of the microfluidic chip determined by the simulation, a quantitative number of nano-microspheres was sucked into the chip inlet, and then the enrichment efficiency of the particles in the chip was observed through a microscope. About 50 nano-microspheres of one particle size were mixed into the aerosol generator each time, all of them were entered from the chip inlet and out, and then the enriched particles were counted by a microscope and the enrichment rate was calculated. The particle size was tested five times, and the statistical boxplot of the enrichment experiment as shown in Figure 10 was obtained. The average enrichment efficiencies of 3 μm, 5 μm, 6 μm, and 8 μm were 87.5%, 82.625%, 82.5%, and 84.87%, respectively. It can be seen from Figure 10 that the enrichment rate obtained in the enrichment experiment is slightly lower than that obtained by simulation, but the enrichment efficiency is stable.

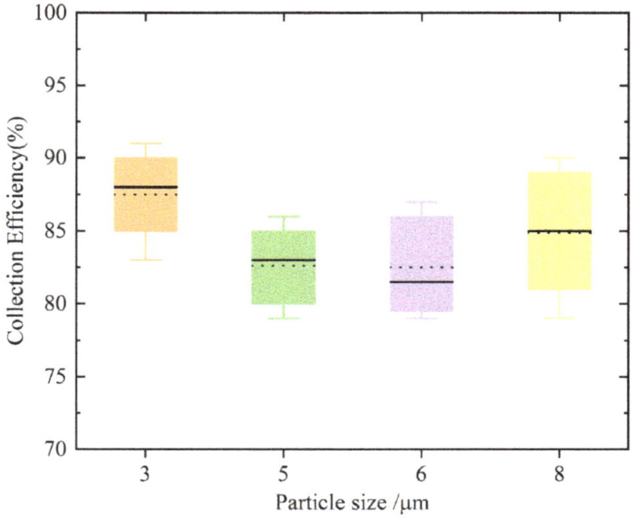

Figure 10. Enrichment Experiment Statistics Box Plot.

3.5. Discussion

Airborne fungal diseases mostly float in the air in the form of spores and spread with the wind before their widespread outbreaks [4,8]. At the time of disease occurrence, the concentration of disease spores in the air is higher than 100 spores /m^3; if timely detection and prevention and control measures are not taken, the disease spores will drift to other areas with the wind and continue to infect other areas. Y. Zhang et al. proposed a deep-learning-based fungal spore detector FSNet for recognition and automatic counting of *Aspergillus glaucus, Penicillium solitum*, and *Aspergillus candidus*, and the experiments demonstrated that FSNet achieved an average precision of 0.9, 0.944, and 0.904 on *Aspergillus glaucus, Penicillium solitum,* and *Aspergillus candidus*, respectively, demonstrating the ability to automate detection of spores in the laboratory [7]. However, the automatic detection of spores based on images cannot accurately identify spores with similar appearance. Aswathi S. et al. were able to differentiate between dead and live *C. sporogenes* spores on media (SBA and TSA) plates using hyperspectral imaging [34]. The use of hyperspectral images and spectral information can accurately identify spores, but the effective reflectance of the hyperspectral spectrum for spores is limited to the spectral band range, and the spores cannot be captured and detected. The method for detecting and identifying disease spores of microfluidic chip combined with Raman microscopy developed in this study can capture spores in the air and then accurately identify the spores by identifying their Raman fingerprints, and this method does not require cumbersome biochemical experiments with low cost.

4. Conclusions

In this study, a two-stage separation and enrichment microfluidic chip with an arc-shaped pretreatment channel was developed and designed to separate and enrich crop disease spores, and then combined with confocal Raman microscopy to conduct Raman fingerprinting of disease conidia. The map was accurately identified. Support vector machine (SVM) and back-propagation artificial neural network (BPANN) were used to identify fungal spore species, and the identification accuracy was 86.32% and 87.61%; the SCARS-SVM model had the best discriminant effect, with a discriminant accuracy of 94.31%. Therefore, the capture and identification method of crop disease spores based on microfluidic chip and micro Raman may become an accurate, convenient, and inexpensive method for detection and identification of fungal spores.

Author Contributions: Conceptualization, X.Z. and F.B.; methodology, X.Z.; software, F.B.; validation, F.B., Y.W. and N.Y.; formal analysis, F.B.; investigation, H.M.; resources, N.Y.; data curation, F.B.; writing—original draft preparation, F.B.; writing—review and editing, X.Z.; visualization, L.H.; supervision, H.M.; project administration, X.Z.; funding acquisition, X.Z. All authors have read and agreed to the published version of the manuscript.

Funding: This research was funded by the Project of Agricultural Equipment Department of Jiangsu University (NZXB20210106); Key Laboratory of Modern Agricultural Equipment and Technology (Jiangsu University), Ministry of Education (Grant No. MAET202111); National Key Research and Development Program for Young Scientists (2022YFD2000013); Key Laboratory of Modern Agricultural Equipment and Technology (Ministry of Education), High-tech Key Laboratory of Agricultural Equipment and Intelligence of Jiangsu Province (Grant No. JNZ201901); Scientific and Technological Project of Henan Province (Grant No. 212102110029); and the National Natural Science Foundation of China (61771224 and 32071905).

Data Availability Statement: Data is contained within the article.

Conflicts of Interest: The authors declare no conflict of interest.

References

1. Savary, S.; Willocquet, L.; Pethybridge, S.J.; Esker, P.; McRoberts, N.; Nelson, A. The global burden of pathogens and pests on major food crops. *Nat. Ecol. Evol.* **2019**, *3*, 430–439. [CrossRef] [PubMed]
2. Ngolong Ngea, G.L.; Qian, X.; Yang, Q.; Dhanasekaran, S.; Ianiri, G.; Ballester, A.R.; Zhang, X.; Castoria, R.; Zhang, H. Securing fruit production: Opportunities from the elucidation of the molecular mechanisms of postharvest fungal infections. *Compr. Rev. Food Sci. Food Saf.* **2021**, *20*, 2508–2533. [CrossRef] [PubMed]
3. Walsh, N.M.; Botts, M.R.; McDermott, A.J.; Ortiz, S.C.; Wuthrich, M.; Klein, B.; Hull, C.M. Infectious particle identity determines dissemination and disease outcome for the inhaled human fungal pathogen Cryptococcus. *PLoS Pathog.* **2019**, *15*, e1007777. [CrossRef] [PubMed]
4. Li, X.; Liu, D.; Yao, J. Aerosolization of fungal spores in indoor environments. *Sci. Total Environ.* **2022**, *820*, 153003. [CrossRef]
5. Lei, Y.; Yao, Z.; He, D. Automatic detection and counting of urediniospores of *Puccinia striiformis* f. sp. tritici using spore traps and image processing. *Sci. Rep.* **2018**, *8*, 13647. [CrossRef]
6. Wang, Y.; Mao, H.; Xu, G.; Zhang, X.; Zhang, Y. A Rapid Detection Method for Fungal Spores from Greenhouse Crops Based on CMOS Image Sensors and Diffraction Fingerprint Feature Processing. *J. Fungi* **2022**, *8*, 374. [CrossRef]
7. Zhang, Y.; Li, J.; Tang, F.; Zhang, H.; Cui, Z.; Zhou, H. An Automatic Detector for Fungal Spores in Microscopic Images Based on Deep Learning. *Appl. Eng. Agric.* **2021**, *37*, 85–94. [CrossRef]
8. Imbusch, F.; Liebe, S.; Erven, T.; Varrelmann, M. Dynamics of cercospora leaf spot disease determined by aerial spore dispersal in artificially inoculated sugar beet fields. *Plant Pathol.* **2021**, *70*, 853–861. [CrossRef]
9. Munir, M.; Wang, H.; Dufault, N.S.; Anco, D.J. Early Detection of Airborne Inoculum of Nothopassalora personata in Spore Trap Samples from Peanut Fields Using Quantitative PCR. *Plants* **2020**, *9*, 1327. [CrossRef]
10. Zhu, Y.; Liu, S.; Li, M.; Liu, W.; Wei, Z.; Zhao, L.; Liu, Y.; Xu, L.; Zhao, G.; Ma, Y. Preparation of an AgNPs@Polydimethylsiloxane (PDMS) multi-hole filter membrane chip for the rapid identification of food-borne pathogens by surface-enhanced Raman spectroscopy. *Spectrochim. Acta Part A Mol. Biomol. Spectrosc.* **2021**, *267*, 120456. [CrossRef]
11. Alsamad, F.; Gobinet, C.; Vuiblet, V.; Jaisson, S.; Piot, O. Towards normalization selection of Raman data in the context of protein glycation: Application of validity indices to PCA processed spectra. *Analyst* **2020**, *145*, 2945–2957. [CrossRef] [PubMed]
12. Huayhongthong, S.; Khuntayaporn, P.; Thirapanmethee, K.; Wanapaisan, P.; Chomnawang, M.T. Raman spectroscopic analysis of food-borne microorganisms. *LWT* **2019**, *114*, 108419. [CrossRef]
13. Breuch, R.; Klein, D.; Siefke, E.; Hebel, M.; Herbert, U.; Wickleder, C.; Kaul, P. Differentiation of meat-related microorganisms using paper-based surface-enhanced Raman spectroscopy combined with multivariate statistical analysis. *Talanta* **2020**, *219*, 121315. [CrossRef] [PubMed]
14. Mi, F.; Hu, C.; Wang, Y.; Wang, L.; Peng, F.; Geng, P.; Guan, M. Recent advancements in microfluidic chip biosensor detection of foodborne pathogenic bacteria: A review. *Anal. Bioanal. Chem* **2022**, *414*, 2883–2902. [CrossRef] [PubMed]
15. Gao, Y.; Pan, X.; Xu, S.; Liu, Z.; Wang, J.; Yu, K.; Wang, C.; Yuan, H.; Wu, S. Fluorescence-enhanced microfluidic sensor for highly sensitive in-situ detection of copper ions in lubricating oil. *Mater. Des.* **2020**, *191*, 108693. [CrossRef]
16. Chen, T.; Sun, J.; Ma, T.; Li, T.; Liu, C.; Zhu, X.; Xue, N. Design and Analysis of Particulate Matter Air-Microfluidic Grading Chip Based on MEMS. *Micromachines* **2019**, *10*, 497. [CrossRef]
17. Wang, P.; Yuan, S.; Yang, N.; Wang, A.; Fordjour, A.; Chen, S. The Collection Method for Crop Fungal Spores Based on an Efficient Microfluidic Device. *Aerosol Air Qual. Res.* **2020**, *20*, 72–79. [CrossRef]
18. Wang, Y.; Zhang, X.; Yang, N.; Ma, G.; Du, X.; Mao, H. Separation-enrichment method for airborne disease spores based on microfluidic chip. *Int. J. Agric. Biol. Eng.* **2021**, *14*, 199–205. [CrossRef]

19. Yang, N.; Chen, C.; Li, T.; Li, Z.; Zou, L.; Zhang, R.; Mao, H. Portable Rice Disease Spores Capture and Detection Method Using Diffraction Fingerprints on Microfluidic Chip. *Micromachines* **2019**, *10*, 289. [CrossRef]
20. Wang, Y.; Mao, H.; Zhang, X.; Liu, Y.; Du, X. A Rapid Detection Method for Tomato Gray Mold Spores in Greenhouse Based on Microfluidic Chip Enrichment and Lens-Less Diffraction Image Processing. *Foods* **2021**, *10*, 3011. [CrossRef]
21. Grewling, L.; Magyar, D.; Chlopek, K.; Grinn-Gofron, A.; Gwiazdowska, J.; Siddiquee, A.; Ianovici, N.; Kasprzyk, I.; Wojcik, M.; Laffersova, J.; et al. Bioaerosols on the atmospheric super highway: An example of long distance transport of Alternaria spores from the Pannonian Plain to Poland. *Sci. Total Environ.* **2022**, *819*, 153148. [CrossRef] [PubMed]
22. Chen, Y.; Dai, L. An Automated Baseline Correction Method Based on Iterative Morphological Operations. *Appl. Spectrosc.* **2018**, *72*, 731–739. [CrossRef]
23. Guo, Z.; Chen, P.; Wang, M.; Zuo, M.; El-Seedi, H.R.; Chen, Q.; Shi, J.; Zou, X. Rapid enrichment detection of patulin and alternariol in apple using surface enhanced Raman spectroscopy with coffee-ring effect. *LWT* **2021**, *152*, 112333. [CrossRef]
24. Beattie, J.R.; Esmonde-White, F.W.L. Exploration of Principal Component Analysis: Deriving Principal Component Analysis Visually Using Spectra. *Appl. Spectrosc.* **2021**, *75*, 361–375. [CrossRef] [PubMed]
25. Bro, R.; Smilde, A.K. Principal component analysis. *Anal. Methods* **2014**, *6*, 2812–2831. [CrossRef]
26. Li, H.; Liang, Y.; Xu, Q.; Cao, D. Key wavelengths screening using competitive adaptive reweighted sampling method for multivariate calibration. *Anal. Chim. Acta* **2009**, *648*, 77–84. [CrossRef] [PubMed]
27. Chauhan, V.K.; Dahiya, K.; Sharma, A. Problem formulations and solvers in linear SVM: A review. *Artif. Intell. Rev.* **2018**, *52*, 803–855. [CrossRef]
28. Marple, V.A.; Chien, C.M. Virtual impactors: A theoretical study. *Environ. Sci. Technol.* **1980**, *14*, 976–985. [CrossRef]
29. Guo, Z.; Wang, M.; Barimah, A.O.; Chen, Q.; Li, H.; Shi, J.; El-Seedi, H.R.; Zou, X. Label-free surface enhanced Raman scattering spectroscopy for discrimination and detection of dominant apple spoilage fungus. *Int. J. Food Microbiol.* **2021**, *338*, 108990. [CrossRef]
30. Wang, F.; Sethiya, P.; Hu, X.; Guo, S.; Chen, Y.; Li, A.; Tan, K.; Wong, K.H. Transcription in fungal conidia before dormancy produces phenotypically variable conidia that maximize survival in different environments. *Nat. Microbiol.* **2021**, *6*, 1066–1081. [CrossRef]
31. Bowman, S.M.; Free, S.J. The structure and synthesis of the fungal cell wall. *Bioessays* **2006**, *28*, 799–808. [CrossRef] [PubMed]
32. Lin, Y.J.; Lin, H.K.; Lin, Y.H. Construction of Raman spectroscopic fingerprints for the detection of Fusarium wilt of banana in Taiwan. *PLoS ONE* **2020**, *15*, e0230330. [CrossRef] [PubMed]
33. Lemma, T.; Wang, J.; Arstila, K.; Hytonen, V.P.; Toppari, J.J. Identifying yeasts using surface enhanced Raman spectroscopy. *Spectrochim. Acta Part A Mol. Biomol. Spectrosc.* **2019**, *218*, 299–307. [CrossRef] [PubMed]
34. Soni, A.; Al-Sarayreh, M.; Reis, M.M.; Brightwell, G. Hyperspectral imaging and deep learning for quantification of Clostridium sporogenes spores in food products using 1D- convolutional neural networks and random forest model. *Food Res. Int.* **2021**, *147*, 110577. [CrossRef] [PubMed]

foods MDPI

Article

Real-Time Grading of Defect Apples Using Semantic Segmentation Combination with a Pruned YOLO V4 Network

Xiaoting Liang [1,2,3], Xueying Jia [1,2,3], Wenqian Huang [1,3], Xin He [1,3], Lianjie Li [1,3], Shuxiang Fan [1,3], Jiangbo Li [1,3], Chunjiang Zhao [1,3] and Chi Zhang [1,3,*]

1 Intelligent Equipment Research Center, Beijing Academy of Agriculture and Forestry Sciences, Beijing 100097, China
2 College of Information Technology, Shanghai Ocean University, Shanghai 201306, China
3 National Research Center of Intelligent Equipment for Agriculture, Beijing 100097, China
* Correspondence: zhangchi@nercita.org.cn

Abstract: At present, the apple grading system usually conveys apples by a belt or rollers. This usually leads to low hardness or expensive fruits being bruised, resulting in economic losses. In order to realize real-time detection and classification of high-quality apples, separate fruit trays were designed to convey apples and used to prevent apples from being bruised during image acquisition. A semantic segmentation method based on the BiSeNet V2 deep learning network was proposed to segment the defective parts of defective apples. BiSeNet V2 for apple defect detection obtained a slightly better result in MPA with a value of 99.66%, which was 0.14 and 0.19 percentage points higher than DAnet and Unet, respectively. A model pruning method was used to optimize the structure of the YOLO V4 network. The detection accuracy of defect regions in apple images was further improved by the pruned YOLO V4 network. Then, a surface mapping method between the defect area in apple images and the actual defect area was proposed to accurately calculate the defect area. Finally, apples on separate fruit trays were sorted according to the number and area of defects in the apple images. The experimental results showed that the average accuracy of apple classification was 92.42%, and the F1 score was 94.31. In commercial separate fruit tray grading and sorting machines, it has great application potential.

Keywords: defective apples; apple grading; deep learning; object detection; semantic segmentation

Citation: Liang, X.; Jia, X.; Huang, W.; He, X.; Li, L.; Fan, S.; Li, J.; Zhao, C.; Zhang, C. Real-Time Grading of Defect Apples Using Semantic Segmentation Combination with a Pruned YOLO V4 Network. *Foods* **2022**, *11*, 3150. https://doi.org/10.3390/foods11193150

Academic Editor: Oscar Núñez

Received: 28 August 2022
Accepted: 4 October 2022
Published: 10 October 2022

1. Introduction

Recently, consumers' awareness of fresh fruit quality is increasing. They often prefer to buy apples with regular shapes, smooth surfaces and no obvious scars and damages. Therefore, it is particularly important to detect and grade apples before they are sent to the market, which will greatly improve the income of fruit farmers. Apples can be classified into different grades based on basic characteristics such as size, shape, color and whether they are defective. However, it is still a challenging task to accurately detect the apple defect area on the automatic sorting line. Especially for expensive fruits, if the number and area of defects are not considered and all defective apples are treated as substandard fruits, it will cause potential economic losses to fruit farmers. Therefore, the detection and grading of apple surface defects is an urgent problem to be solved for expensive apple grading and sorting.

Mizushima et al. [1] applied a linear support vector machine (SVM) and Otsu method to classify apples. First, the optimal classification hyperplane was calculated, and then the color image was gray scaled with SVM. The optimal threshold near the fruit boundary was obtained by the Otsu method. Finally, apples were eventually divided into three commercial grades. Jawale et al. [2] proposed the K-means clustering method to segment the image. Then, an artificial neural network (ANN) combined with color and texture features was used to separate the defective apples. Mohammadi et al. [3] used a simple threshold to extract gray-scale images. Then, the shape features such as roundness of the segmented apples were obtained and used to detect defects. Nosseir [4] proposed an algorithm to identify rotten fruit by extracting RGB value. The recognition accuracy of the method was 96.00%. Wenzhuo Zhang et al. [5] proposed an apple suspicious defect detection method based on a multivariable image analysis strategy. The FCM-NPGA algorithm was used to segment the suspicious apple defect image. The overall detection accuracy was 98%. Chi Zhang et al. [6] used NIR-coded structured light and fast lightness correction to automatically detect defective apples. Defective regions or stem/calyx regions can be correctly distinguished. The identification rate of defective apples with this method was 90.2%. Integrating the four characteristics of apple size, color, shape and surface defects, the apples were divided into three levels by support vector machine (SVM). The detection accuracy of surface defects based on a single index was 95.85%, while the average classification accuracy of apple surface defects based on multiple features was 95.49%.

The feature extraction method is the key to the accuracy of fruit detection for traditional machine vision technology. However, some methods require complex acquisition systems, and some may not be used in online applications [7]. Recently, multispectral imaging (MSI) and hyperspectral imaging (HSI) systems have been applied to nondestructive detection of fruits such as apples, oranges, etc. [3,8]. However, due to the time consumption of image acquisition and the high price of the HSI camera, the practical application of HSI was limited [9,10]. Huang et al. [11] used principal component analysis (PCA) to detect apple defects in hyperspectral images. However, the classification accuracy in the online test was 74.6%. In addition, the research based on a laboratory MSI system could only deal with defect detection under static conditions, which was difficult to apply to online detection.

Due to the fast detection speed and low cost of the RGB color camera, traditional machine vision using an RGB color camera had obvious advantages in online fruit grading and sorting based on color, size, shape and defect compared with other nondestructive testing technologies [12]. In recent years, deep learning has been widely used in agriculture, industry, medicine and other fields. It automatically learned image features from the input image and the key features with fewer human factors were extracted for subsequent tasks [13]. Machine vision based on an RGB color camera combined with various deep learning models greatly improved the online grading accuracy of apples.

For postharvest quality grading of ordinary apples, it is sufficient to divide apples into normal and defective apples without locating the defects of each apple [10]. Therefore, Yujian Xin et al. [14] compared the detection results of SVM, Fast RCNN, YOLOv2 and YOLOv3 models on apple images. The YOLOv3 model had the best effect on apple defect detection. The average detection time of an apple image was 1.12 s, and the F1 score was 92.35%. Paolo et al. [15] regarded apple defects as object detection problems. After comparing a single shot detector (SSD) with YOLOv3, the YOLOv3 model was trained using a dataset containing healthy and defective apples to detect which apples were healthy. The overall mAP was less than 74%. Guangrui Hu et al. [16] used the TensorFlow deep learning framework and SSD deep learning algorithm to identify apple surface defects. Yanfei Li et al. [17] proposed a fast classification model of apple quality based on a convolutional neural network (CNN) and compared it with the Google InceptionV3 model and HOG/GLCM + SVM. It was concluded that the accuracy of apple quality classification was 95.33%. Fan et al. [18] compressed the depth and width of the YOLO V4 network through channel pruning and layer pruning, which reduced the inference time and model size of the network by 10.82 ms and 241.24 MB, respectively, and increased the mAP to 93.74%. This method was suitable for the defect identification of different varieties of apples. Zhipeng Wang et al. [7] proposed an object-detection algorithm based on YOLOv5. The real-time detection of apple stem/calyx could be realized, and the detection accuracy was 93.89%.

However, current research on apple surface defect detection has either been on the condition of a static environment or based on online detection using a roller conveyor. Although sorting machines with a roller conveyor have fast sorting speed, it is easy to cause mechanical damage to the apple and reduces the quality of the apple when the apples are rotated with the roller. The widely used roller conveyor sorting equipment requires that the fruit to be sorted has high hardness. Although image processing techniques are applied to sorting fruits, sorting fruits according to the number and area of surface defects is still a difficult problem. For fragile fruits with low hardness, it is easy to cause damage and economic losses when sorting with chain transmission equipment. Fruits with higher prices are also likely to cause potential economic losses in the process of sorting with a roller conveyor. Therefore, a fruit sorting machine based on separate fruit trays was designed which could protect the apples from damage. The separate fruit tray has been especially suitable for online grading of high-quality apples. With the improvement of classification requirements, it was not only necessary to determine whether the apple's surface had defects but also to identify the number and area of apple surface defects. For expensive high-end fruits, if both slight defects and severe defects were considered as equal defective fruits and discarded, this would cause economic losses to farmers, so it is necessary to grade fruits according to the number and size of defects. Due to the curvature of the fruit's shape, the area of defect in the image would be compacted compared with actual defect area. Therefore, it was necessary to further accurately grade the defective apples in high-quality apples according to the number and area of defects, so as to reduce the economic losses of fruit farmers.

In this paper, a defect grading method based on deep learning is proposed to identify the number and area of defects in an apple image. The specific objectives were: (1) using the BiSeNet V2 network to build a defect detection and segmentation model, (2) using the YOLO V4 network to correct the results of BiSeNet V2 detection, (3) building the corresponding relationship between the number of pixels in the defect area of an apple image and the actual defect area, and (4) grading the defective apple according to the defect area and quantity.

2. Materials and Methods

2.1. Samples

The samples were composed of 180 defective Fuji apples and 50 healthy Fuji apples. Apples with different degrees of defects and healthy apples were picked in November 2021 from a commercial orchard in Beijing. Before capturing apple images, mud points on the apples' surface were washed to avoid mistaking them as defects.

2.2. Computer Vision System

The computer vision system was composed of industrial control computer, RGB camera (acA1920-40 gc, Basler, German), lens (M0814-MP2 8 mm, Computar, Japan), hemispherical lighting hood and the outermost light chamber (Figure 1). There was a circular opening at the top of the hemispherical lighting hood and a circle of light-emitting diode (LED) lights at the bottom of the hemispherical lighting hood. All components (except industrial control computer) were fixed in the light chamber.

The quality of apple images was directly related to the detection accuracy of apple defects. It was significant to capture an image without any light spots. Direct illumination would bring about obvious bright spots on the apple. At the same time, the central regions of the apple images were bright, and the surrounding regions were dark, which increased the difficulty of accurate detection. Image quality, related to the performance of the illumination system, would affect the detection results of apple defects. It was quite important to adopt a suitable illumination system. Therefore, a hemispherical lighting hood with LED light source (wavelength range between 500 nm and 630 nm) was applied to realize the irradiation effect of diffuse reflection in this study.

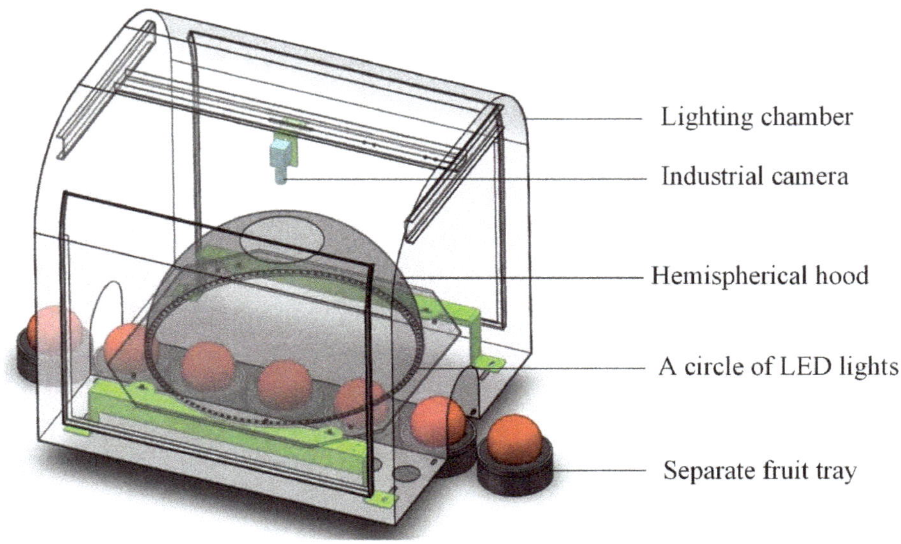

Figure 1. Image acquisition system.

The RGB camera (1920 × 1200 pixels) was installed directly above the hemispherical hood. The apple on the separate fruit tray continuously transmitted under the camera. Apple images could be captured through the circular opening at the top of the hemispherical lighting hood by the camera. In total, 112 LED lamp beads were built at the bottom of the hemispherical lighting hood to form a circular light source. The power of a single LED lamp bead is 3 W, and the color temperature is 6500 k. The LED lights were controlled by the hardware trigger in the control unit. The output voltage of LED power supply is continuously adjustable from 13 V to 24 V, and the light intensity can be adjusted by manually rotating the button of LED power control unit. When an apple passed, the LED lights were on and off for the rest of the time. White diffuse reflective coating was painted on the inner surface of the lighting hood and the reflectivity was 99%, which could obtain uniform illumination. Therefore, the apple images in this study do not need corrected brightness.

The following frameworks were used to obtain the segmentation model and detection model, respectively, in this study: PaddleSeg-based framework (Baidu, China) of version 2.1 for BiSeNet V2 and Darknet-based framework (open-source framework) for YOLO V4, together with Python version 3.7. All experiments were performed on a 64 bits Intel Core i7-6700 CPU with 3.4 GHz and 32 GB RAM memory. One graphics processing unit (GPU), GeForce GTX 2080 with 8 GB of memory under CUDA version 10.1, was employed in this study. The operating system was Windows version 10. C++ language was used to realize online deployment.

2.3. Image Dataset

The apple images used in this research were captured by the machine vision system, as shown in Figure 1. Before capturing the apple images, the apples were put on the separate fruit tray, and the separate fruit tray moved with the conveyor belt. When the apples passed through the lighting chamber, the camera on the top of the lighting chamber would automatically capture the apple image directly under the control of hardware trigger signal. Then, grading software read the apple image from the camera buffer and saved the image. Three thousand apple images were finally obtained as the dataset of this research. The size of all apple images was 400 pixels × 336 pixels. Before training, the resize function of openCV was used to resize the input images into 512 × 512 pixels.

An open-source annotation tool-LabelMe-was used to semantically label the captured apple defect images and establish a standard semantic label dataset. Meanwhile, Labe-lImg was used to mark the stem, calyx and defect regions in the apple images. In total, 2400 images were selected as the training set of the network and the remaining 600 as the validation set.

2.4. Apple Surface Defect Detection Based on BiSeNet V2

In order to obtain an optimal lightweight network model to reduce the network parameters, many researchers were looking for a balance among the amount of computation, parameters and accuracy, hoping to use as few computations and parameters as possible to obtain high accuracy of the detection model [19]. In the field of semantic segmentation, reducing the image size or reducing the complexity of the model could decrease the computation cost caused by semantic segmentation.

Reducing the image size could directly reduce the amount of computation, but the image would lose many details, which would affect the image accuracy. In addition, reducing the complexity of the model would weaken the feature extraction ability of the model, which would affect the segmentation accuracy. Therefore, it was quite challenging to apply lightweight model in semantic segmentation task while taking into account accuracy and real-time performance.

The BiSeNet network could basically balance the relationship between real-time performance and accuracy [20]. So, it was used in this research, and the architecture of it is shown in Figure 2.

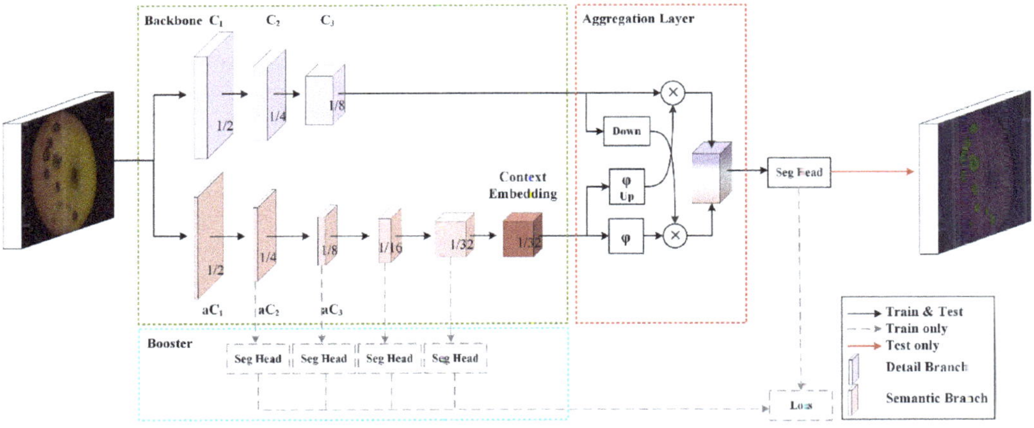

Figure 2. The architecture of BiSeNet V2 network.

The BiSeNetV2 network is divided into three main components: the two-pathway backbone (green dashed box) with a detail branch (the purple cubes), a semantics branch (the pink cubes), the booster component (blue dashed box) and the aggregation layer (red dashed box). C1, C2 and C3 indicate the channels of the detail branch, respectively. The context embedding block as the output of the semantics branch is in the last stage. Down and up represent the down-sampling and the up-sampling operation, respectively. The sigmoid function and the elementwise product were represented by φ and $\otimes$, respectively.

Shallow layers and wide channel dimensions are the characteristics of the detail branch, which have a small receptive field of spatial detail used to generate high-resolution feature representation and capture low-level detail. The semantic branch with deep layers and narrow channel dimensions has a large receptive field for the categorical semantics to capture high-level semantics. The gaps between the semantic and resolution were compensated by the aggregation layer. The initialization parameters of BiSeNet V2 network are shown in Table 1.

Table 1. The initialization parameters of BiSeNet V2 network.

Input Size of Images/Pixel	Batch Size	Initial Learning Rate	Iterations
512 × 512	4	1.1×10^{-3}	1000

Because defects were considered as the region of interest in apple images and in order to ensure the real-time detection, apple images were only segmented into defect region and background region. The segmentation result based on BiSeNetV2 used binary image I_B to present. The gray value of defect region was set as B_V, where B_V was not equal to 0. The gray value of background region was set as 0. In practical application, there might be multiple defect regions in apple images. So, R_B ($R_B \in \{R_{b1}, R_{b2}, \ldots R_{bn}\}$) was used to store the position values of different defect regions, where n was the total number of defects obtained using BiSeNetV2 model in apple image.

The overall goal of this study was to quickly and accurately realize the online grading of defective apples. Therefore, it was necessary to further calculate the area and the number of defects of defective apples. Finally, the grade of apple could be determined according to the comparison between the defect information and the grading standard.

2.5. The Correction of Apple Defect Detection Based on Pruned YOLO V4 Network

Although as a lightweight semantic segmentation model BiSeNet V2 could realize real-time semantic segmentation, it might incorrectly segment the apple stem/calyx region as defect region. Therefore, the object detection model was further used to accurately determine the location of the defect region.

The result of object-detection algorithm required not only identifying the object category in the pictures but also marking the position parameters of the objects. Among them, RCNN, Fast RCNN, SPP-Net [21] and Faster RCNN [22] could be divided into two main parts: region proposal and extraction regions. Therefore, YOLO model had less computation and was faster than two parts methods, as YOLO model replaced numerous regions through grid division and anchor method. YOLO V4 [23] model was implemented based on Darknet framework, which could easily and flexibly use C++ language to deploy the trained network model in practical application. Therefore, a defect detection model based on the YOLO V4 was proposed to identify the defect region in RGB apple images.

The YOLO V4 object-detection algorithm was an improved version of YOLO V3 [24]. Compared with the YOLO V3 object-detection algorithm, YOLO V4 improved the speed and accuracy of real-time detection of the algorithm [25].

CSP (Cross Stage Partial) module could improve the learning ability of the network. CBL module was composed of the Convolution, batch normalization and Leaky_ReLU and CBM module was composed of the Convolution, batch normalization and Mish [26]. These two modules were used to extract input image features. SPP (Spatial pyramid pooling) module used the max-pooling of different scales to pool the input feature layers and then stacked them, which could greatly increase the receptive field.

Recently, YOLO V4 has been used for defect detection of a variety of objects. With the proposal of YOLOv5 and YOLOX, many researchers focus on the newly proposed network, but from the perspective of practical application, YOLO V4 is easier to deploy and realize the online detection of apple defects. The architecture of YOLO V4 network is shown in the Figure 3.

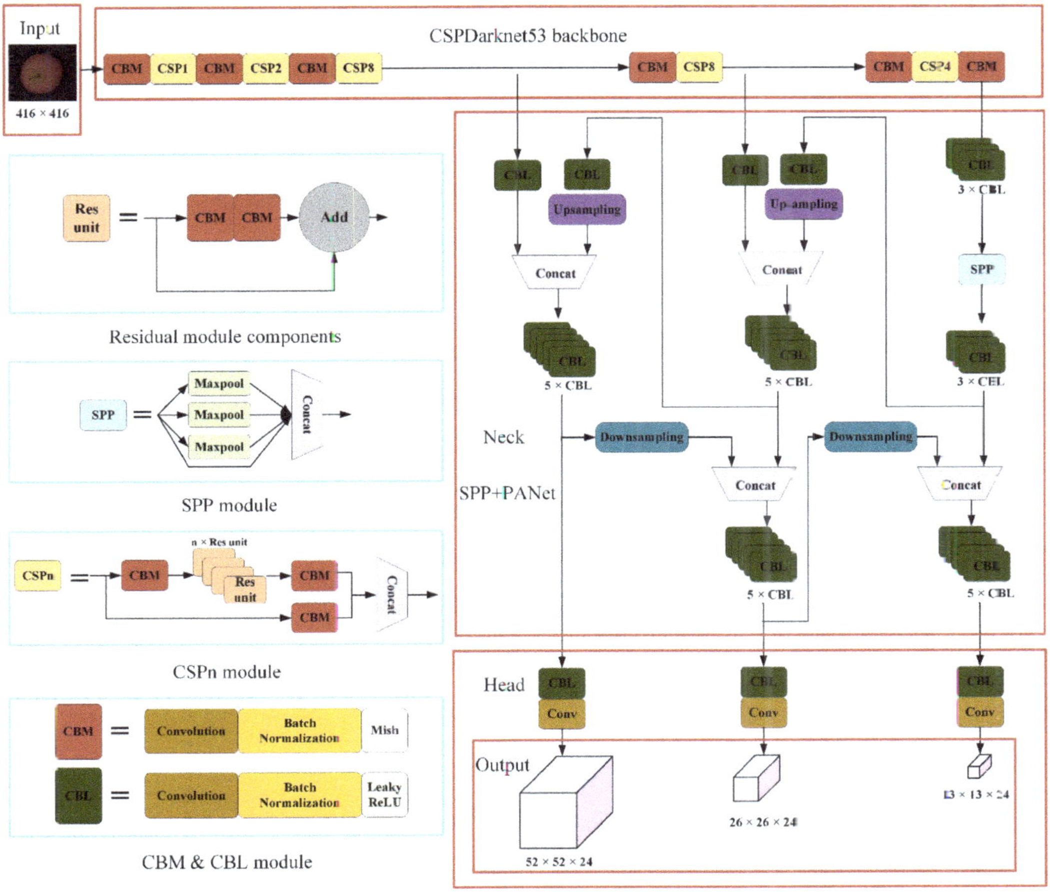

Figure 3. The architecture of YOLO V4 network.

In order to realize the real-time detection of apple surface defects, it was necessary to deploy the YOLO V4 model into the apple automatic grading system. After experiment comparison, it was found that a large number of network parameters, network layers and complex structure of YOLO V4 lead to excessive calculation. It could not meet the real-time requirements of apple surface defect detection. Considering the memory occupation of the YOLO V4 and ensuring the real-time and stability in the detection process, a lightweight model, pruned YOLO V4, was obtained by compressing the YOLO V4 network based on RGB images of apples. Model pruning method could achieve the best balance between model detection speed and detection accuracy. It was also a method to automatically obtain the simplest network structure of the original model. The pruned YOLO V4 model could be deployed on the Windows 10 operating system and hardware and realized the real-time detection of apple surface defects.

In order to obtain pruned YOLO V4, firstly, the sparsity training was introduced into the YOLO V4 network. The scale factors were sorted after sparsity training, and the maximum scale factor meeting the requirements of pruning rate was set as the threshold. Then, it deleted the channels that were less than the threshold. High contribution channels were retained, and low contribution channels were deleted according to the scale factor. So, the channel pruning was completed. However, the object detection model still could not meet the requirements of real-time detection after the channel pruning of the model, which compressed the network width. So, the depth of the network model was further compressed using the layer-pruning method. The mean value of the scale factor of each layer was sorted. The layer with the lower mean value was selected for pruning, which completed the layer pruning. After completing the channel pruning and layer pruning of the YOLO V4 model, the accuracy of the model may decline. Fine-tune operation was used to improve the detection accuracy of the pruned model. Finally, a pruned YOLO V4 model for defect detection was obtained. The result of pruning is shown in Figure 4.

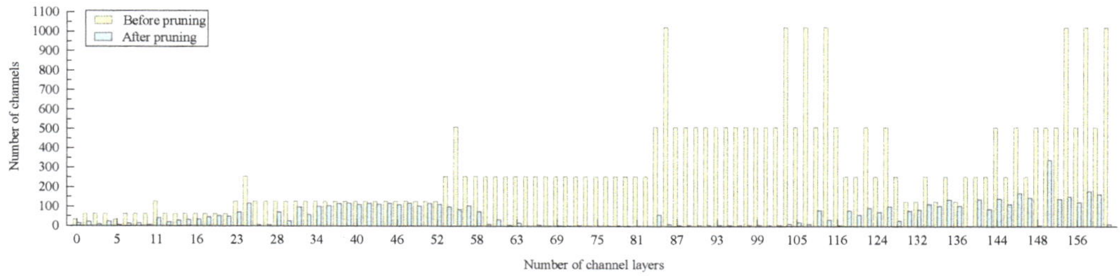

Figure 4. The channel changes of each layer of YOLO V4 model before and after pruning.

Pruned YOLO V4 network could accurately locate the location of apple defects and identify the apple stem/calyx. For the defection of regions identified by the pruned YOLO V4 network, the corresponding location information of each defect region was compared with the binary image I_B generated by BiSeNet V2 network. For the position where the gray value Bv was in the defect area determined by pruned YOLO V4 network, it was determined as the defect area. Finally, the defect region in apple image was segmented accurately by combining BiSeNet V2 network and pruned YOLO V4 network.

2.6. Defect Area Correction

If the extent of surface defects of expensive fruits is not graded, all fruits containing defects could be sorted as substandard fruits, which would cause serious economic losses to fruit farmers. So, some grading standards of fruits classify fruits according to extent of defects. For example, the number and area of apple surface defects under different grades were restricted in the local standard for apple grading in Beijing, China. Therefore, the number and area of the defects in defective apples needed to be accurately calculated.

The surface of the apple has a certain curvature because of the similarity between an apple and a sphere. When the apple was placed on the separate fruit tray and the industrial camera captured image of the apple, the defects in different areas of the outer surface of the apple would be scaled to varying degrees. Therefore, the actual area of defect might be different from the region of defect in apple images. Thus, projection method was presented to provide a solution for building the relationship between actual area of defect and defect region in apple images.

In order to eliminate the influence of surface curvature on the defect area in the apple image, it was necessary to correct the number of pixels in the defect area of the apple image. Firstly, apple models of different sizes were obtained by 3D printing, referring to the characteristics of apple surface changes in orchards. There were 12 apple models with horizontal diameter from 68 mm to 90 mm, as shown in Figure 5. In order to establish the corresponding relationship between the number of real pixels and defective pixels in the apple image, a series of black square labels with a side length of 3 mm was printed and pasted on the surface of apple models of different sizes to simulate the change of defect area at different positions.

Figure 5. The apple models with labels.

Then, each apple model was put under the camera and the images of the apple models were captured statically. The actual pixel value of the squares and their pixel value in the image were different because of the change of curvature and the different distance from each square to the center of the apple. Therefore, it was necessary to establish the function relationship between the three variables, namely the number Z of real pixels, the distance d from the defect region to the center of the apple and the r representing the radius of the apple. The function relationship could be represented as $Z = F(d, r)$. In order to obtain the expression of the function, the number of pixels in the defect area of apple model image at different positions was recorded manually for apple models with different sizes. Then, the dataset (d_w, r_w, z_w) corresponding to the three variables was generated.

So, given a dataset (d_w, r_w, z_w), w = 1, 2, 3,..., n. The bivariate polynomial function $F(d, r)$ based on the dataset could be expressed as:

$$F(d,r) = \sum_{ij=1,1}^{p,q} g_{ij} d^{i-1} r^{j-1} = \sum_{i=1}^{p} \sum_{j=1}^{q} g_{ij} d^{i-1} r^{j-1} \tag{1}$$

Let

$$d = \begin{bmatrix} 1 \\ d \\ d^2 \\ \vdots \\ d^p \end{bmatrix}, r = \begin{bmatrix} 1 \\ r \\ r^2 \\ \vdots \\ r^p \end{bmatrix}, G = \begin{bmatrix} g_{11} & \cdots & g_{1q} \\ \vdots & \ddots & \vdots \\ g_{p1} & \cdots & g_{pq} \end{bmatrix} \tag{2}$$

Then, the function could be expressed as

$$F(d,r) = d^{\mathrm{T}} \mathrm{G} r \tag{3}$$

The goal of fitting was to obtain the parameter matrix G. To obtain the parameter matrix G, a multivariate function with respect to the parameter g_{ij} was constructed:

$$L\left(g_{11}, \cdots, g_{pq}\right) = \sum_{w=1}^{n} [F(d_w, r_w) - z_w]^2 = \sum_{w=1}^{n} \left(\sum_{i=1}^{p} \sum_{j=1}^{q} g_{ij} d^{i-1} r^{j-1} - z_w \right)^2 \tag{4}$$

The point (g_{11},..., g_{pq}) was the minimum point of the multivariate function L (g_{11},..., g_{pq}), and z_w was the number of actual pixels, so the point (g_{11},..., g_{pq}) must satisfy the equation:

$$\frac{\partial L}{\partial g_{ij}} = 2 \sum_{w=1}^{n} \left[d_w^{i-1} r_w^{j-1} F(d_w, r_w) - d_w^{i-1} r_w^{j-1} z_w \right] = 0 \tag{5}$$

So, the following equation could be obtained:

$$\sum_{w=1}^{n} d_w^{i-1} r_w^{j-1} F(d_w, r_w) = \sum_{w=1}^{n} d_w^{i-1} r_w^{j-1} z_w \tag{6}$$

According to Equation (1), there were:

$$\sum_{w=1}^{n} d_w^{i-1} r_w^{j-1} z_w \sum_{\alpha=1}^{p} \sum_{\beta=1}^{q} g_{\alpha\beta} d_w^{\alpha-1} r_w^{\beta-1} = \sum_{w=1}^{n} d_w^{i-1} r_w^{j-1} z_w \tag{7}$$

$$\sum_{\alpha\beta=1,1}^{p,q} \left[g_{\alpha\beta} \sum_{w=1}^{n} \left(d_w^{\alpha-1} r_w^{\beta-1} d_w^{i-1} r_w^{j-1} \right) \right] d_w^{i-1} r_w^{j-1} z_w = \sum_{w=1}^{n} d_w^{i-1} r_w^{j-1} z_w \tag{8}$$

Let $u_{\alpha\beta}(i,j) = \sum_{w=1}^{n} \left(d_w^{\alpha-1} r_w^{\beta-1} d_w^{i-1} r_w^{j-1} \right)$ and $v(i,j) = \sum_{w=1}^{n} d_w^{i-1} r_w^{j-1} z_w$
So, Equation (8) can be rewritten in matrix form:

$$\begin{bmatrix} u_{11}(1,1) & \cdots & u_{pq}(1,1) \\ \vdots & \ddots & \vdots \\ u_{11}(p,q) & \cdots & u_{pq}(p,q) \end{bmatrix} \begin{bmatrix} g_{11} \\ \vdots \\ g_{pq} \end{bmatrix} = \begin{bmatrix} v(1,1) \\ \vdots \\ v(p,q) \end{bmatrix} \tag{9}$$

Equation (9) could be rewritten as the form Ug = V, where U is matrix with $pq \times pq$, and V is a column vector with length pq. The column vector g could be calculated. Then, g was transformed into the parameter matrix G. So, the function $F(d,r)$ could be determined using matrix G.

The object distance between the apple and the camera lens would change due to the different size of the apple, and it would affect the conversion between the real pixels and real area corresponding to the real pixels. In order to determine the corresponding relationship between the number of defective pixels and the actual defect areas under different apple sizes, every model apple with a certain size was used to determine the calibration coefficient c(r) between the number of pixels and the real areas S, where r was the radius of apple. For c(r)∈C, C was composed of 20 calibration coefficients. The final area projection equation was:

$$\mathrm{S} = \mathrm{c}(r) \sum_{i \in S_\mathrm{I}} F(d_i, r) \tag{10}$$

where i represents the pixel located in the defect region S_I in the apple image.

According to Equation (1) to Equation (9), the actual number of pixels corresponding to the defect in the image could be determined. Then, the actual area of the defect could be obtained according to Equation (10). So, the grade of defective apple could be determined according to the apple grading standard.

2.7. Evaluation Metrics of the Model Performance

Several indicators [27] were used to evaluate the performance of the proposed model, such as accuracy (A), pixel accuracy (PA), mean intersection over union (MIoU), mean pixel accuracy (MPA), recall (R), precision (P) and F1 value, where TP, FP, TN and FN represented true positive, false positive, true negative and false negative, respectively.

$$\mathrm{A} = \frac{\mathrm{TP} + \mathrm{TN}}{\mathrm{TP} + \mathrm{TN} + \mathrm{FP} + \mathrm{FN}} \tag{11}$$

$$\mathrm{PA} = \frac{\sum_{i=0}^{k} p_{ii}}{\sum_{i=0}^{k} \sum_{j=0}^{k} p_{ij}} \tag{12}$$

$$\mathrm{MPA} = \frac{1}{k+1} \sum_{i=0}^{k} \frac{p_{ii}}{\sum_{j=0}^{k} p_{ij}} \tag{13}$$

$$\mathrm{MIoU} = \frac{1}{k+1} \sum_{i=0}^{k} \frac{p_{ii}}{\sum_{j=0}^{k} p_{ij} + \sum_{j=0}^{k} p_{ji} - p_{ii}} \quad (14)$$

$$\mathrm{R} = \frac{\mathrm{TP}}{\mathrm{TP} + \mathrm{FN}} \quad (15)$$

$$\mathrm{P} = \frac{\mathrm{TP}}{\mathrm{TP} + \mathrm{FP}} \quad (16)$$

$$\mathrm{F1} = \frac{2\mathrm{P} \cdot \mathrm{R}}{\mathrm{P} + \mathrm{R}} \quad (17)$$

Assume that there were k + 1 classes (0... k) in the dataset and 0 usually represented the background. p_{ij} indicated that it was originally class i and was predicted to be class j, and p_{ji} indicated that it was originally class j but was predicted to be class i. Pixel accuracy (PA) refers to the proportion of pixels predicted correctly in the total pixels. Mean pixel accuracy (MPA) was an improvement on PA. It calculated PA for each class and then averaged PA for all classes.

3. Results and Discussion

In order to quickly and accurately realize the online classification of defective apples, the number and the area of defects needed to be calculated after apple defects were detected. Therefore, three semantic segmentation methods including DAnet [28], Unet [29] and BiSeNet V2 were compared. The detection results of the semantic segmentation for comparison are shown in Figure 6. In Figure 6, the green mark was used to label the pixels of the defect area detected by the semantic segmentation methods. Using the DAnet and Unet networks, the stem/calyx region was more likely to be wrongly segmented as a defective region, while the BiSeNet V2 network had a higher segmentation accuracy than other networks.

The performance comparison of different semantic segmentation models is shown in Table 2. It was observed in the results presented in Table 2 that the mean pixel accuracy (MPA) of the three semantic segmentation methods for apple defect detection were up to 99%. BiSeNet V2 for apple defect detection obtained a slightly better result in MPA with a value of 99.66%, which was 0.14 and 0.19 percentage points higher than DAnet and Unet, respectively. In addition, the mean intersection over union (MIoU) of the semantic segmentation method based on BiSeNet V2 for apple defect detection was 80.46%, which was 6.38 and 6.53 percentage points higher than DAnet and Unet, respectively. The results showed that BiSeNet V2 had a better ability to identify apple surface defects that DAnet and Unet failed to identify. DAnet, Unet and BiSeNet V2 took 37.40 ms, 22.64 ms and 9.00 ms, respectively, for a single image. Inference time is an important factor in evaluating online detection models. BiSeNet V2 took the shortest time, which was 75.94% and 60.25%, shorter than DAnet and Unet, respectively. Meanwhile, BiSeNet V2 had a smaller model size than other models. After comparing the pixel accuracy, inference time, parameter quantity and model size of the models, BiSeNet V2 could give consideration to higher segmentation accuracy and real-time performance. Therefore, the BiSeNet V2 model could meet the actual requirement of apple defect online detection.

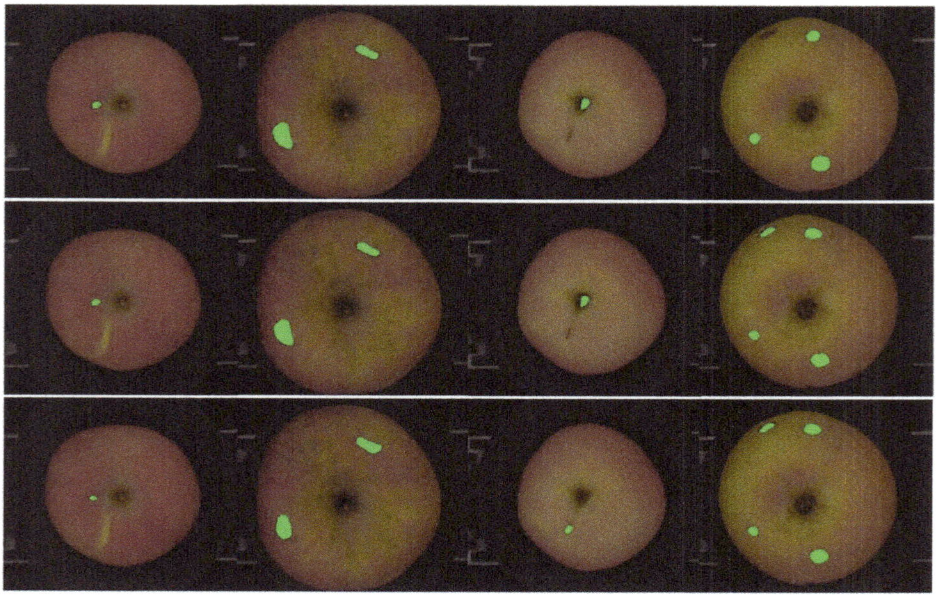

Figure 6. The segmentation results of the Unet, DAnet and BiSeNet V2 networks.

Table 2. The comparison of different semantic segmentation models.

Models	MIoU/%	MPA/%	Inference Time/ms	Parameters/MB	Model Size/MB
DAnet	74.08	99.52	37.40	45.31	181.30
Unet	73.93	99.47	22.64	12.78	51.15
BiSeNet V2	80.46	99.66	9.00	2.22	9.67

3.1. Analysis of Improvement Using Pruned YOLO V4

In the above discussion, the BiSeNet V2 network exhibits better accuracy and faster detection speed, but in practical applications, there is still mis-segmentation as shown in Figure 7. The pixels in stem/calyx regions were mistakenly identified as defects, respectively, using BiSeNet V2. In Figure 7, the green mark in the first row shows the defective parts. The pruned YOLO V4 network with higher accuracy could be used to solve this problem. The pruned YOLO V4 model was used to process the apple images after semantic segmentation. The detection results are shown in the second row of Figure 7. A green bounding box was used to label defect regions. A purple bounding box and yellow bounding box were used to label calyx and stem regions, respectively. The apple stem/calyx region and defect region in the images of the second row of Figure 7 were identified accurately. Finally, comparing the result of semantic segmentation with the result of object detection, the defect area confirmed by the two results at the same time was determined as the true defect area as shown in the last row of Figure 7. Finally, by combining BiSeNet V2 and the pruned YOLO V4 network, the defect region in apple images was obtained accurately. Therefore, the combination of BiSeNet V2 and YOLO V4 could improve the segmentation results of defect regions in apple images.

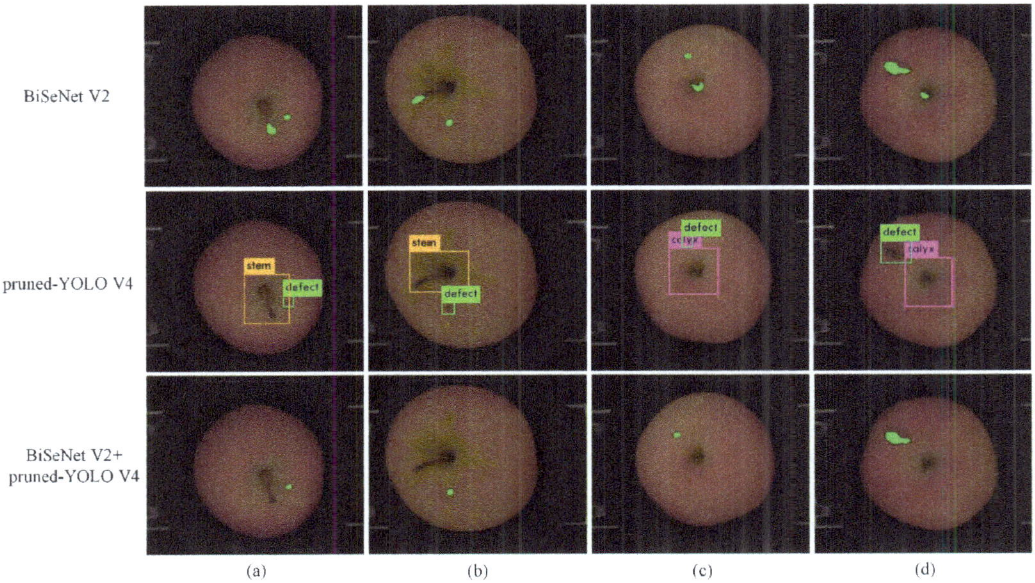

Figure 7. Comparison between segmentation results and detection results.((**a**,**b**) are images taken with stem upward. (**c**,**d**) are images taken with calyx upward.)

3.2. Results Analysis of Defect Area Calculation

According to the method of defect area correction, the calculation of apple defect area was tested. Apples with defects of different sizes in different regions were tested, respectively. Each defect was tested ten times. Then, the average value of each defect was calculated.

When the defect was located in region A, as shown in Figure 8, the defect area calculated according to the method of defect area correction was compared with the actual defect area. Three apples with defects of different sizes in region A were selected, and the defect areas on each apple were different. The actual number of pixels was calculated in the defect region according to Equation (1). Then, the area of the defect region was computed according to Equation (10). Finally, the result was compared with the actual defect area. The measurement process was repeated ten times for each apple. Then, the average value of each area was calculated. Similarly, the calculated results of three apples with defects of different sizes in region B and region C are shown in Figure 9. It could be concluded that whether the defect was located in region A, region B or region C, the difference between the calculated defect area and the actual defect area were less than 2.23 mm^2. When the defect was located in region C, the mean square error between the calculated area and the actual area was between 3.03 and 3.22, which was higher than that of defects in region A and region B. It could be concluded from Table 3 that when the defect area was large, the error of the defect area obtained was also large. This was mainly because when the samples used for training the semantic segmentation model were marked manually, it was difficult to accurately mark the edge of the apple defect region. Therefore, there might be a certain transition area between the defect edge region and the normal peel region, which led to the error in the calculated result of the defect area. Meanwhile, the sensor of the industrial camera used in the machine vision system was a CMOS chip, which led to differences in the images captured every time, even in the same conditions. This further leads to the error of defect area calculation.

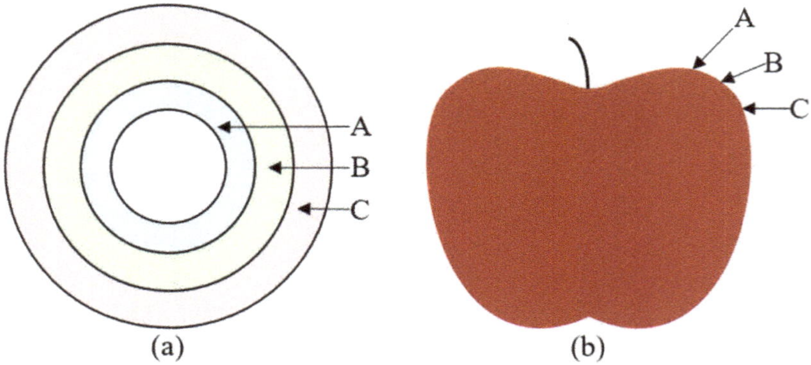

Figure 8. Schematic diagram of defect location. (A,B,C represent different areas of the apple. (**a**) is a diagram of the top view of the apple and (**b**) is a diagram of the front view of the apple).

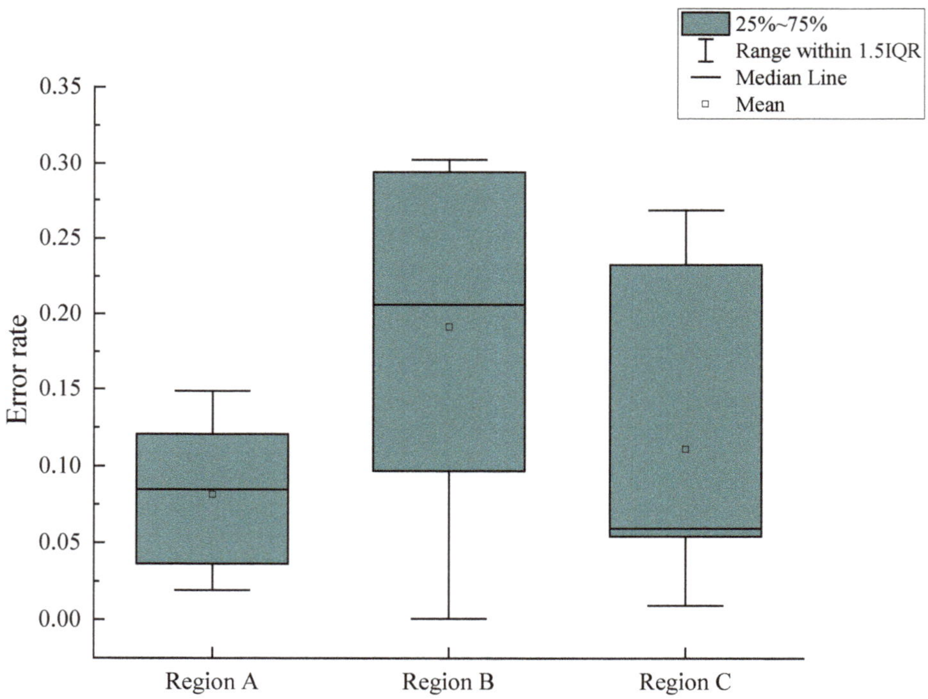

Figure 9. Boxplot of the error rate between calculated and actual defect area at different regions of fruit.

Table 3. Calculation results of defect area.

Defect Position	The Pixels of the Defect	Calculated Defect Area/mm^2	Actual Defect Area/mm^2	Root Mean Square Error
A	389.47	39.79	38.48	2.38
	243.55	24.58	22.90	2.76
	186.92	19.08	18.10	1.85
B	370.41	37.75	36.32	1.36
	218.63	21.86	19.63	2.13
	157.68	15.77	16.62	2.92
C	323.90	32.39	33.18	3.16
	221.58	22.16	20.43	3.22
	135.27	13.53	12.56	3.03

3.3. Results of the Defective Apple Grading

In order to verify the detection effect of the proposed method of defective apple grading, 68 first-class apples, 64 second-class apples and 62 third-class apples were purchased in a supermarket and selected for testing referring to the grading standard of apples in Beijing. The experimental results are shown in Table 4 and Figure 10.

Table 4. The average of detection results of three grades.

Defect Level	Precision/%	Recall/%	Accuracy/%	F1/%
First class	95.59%	95.59%	-	-
Second class	92.06%	90.63%	-	-
Third class	95.24%	96.77%	-	-
Total	94.30%	94.33%	92.42%	94.31%

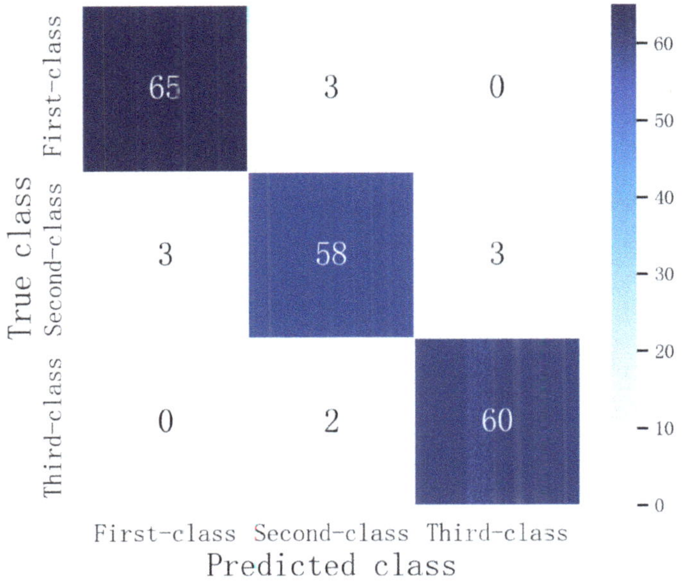

Figure 10. The confusion matrix of defective apple grading.

As shown in Table 4, the precision and recall in the three grades of apples were above 90.63%, and the overall precision and recall were 94.30% and 94.33%, respectively. The detection accuracy of apples was 92.42%, and the F1 value was 94.31%. Among the three grades, the apple grade identification with the highest precision was the first-class apples (95.59%), and the ones with lowest precision were second-class apples (92.06%). Due to the error that would occur when the area values of defect were close to the junction of two adjacent grades, first class and third class might be misclassified into second class, and second-class could also be misclassified into first class and third class. Meanwhile, if the defect was located at the edge of apple, it was sometimes incorrectly detected as the background by the object-detection algorithm and semantic segmentation model, which would also lead to the reduction of classification accuracy.

4. Conclusions

In this paper, a grading method of defective apples was proposed and applied to the separate fruit tray sorting machine. The BiSeNet V2 network and pruned YOLO V4 network were combined to extract the defect regions in apple images. The BiSeNet V2 network was utilized to determine the latent location of defect regions. The pruned YOLO V4 network was used to remove the non-defective region. A projection algorithm was proposed to build the corresponding relationship between the defect area in the image and the actual defect area on the apple's surface. After the two deep learning models were deployed using C++ language, the average accuracy and the F1 score of defective apple grading in the online test were 92.42% and 94.31%, respectively.

The overall results denoted that the proposed method has potential to be implemented in commercial fruit-grading machines. Meanwhile, the proposed method has the potential for being extended to other fruit. Because separate fruit tray grading equipment in the market can only capture the upper surface of the fruit, we are developing a flexible air suction device to assist the camera with capturing the full surface image of the fruit. Future work will focus on improving the segmentation accuracy of defects and the projection accuracy of the defect area for improving the accuracy of grading defective apples.

Author Contributions: Conceptualization, X.L. and C.Z. (Chi Zhang); methodology, X.L.; software, C.Z. (Chi Zhang); validation, X.J., L.L. and X.L.; formal analysis, X.L.; investigation, C.Z. (Chi Zhang) and X.H.; resources, W.H.; data curation, X.J.; writing—original draft preparation, X.L.; writing—review and editing, C.Z. (Chi Zhang), S.F. and C.Z. (Chunjiang Zhao); visualization, C.Z. (Chi Zhang); supervision, J.L.; project administration, C.Z. (Chi Zhang.); funding acquisition, S.F. All authors have read and agreed to the published version of the manuscript.

Funding: This research was funded by the National Natural Science Foundation of China (NSFC no. 31871523) and the Young Elite Scientists Sponsorship Program by CAST (2019QNRC001).

Informed Consent Statement: Not applicable.

Data Availability Statement: The data presented in this study are available on request from the corresponding author. The data are not publicly available since future studies are related to current data.

Conflicts of Interest: The authors declare no conflict of interest.

References

1. Mizushima, A.; Lu, R. An image segmentation method for apple sorting and grading using support vector machine and otsu's method. *Comput. Electron. Agric.* **2013**, *94*, 29–37. [CrossRef]
2. Jawale, D.; Deshmukh, M. Real time automatic bruise detection in (Apple) fruits using thermal camera. In Proceedings of the 2017 International Conference on Communication and Signal Processing (ICCSP), Melmaruvathur, India, 6–8 April 2017; pp. 1080–1085.
3. Mohammadi Baneh, N.; Navid, H.; Kafashan, J. Mechatronic components in apple sorting machines with computer vision. *J. Food Meas. Charact.* **2018**, *12*, 1135–1155. [CrossRef]
4. Nosseir, A.; Ahmed SE, A. Automatic Classification for Fruits' Types and Identification of Rotten Ones using k-NN and SVM. *Int. J. Online Biomed. Eng.* **2019**, *15*, 47–61. [CrossRef]

5. Zhang, W.; Hu, J.; Zhou, G.; He, M. Detection of apple defects based on the FCM-NPGA and a multivariate image analysis. *IEEE Access* **2020**, *8*, 38833–38845. [CrossRef]
6. Zhang, C.; Zhao, C.; Huang, W.; Wang, Q.; Liu, S.; Li, J.; Guo, Z. Automatic detection of defective apples using NIR coded structured light and fast lightness correction. *J. Food Eng.* **2017**, *203*, 69–82. [CrossRef]
7. Wang, Z.; Jin, L.; Wang, S.; Xu, H. Apple stem/calyx real-time recognition using YOLO-v5 algorithm for fruit automatic loading system. *Postharvest Biol. Technol.* **2022**, *185*, 111808. [CrossRef]
8. Keresztes, J.C.; Goodarzi, M.; Saeys, W. Real-time pixel based early apple bruise detection using short wave infrared hyperspectral imaging in combination with calibration and glare correction techniques. *Food Control* **2016**, *66*, 215–226. [CrossRef]
9. Fan, S.; Li, C.; Huang, W.; Chen, L. Detection of blueberry internal bruising over time using NIR hyperspectral reflectance imaging with optimum wavelengths. *Postharvest Biol. Technol.* **2017**, *134*, 55–66. [CrossRef]
10. Fan, S.; Li, J.; Zhang, Y.; Tian, X.; Wang, Q.; He, X.; Huang, W. On line detection of defective apples using computer vision system combined with deep learning methods. *J. Food Eng.* **2020**, *286*, 110102. [CrossRef]
11. Huang, W.; Li, J.; Wang, Q.; Chen, L. Development of a multispectral imaging system for online detection of bruises on apples. *J. Food Eng.* **2015**, *146*, 62–71. [CrossRef]
12. Zhang, B.; Huang, W.; Li, J.; Zhao, C.; Fan, S.; Wu, J.; Liu, C. Principles, developments and applications of computer vision for external quality inspection of fruits and vegetables: A review. *Food Res. Int.* **2014**, *62*, 326–343. [CrossRef]
13. Yang, J.; Xu, J.; Zhang, X.; Wu, C.; Lin, T.; Ying, Y. Deep learning for vibrational spectral analysis: Recent progress and a practical guide. *Anal. Chim. Acta* **2019**, *1081*, 6–17. [CrossRef] [PubMed]
14. Xin, Y.; Ma, S.; Wei, Y.; Hu, J.; Ding, Z.; Wang, F. Detection of Apple Surface Defect Based on YOLOv3. In Proceedings of the 2021 ASABE Annual International Virtual Meeting, Virtual, 12–16 July 2021; American Society of Agricultural and Biological Engineers: St. Joseph, MI, USA, 2021; p. 1. [CrossRef]
15. Valdez, P. Apple defect detection using deep learning based object detection for better post harvest handling. *arXiv* **2020**, arXiv:2005.06089.
16. Hu, G.; Zhang, E.; Zhou, J.; Zhao, J.; Gao, Z.; Sugirbay, A.; Chen, J. Infield Apple Detection and Grading Based on Multi-Feature Fusion. *Horticulturae* **2021**, *7*, 276. [CrossRef]
17. Li, Y.; Feng, X.; Liu, Y.; Han, X. Apple quality identification and classification by computer vision based on deep learning. *Sci. Rep.* **2021**, *11*, 16618. [CrossRef] [PubMed]
18. Fan, S.; Liang, X.; Huang, W.; Zhang, V.J.; Pang, Q.; He, X.; Zhang, C. Real-time defects detection for apple sorting using NIR cameras with pruning-based YOLOV4 network. *Comput. Electron. Agric.* **2022**, *193*, 106715. [CrossRef]
19. Nagaraju, Y.; Hegde, S.U.; Stalin, S. Fine-tuned mobilenet classifier for classification of strawberry and cherry fruit types. In Proceedings of the 2021 International Conference on Computer Communication and Informatics (ICCCI), Coimbatore, India, 27–29 January 2021; pp. 1–8. [CrossRef]
20. Yu, C.; Wang, J.; Peng, C.; Gao, C.; Yu, G.; Sang, N. Bisenet: Bilateral segmentation network for real-time semantic segmentation. In Proceedings of the European Conference on Computer Vision (ECCV), Munich, Germany, 8–14 September 2018; pp. 325–341.
21. He, K.; Zhang, X.; Ren, S.; Sun, J. Spatial pyramid pooling in deep convolutional networks for visual recognition. *IEEE Trans. Pattern Anal. Mach. Intell.* **2015**, *37*, 1904–1916. [CrossRef]
22. Ren, S.; He, K.; Girshick, R.; Sun, J. Faster R-CNN: Towards real-time object detection with region proposal networks. *IEEE Trans. Pattern Anal. Mach. Intell.* **2017**, *39*, 1137–1149. [CrossRef]
23. Bochkovskiy, A.; Wang, C.Y.; Liao, H.Y.M. Yolov4: Optimal speed and accuracy of object detection. *arXiv* **2020**, arXiv:2004.10934.
24. Redmon, J.; Farhadi, A. Yolov3: An incremental improvement. *arXiv* **2018**, arXiv:1804.02767.
25. Molchanov, P.; Mallya, A.; Tyree, S.; Frosio, I.; Kautz, J. Importance estimation for neural network pruning. In Proceedings of the IEEE/CVF Conference on Computer Vision and Pattern Recognition, Long Beach, CA, USA, 15–20 June 2019; pp. 11264–11272.
26. Misra, D. Mish: A self regularized non-monotonic neural activation function. *arXiv* **2019**, arXiv:1908.08681.
27. Garcia-Garcia, A.; Orts-Escolano, S.; Oprea, S.; Villena-Martinez, V.; Garcia-Rodriguez, J. A review on deep learning techniques applied to semantic segmentation. *arXiv* **2017**, arXiv:1704.06857.
28. Fu, J.; Liu, J.; Tian, H.; Li, Y.; Bao, Y.; Fang, Z.; Lu, H. Dual attention network for scene segmentation. In Proceedings of the IEEE/CVF Conference on Computer Vision and Pattern Recognition, Long Beach, CA, USA, 15–20 June 2019; pp. 3146–3154.
29. Ronneberger, O.; Fischer, P.; Brox, T. U-net: Convolutional networks for biomedical image segmentation. In Proceedings of the International Conference on Medical Image Computing and Computer-Assisted Intervention, Munich, Germany, 5–9 October 2015; pp. 234–241.

foods

MDPI

Article

Real-Time Monitoring of the Quality Changes in Shrimp (*Penaeus vannamei*) with Hyperspectral Imaging Technology during Hot Air Drying

Wenya Xu, Fan Zhang, Jiarong Wang, Qianyun Ma, Jianfeng Sun, Yiwei Tang, Jie Wang and Wenxiu Wang *

College of Food Science and Technology, Hebei Agricultural University, Baoding 071000, China
* Correspondence: cauwwx@hebau.edu.cn; Tel.: +86-0312-7520197

Abstract: Hot air drying is the most common processing method to extend shrimp's shelf life. Real-time monitoring of moisture content, color, and texture during the drying process is important to ensure product quality. In this study, hyperspectral imaging technology was employed to acquire images of 104 shrimp samples at different drying levels. The water distribution and migration were monitored by low field magnetic resonance and the correlation between water distribution and other quality indicators were determined by Pearson correlation analysis. Then, spectra were extracted and competitive adaptive reweighting sampling was used to optimize characteristic variables. The grey-scale co-occurrence matrix and color moments were used to extract the textural and color information from the images. Subsequently, partial least squares regression and least squares support vector machine (LSSVM) models were established based on full-band spectra, characteristic spectra, image information, and fused information. For moisture, the LSSVM model based on full-band spectra performed the best, with residual predictive deviation (RPD) of 2.814. For L^*, a^*, b^*, hardness, and elasticity, the optimal models were established by LSSVM based on fused information, with RPD of 3.292, 2.753, 3.211, 2.807, and 2.842. The study provided an in situ and real-time alternative to monitor quality changes of dried shrimps.

Keywords: shrimp; hot air drying; quality change; hyperspectral images; low field magnetic resonance

Citation: Xu, W.; Zhang, F.; Wang, J.; Ma, Q.; Sun, J.; Tang, Y.; Wang, J.; Wang, W. Real-Time Monitoring of the Quality Changes in Shrimp (*Penaeus vannamei*) with Hyperspectral Imaging Technology during Hot Air Drying. *Foods* **2022**, *11*, 3179. https://doi.org/10.3390/foods11203179

Academic Editor: James Carson

Received: 25 August 2022
Accepted: 7 October 2022
Published: 12 October 2022

1. Introduction

Shrimp (*Penaeus vannamei*) harvesting is one of the most economically significant fishing activities in China attracting attention from consumers due to the high protein content and rich nutritional composition of shrimp [1,2]. According to the China Fisheries Statistical Yearbook, the *Penaeus vannamei* aquaculture production in China was 1.1977 millions of tons in 2020. However, the shrimp harvest suffers from rapid deterioration due to biochemical reactions and microbial activity after death [3–5], which directly affect its shelf life. Hot air drying, a common and practical method of drying seafood, can prolong the shelf life of the shrimp harvest [6–8]. As a foodstuff, dried shrimp has the advantages of a unique flavor, rich nutrition, easy storage, and high consumer demand [9,10]. However, drying is a complex process involving water evaporation, protein degradation and denaturation, and the formation of flavor compounds [11,12]. Ineffective drying can adversely impact the color, texture, and nutrition attributes of the dried shrimp product [13]. Therefore, it is imperative to monitor and control critical quality parameters during the drying process to ensure consistency among batches, as well as uniformity of the end-product.

Current analytical methods employed to measure these quality characteristics in factories, such as oven drying and texture profile analysis (TPA), are time-consuming, destructive, cumbersome, and restricted to off-line usage [14–16]. Therefore, it is necessary to develop an effective, rapid, nondestructive, and real-time detection method for dried shrimp quality control. With the development of optical and spectroscopic technologies, hyperspectral imaging (HSI) has been successfully applied to evaluate food safety and quality, monitor food preparation processes, and identify adulteration [17,18]. HSI simultaneously captures both spectral and spatial information of a sample by integrating spectroscopic and computer vision or imaging techniques into one system [19–21]. Another unique characteristic is that HSI generates a visual distribution map of reference values to enable the prediction and quantification of internal sample constituents, as well as the simultaneous determination of their location on the sample surface [22,23].

Based on these advantages, HSI has been applied to monitor quality changes in meat, fruit, vegetable, and cereal foods during drying. For example, Sun et al. used HSI to monitor the moisture contents of scallops during drying, and reported a model prediction accuracy of greater than 0.9 [24]. Moreover, Netto et al. used HSI to evaluate the water uniformity of the melon drying process under different pretreatments by visualizing the moisture content in the samples [25]. However, most existing studies only employ spectral information for quality indicator evaluation, ignoring image information, such as color and texture in their modeling. To improve prediction accuracy, the importance of combining spectral and spatial HIS information has been emphasized by several researchers. This technique has been used to discriminate between different breeds of chicken [26], predict the storage time and moisture content of cooked beef [19], and assess the fat and moisture contents of salmon [27]. The results indicate that a combination of spectral and spatial HSI data is more comprehensive and intuitional than conventional analyses. Furthermore, considering that the shrimp drying process involves color and texture changes, it is crucial to include image information in the spectral model for quality control. To the best of our knowledge, there are no previous data fusion studies on the visualization of moisture and other quality indicators in dried shrimp. Additionally, previous studies only predicted moisture and other quality indicator contents, neglecting the link between moisture distribution and other quality characteristics, which may clarify the mechanisms governing shrimp quality changes during the drying process.

Therefore, the purpose of the current study is to explore the correlation between shrimp water distribution state and other quality indices and combine spatial and spectral information of the hypercube to measure shrimp quality changes during the drying process. The specific objectives are: (1) to quantify changes in shrimp during hot air drying through moisture content measurement, color properties (L^*, a^*, b^*) analysis, and texture profile analysis (hardness, adhesiveness, elasticity, stickiness, and chewiness); (2) to monitor the dynamic water sate and water migration by low field magnetic resonance (LF-NMR) and determine the correlation between water distribution and other quality indicators by Pearson correlation analysis; (3) to acquire hyperspectral reflectance images of shrimps at different drying stages, as well as spectral data and color and textural features from the region of interest (ROI); (4) to establish partial least squares regression (PLSR) and least squares support vector machine (LSSVM) models based on spectral, image, and fusion information; and (5) to visualize shrimp quality at the pixel level using the optimal models.

2. Materials and Methods

2.1. Sample Preparation and Drying Experiments

Live shrimp (*Penaeus vannamei*), each weighing approximately 15 ± 3 g, were purchased from a local market in Baoding, China. The live shrimp were transported to the laboratory within 30 min and stored in ice water, until their death. The shrimp were boiled in salt water with a mass fraction of 3% for 2 min and removed to dry the surface moisture. Then, they were used for constant drying experiments at 55 °C using an electrical blast drying oven. When the moisture content had been reduced by approximately 35%, the taste of the dried shrimp was optimal. The total dry processing time was 12 h, and shrimp sample were collected after boiling and drying for 1, 2, 3, 4, 5, 6, 7, 8, 9, 10, 11, and 12 h. A total of 16 samples were collected at each sampling point, half were used for color, moisture determination, and hyperspectral measurement, and the other half were used for texture analysis, LF-NMR analysis, magnetic resonance imaging (MRI) measurement, and hyperspectral measurement. Thus, 104 samples (8 samples × 13 drying levels) were involved for each quality indicator prediction.

2.2. Quality Indicators Determination

2.2.1. Moisture Content Measurement

The direct drying method found in the *National Standards of China (GB 5009.3-2016)* was used to calculate the moisture content of the shrimp samples. The specific steps employed were as follows. First, the glass flat weighing bottles were placed in a 105 °C oven to dry to a constant weight. Second, 3.0 g of shrimp samples for each drying period were placed into weighing bottles and the total mass of the bottles and shrimp samples were accurately weighed, noted as m_1. Third, the bottles were placed in an oven at 105 °C for drying and removed after 1 h, then weighed after cooling in the desiccator for half an hour. These steps were repeated until the total mass was not changing and the final mass was weighed accurately and noted as m_2. The moisture content of the shrimp samples at each time point during the drying process was calculated as follows:

$$X = \frac{m_1 - m_2}{m_1 - m_3} \times 100\% \quad (1)$$

where X (unit: g/100 g) indicates the moisture content of the shrimp samples at each time point during the drying process; m_1 and m_2 (unit: g) indicate the mass of the weight bottles and samples before and after drying, respectively; and m_3 (unit: g) indicates the mass of the weight bottles.

2.2.2. Color Analysis

The color of the shrimp samples was evaluated using a CR-400 color difference meter (Konica Minolta Co., Ltd., Tokyo, Japan) after equilibration to room temperature. The second abdominal segment of the shrimp was used for color measurement. The color differences were analyzed using lightness (L^*), green to red (a^*), and blue to yellow (b^*). All experiments were conducted eight times.

2.2.3. Texture Profile Analysis (TPA)

The shrimp with the head and shell removed were subjected to texture analysis. Texture variables, including hardness, elasticity, stickiness, adhesiveness, and chewiness, were obtained using Texture Expert software (TMS-Pro, Food Technology Corporation, Sterling, VA, USA) The measurement parameters were set to TPA mode; the probe type was P/5, compression ratio was 45%, detection rate was 30 mm/min, shape variable was 60%, minimum force was 0.5 N, and return distance was 2.5 cm. Each shrimp sample was measured eight times at each point during the drying processes.

2.3. LF-NMR Transverse Relaxation Measurements

The relaxation measurements were performed on a Meson NMI20-040H-I LF-NMR analyzer (NMI20-040H-I, NIUMAG Electronic Technology Co., Ltd., Shanghai, China) with a magnetic field strength of 0.5 T and corresponding resonance frequency for protons of 20 MHz. The shrimp samples were placed in a cylindrical glass tube, and a 30-mm diameter radio frequency coil was used to collect Carr–Purcell–Meiboom–Gill sequence (CPMG) decay signals, with a π-value (the time between pulses 90 and 180) of 200 μs; the lengths of these two pulses were 9.52 μs and 18.48 μs, respectively. The repetition time between two scans was 1500 ms. Distributed multiexponential fitting analysis was performed on the T_2 relaxation data using MultiExp Inv Analysis software (NIUMAG Electronic Technology Co., Ltd., Shanghai, China). The T_2 relaxation spectra were obtained from this analysis; the lateral and vertical axes represent the relaxation time and signal intensity, respectively (corresponding to the proportion of water molecules exhibited at that relaxation time).

2.4. MRI Analysis

MRI was also performed using a Meson NMI20-040H-I LF-NMR analyzer ((NMI20-040H-I, NIUMAG Electronic Technology Co., Ltd., Shanghai, China)) equipped with a 60-mm radio frequency coil. A spin echo (SE) sequence was applied to obtain T_2 weighted images of the shrimp. The following scanning parameters were used: field of view (FOV) = 100 mm × 100 mm, slice width = 1.1 mm, slice gap = 1.1 mm, average = 8, read size = 256, phase size = 192, T_2 weighted image echo time (TE) = 20 ms, and repetition time (TR) = 500 ms.

2.5. Hyperspectral Image Acquisition and ROI Selection

Images of shrimp were acquired using a pushbroom HSI system in the reflectance mode. The system consisted of 4 components: a charge-coupled device (CCD) camera (FX 10, Specim Ltd., Helsinki, Finland) with a resolution of 1024 pixels in the spatial dimension and 224 bands in the spectral dimension, hyperspectral imaging workstation with a spectral range of 400–1000 nm, 2 halogen lamps, and computer with hyperspectral image analysis software. The spectral resolution was 5.5 nm, and the imaging speed of full band acquisition was 330 Frames Per Second (FPS). Before the experiment, the instrument was preheated for 30 min to ensure its stability. The samples were placed on a mobile platform for image acquisition. To prevent image oversaturation, it was necessary to set the speed of the moving platform, camera exposure time, and acquisition distance in advance; after repeated testing, these three parameters were set to 7.5 mm/s, 50 ms, and 30 cm, respectively. Simultaneously, black and white correction of the acquired hyperspectral image was conducted to reduce the influence of the dark current of the CCD camera and uneven brightness of the light source. The correction formula is given by:

$$R = \frac{R_o - R_d}{R_w - R_d} \quad (2)$$

where R_o represents the original spectral image, R_w represents the whiteboard image, and R_d represents the darkfield image.

ROI spectral extraction of the hyperspectral image was performed using ENVI 5.2 software (Exelis Visual Information Solutions Co., Boulder, CO, USA). All pixels, except those corresponding to the shrimp head and tail, were selected to contain as much information as possible about the sample. As the collected spectral images were clear in all bands, the entire spectral range with 224 bands was retained for analysis.

2.6. Spectral Pre-Processing and Optimal Wavelengths Selection

Spectral preprocessing involves the use of appropriate mathematical analyses to correct random noise in the spectra and light scattering generated by the instruments, which is helpful for highlighting valuable spectral information [28]. In current work, Savitzky–Golay smoothing and standard normal variable transformation (SNV) method were employed to remove the interference information from the spectra. Meanwhile, among the collected spectral information, the spectral data of certain bands could be explained or replaced by those of other bands. This situation caused a large amount of redundant information in the spectrum. Owing to the existence of redundant information, the prediction accuracy of the established model decreased; as the computational burden increased, the computational speed decreased. To overcome these problems, it was important to select a small set of optimal wavelengths that reflected the changes in quality to establish the model. The competitive adaptive reweighting sampling (CARS) method was used to select the optimal wavelengths in this study.

2.7. Image Color and Texture Information Extraction

Compared with traditional spectroscopic methods, HSI has the advantage of providing abundant image information related not only to size and shape, but also color and textural features. Color moments represent a simple and effective means of representing the color features, with the first, second, and third order moments commonly used to express the color distribution of images. Because of its advantages of low feature vector dimensionality and no color space quantization, color moments are effective for characterizing color distributions in images [29]. In this study, we extracted the *RGB* (red, green, and blue) images synthesized from the hyperspectral images at 647 nm, 550 nm and 460 nm using ENVI 5.2 (Exelis Visual Information Solutions Co., Boulder, CO, USA), and the first-order moment and second-order moment information were calculated. Then, the *RGB* tricolor was transformed into HSV (hue, saturation, and value) mode, which is based on the intuitive properties of color, to extract three additional feature variables. Finally, nine color features were obtained to reflect the image difference of samples with different drying levels.

The gray-level co-occurrence matrix (GLCM) method was used to extract the texture information about the shrimp images. Four special mutually independent features of contrast, correlation, energy, and homogeneity were used to describe the co-occurrence matrix data in four orientations of 0°, 45°, 90°, and 135°, and the distance of each pixel pair was set to 1. The contrast value expresses local variations in the gray levels of the GLCM, the correlation measures the image linearity among pixels, the homogeneity measures the density of the distribution of elements in the GLCM to its diagonal, and the energy measures the textural uniformity of the image [30]. All textural values based on the different directions were then averaged into one value representing the textural features of the sample for subsequent analysis. Before constructing the texture matrix, principal component analysis (PCA) was performed to select the optimal characteristic images [31]. The implementation procedures for PC images were performed using the ENVI 5.2 software (Exelis Visual Information Solutions Co., Boulder, Colorado, USA), and the color and texture feature extraction were performed in Matlab 2012a (MathWorks Co., Natick, MA, USA).

2.8. Quantitative Analysis Models

In this study, PLSR and LSSVM techniques were compared to establish the quantitative relationships between spectroscopic data and image information and the measured moisture content, *L**, *a**, *b**, hardness, and elasticity during the drying process. The 104 samples were divided 3:1 into calibration and prediction sets for L and LSSVM modeling. PLSR is an effective multivariate regression method that enables regression modeling of multiple independent variables; it is particularly effective when the variables are highly linearly correlated [32]. The LSSVM technique can be applied to both linear and nonlinear regression models. For nonlinear regression problems, the LSSVM approach first performs nonlinear mapping from the input space onto a high-dimensional feature space using a nonlinear kernel function. This method then performs linear regression in the same feature space, which can be used to solve linear regression problems [33]. The predicted results were compared with the actual values, and the model performance was evaluated in terms of the correlation coefficient (*R*), root-mean-squared error of calibration set (RMSEC) and root-mean-squared error of prediction set (RMSEP), and residual predictive deviation (RPD). The afore-mentioned data analyses were implemented using Matlab2012a (MathWorks Co., Natick, MA, USA).

2.9. Visualization of Shrimp Quality Indicators

The advantage of HSI is its ability to transfer multivariate spectral data in a pixel-wise manner by inputting the spectra in each pixel into an established calibration model. In this study, we selected the final optimal models of moisture content, *L**, *a**, *b**, hardness, and elasticity for visualization by pseudo-color data processing. All visualization steps were executed in Matlab2012a (MathWorks Co., Natick, MA, USA). The key steps of the analysis procedure are summarized in Figure 1.

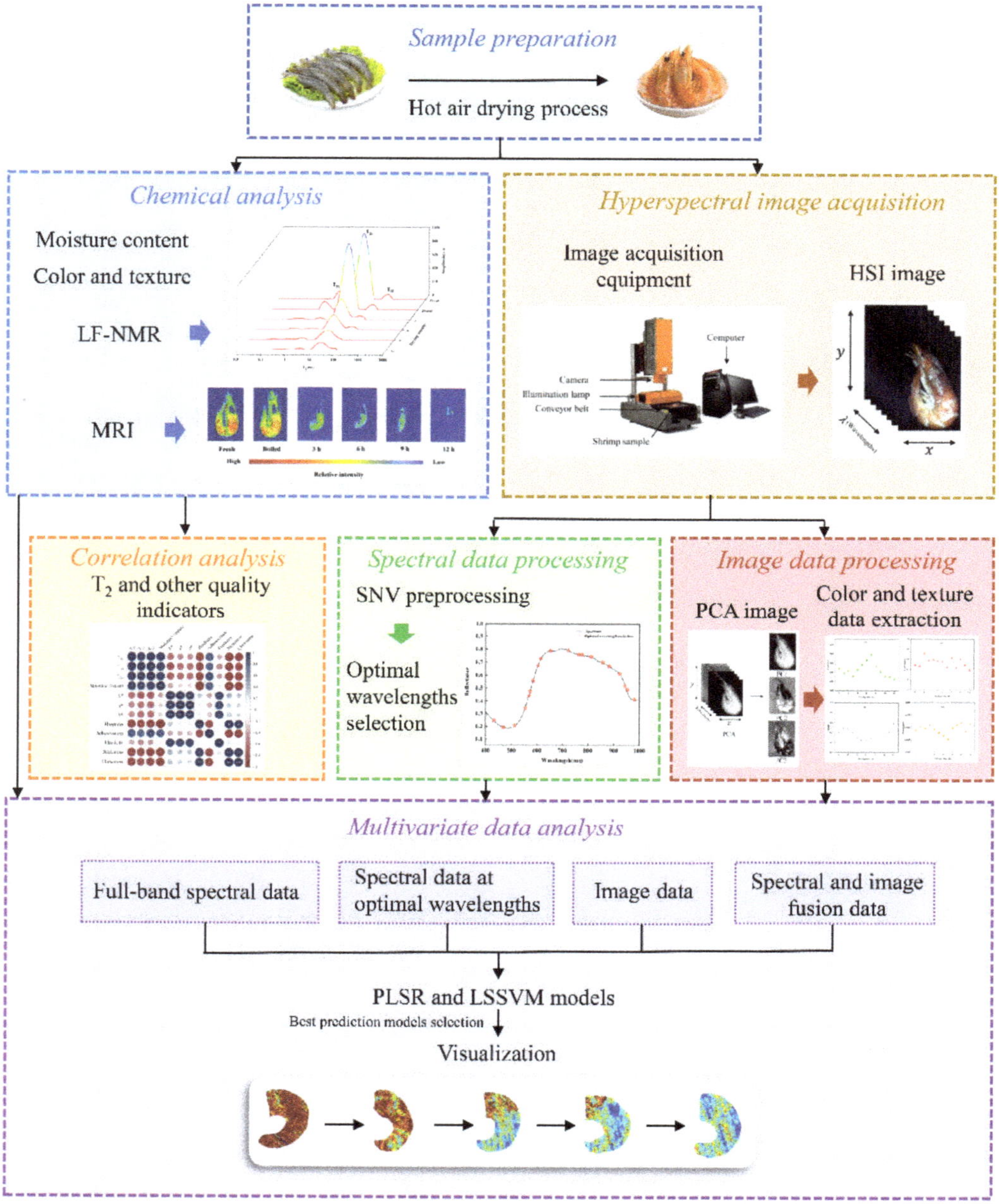

Figure 1. Flowchart of data analyses.

2.10. Statistical Analysis

The physicochemical data were statistically analyzed using the Statistical Package for the Social Sciences (SPSS) version 18.0 software package (SPSS Inc., Chicago, IL, USA). Data are expressed as mean ± standard deviation (SD), and significance was defined as

$p < 0.05$. The correlation between the LF-NMR results and the physicochemical parameters was determined by Pearson correlation analysis.

3. Results and Discussion

3.1. Quality Indicators Analysis

3.1.1. Moisture Content Analysis

The changes in the moisture content and drying rate of shrimp during the hot air drying process are shown in Figure 2. The moisture content of the fresh shrimp was 75.87%. The drying endpoint is 12 h at which point the moisture content decreases to 35.02%. It can be seen from Figure 2 that within 2 h of drying, the moisture content decreases at a slower rate; the drying rate at this point is 3.74% w.b h^{-1}; during 2–8 h, the moisture content decreases at a faster rate and the drying rate reaches a maximum of 5.72% w.b h^{-1} at 4 h; and after 8 h, the drying rate decreases slowly. The reason for this phenomenon could be that, in the early drying stage, due to the high moisture content of shrimp, the oven space was saturated, and the moisture on the surface of the shrimp could not evaporate in time [34]. This situation increased the humidity in the oven. With further hot air drying, the protein denatured because of the heat, which reduced the interaction between matter and water. The release of water and increase in the drying rate could also have been due to fiber shrinkage, leading to decreased intracellular spaces and thus facilitating evaporation. Similar results were obtained by Sun et al., who found that the moisture content of scallops decreased by approximately 50% during drying at 55 °C for 5 h and that the decrease in moisture content was mainly associated with free water migration [24]. When the moisture on the surface of the shrimp evaporated, the free water in the body evaporated to a certain extent, and the remaining bound water could not easily flow and evaporate, resulting in a slow decline in the moisture content and drying rate. Shi et al. found that the decrease in the moisture content of beef jerky with increased drying time and temperature was related to the degree of moisture migration [35]. The current results corroborate these findings.

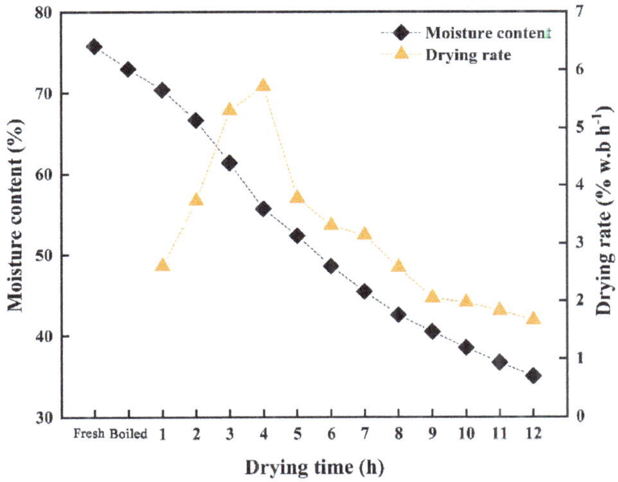

Figure 2. Averaged measured moisture contents of all samples at different drying levels.

3.1.2. TPA

Texture analysis of the shrimp was performed during the drying process, and the results are shown in Table 1. The TPA parameters include hardness, adhesiveness, elasticity, stickiness, and chewiness. As shown in Table 1, the hardness of the shrimp samples significantly increased with increasing drying time ($p < 0.05$) due to the change of drying rate and the shrinkage of shrimp muscle fibers during the drying process. Latorre et al.

explained that the dissociation of actin and myosin, disintegration of muscle fibers, and myofibril dissociation lead to the formation of small fragments and increase the hardness of the disordered structure of the muscle fiber [36]. The shrimp showed the least adhesiveness after drying for 12 h, which highlighted their improved fragility. The elasticity of shrimp firstly increased and then decreased during drying and reached its highest value at 7 h of drying. The changes in elasticity were related to the contraction of the muscle fiber. As muscle fiber contracts, the muscle proteins form a dense reticular structure, which is prone to irreversible deformation when the muscle tissue is extruded, and the spatial structure of muscle proteins is small, resulting in reduced elasticity [37]. In addition, the stickiness and chewiness of shrimp samples increase with increasing drying time. Chewiness and stickiness are parameters used in comprehensive analysis. Chewiness represents the energy required to chew solid samples, whereas stickiness represents the energy required to separate food from its contact material. As hardness and elasticity showed significant changes during drying, they were chosen as representative indicators of TPA for further modeling.

Table 1. Effects of hot air drying on the texture of shrimp.

Drying Time (h)	Hardness (N)	Adhesiveness (N)	Elasticity (mm)	Stickiness (mJ)	Chewiness (mJ)
Fresh	127.96 ± 15.44^{f}	1.90 ± 0.31^{a}	0.83 ± 0.13^{f}	2.84 ± 0.58^{h}	2.38 ± 0.74^{h}
Boiled	144.26 ± 16.24^{f}	1.48 ± 0.17^{c}	$1.96 \pm 0.23^{c,d,e}$	12.23 ± 1.99^{g}	24.37 ± 6.85^{g}
1	$159.64 \pm 28.24^{e,f}$	1.45 ± 0.18^{c}	$1.90 \pm 0.22^{d,e}$	$14.61 \pm 3.32^{f,g}$	$28.44 \pm 9.41^{f,g}$
2	$168.89 \pm 34.57^{e,f}$	1.36 ± 0.14^{c}	$1.99 \pm 0.21^{c,d,e}$	$17.06 \pm 3.96^{e,f,g}$	$34.63 \pm 11.19^{f,g}$
3	$162.99 \pm 20.22^{e,f}$	1.14 ± 0.11^{d}	$1.87 \pm 0.16^{d,e}$	$14.55 \pm 1.24^{f,g}$	$27.21 \pm 4.07^{f,g}$
4	$197.26 \pm 45.33^{d,e}$	1.48 ± 0.19^{c}	1.74 ± 0.18^{e}	$19.40 \pm 4.70^{d,e,f}$	$34.44 \pm 11.17^{f,g}$
5	$231.38 \pm 41.98^{c,d}$	1.37 ± 0.11^{c}	$1.89 \pm 0.18^{d,e}$	$20.74 \pm 4.40^{c,d,e,f}$	$39.41 \pm 10.75^{e,f,g}$
6	245.64 ± 52.16^{c}	1.67 ± 0.32^{b}	$1.91 \pm 0.31^{d,e}$	$21.20 \pm 6.82^{c,d,e,f}$	$41.67 \pm 18.05^{d,e,f,g}$
7	$233.56 \pm 56.55^{c,d}$	$0.25 \pm 0.06^{f,g}$	$2.09 \pm 0.20^{c,d}$	$22.15 \pm 5.17^{c,d,e}$	$46.59 \pm 12.44^{c,d,e,f}$
8	$265.54 \pm 39.32^{b,c}$	1.00 ± 0.12^{d}	$2.05 \pm 0.40^{c,d}$	26.55 ± 5.60^{c}	$55.78 \pm 21.14^{c,d,e}$
9	$267.03 \pm 58.75^{b,c}$	$0.33 \pm 0.06^{e,f,g}$	$1.36 \pm 0.23^{a,b}$	26.61 ± 6.87^{c}	63.47 ± 19.29^{c}
10	$252.94 \pm 44.79^{b,c}$	$0.41 \pm 0.10^{e,f}$	$1.23 \pm 0.34^{b,c}$	$25.95 \pm 8.07^{c,d}$	$60.13 \pm 25.82^{c,d}$
11	296.95 ± 63.08^{b}	0.47 ± 0.08^{e}	1.51 ± 0.26^{a}	37.14 ± 10.70^{b}	92.36 ± 24.28^{b}
12	344.78 ± 44.22^{a}	0.20 ± 0.02^{g}	1.56 ± 0.35^{a}	45.01 ± 9.78^{a}	116.50 ± 34.33^{a}

Note: All data are presented as mean ± standard error. Mean values with different letters within each line are significantly different ($p < 0.05$) with respect to processing.

3.1.3. Color Analysis

The market value of shrimp depends on the visual appearance of their body color, which is attributed to the presence of astaxanthin [38]. This carotenoid pigment is responsible for orange red tissue pigmentation in shrimp meat. Table 2 shows the color differences of the shrimp. The L^* value of shrimp increases from 40.71 when fresh to 63.85 after boiling ($p < 0.05$), which may be due to the increase in heat during boiling, resulting in protein accumulation and an increase in opacity. However, the L^* value of the shrimp decreases with more drying ($p < 0.05$). The blackening of dried shrimp is attributed to the Maillard reaction during drying [39]. Moreover, a^* and b^* exhibit similar trends throughout drying. The a^* and b^* values of dried shrimp are significantly higher than those of fresh shrimp ($p < 0.05$). The formation of redness upon the exposure of shrimp meat to heat is a result of the release of astaxanthin owing to the breakdown of carotene protein during denaturation. There are slight decreases in a^* and b^* values in the late drying period, which may be due to a slower drying rate and longer drying time, resulting in the slight damage of astaxanthin from the extension of hot air-drying. Regarding ΔE values, the results for $\Delta E > 12$ show that the color of shrimp during drying is notably different from that of fresh shrimp.

Table 2. Effects of hot air drying on the color of shrimp.

Drying Time (h)	*L**	*a**	*b**	Δ*E*
Fresh	40.71 ± 1.22 [f]	1.22 ± 0.55 [f]	3.08 ± 1.39 [h]	-
Boiled	63.85 ± 1.22 [a]	23.38 ± 2.30 [a]	29.03 ± 2.46 [a]	41.304 ± 2.34 [a]
1	63.89 ± 1.13 [a]	21.33 ± 1.79 [b]	25.85 ± 1.96 [b]	38.261 ± 1.97 [b]
2	63.68 ± 1.17 [a]	20.02 ± 1.39 [b,c]	24.19 ± 2.08 [b,c,d]	36.460 ± 2.12 [b,c]
3	63.23 ± 2.17 [a,b]	19.35 ± 1.87 [c]	25.16 ± 1.84 [b,c]	36.493 ± 1.40 [b,c]
4	63.12 ± 1.61 [a,b,c]	20.24 ± 1.57 [b,c]	24.09 ± 1.65 [b,c,d]	35.158 ± 2.26 [c]
5	61.79 ± 0.88 [b,c]	17.56 ± 1.61 [d]	24.15 ± 1.68 [b,c,d]	34.043 ± 1.41 [d]
6	62.04 ± 1.52 [b,c]	15.86 ± 2.74 [d,e]	21.94 ± 1.37 [e,f,g]	32.144 ± 1.49 [e,f]
7	61.62 ± 2.24 [c,d]	16.60 ± 2.40 [d,e]	23.35 ± 1.79 [c,d,e]	33.025 ± 2.62 [d,e]
8	60.23 ± 1.28 [d,e]	15.99 ± 1.95 [d,e]	22.25 ± 1.77 [d,e,f]	31.152 ± 2.09 [e,f,g]
9	59.59 ± 1.58 [e]	16.93 ± 1.28 [d,e]	21.78 ± 1.79 [e,f,g]	30.938 ± 1.52 [f,g]
10	59.63 ± 1.68 [e]	15.56 ± 1.06 [d,e]	20.36 ± 1.75 [f,g]	29.433 ± 1.60 [g]
11	59.93 ± 0.56 [e]	15.44 ± 1.73 [e]	19.83 ± 2.56 [g]	29.289 ± 1.80 [g]
12	59.58 ± 0.79 [a]	16.08 ± 1.31 [d,e]	20.25 ± 2.42 [f,g]	29.617 ± 1.28 [g]

Note: All data are presented as mean ± standard error. Mean values with different letters within each line are significantly different ($p < 0.05$) with respect to processing. "-"represents the blank.

3.2. LF-NMR Analysis

LF-NMR spectroscopy measures the absorption of radio frequency resonance in presence of an external magnetic field [30]; thus, the spin–spin relaxation time (T_2) is closely related to the water state and dynamics in foods. Protons of all substances are surrounded by a small magnetic field; thus, each proton creates a tiny magnetic field that is affected by the magnetic field of other protons [40]. Therefore, as the T_2 of a sample is small or large when the distance between protons is relatively small or large, respectively, T_2 value analysis is a fast and effective method that allows to identify changes in moisture content and status, and reflects (to some extent) the micro-molecular structure of a sample [41,42]. Since water can alter the interaction between the different components of foods, drying can significantly modify the microstructure of foods. Herein, the T_2 signal amplitude of shrimp at different drying stages was measured to characterize the change of water state (Figure 3a). To better investigate the water state in the different samples, the relaxation times T_{21}, T_{22}, and T_{23} of shrimp were defined as bound water that was tightly attached to macromolecules when T_{21} was 0.01–10 ms, immobilized water entrapped within the extra-myofibrillar lattice when T_{22} was 10–100 ms, and free water when T_{23} was 100–10,000 ms, respectively. Noteworthily, the levels of bound water, immobilized water, and free water quickly decreased, as denoted by the shift of the main peaks and signal amplitudes to the left direction with increased drying time. These results indicate that the remaining water molecules within the shrimp samples form strong adsorption connections with the dry matter. The strongest T_2 signal amplitudes were observed in fresh and boiled shrimp, mainly due to their free and immobilized water, whereas the signals of free water gradually disappeared and those of bound and immobilized water decreased as drying proceeded (Figure 3a). Moreover, the relaxation times of T_{21} and T_{22} decreased from 3.05 to 1.52 and 28.48 to 14.17, respectively, and T_{23} became 0 ms after 9 h of drying (Figure 3b), which indicates that the free water is the main moisture lost during drying. Therefore, the LF-NMR results revealed that the mobility of the bound, immobilized, and free water molecules is reduced due to shrimp muscle contraction and the marked evaporation of free water during the drying process.

3.3. MRI Analysis

The hydrogen proton MRI has been used as a noninvasive method to evaluate the distribution of moisture content within food products [43]. The T_2 weighted images taken at the transverse geometric center of each sample during the drying process revealed the distribution of the water within high-mobility protons (Figure 4). With increasing drying time, a continuous decrease is observed in the size of the brighter regions, suggesting the loss of a longer relaxation signal of water during drying. In addition, a decrease in the signal intensity from the external surface to the inner region is evident. Similar phenomenon was also observed by Ling et al. who found that the red region gradually changed to blue, and the color and size of the blue region remarkably decreased from the exterior to the interior part with increasing drying time of shrimp [39]. These results confirmed that the water relaxation signal gradually weakens and the water content continuously decreases during the drying process, in agreement with the above-described changes observed in moisture content in shrimp during drying.

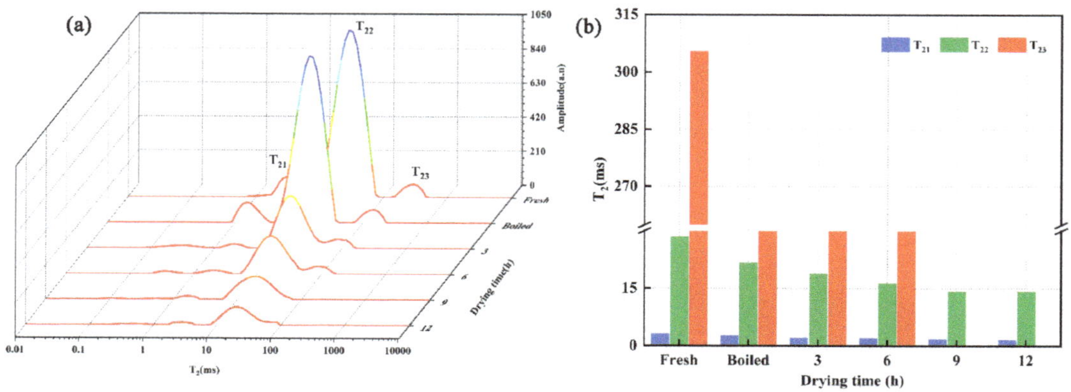

Figure 3. (**a**) Distribution of T_2 relaxation spectra and (**b**) change of T_2 relaxation times obtained by multi-exponential fitting of the continuously distributed Carr–Purcell-Meiboom-Gill relaxation curve of different shrimp samples during drying.

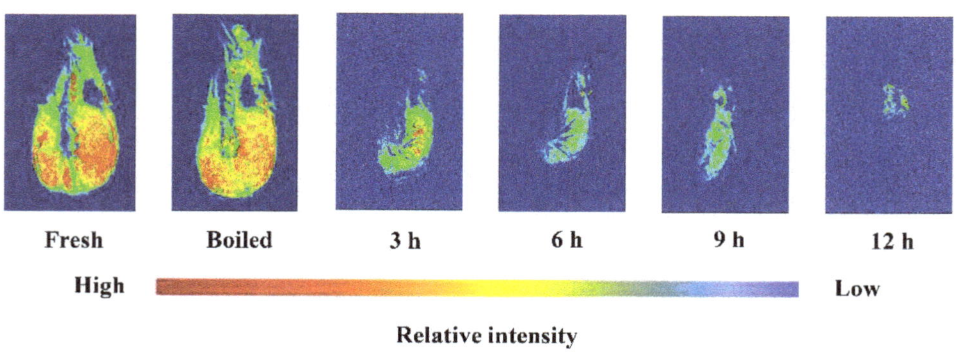

Figure 4. T_2 weighted MRI images of shrimp dried by hot air drying at different levels.

3.4. Correlation between LF-NMR and Physicochemical Properties

As a rapid, noninvasive method, LF-NMR relaxation is often applied to investigate water mobility in materials and foods [44]. However, the correlation between water distribution state and the physicochemical parameters of shrimp during the drying process needs deeper exploration. Therefore, the relationship between T_2 relaxation times (T_{21}, T_{22}, and T_{23}) and shrimp physicochemical properties was determined by Pearson correlation analysis (Figure 5). The results indicated good correlations between LF-NMR data and shrimp moisture content, hardness, adhesiveness, stickiness, and chewiness. Specifically, moisture content was significantly positively correlated with T_{21} ($R = 0.943$), T_{22} ($R = 0.914$), and T_{23} ($R = 0.903$), which may be explained by the substantial effect of moisture on the proteins and myofibril in shrimps. This was similar to the findings of Cheng et al., who also reported a positive correlation between the decrease in moisture content and the change in relaxation times [42]. Regarding shrimp texture, its hardness, stickiness, and chewiness were negatively correlated with T_{21} ($R = -0.877$, $R = -0.889$, $R = -0.852$) and T_{23} ($R = -0.846$, $R = -0.875$, $R = -0.844$), whereas adhesiveness was positively correlated with T_{21} ($R = 0.832$) and T_{23} ($R = 0.872$). These results agree with those of Wang et al., who reported that T_{22} was highly correlated ($p < 0.01$) with hardness, elasticity, and chewiness, thereby consequently leading to moisture changes that will affect muscle fiber contraction and alter the texture of shrimp meat [44]. However, the color indicators exhibited a weaker correlation with the LF-NMR, which may be due to the fact that the color change is mainly caused by fat and pigmentation, and is weakly related with water splitting. In summary, the strong correlations between LF-NMR data and shrimp moisture content and texture properties indicate the potential of LF-NMR as a fast and nondestructive alternative method of detecting quality changes during shrimp drying.

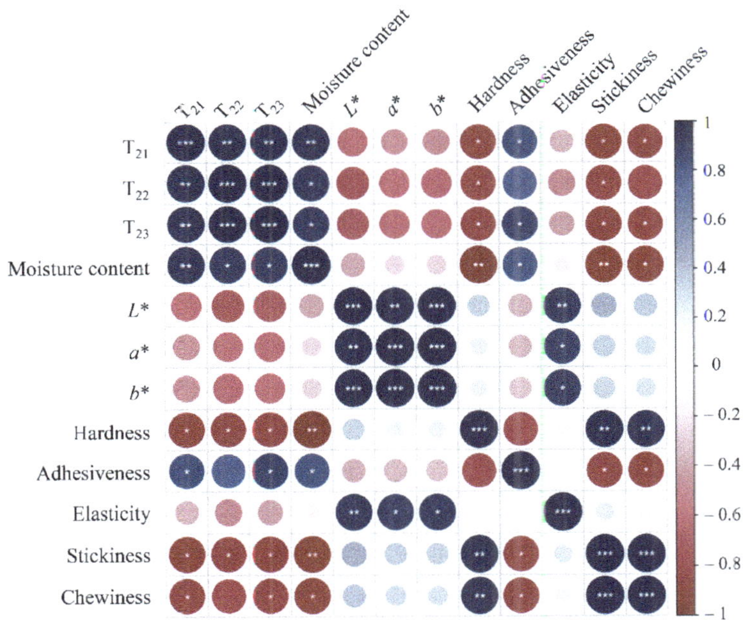

Figure 5. Correlation analysis between LF-NMR relaxation parameters and moisture content, color, texture of shrimp during drying.

3.5. Analysis of Modeling Results Based on Spectral Information

3.5.1. Spectral Characteristics of Drying Processes

Figure 6 shows the average spectra of the ROIs in the shrimp samples. The spectral reflectance curves of the shrimp samples with different drying levels are smooth and exhibit the same trends across the entire wavelength region. As shown in Figure 6a, a prominent absorption peak is centered at approximately 480 nm, which is probably due to the presence of astaxanthin in the shrimp [45]. Astaxanthins present in the dermis of the carapace are bound to proteins, and when shrimp are heated at high temperatures, astaxanthin detaches from the proteins, causing red astaxanthin to become present. Another intense absorption peak occurred at approximately 960 nm, which was attributed to water absorption corresponding to the second overtone of O–H stretching [24]. Because water is the main component of shrimp, it absorbs the radiation of light waves and dominates the spectral characteristics between 950 and 1000 nm. Figure 6b shows the representative reflectance spectra of boiled and processed shrimp at different drying times (3, 6, 9, and 12 h). Over the wavelength region of 400–920 nm, the reflectance of boiled shrimp was greater than that in the dried samples, and the reflectance of dried shrimp decreased as the drying time increased. This phenomenon is related to moisture changes during shrimp-drying, especially to the mechanism of vapor diffusion [46]. Changes in muscle tissue and pigmentation during drying also contribute to this phenomenon.

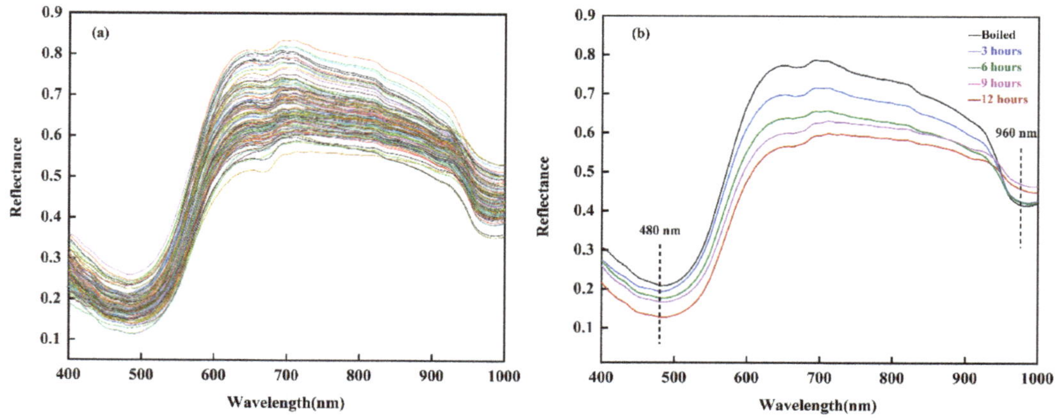

Figure 6. (**a**) Mean reflectance spectra of the ROIs in shrimp samples with different drying levels and (**b**) reflectance spectra at different drying times (boiled, 3, 6, 9, and 12 h).

3.5.2. Prediction Models Using Whole Spectra

After spectral pretreatment, PLSR and LSSVM calibration models were established using the mean spectra from 400–1000 nm (224 bands) to predict quality changes in shrimp during drying. The main statistical parameters used to evaluate model performance are shown in Table 3. The two models exhibited reasonable and similar performance. For shrimp moisture content, both PLSR and LSSVM models yielded satisfactory results with $R_p > 0.92$ and RPD > 2.5. Both models performed well for the prediction set, with RPD values of 2.623 and 2.814, respectively, indicating that the LSSVM model is superior. For shrimp color (L^*, a^*, and b^*), the R_p values of L^*, a^*, and b^* obtained with the LSSVM model were 0.898, 0.919, and 0.906, respectively, showing excellent accuracy. Compared with the LSSVM results, the R_p values of L^*, a^*, and b^* obtained with the PLSR model were 0.853, 0.887, and 0.891, indicating a decrease of 0.045, 0.032, and 0.015, respectively. The performances of the PLSR and LSSVM models were much better than those obtained in a previous study by Wu et al. in which low R_p values of 0.864, 0.736, and 0.798 were achieved respectively for L^*, a^*, and b^* prediction in salmon [47]. Significant correlations between the color parameters (L^*, a^*, and b^*) and reflectance spectra could imply that the color changes indicate the shrimp chemical composition that indirectly influences the reflectance spectra. Compared to the PLSR model, the RMSEP for hardness and elasticity decreased from 32.663 N to 20.486 N and from 0.181 mm to 0.151 mm, respectively, in the LSSVM model, whereas RPD increased from 2.162 to 2.226 and from 2.118 to 2.208, respectively. These findings prove that the LSSVM model is more effective in terms of hardness and elasticity prediction, and demonstrate the potential of using HSI to estimate shrimp quality during the drying process.

3.5.3. Prediction Models Using Characteristic Wavelengths

As multivariable (high-dimensional) data are extracted from hyperspectral images; they contain many inter-band correlations, resulting in long data processing times and low accuracy and robustness of the models [48,49]. After the SNV spectral pretreatment, the CARS algorithm was employed to identify the optimal wavelengths that carry the most information, which is useful for determining the moisture content, L^*, a^*, b^*, hardness, and elasticity The number of Monte Carlo sampling runs was set to 1000, and the number of selected wavelengths was determined by 10-fold cross-validation. As a result, 42, 25, 39, 20, 29, and 18 optimal wavelengths were selected from the 400–1000 nm range, which occupied <19% of the entire wavelength range (224).

Table 3. Prediction models for moisture content, L^*, a^*, b^*, hardness, and elasticity values using 224 wavelengths.

Parameters	Pre-Processing	Model	Calibration Set		Prediction Set		RPD
			R_c	RMSEC	R_p	RMSEP	
Moisture content	SNV	PLSR	0.929	4.369	0.925	4.512	2.623
	SNV	LSSVM	0.959	3.378	0.938	4.312	2.814
L^*	SNV	PLSR	0.891	0.975	0.853	1.257	1.944
	SNV	LSSVM	0.906	1.002	0.898	1.031	1.958
a^*	SNV	PLSR	0.905	1.010	0.887	1.249	2.016
	SNV	LSSVM	0.937	0.998	0.919	1.181	2.246
b^*	SNV	PLSR	0.937	1.045	0.891	1.325	1.894
	SNV	LSSVM	0.940	0.875	0.906	0.945	2.065
Hardness	SNV	PLSR	0.957	16.545	0.941	32.663	2.162
	SNV	LSSVM	0.968	12.758	0.915	20.486	2.226
Elasticity	SNV	PLSR	0.937	0.116	0.928	0.181	2.118
	SNV	LSSVM	0.958	0.073	0.910	0.151	2.208

Based on the identified optimal wavelengths, simplified PLSR (CARS-PLSR) and LSSVM (CARS-LSSVM) models were established for the prediction of quality parameters of shrimp during the drying processes, and the results are presented in Figure 7. Compared with the PLSR and LSSVM models based on full spectra, the CARS-PLSR and CARS-LSSVM models achieved a better prediction result for all quality indicators (L^*, a^*, b^*, hardness, and elasticity) except moisture content, which could be attributed to the selection of effective wavebands during optimal wavelength selection in the CARS method. For shrimp moisture content, the RPD based on the characteristic wavelengths model was slightly lower than that determined using the full spectra because the process of filtering the characteristic wavelengths misses some important information. For shrimp color and texture, the prediction results of the characteristic wavelengths models were significantly improved, and the LSSVM models results were better than the PLSR model results. The RPD of the LSSVM model reached 2.541, 2.550, and 2.795 for L^*, hardness, and elasticity, respectively. Overall, it is reasonable to select the optimal wavelengths by employing the CARS method, which removed approximately 80% of the wavebands, significantly decreasing the data processing time and increasing the working efficiency. The newly developed model based on optimal wavelengths exhibits a powerful ability to predict the quality parameters of shrimp during drying.

3.6. Analysis of Modeling Results Based on Image Information

3.6.1. Color Feature Information Extraction

The hyperspectral images at 647 nm, 550 nm, and 460 nm were used to synthesize *RGB* images as the target images for color feature extraction. The first- and second-order moment statistics for the *R*, *G*, and *B* components were calculated and listed in Table 4. Owing to the large amount of data, the color moment information of the eight samples was averaged. The first-order moment represents the average strength of the color component, whereas the second-order moment represents the color variance (i.e., non-uniformity) [29]. As shown in Table 4, the first-order moments show an overall increasing trend, and the second-order moments exhibit a decreasing trend; it indicates that the average intensity of the image color increases, and the color distribution becomes more uniform. These characteristics may be due to the oxidation of astaxanthin in shrimp with increasing drying time, resulting in a darker color. Because the *RGB* color space does not match human color perception, this space was converted into a visual-perception-oriented HSV space to calculate the histogram and quantify information. The mean grayscale values of the H, S, and V components are listed in Table 4. As the drying time increases, the overall S and V values increase, whereas the difference in H is small, indicating that shrimp images with different degrees of drying show less variation in hue.

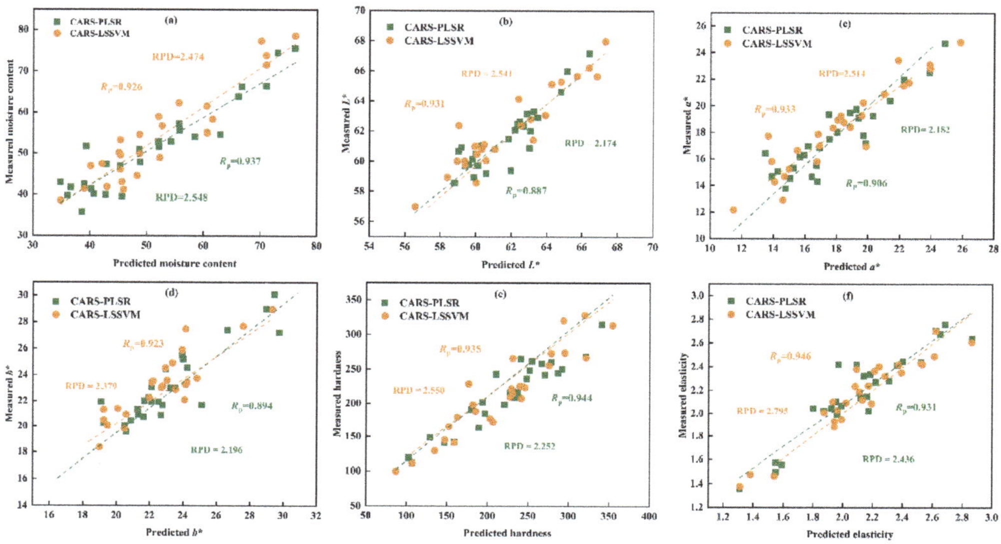

Figure 7. Comparison of CARS-PLSR and CARS-LSSVM in terms of (**a**) moisture content, (**b**) L^*, (**c**) a^*, (**d**) b^*, (**e**) hardness, and (**f**) elasticity based on quantitative analysis models in shrimp during drying.

Table 4. Extracted image feature information of color.

Drying Times (h)	First Order Moments			Second Order Moments			*H*	*S*	*V*
	R	*G*	*B*	*R*	*G*	*B*			
Boiled	62.284	56.703	62.217	38.837	35.045	27.693	0.487	0.595	0.307
1	61.977	55.456	60.337	37.209	34.035	27.374	0.482	0.605	0.301
2	60.761	57.174	64.857	36.829	33.947	26.746	0.496	0.605	0.311
3	61.490	55.945	62.807	38.330	34.523	27.186	0.495	0.615	0.309
4	60.892	55.974	62.138	37.353	33.333	25.926	0.488	0.615	0.308
5	61.110	56.265	62.507	35.692	32.388	25.120	0.490	0.607	0.309
6	63.739	58.015	62.952	34.702	31.067	23.947	0.479	0.613	0.318
7	65.538	59.785	62.133	35.201	30.905	24.029	0.459	0.609	0.322
8	60.655	55.422	60.712	36.566	31.897	24.553	0.481	0.623	0.308
9	64.497	57.990	61.884	36.441	31.830	24.614	0.479	0.600	0.317
10	63.540	58.484	63.141	38.164	32.466	24.667	0.476	0.622	0.322
11	63.438	57.540	62.576	36.181	31.338	23.793	0.483	0.621	0.321
12	68.387	61.024	64.176	36.032	31.173	23.693	0.471	0.609	0.334

3.6.2. Texture Feature Information Extraction

As important as visual characteristics, texture features can also reflect differences in the chemical composition and structure of foods [50]. In this study, PCA was conducted for each individual image to evaluate the spatial variability of the samples; the top three principal component images (PC1, PC2, and PC3) with a cumulative contribution of 99.58% were selected for GLCM to obtain the contrast, correlation, energy, and homogeneity. The PCA process and average trends of the four textural features of the eight samples with different drying times are shown in Figure 8. It was clear that the contrast of the samples differed with increasing drying time, as denoted by the large differences in the gray value of the images, firstly exhibiting an increasing trend followed by a decrease in contrast (Figure 8a), which may be related to changes in the muscle texture during the shrimp-drying process. The correlation varies less (Figure 8b), fluctuating from 0.7 to 0.9, indicating that the texture uniformity of shrimp images with different drying levels is similar. As the drying time increases, the energy firstly decreases and then increases (Figure 8c). Homogeneity shows an opposite trend, reaching a minimum value at the seventh hour of drying (Figure 8d).

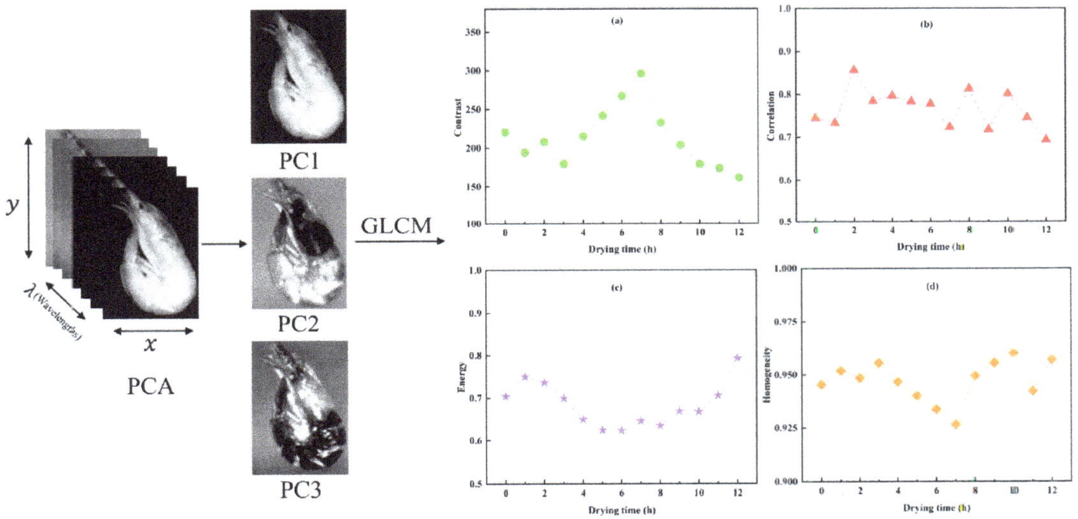

Figure 8. PCA process and texture features of shrimp samples, (**a**–**d**) stand for the change of contrast, correlation, energy, and homogeneity, respectively.

3.6.3. Image Information Modeling Results

To verify whether the color and texture features of the hyperspectral images can be used to predict the quality indicators of shrimp during drying, nine color variables and four texture variables were selected and used to construct PLSR and LSSVM prediction models. The color variables were used to predict L^*, a^*, and b^*, the texture variables were used to predict hardness and elasticity, and 13 integration variables were used to predict moisture content. The PLSR and LSSVM model results based on image information are presented in Table 5. The LSSVM model yielded better predictions than the PLSR model. Specifically, the LSSVM model results for color were good, with RPD values of 1.642, 1.510, and 1.544 for L^*, a^*, and b^*, indicating that images can be used to predict shrimp color. However, the hardness and elasticity predictions were relatively poor, which may be because the amount of extracted textural information was not sufficient to accurately reflect shrimp hardness and elasticity. Overall, the models based on hyperspectral image information were inferior to those based on spectral data, which highlights the inadequacy of using only external image features to predict the quality indicators of shrimp during drying.

3.7. Analysis of Modeling Results Based on Fusion Mapping Feature Information

To further verify whether the integration of the image and spectral data from shrimp samples could optimize the prediction model and improve the accuracy for moisture content, color (L^*, a^* and b^*), and texture (hardness and elasticity), the variables from the optimal spectra and HSI color and texture information were integrated by feature-level fusion using the normalization technique. Thus, fusion data comprising the optimal wavelength of each indicator and 13 color and texture features were used to establish new PLSR and LSSVM models. The prediction results of full bands, characteristic bands, and fusion information are given and compared in Figure 9. Regarding shrimp moisture content (Figure 9a), the fusion models achieved limited improvement. The LSSVM model using full-band spectral information exhibited the best performance for dried shrimp (R_c = 0.959; R_p = 0.938; RPD = 2.814). For L^*, a^*, and b^* (Figure 9b–d), the fusion-based PLSR and LSSVM models exhibited substantial improvement. The LSSVM model was superior to the PLSR model, with RPD values for L^*, a^*, and b^* of 3.292, 2.753, and 3.211, indicating an increase in the prediction performance of 0.866, 0.172, and 0.859 than the PLSR model, respectively. For hardness and elasticity (Figure 9e,f), the fusion-based LSSVM model also showed excellent results compared to the fusion-based PLSR model, with the RPD values increasing from 2.612 to 2.807 and from 2.717 to 2.842, respectively. Thus, combining the internal components and external attributes of shrimp can more fully explain the color and texture changes of shrimp during drying, leading to better prediction results.

Table 5. Results of PLSR and LSSVM models based on image information.

Parameters	PLSR			LSSVM		
	R_p	RMSEP	RPD	R_p	RMSEP	RPD
Moisture content	0.695	9.569	1.065	0.730	8.564	1.197
L^*	0.690	2.323	1.243	0.798	1.845	1.642
a^*	0.701	2.570	1.317	0.762	1.901	1.510
b^*	0.655	2.609	1.287	0.794	2.010	1.544
Hardness	0.591	46.198	1.088	0.685	40.103	1.395
Elasticity	0.581	0.332	1.192	0.698	0.207	1.404

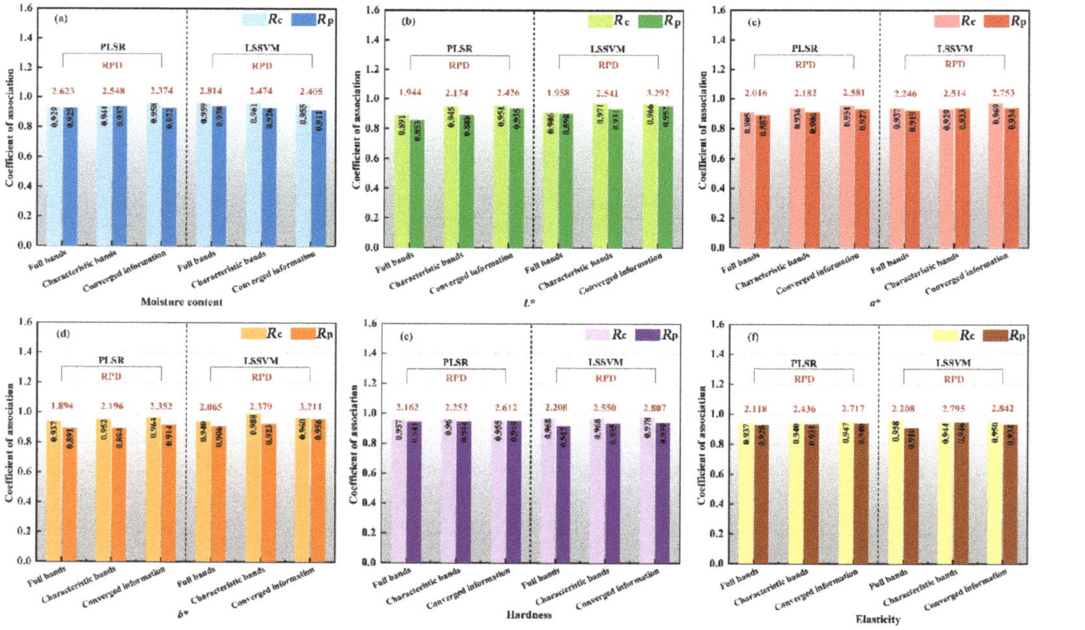

Figure 9. Comparison results of (**a**) moisture content, (**b**) L^*, (**c**) a^*, (**d**) b^*, (**e**) hardness, and (**f**) elasticity models.

3.8. Visualization of Quality Indicators

A unique advantage of HSI technology compared with traditional spectroscopy or computer imaging technology is visualization of the prediction index of tested samples. Figure 10 visualizes the moisture content, L^*, a^*, b^*, hardness, and elasticity of shrimp generated by the optimal model selected from the modeling results. In the maps, the distribution of shrimp moisture content is expressed by a linear color bar ranging from blue (low value) to red (high value). The boiled shrimp have a high moisture content of 73.02%. The moisture content of the samples then gradually decreases with drying time to a final value of 35.02%. As for shrimp color, L^*, a^*, and b^* values tend to decrease during the drying process. Although this difference cannot be observed by visual inspection, the spatial distribution of color features within the shrimp was detected in the final distribution map generated by analyzing the hyperspectral image of the sample. Furthermore, the visualization maps show a clear increase in the hardness of shrimp, whereas the distribution of elasticity is more complex. Thus, the distribution maps of shrimp moisture content, color, and texture provide an intuitive analysis of changes in the quality reference values for dried shrimp, which are unlikely to be observed by the naked eye or an RGB image.

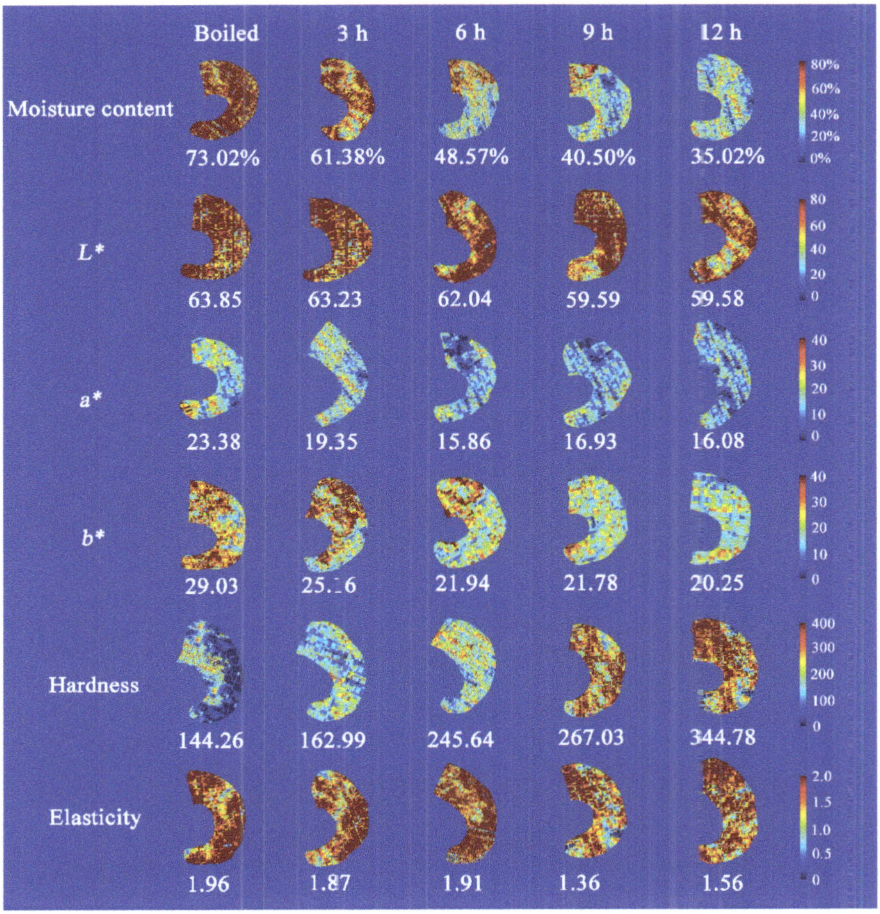

Figure 10. Moisture content, color (L^*, a^*, and b^*) and texture (hardness, and elasticity) visualization map of shrimp at different drying times (boiled, 3, 6, 9 and 12 h).

In more detail, the moisture content distribution is non-uniform and asymmetric. This may be attributed to complex changes in protein decomposition, lipid oxidation, etc. Furthermore, drying temperature and time may accelerate the degradation of ruptured tissue and cells in meat, leading to further uneven water loss [51]. Following shrimp drying, the L^* value decreases with the oxidization of myoglobin and hemoglobin into metmyoglobin and methemoglobin [52]. In addition, the color of shrimp becomes orange, with yellow or orange-red colors resulting from the oxidization reaction and the presence of astaxanthin. The hardness and elasticity of shrimp exhibits a non-uniform distribution that is related to the distribution of fat, pigments, and collagen [53]. In summary, HSI combined with data fusion can achieve the nondestructive detection and visualization of shrimp color and texture during drying. Specifically, the distribution maps of quality indicators generated using HSI clarify the location and movement of water, color, and textures through the shrimp samples during the hot air drying process. Such maps help consumers intuitively understand the dynamic changes in shrimp quality and the shelf life of dried shrimp production. Thus, we present a valid alternative to traditional methods of monitoring shrimp drying that has substantial potential for further development and can be applied to detect freshness or other indexes during aquatic production.

4. Conclusions

In this study, we described the changes in shrimp quality, evaluated the correlation between shrimp water distribution state and other quality indices, and combined spectral and image information of the hypercube to monitor shrimp quality changes during the drying process. Throughout the process, the moisture content showed a downward trend, the hardness and elasticity reached 344.78 N and 1.56 mm, respectively, and the color turned bright yellow at the end of drying. Significant correlations between the moisture content, TPA parameters (hardness, adhesiveness, elasticity, stickiness, and chewiness), and LF-NMR parameters (T_{21}, T_{22}, and T_{23}) were observed. The HSI system in the spectral range of 400–1000 nm was used to monitor the quality changes (moisture content, L^*, a^*, b^*, hardness, and elasticity) of shrimps during drying. The results demonstrate the following: first, the ability of the HSI method to evaluate the quality changes of shrimps during drying; second, the optional wavelengths selected by the CARS method carried the most effective information, which reduced the spectral dimension and accelerated the calibration process; and finally, the spectral information model predicts better than the image color and texture information model, and the LSSVM built by combining image information with spectral information in characteristic bands has powerful and accurate prediction capabilities. Thus, HSI can be utilized to visualize the quality changes in shrimps in a pixel-wise manner both quantitatively and automatically, reducing the overall production cost, saving time, and avoiding subjectivity and discrepancies.

Author Contributions: Conceptualization, W.X. and W.W.; writing—original draft preparation, W.X.; methodology, F.Z.; software, W.X.; validation, J.W. (Jiarong Wang) and Y.T.; formal analysis, J.W. (Jiarong Wang); investigation, Q.M.; resources, W.W.; writing—review and editing, J.S.; visualization, J.W. (Jie Wang); funding acquisition, W.W. All authors have read and agreed to the published version of the manuscript.

Funding: This research was supported by the Natural Science Foundation, grant number 32001732, and the National Key R&D Program of China, grant number 2018YFD0901004.

Data Availability Statement: The data presented in this study are available on request from the corresponding author.

Acknowledgments: We appreciate the research was funded by the Natural Science Foundation, grant number 32001732, and the National Key R&D Program of China, grant number 2018YFD0901004.

Conflicts of Interest: The authors declare no conflict of interest.

References

1. Cen, S.J.; Fang, Q.; Tong, L.; Yang, W.G.; Zhang, J.J.; Lou, Q.M.; Huang, T. Effects of chitosan-sodium alginate-nisin preservatives on the quality and spoilage microbiota of *Penaeus vannamei* shrimp during cold storage. *Int. J. Food Microbiol.* **2021**, *349*, 109227. [CrossRef]
2. Ambasankar, K.; Dayal, J.S.; Vasagam, K.P.K.; Sivaramakrishnan, T.; Sandeep, K.P.; Panigrahi, A.; Raja, R.A.; Burri, L.; Vijayan, K.K. Growth, fatty acid composition, immune-related gene expression, histology and haematology indices of *Penaeus vannamei* fed graded levels of Antarctic krill meal at two different fishmeal concentrations. *Aquaculture* **2022**, *553*, 738069. [CrossRef]
3. Liang, R.; Lin, S.Y.; Chen, D.; Sun, N. Differentiation of *Penaeus vannamei* from different thermal processing methods in physico-chemical, flavor and sensory characteristics. *Food Chem.* **2022**, *378*, 132092. [CrossRef]
4. Xu, W.Y.; Ma, Q.Y.; Sun, J.F.; Li, Y.L.; Wang, J.; Tang, Y.W.; Liu, Y.Q.; Mu, J.L.; Wang, W.X. Changes in quality characteristics of shrimp (*Penaeus chinensis*) during refrigerated storage and their correlation with protein degradation. *J. Food Compos. Anal.* **2022**, *114*, 104773. [CrossRef]
5. Ge, Y.J.; Li, Y.; Wu, T.T.; Bai, Y.; Yuan, C.H.; Chen, S.G.; Gakushi, I.; Hu, Y.Q. The preservation effect of CGA-Gel combined with partial freezing on sword prawn (*Parapenaeopsis hardwickii*). *Food Chem.* **2020**, *313*, 126078. [CrossRef] [PubMed]
6. Li, D.Y.; Xie, H.K.; Liu, Z.Y.; Li, A.; Li, J.X.; Liu, B.; Liu, X.Y.; Zhou, D.Y. Shelf life prediction and changes in lipid profiles of dried shrimp (*Penaeus vannamei*) during accelerated storage. *Food Chem.* **2019**, *297*, 124951. [CrossRef]
7. Zhu, Y.; Chen, X.T.; Pan, N.; Liu, S.J.; Su, Y.C.; Xiao, M.T.; Shi, W.Z.; Liu, Z.Y. The effects of five different drying methods on the quality of semi-dried *Takifugu obscurus* fillets. *LWT-Food Sci. Technol.* **2022**, *161*, 113340. [CrossRef]
8. Wu, T.; Mao, L.C. Influences of hot air drying and microwave drying on nutritional and odorous properties of grass carp (*Ctenopharyngodon idellus*) fillets. *Food Chem.* **2008**, *110*, 647–653. [CrossRef]
9. Li, D.Y.; Yuan, Z.; Liu, Z.Q.; Yu, M.M.; Guo, Y.; Liu, X.Y.; Zhang, M.; Liu, H.L.; Zhou, D.Y. Effect of oxidation and maillard reaction on color deterioration of ready-to-eat shrimps during storage. *LWT-Food Sci. Technol.* **2020**, *131*, 109696. [CrossRef]
10. Hu, M.Y.; Wang, S.Y.; Liu, Q.; Cao, R.; Xue, Y. Flavor profile of dried shrimp at different processing stages. *LWT-Food Sci. Technol.* **2021**, *146*, 111403. [CrossRef]
11. Zhang, D.; Ji, H.W.; Liu, S.C.; Gao, J. Similarity of aroma attributes in hot-air-dried shrimp (*Penaeus vannamei*) and its different parts using sensory analysis and GC–MS. *Food Res. Int.* **2020**, *137*, 109517. [CrossRef]
12. Li, X.J.; Luan, A.N.; Li, X.H.; Wang, F.X.; Huang, Y.Q.; Li, A.P.; Liu, Y.L. Protein degradation and aggregation in silver carp (*Hypophthalmichthys molitrix*) muscle during hot air drying. *LWT-Food Sci. Technol.* **2022**, *163*, 113540. [CrossRef]
13. Li, D.Y.; Li, N.; Dong, X.H.; Tan, Z.F.; Na, X.K.; Liu, X.Y.; Zhou, D.Y. Effect of phytic acid combined with lactic acid on color and texture deterioration of ready-to-eat shrimps during storage. *Food Chem.* **2022**, *396*, 133702. [CrossRef] [PubMed]
14. Chen, X.X.; Jiao, Y.L.; Liu, B.; Chao, W.H.; Duan, X.C.; Yue, T.L. Using hyperspectral imaging technology for assessing internal quality parameters of persimmon fruits during the drying process. *Food Chem.* **2022**, *386*, 132774. [CrossRef]
15. Yu, D.W.; Feng, T.Y.; Jiang, Q.X.; Yang, F.; Gao, P.; Xu, Y.S.; Xia, W.S. The change characteristics in moisture distribution, physical properties and protein denaturation of slightly salted silver carp (*Hypophthalmichthys molitrix*) fillets during cold/hot air drying processing. *LWT-Food Sci. Technol.* **2021**, *137*, 110466. [CrossRef]
16. Elmas, F.; Bodruk, A.; Koprüalan, O.; Arıkaya, S.; Koca, N.; Serdaroglu, F.M.; Kaymak-Ertekin, F.; Koç, M. The effect of pre-drying methods on physicochemical, textural and sensory characteristics on puff dried Turkey breast meat. *LWT-Food Sci. Technol.* **2021**, *145*, 111350. [CrossRef]
17. Aviara, N.A.; Liberty, J.T.; Olatunbosun, O.S.; Shoyombo, H.A.; Oyeniyi, S.K. Potential application of hyperspectral imaging in food grain quality inspection, evaluation and control during bulk storage. *J. Agric. Food Res.* **2022**, *8*, 100288. [CrossRef]
18. Xu, J.L.; Riccioli, C.; Sun, D.W. Development of an alternative technique for rapid and accurate determination of fish caloric density based on hyperspectral imaging. *J. Food Eng.* **2016**, *190*, 185–194. [CrossRef]
19. Yang, D.; He, D.D.; Lu, A.X.; Ren, D.; Wang, J.H. Combination of spectral and textural information of hyperspectral imaging for the prediction of the moisture content and storage time of cooked beef. *Infrared Phys. Technol.* **2017**, *83*, 206–216. [CrossRef]
20. Yu, H.D.; Qing, L.W.; Yan, D.T.; Xia, G.H.; Zhang, C.H.; Yun, Y.H.; Zhang, W.M. Hyperspectral imaging in combination with data fusion for rapid evaluation of tilapia fillet freshness. *Food Chem.* **2021**, *348*, 129129. [CrossRef] [PubMed]
21. Kamruzzaman, M.; Makino, Y.; Oshita, S. Online monitoring of red meat color using hyperspectral imaging. *Meat Sci.* **2016**, *116*, 110–117. [CrossRef]
22. Lohumi, S.; Lee, S.; Lee, H.; Kim, M.S.; Lee, W.H.; Cho, B.K. Application of hyperspectral imaging for characterization of intramuscular fat distribution in beef. *Infrared Phys. Technol.* **2016**, *74*, 1–10. [CrossRef]
23. Chen, Q.S.; Zhang, Y.H.; Zhao, J.W.; Hui, Z. Nondestructive measurement of total volatile basic nitrogen (TVB-N) content in salted pork in jelly using a hyperspectral imaging technique combined with efficient hypercube processing algorithms. *Anal. Methods* **2013**, *5*, 6382. [CrossRef]
24. Sun, J.L.; Zhang, X.Y.; Qiu, Z.J.; Zhu, X.Y.; Zhang, T.; Yang, J.X.; Zhang, X.; Lv, Y.; Wang, H.H. Hyperspectral data for predicting moisture content and distribution in scallops during continuous and intermittent drying. *Dry Technol.* **2020**, *40*, 924–937 [CrossRef]
25. Netto, J.M.S.; Honorato, F.A.; Azoubel, P.M.; Kurozawa, L.E.; Barbin, D.F. Evaluation of melon drying using hyperspectral imaging technique in the near infrared region. *LWT-Food Sci. Technol.* **2021**, *143*, 111092. [CrossRef]

26. Xiong, Z.J.; Sun, D.W.; Pu, H.B.; Zhu, Z.W.; Luo, M. Combination of spectra and texture data of hyperspectral imaging for differentiating between free-range and broiler chicken meats. *LWT-Food Sci. Technol.* **2015**, *60*, 649–655. [CrossRef]
27. Zhang, H.L.; Zhang, S.A.; Chen, Y.; Luo, W.; Huang, Y.F.; Tao, D.; Zhana, B.S.; Liu, X.M. Non-destructive determination of fat and moisture contents in Salmon (*Salmo salar*) fillets using near-infrared hyperspectral imaging coupled with spectral and textural features. *J. Food Compos. Anal.* **2020**, *92*, 103567. [CrossRef]
28. Zhang, F.; Kang, T.H.; Sun, J.F.; Wang, J.; Zhao, W.; Gao, S.; Wang, W.X.; Ma, Q.Y. Improving TVB-N prediction in pork using portable spectroscopy with just-in-time learning model updating method. *Meat Sci.* **2022**, *188*, 108801. [CrossRef] [PubMed]
29. Ma, J.; Pu, H.B.; Sun, D.W.; Gao, W.H.; Qu, J.H.; Ma, K.Y. Application of Vis–NIR hyperspectral imaging in classification between fresh and frozen-thawed pork *Longissimus Dorsi* muscles. *Int. J. Refrig.* **2015**, *50*, 10–18. [CrossRef]
30. Antequera, T.; Caballero, D.; Grassi, S.; Uttaro, B.; Perez-Palacios, T. Evaluation of fresh meat quality by Hyperspectral Imaging (HSI), Nuclear Magnetic Resonance (NMR) and Magnetic Resonance Imaging (MRI): A review. *Meat Sci.* **2021**, *172*, 108340. [CrossRef]
31. Jiang, H.Z.; Yuan, W.D.; Ru, Y.; Qing Chen, Q.; Wang, J.P.; Zhou, H.P. Feasibility of identifying the authenticity of fresh and cooked mutton kebabs using visible and near-infrared hyperspectral imaging. *Spectrochim. Acta Part A Mol. Biomol. Spectrosc.* **2022**, *282*, 121689. [CrossRef] [PubMed]
32. He, H.J.; Wu, D.; Sun, D.W. Non-destructive and rapid analysis of moisture distribution in farmed Atlantic salmon (*Salmo salar*) fillets using visible and near-infrared hyperspectral imaging. *Innov. Food Sci. Emerg.* **2013**, *18*, 237–245. [CrossRef]
33. Yu, X.J.; Yu, X.; Wen, S.T.; Yang, J.Q.; Wang, J.P. Using deep learning and hyperspectral imaging to predict total viable count (TVC) in peeled Pacifc white shrimp. *J. Food Meas. Charact.* **2019**, *13*, 2082–2094. [CrossRef]
34. Sanchez-Torres, E.A.; Abril, B.; Benedito, J.; Bon, J.; Toldra, M.; Pares, D.; García-Perez, J.V. Airborne ultrasonic application on hot air-drying of pork liver. Intensification of moisture transport and impact on protein solubility. *Ultrason. Sonochem.* **2022**, *86*, 106011. [CrossRef]
35. Shi, S.; Feng, J.; An, G.; Kong, B.H.; Wang, H.; Pan, N.; Xia, X.F. Dynamics of heat transfer and moisture in beef jerky during hot air drying. *Meat Sci.* **2021**, *182*, 108638. [CrossRef]
36. Latorre, M.E.; Velazquez, D.E. Effects of thermal treatment on collagen present in bovine *M. Semitendinosus* intramuscular connective tissue. Analysis of the chemical, thermal and mechanical properties. *Food Struct.* **2021**, *27*, 100165. [CrossRef]
37. Fulladosa, E.; Guerrero, L.; Illana, A.; Olmos, A.; Coll-Brasas, E.; Gou, P.; Muñoz, I.; Arnau, J. Instrumental texture analysis on the surface of dry-cured ham to define the end of the process. *Meat Sci.* **2021**, *172*, 108334. [CrossRef]
38. Wang, S.N.; Das, A.K.; Pang, J.; Liang, P. Real-time monitoring the color changes of large yellow croaker (*Larimichthys crocea*) fillets based on hyperspectral imaging empowered with artificial intelligence. *Food Chem.* **2022**, *382*, 132343. [CrossRef]
39. Ling, J.G.; Xuan, X.T.; Yu, N.; Yan Cui, Y.; Shang, H.T.; Liao, X.J.; Lin, X.D.; Yu, J.F.; Liu, D.H. High pressure-assisted vacuum-freeze drying: A novel, efficient way to accelerate moisture migration in shrimp processing. *J. Food Sci.* **2020**, *85*, 1167–1176. [CrossRef]
40. Yu, Y.; Tang, M.; Dai, H.J.; Feng, X.; Ma, L.; Zhang, Y.H. Dominating roles of protein conformation and water migration in fish muscle quality: The effect of freshness and heating process. *Food Chem.* **2022**, *388*, 132881. [CrossRef]
41. Qiao, S.C.; Tian, Y.W.; Wang, Q.H.; Song, S.Y.; Song, P. Nondestructive detection of decayed blueberry based on information fusion of hyperspectral imaging (HSI) and low-Field nuclear magnetic resonance (LF-NMR). *Comput. Electron. Agric.* **2021**, *184*, 106100.
42. Cheng, S.S.; Tang, Y.Q.; Zhang, T.; Song, Y.K.; Wang, X.H.; Wang, H.H.; Wang, H.T.; Tan, M.Q. An approach for monitoring the dynamic states of water in shrimp during drying process with LF-NMR and MRI. *Dry Technol.* **2017**, *36*, 841–848. [CrossRef]
43. Wang, H.H.; Wang, R.Y.; Song, Y.K.; Kamal, T.; Lv, Y.; Zhu, B.W.; Tao, X.H.; Tian, M.Q. A fast and non-destructive LF-NMR and MRI method to discriminate adulterated shrimp. *J. Food Meas. Charact.* **2018**, *12*, 1340–1349. [CrossRef]
44. Wang, X.; Xie, X.R.; Zhang, T.; Zheng, Y.; Guo, Q.Y. Effect of edible coating on the whole large yellow croaker (*Pseudosciaena crocea*) after a 3-day storage at −18 °C: With emphasis on the correlation between water status and classical quality indices. *LWT-Food Sci. Technol.* **2022**, *163*, 113514. [CrossRef]
45. Sun, D.W.; Weng, H.Y.; He, X.T.; Li, L.; He, Y.; Cen, H.Y. Combining near-infrared hyperspectral imaging with elemental and isotopic analysis to discriminate farm-raised pacific white shrimp from high-salinity and low-salinity environments. *Food Chem.* **2019**, *299*, 125121. [CrossRef]
46. Hua, M.H.; Dong, Q.L.; Liu, B.L.; Opara, U.L. Prediction of mechanical properties of blueberry using hyperspectral interactance imaging. *Postharvest Biol. Technol.* **2016**, *115*, 122–131. [CrossRef]
47. Wu, D.; Sun, D.W.; He, Y. Application of long-wave near infrared hyperspectral imaging for measurement of color distribution in salmon fillet. *Innov. Food Sci. Emerg.* **2012**, *16*, 361–372. [CrossRef]
48. Yang, C.S.; Zhao, Y.; An, T.; Liu, Z.Y.; Jiang, Y.W.; Li, Y.Q.; Dong, C.W. Quantitative prediction and visualization of key physical and chemical components in black tea fermentation using hyperspectral imaging. *LWT-Food Sci. Technol.* **2021**, *141*, 110975. [CrossRef]
49. Jiang, X.N.; Tian, J.P.; Huang, H.P.; Hu, X.J.; Han, L.P.; Huang, D.; Luo, H.B. Nondestructive visualization and quantification of total acid and reducing sugar contents in fermented grains by combining spectral and color data through hyperspectral imaging. *Food Chem.* **2022**, *386*, 132779. [CrossRef]
50. Wang, C.X.; Wang, S.L.; He, X.G.; Wu, L.G.; Li, Y.L.; Guo, J.H. Combination of spectra and texture data of hyperspectral imaging for prediction and visualization of palmitic acid and oleic acid contents in lamb meat. *Meat Sci.* **2020**, *169*, 108194. [CrossRef]

51. Cheng, J.H.; Sun, D.W.; Han, Z.; Zeng, X.A. Texture and structure measurements and analyses for evaluation of fish and fillet freshness quality: A review. *Compr. Rev. Food Sci. Food Saf.* **2014**, *13*, 52–61. [CrossRef] [PubMed]
52. Wu, L.L.; Pu, H.B.; Sun, D.W. Novel techniques for evaluating freshness quality attributes of fish: A review of recent developments. *Trend Food Sci. Technol.* **2019**, *83*, 259–273. [CrossRef]
53. Wu, D.; Sun, D.W.; He, Y. Novel non-invasive distribution measurement of texture profile analysis (TPA) in salmon fillet by using visible and near infrared hyperspectral imaging. *Food Chem.* **2014**, *145*, 417–426. [CrossRef] [PubMed]

foods

MDPI

Article

Detection of Soluble Solids Content in Different Cultivated Fresh Jujubes Based on Variable Optimization and Model Update

Haixia Sun, Shujuan Zhang *, Rui Ren, Jianxin Xue and Huamin Zhao

College of Agricultural Engineering, Shanxi Agricultural University, Jinzhong 030801, China
* Correspondence: zsj2021@sxau.edu.cn; Tel.: +86-139-3549-1091

Abstract: To solve the failure problem of the visible/near infrared (VIS/NIR) spectroscopy model, soluble solids content (SSC) detection for fresh jujubes cultivated in different modes was carried out based on the method of variable optimization and model update. Iteratively retained informative variables (IRIV) and successive projections algorithm (SPA) algorithms were used to extract characteristic wavelengths, and least square support vector machine (LS-SVM) was used to establish detection models. Compared with IRIV, IRIV-SPA achieved better performance. Combined with the offset properties of the wavelength, repeated wavelengths were removed, and wavelength recombination was carried out to create a new combination of variables. Using these fused wavelengths, the model was recalibrated based on the Euclidean distance between samples. The LS-SVM detection model of SSC was established using the update method of wavelength fusion-Euclidean distance. Good prediction results were achieved using the proposed model. The determination coefficient (R^2), root mean square error (RMSE), and residual predictive deviation (RPD) of the test set on SSC of fresh jujubes cultivated in the open field were 0.82, 1.49%, and 2.18, respectively. The R^2, RMSE, and RPD of the test set on SSC of fresh jujubes cultivated in the rain shelter were 0.81, 1.44%, and 2.17, respectively. This study realized the SSC detection of fresh jujubes with different cultivation and provided a method for the establishment of a robust VIS/NIR detection model for fruit quality, effectively addressing the industry need for identifying jujubes grown in the open field.

Keywords: cultivation; visible/near infrared spectrum; fresh jujube; model update; variable fusion

Citation: Sun, H.; Zhang, S.; Ren, R.; Xue, J.; Zhao, H. Detection of Soluble Solids Content in Different Cultivated Fresh Jujubes Based on Variable Optimization and Model Update. *Foods* **2022**, *11*, 2522. https://doi.org/10.3390/foods11162522

Academic Editors: Zhiming Guo, Zhao Zhang and Dong Hu

Received: 6 July 2022
Accepted: 18 August 2022
Published: 20 August 2022

1. Introduction

Containing various types of ingredients (such as sugars, vitamin C, and minerals), "Huping" jujube has high nutritional and medicinal values. The content of soluble solids (SSC) is an important evaluation index for the internal quality of fruit and vegetables, which is closely related to improving the added value of products and meeting consumer needs [1,2]. In traditional detection of SSC, destructive or invasive methods (e.g., refractometers) were used, which damaged the integrity of the sample and were cumbersome, time-consuming, and laborious to perform. This destructive approach is unfavorable for large-scale collection, implementation assessment, and industrial applications. Therefore, it is important to achieve rapid, non-destructive detection of SSCs to support the quality assessment and grading of agricultural products.

Visible/near infrared spectroscopy (VIS/NIR) [3,4] uses absorption characteristics of the frequency doubling and combined frequency absorption of hydrogen-containing groups (such as C-H, N-H, and O-H) to obtain characteristic information of samples, which realizes the detection of key chemical components and physical properties. Compared with traditional detection methods, VIS/NIR technology requires little or no sample preparation and has the characteristics of rapidity, non-destructiveness, real-time application, and low cost. It has been widely applied in the quality detection of agricultural products, such as fruits [5], vegetables [6], cereals [7], and pulses [8]. VIS/NIR spectroscopy is multivariate and contains multiple overlapping peaks related to compounds such as water, sugars, and proteins. The prediction accuracy of the VIS/NIR model was affected by some conditions such as samples (for example, maturity, variety, season, year, and batch) [9–11], instruments [12,13], and environment (for example, temperature) [14,15]. Developed models based on VIS/NIR spectral data were generally applicable to the quality detection of samples in a single condition. There is some variability in measured values under the new conditions. Modeling based on the data from the first condition does not involve this variability, and these models are usually not robust for the actual variability. For samples of different conditions, models built with a single condition perform poorly, and the bias and error are generally high.

The damage [16], pest [17], crack [18], SSC [19], and hardness [20] have been carried out in the quality detection of fresh jujubes using VIS/NIR spectroscopy. Those quality detections were implemented in the open-field cultivation mode, and the predicted samples had similar characteristics to those modeling samples. In addition to the open-field cultivation mode, there is also a rain-shelter cultivation mode that adopts the method of building a rainproof shed in the actual "Huping" jujube cultivation [21]. The rain-shelter method can avoid direct contact between rainwater and jujube fruit; reduce the impact of cracking, diseases, and insect pests on jujubes; and have good ventilation performance. Due to the differences in temperature, humidity, and solar radiation between rain-shelter and open-field cultivation, various internal component contents of different cultivated fruit are different, such as pear [22,23], cherry [24], and grape [25]. Inside the samples, the chemical composition is related to its optical absorption properties, and the physical structure is related to its scattering properties. Changes in the texture and internal component contents lead to different optical responses, which would affect the performance of the model built in spectral detection [26,27]. In the above studies of quality detection, spectral detection models were mainly developed for samples cultivated in open fields. However, the analysis of this model prediction performance for fresh jujubes from different cultivation modes is rarely reported.

Several studies have been reported to address the poor performance of models constructed from a single condition. Mishra et al. [28] updated the NIR detection models of the moisture content and SSC for pears, which significantly improved the prediction results for samples of different batches. Sun et al. [29] pointed out that temperature had an influence on the spectral detection model of mango dry matter content and established a robust prediction model using the method of temperature correction. For mango dry matter content based on multi-season, multi-variety, and multi-growing regions, Anderson et al. [30] established a robust prediction model. In order to reduce the influence of instruments, seasons, and temperature changes on the fruit NIR detection model in the study of Mishra et al. [31], calibration models that preserved performance under new conditions were established. The above studies showed that reference measurements from new conditions were generally required in order to compensate for external influences. Model updating was an important method of resolving poor performance when the VIS/NIR model was applied to new conditions. New samples were required in the model update, and the method for determining the number of new samples needed to be investigated. In addition, a reasonable selection of variables could reduce the influence of interfering information and improve model performance due to the high dimensionality and overlapping peaks of VIS/NIR data. There are few studies incorporate wavelength offset properties into the selection of variables.

Therefore, the objective of this study was to develop a robust model for the SSC detection of fresh jujubes from different cultivation modes based on VIS/NIR spectroscopy. To achieve this aim in this study, the IRIV-SPA was used to select characteristic wavelengths, and a new combination of variables was established in combination with wavelength position offset properties. A model update using wavelength fusion-Euclidean distance was proposed to re-calibrate the SSC model. The proposed method achieves the SSC prediction of fresh jujubes from different cultivation modes and improves the generalizability and stability of the model.

2. Materials and Methods

2.1. Sample Collection

The two methods are applied to the planting of "Huping" jujube. The rain shelter was built in the immature period of jujube fruit. The top is covered with plastic film, and the surrounding is ventilated. The full-maturity "Huping" jujubes cultivated in the open field and the rain shelter were collected from an orchard in Taigu, China, respectively. On the same day of collection, samples were transported to the laboratory, cleaned, and placed for four hours to return to room temperature. In this study, a total of 300 intact samples (150 rain-shelter samples and 150 open-field samples) were selected. For each cultivation, the KS algorithm [32] was used to divide the data set into a calibration set (114 samples) and a prediction set (36 samples) with a ratio of 3:1.

2.2. Spectrum Acquisition and SSC Determination

Spectrum data were collected using an ASD Fieldspec3 spectrometer (analytical spectral device, Longmont, CO, USA) with spectral resolutions of 3 nm@350~1000 nm and 10 nm@1000~2500 nm and a spectral range of 350~2500 nm. The spectral curves of fresh jujubes cultivated in the open field and the rain shelter are shown in Figure 1. Although there were some differences in reflectance values of fresh jujubes between the two cultivation modes, the curve-change trends of the two cultivation methods were relatively similar. There were obvious absorption peaks related to the stretching vibration of O-H bond at the vicinity of 970 nm and 1400 nm. The signal-to-noise ratio was low, and the noise was loud at 350~450 nm and 2400~2500 nm. Therefore, the spectral information of 450~2400 nm was selected for subsequent analysis.

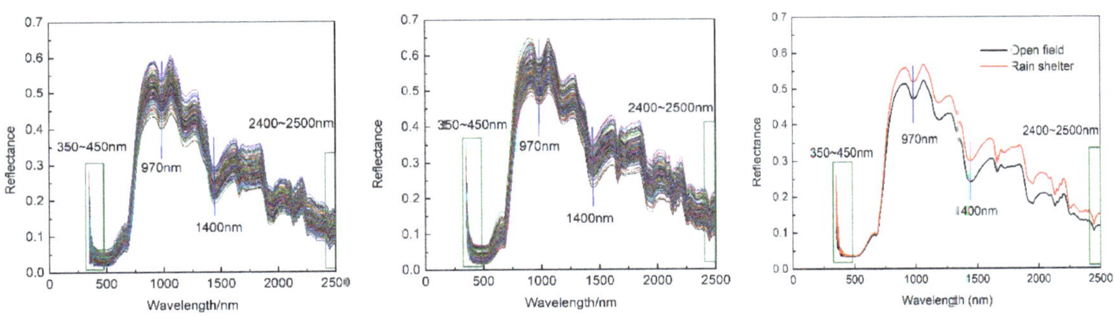

Figure 1. Mean spectrum of fresh jujubes. (**a**) Spectral curves of open-field cultivation, (**b**) spectral curves of rain-shelter cultivation, and (**c**) average spectrum.

The SSC of each sample was measured using a hand-held refractometer. The statistical values are shown in Table 1. The SSCs of fresh jujubes cultivated in the open field and rain shelter were 21.2–35.5% and 21.8–37.4%, respectively.

Table 1. Statistics of soluble solids content (%).

Cultivation	Data Set	Maximum	Minimum	Mean	Standard Deviations
Open field	Total samples	35.5	21.2	26.59	2.78
	Calibration set	35.5	21.2	26.55	2.86
	Prediction set	32.5	22	26.69	2.57
Rain shelter	Total samples	37.4	21.8	28.81	3.10
	Calibration set	37.4	21.8	28.75	3.18
	Prediction set	34.2	24.3	28.97	2.84

2.3. Data Analysis

In this study, the baseline and Savitzky Golay (SG) were used for the preprocessing, respectively. Spectral preprocessing was performed using Unscrambler X10.1 software. The IRIV and SPA were used to extract characteristic wavelengths, and the LS-SVM was adopted to build the model. LS-SVM [33,34] follows the principle of structural risk minimization and transforms the convex quadratic programming problem of traditional support vector machines into the problem of solving a system of linear equations, which reduces the computational complexity. Variable extraction of IRIV and SPA and modeling of LS-SVM were carried out using MATLAB R2020a. The determination coefficient (Rc^2) and the root mean square error (RMSEC) of the calibration set, the determination coefficient (Rp^2) and the root mean square error (RMSEP) of the prediction set, and the residual predictive deviation (RPD) were used to evaluate the model performance.

2.4. Data Processing Method

2.4.1. Wavelength Extraction Method

The SPA [35,36] eliminates the influence of collinearity among various variables, reduces the overlap of effective information, and accelerates the modeling speed.

The IRIV method [37] utilizes random combinations of variables and interactions between variables to select variables based on binary matrix rearrangement filters. Based on the model cluster analysis method, the difference of mean values (DMEAN) and the P value were calculated to determine the class of each variable. The classification rules of variables are shown in Table 2.

Table 2. Variable classification rules of iteratively retains informative variables (IRIV) [37].

Variable Class	Classification Rules
Interfering variable	$DMEANi > 0$, $Pi < 0.05$
Uninformative variable	$DMEANi > 0$, $Pi > 0.05$
Strongly informative variable	$DMEANi < 0$, $Pi < 0.05$
Weakly informative variable	$DMEANi < 0$, $Pi > 0.05$

2.4.2. Model Update Method

To improve the applicability of the model, a sample addition algorithm was proposed to update the model. Combined with wavelength position offset properties, wavelength fusion was performed. The KS algorithm calculates the Euclidean distance and adds the two samples with the largest distance to the calibration set. Then, the distance minimum (Dm) value between each remaining sample and the selected calibration set sample is calculated. The sample with the largest Dm is added to the calibration set until the calibration set reaches the specified number of samples. Therefore, samples with large spectral differences between different cultivation modes can be extracted for use as a calibration set using the KS algorithm. In this study, based on the Euclidean distance between samples, the KS algorithm was used to select new samples in turn, and the RMSECV of the established PLSR model was calculated.

3. Results and Discussion

3.1. Establishment of SSC Detection Model for Fresh Jujubes in Open-Field Cultivation

3.1.1. SSC Detection Models Using Full Wavelengths

Open-field cultivation is the main mode in the planting of jujube. The spectral information of the open-field cultivation samples was pre-processed using baseline and SG. Based on the spectrum with no-pretreatment and pretreatment, LS-SVM models of the SSC were established, and fresh jujubes from two cultivation modes were predicted. The prediction results are shown in Table 3.

Table 3. Prediction results of SSC using different pre-processing methods.

Pretreatment	Prediction Set	Rc^2	RMSEC (%)	Rp^2	RMSEP (%)	RPD
No-pretreatment	Open field Rain shelter	0.84	1.15	0.80 0.59	1.14 2.54	2.25 1.12
Baseline	Open field Rain shelter	0.83	1.24	0.67 0.47	1.48 2.77	1.74 1.03
SG	Open field Rain shelter	0.84	1.15	0.80 0.58	1.14 2.54	2.25 1.12

In Table 3, when built models with the samples cultivated in the open field were used to predict the samples of the same cultivation methods, results of no-pretreatment and SG pretreatment were similar (Rp^2 = 0.80, RMSEP = 1.14%, RPD = 2.25) and better than the results of baseline. The prediction ability of the three methods for rain-shelter samples was not ideal (Rp^2 = 0.47~0.59, RMSEP = 2.54~2.77%, RPD = 1.03~1.12), which indicated that the sharing ability of the model needs to be improved. On the whole, compared with the results of baseline and SG, the prediction ability of the constructed model with no-pretreatment was better. Therefore, spectral data without preprocessing were used for analysis in the following study.

3.1.2. Establishment of an SSC Model Using the IRIV-SPA

Based on the spectral data of fresh jujubes in the open field, the IRIV algorithm with an inverse elimination was used to select variables. The number of cross-validation, maximum principal component, and iteration were set as 10, 20, and 8, respectively. In this IRIV iterative process, the change curve of the retained variable number is shown in Figure 2. As the iteration number increased, the number of retained variables decreased, and the downward trend gradually flattened. Overall, 87 wavelength variables were retained at the 8th iteration. The DMEAN and P−values of retained variables are shown in Figure 3.

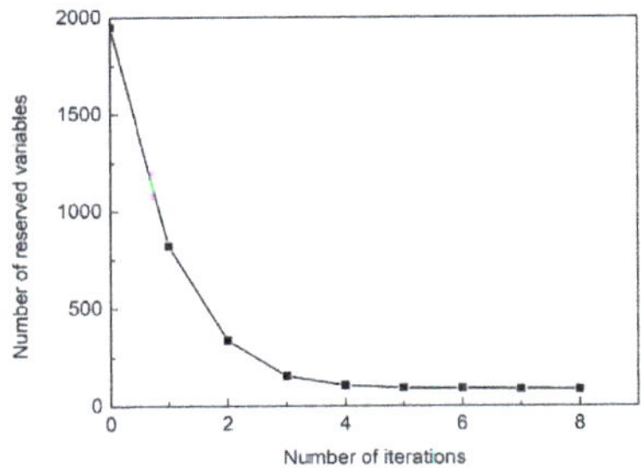

Figure 2. Number of retained variables.

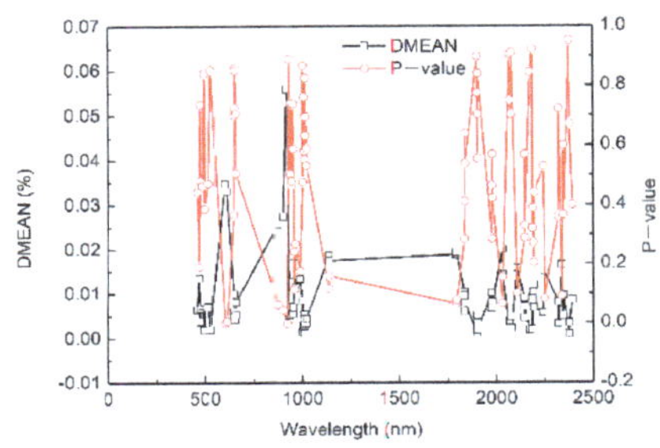

Figure 3. DMEAN and P−values of the nonparametric Mann–Whitney U test on variable.

Combining Figure 3 and the rules in Table 2, the variable type was divided. The selected variables and types are shown in Figure 4. In total, 4 strong informative variables and 83 weak informative variables were selected from 1951 variables, respectively. An inverse elimination was performed, and 71 characteristic wavelengths were preserved.

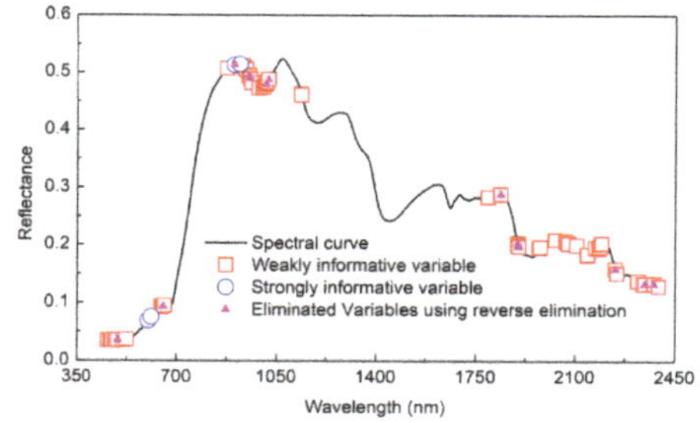

Figure 4. Selection of characteristic wavelength using IRIV based on open-field cultivation.

The number of selected characteristic wavelengths using the IRIV algorithm was high. For further data dimensionality reduction, SPA was used to perform an extraction of characteristic wavelengths for the second time based on 71 extracted characteristic wavelengths using IRIV. When the RMSE was 1.0257%, 10 characteristic wavelengths were extracted. According to the degree of importance, the extracted wavelengths using IRIV-SPA were 957, 1008, 2339, 920, 2248, 2394, 1137, 1976, 647, and 602 nm in turn.

Based on extracted characteristic wavelengths using IRIV and IRIV-SPA, LS-SVM was used to establish SSC detection models. The SSC of fresh jujubes from two cultivation modes was predicted. The SSC-predicted results are shown in Table 4.

Table 4. Prediction results of SSC using IRIV and IRIV-SPA.

Variable Selection Methods	Number of Wavelengths	Prediction Set	Rc^2	RMSEC (%)	Rp^2	RMSEP (%)	RPD
IRIV	71	Open field Rain shelter	0.93	0.76	0.85 0.71	1.02 2.50	2.52 1.14
IRIV-SPA	10	Open field Rain shelter	0.82	1.23	0.79 0.65	1.20 3.33	2.14 0.85

Compared with the built model using full wavelengths (in Table 3), the IRIV-LS-SVM model improved the prediction ability. For the SSC of open-field samples, the Rp^2 (from 0.80 to 0.85) and the RPD (from 2.25 to 2.52) were increased, and the RMSEP decreased from 1.14% to 1.02%. For the SSC of rain-shelter samples, the Rp^2 (from 0.59 to 0.71) and the RPD (from 1.12 to 1.14) were increased, and the RMSEP decreased from 2.54% to 2.50%. For the prediction results of open-field samples, the IRIV-SPA-LS-SVM model and the full-wavelength LS-SVM model were basically the same. For the prediction results of samples cultivated in the rain shelter, the IRIV-SPA-LS-SVM model was slightly worse than the full-wavelength LS-SVM model. Based on the IRIV-LS-SVM and IRIV-SPA-LS-SVM models of fresh jujubes cultivated in the open field, the SSC of samples in the same cultivation was well-predicted, but the SSC of samples in the rain-shelter cultivation was poorly predicted. Therefore, the cultivation mode has a certain influence on the SSC detection model, and the model needs to be further optimized to improve the predictive ability. Compared with IRIV, the number of extracted characteristic wavelengths using IRIV-SPA was significantly reduced (from 71 to 10) on the premise of ensuring the model performance. The IRIV-SPA algorithm achieved a better comprehensive ability and was used to select the characteristic wavelength in the following research.

3.2. Update of SSC Detection Model

3.2.1. Variable Optimization

Based on the spectral information of samples cultivated in the rain shelter, the selection of SSC characteristic wavelengths using IRIV is shown in Figure 5. Thirteen strongly informative variables and ninety-three weakly informative variables were selected. After reverse elimination of variables, the final number of optimal characteristic wavelengths was 73.

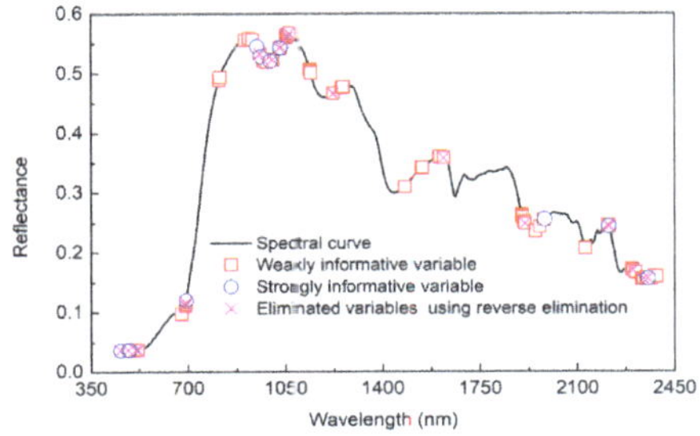

Figure 5. Selection of characteristic wavelength using IRIV based on rain-shelter cultivation.

Based on the extracted characteristic wavelengths using IRIV, SPA was used to extract the characteristic wavelengths for the second time. Ten characteristic wavelengths were extracted when the RMSE was 1.0257%. According to the importance, the extracted characteristic wavelengths using IRIV-SPA were 1257, 962, 905, 1137, 2337, 2300, 1541, 2378, 2386, 1947, 1907, 1480, 1058, 2128, 811, and 693 nm in turn.

For fresh jujubes from two cultivation modes, the selected characteristic wavelengths using IRIV-SPA are shown in Figure 6. There was a certain difference between the extracted characteristic wavelengths using the two cultivation modes. For the characteristic wavelengths extracted from a single cultivation mode, it was difficult to cover up the characteristic information of another cultivation mode.

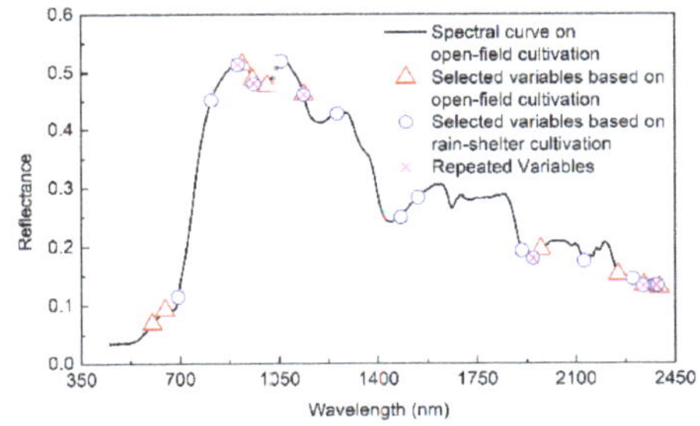

Figure 6. Fusion of characteristic wavelength.

Therefore, a new variable combination that integrated the extracted characteristic wavelengths of open-field and rain-shelter cultivation was proposed. In Figure 6, there were also the same and similar wavelengths between the characteristic wavelengths of those two cultivation modes. The selected variables of the two cultivation modes were added, redundant repeat variables were removed from the added characteristic variables, and the remaining variables were used as the fused characteristic wavelengths. Due to differences in the physicochemical properties of the sample, the external environment, and other factors, there would be a certain positional shift between wavelengths [38–40]. In this study, the wavelength corresponding to the position shift in the range of (−30 nm, 30 nm) was used as a repeated variable. Only one variable remained among the repeated variables, and redundant variables were removed. The extracted SSC characteristic wavelengths after fusion (in Figure 6) were 602, 647, 693, 811, 920, 957, 1008, 1058, 1137, 1257, 1480, 1541, 1907, 1976, 2128, 2248, 2300, 2339, and 2394 nm.

3.2.2. Model Update

Because of the difference between the spectral curves of the two modes, the Euclidean distance between full wavelength spectrum of samples from the rain shelter. The KS algorithm was used to sequentially select samples from the calibration set of fresh jujubes cultivated in the rain shelter, sequentially. The new selected samples were added to the calibration set of fresh jujubes cultivated in the open field to form an updated calibration set, sequentially. PLSR was adopted to establish SSC detection models based on the updated calibration set, and the minimum value of RMSECV was used as the rule for selecting samples. The changing curve of RMSECV for SSC is shown in Figure 7. The minimum value of RMSECV was 1.33%. Correspondingly, 33 samples were selected from the calibration set of fresh jujubes cultivated in the rain shelter.

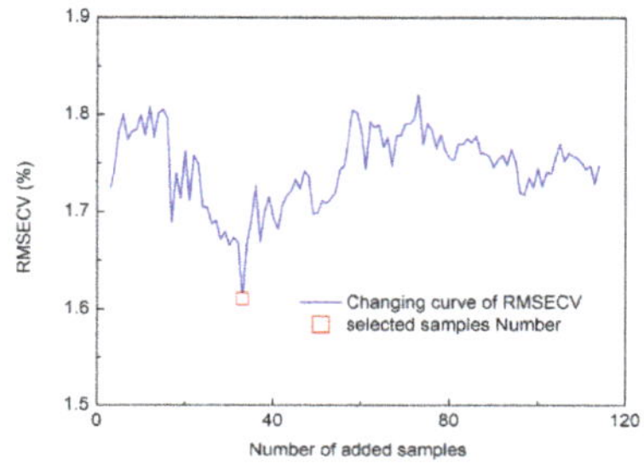

Figure 7. RMSECV distribution in different numbers of samples.

Thirty-three selected samples from rain shelter cultivation were combined with the calibration set from open-field cultivation (114 samples) to form an updated calibration set (147 samples). Based on the fused characteristic wavelengths, the original calibration set and the updated calibration set were used to establish LS-SVM detection models, respectively. The predicted results are shown in Table 5.

Table 5. Prediction results of SSC after model update.

Model Update	Prediction Set	Rc^2	RMSEC (%)	Rp^2	RMSEP (%)	RPD
No update	Open field Rain shelter	0.82	1.23	0.79 0.65	1.20 3.33	2.14 0.85
Wavelength fusion	Open field Rain shelter	0.85	1.10	0.80 0.69	1.17 2.96	2.20 0.96
Wavelength fusion-Euclidean distance	Open field Rain shelter	0.88	1.02	0.79 0.81	1.17 1.35	2.20 2.10

For the prediction ability (in Table 5) of fresh jujubes from two cultivation modes, the established model based on wavelength fusion was better than the model established before updating. This indicated that the updated characteristic wavelengths after wavelength fusion did not interfere with the performance of the established model on the open-field cultivation. For the SSC prediction results of fresh jujubes cultivated in the open field, the two update methods were good and similar. For the SSC prediction results of fresh jujubes cultivated in the rain shelter, the updated model based on wavelength fusion-Euclidean distance (Rp^2 = 0.81, RMSEP = 1.35%, RPD = 2.10) was significantly better than the updated model based on wavelength fusion (Rp^2 = 0.69, RMSEP = 2.96%, RPD = 0.96). Compared with the prediction performance before the model update, the ability of the updated LS-SVM model with the wavelength fusion-Euclidean distance was significantly improved. For fresh jujubes cultivated in open field, the Rp^2 (0.79) was the same, the RPD increased from 2.14 to 2.20, and the RMSEP decreased from 1.20% to 1.17%. For fresh jujubes cultivated in the rain shelter, the Rp^2 (from 0.65 to 0.81) and the RPD (from 0.85 to 2.10) were significantly increased, and the RMSEP (from 3.33% to 1.35%) decreased significantly. Therefore, the LS-SVM model based on the updated method of wavelength fusion-Euclidean distance achieved the best SSC prediction for fresh jujubes in both cultivation modes.

To validate the performance of the model based on the updated method of fusion wavelength-Euclidean distance, 50 samples from open-field cultivation and 50 samples from rain-shelter cultivation were collected for testing. To better show the detection results, the SSC-predicted and true values of samples from two cultivation modes are shown in Figure 8.

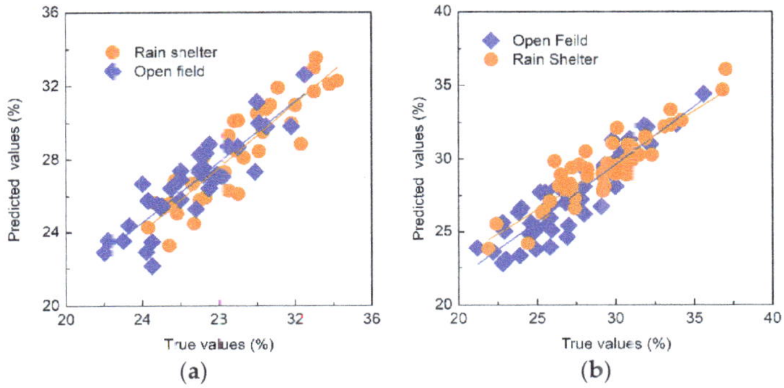

Figure 8. Detection results of SSC after model update using wavelength fusion-Euclidean distance. (**a**) Prediction set results; (**b**) test set results.

It was shown that the LS-SVM model using the update method of wavelength fusion-Euclidean distance achieved good prediction and test results for the SSC of fresh jujubes from both cultivation modes in Figure 8. The R^2, RMSE, and RPD of the test set were 0.82, 1.49%, and 2.18 for the SSC of "Huping" jujubes from open-field cultivation, respectively. The R^2, RMSE, and RPD of the test set were 0.81, 1.44%, and 2.17 for the SSC of "Huping" jujubes from rain-shelter cultivation, respectively.

In the field of fruit quality detection, a common problem was the failure of VIS/NIR spectral models. The established model has good prediction results under a single condition, but the model fails under new conditions with some variability. In the actual production of "Huping" jujube, there are two cultivation modes (open-field cultivation and rain-shelter cultivation). The established SSC detection model based on samples cultivated in open-field cultivation failed to predict samples cultivated in rain-shelter cultivation. In the SSC detection, there was some difference between the extracted characteristic wavelengths from the open-field samples and those from the rain-shelter samples. The dimensionality of the visible/NIR spectra was high and a direct replication of the VIS/NIR spectrum resulted in redundant information, which would affect model performance. In this study, the IRIV-SPA was used to preferentially select feature wavelengths that removed the effects of interfering information and uninformative variables. It was ensured that valid information was extracted, while the dimensionality was reduced. At the same time, variable recombination combined with wavelength position shift theory was used for variable selection. The preferred fusion wavelengths covered the variable information under the new conditions, which increased the coverage of feature information and did not interfere with the modeling ability of the original variables because of the new variables. When new samples were introduced based on Euclidean distances, the variability under the new conditions was increased. These variables were involved in the modeling when the model was recalibrated, which improved the accuracy and robustness of the model. Therefore, the wavelength fusion-Euclidean distance update method achieved good SSC prediction results for fresh jujube from two cultivation modes synchronously. The proposed method is an effective model updating method, which provides methods for the establishment of a robust VIS/NIR detection model and ideas for the online detection of agricultural product quality based on VIS/NIR spectroscopy.

4. Conclusions

In this study, the SSC detection of fresh jujubes cultivated in different modes (open-field cultivation and rain-shelter cultivation) was carried out based on VIS/NIR spectroscopy using variable selection and model updating. Based on the full-wavelength and the extracted characteristic wavelengths using IRIV and IRIV-SPA, the established LS-SVM models all achieved good predictions for the SSC of fresh jujubes cultivated in the open field, but the prediction results for samples cultivated in the rain shelter were all unsatisfactory. Compared with the IRIV algorithm, the IRIV-SPA algorithm achieved better performance. The extracted characteristic wavelengths of the two cultivation modes using IRIV-SPA were fused together. Combining the wavelength shift characteristics of the VIS/NIR spectrum, the repeated wavelengths were eliminated to form a new variable combination. Variable selection using wavelength fusion improved the SSC prediction results, but the degree of improvement in the SSC prediction of samples from rain-shelter cultivation needed to be increased. The method of adding samples under new conditions was applied for the model update. The updated LS-SVM model using the wavelength fusion-Euclidean distance achieved the best prediction results for SSC of fresh jujubes cultivated in the open field (Rp^2 = 0.79, RMSEP = 1.17%, RPD = 2.20) and the rain shelter (Rp^2 = 0.81, RMSEP = 1.35%, RPD = 2.10). The test results showed that the R^2, RMSE, and RPD for the SSC of "Huping" jujubes from open-field cultivation were 0.82, 1.49%, and 2.18, respectively. The R^2, RMSE, and RPD for the SSC of "Huping" jujubes from rain-shelter cultivation were 0.81, 1.44%, and 2.17, respectively. The method proposed in this study realizes the SSC detection of different cultivated fresh jujubes, provides a method for the establishment of a robust VIS/NIR detection model, and provides a basis for the online detection of fruit quality. In the future, a production line for the quality detection of fresh jujubes will be developed and optimized based on VIS/NIR spectroscopy.

Author Contributions: Conceptualization, H.S.; methodology, H.S. and R.R.; software, H.S., J.X., and H.Z.; writing—original draft, H.S. and R.R.; writing—review and editing, H.S. and S.Z. All authors have read and agreed to the published version of the manuscript.

Funding: This research was funded by Applied Basic Research Project of Shanxi Province (Project No. 201901D211359), Award-funded Scientific Research Projects for Outstanding Doctors to Work in Shanxi Province (Project No. SXYBKY2019049), Science and Technology Innovation Fund Project of Shanxi Agricultural University (Project No. 2020BQ02), The Key Research and Development Program of Shanxi Province (Project No. 201903D221027), and National Natural Science Foundation of China (31801632).

Institutional Review Board Statement: Not applicable.

Data Availability Statement: Data are contained within the article.

Conflicts of Interest: The authors declare no conflict of interest.

References

1. Escribano, S.; Biasi, W.V.; Lerud, R.; Slaughter, D.C.; Mitcham, E.J. Non-destructive prediction of soluble solids and dry matter content using NIR spectroscopy and its relationship with sensory quality in sweet cherries. *Postharvest Biol. Technol.* **2017**, *128*, 112–120. [CrossRef]
2. Peng, Y.; Lu, R. Analysis of spatially resolved hyperspectral scattering images for assessing apple fruit firmness and soluble solids content. *Postharvest Biol. Technol.* **2008**, *48*, 52–62. [CrossRef]
3. Posom, J.; Klaprachan, J.; Rattanasopa, K.; Sirisomboon, P.; Saengprachatanarug, K.; Wongpichet, S. Predicting Marian Plum Fruit Quality without Environmental Condition Impact by Handheld Visible-Near-Infrared Spectroscopy. *ACS Omega* **2020**, *5*, 27909–27921. [CrossRef]

4. Guo, Z.; Wang, Q.; Song, Y.; Zou, X.; Cai, J. Research progress of sensing detection and monitoring technology for fruit and vegetable quality control. *Smart Agric.* **2021**, *3*, 14–28. [CrossRef]
5. Minas, I.S.; Blanco-Cipollone, F.; Sterle, D. Accurate non-destructive prediction of peach fruit internal quality and physiological maturity with a single scan using near infrared spectroscopy. *Food Chem.* **2021**, *335*, 127626. [CrossRef]
6. Nakajima, S.; Genkawa, T.; Miyamoto, A.; Ikehata, A. Useful tissues in cabbage head for freshness evaluation with visible and near infrared spectroscopy. *Food Chem.* **2021**, *339*, 128058. [CrossRef]
7. Zareef, M.; Arslan, M.; Hassan, M.M.; Ahmad, W.; Ali, S.; Li, H.; Ouyang, Q.; Wu, X.; Hashim, M.M.; Chen, Q. Recent advances in assessing qualitative and quantitative aspects of cereals using nondestructive techniques: A review. *Trends Food Sci. Technol.* **2021**, *116*, 815–828. [CrossRef]
8. Hang, J.; Shi, D.; Neufeld, J.; Bett, K.E.; House, J.D. Prediction of protein and amino acid contents in whole and ground lentils using near-infrared reflectance spectroscopy. *LWT* **2022**, *165*, 113669. [CrossRef]
9. Li, X.; Li, Z.; Yang, X.; He, Y. Boosting the generalization ability of Vis-NIR-spectroscopy-based regression models through dimension reduction and transfer learning. *Comput. Electron. Agric.* **2021**, *186*, 106157. [CrossRef]
10. Nordey, T.; Joas, J.; Davrieux, F.; Chillet, M.; Léchaudel, M. Robust NIRS models for non-destructive prediction of mango internal quality. *Sci. Hortic.* **2017**, *216*, 51–57. [CrossRef]
11. Teh, S.L.; Coggins, J.L.; Kostick, S.A.; Evans, K.M. Location, year, and tree age impact NIR-based postharvest prediction of dry matter concentration for 58 apple accessions. *Postharvest Biol. Technol.* **2020**, *166*, 111125. [CrossRef]
12. Sun, H.; Zhang, S.; Xue, J.; Zhao, X.; Xing, S.; Chen, C.; Li, C. Model transfer method of fresh jujube soluble solids detection using variables optimization and correction algorithms. *Spectrosc. Spectr. Anal.* **2019**, *39*, 1041–1046. [CrossRef]
13. Mishra, P.; Marini, F.; Brouwer, B.; Roger, J.; Biancolillo, A.; Woltering, E.; Echtelt, E. Sequential fusion of information from two portable spectrometers for improved prediction of moisture and soluble solids content in pear fruit. *Talanta* **2021**, *223*, 121733. [CrossRef]
14. Acharya, U.K.; Walsh, K.B.; Subedi, P.P. Robustness of Partial Least-Squares Models to Change in Sample Temperature: II. Application to Fruit Attributes. *J. Near Infrared Spectrosc.* **2014**, *22*, 287–295. [CrossRef]
15. Isabel, M.; Antolin, G.; Debán, L.; Pardo, R. Assessing the influence of temperature on NIRS prediction models for the determination of sodium content in dry-cured ham slices. *Food Chem.* **2018**, *257*, 237–242. [CrossRef]
16. Zhang, S.; Zhang, H.; Zhao, Y.; Guo, W.; Zhao, H. A simple identification model for subtle bruises on the fresh jujube based on NIR spectroscopy. *Math. Comput. Model.* **2013**, *58*, 545–550. [CrossRef]
17. Wang, J.; Nakano, K.; Ohashi, S.; Kubota, Y.; Takizawa, K.; Sasaki, Y. Detection of external insect infestations in jujube fruit using hyperspectral reflectance imaging. *Biosyst. Eng.* **2011**, *108*, 345–351. [CrossRef]
18. Yu, K.; Zhao, Y.; Li, X.; Shao, Y.; Zhu, F.; He, Y. Identification of crack features in fresh jujube using Vis/NIR hyperspectral imaging combined with image processing. *Comput. Electron. Agric.* **2014**, *103*, 1–10. [CrossRef]
19. Sun, H.; Zhang, S.; Chen, C.; Li, C.; Xing, S.; Liu, J.; Xue, J. Detection of the Soluble Solid Contents from Fresh Jujubes during Different Maturation Periods Using NIR Hyperspectral Imaging and an Artificial Bee Colony. *J. Anal. Methods Chem.* **2019**, *2019*, 5032950. [CrossRef]
20. Liu, J.; Zhang, S.; Sun, H.; Xue, J.; Zhao, X. A kinetic model of hardness in storage periods of fresh jujubes at room temperature using two dimensional correlation spectroscopy. *Spectrosc. Spectr. Anal.* **2018**, *38*, 813–817. [CrossRef]
21. Yin, X.; Liu, Y.; Du, Y.; Gao, P.; Wen, P.; Zhang, P. Effect of the anti-cracking treatment on the fruit development and quality of 'Huping' jujube. *J. Henan Agric. Univ.* **2020**, *54*, 582–588. [CrossRef]
22. Zhang, Q.; Pang, X.; Chen, X.; Ye, J.; Lin, S.; Jia, X. Rain-shelter cultivation influence rhizosphere bacterial community structure in pear and its relationship with fruit quality of pear and soil chemical properties. *Sci. Hortic.* **2020**, *269*, 109419. [CrossRef]
23. Chen, X.; Zhang, Q.; Zeng, S.; Chen, Y.; Guo, Y.; Huang, X. Rhizosphere soil affects pear fruit quality under rain-shelter cultivation. *Can. J. Plant Sci.* **2020**, *100*, 683–691. [CrossRef]
24. Tian, T.; Qiao, G.; Deng, B.; Wen, Z.; Hong, Y.; Wen, X. The effects of rain shelter coverings on the vegetative growth and fruit characteristics of Chinese cherry (*Prunus pseudocerasus* Lindl.). *Sci. Hortic.* **2019**, *254*, 228–235. [CrossRef]
25. Li, X.; He, F.; Wang, J.; Li, Z.; Pan, Q. Simple rain-shelter cultivation prolongs accumulation period of anthocyanins in wine grape berries. *Molecules* **2014**, *19*, 14843–14861. [CrossRef]
26. Sun, C.; Aernouts, B.; Saeys, W. Effects of harvest time, fruit size and cultivar on the bulk optical properties of Satsuma mandarin. *Postharvest Biol. Technol.* **2021**, *175*, 111412. [CrossRef]
27. Cen, H.; Lu, R.; Mendoza, F.; Beaudry, R.M. Relationship of the optical absorption and scattering properties with mechanical and structural properties of apple tissue. *Postharvest Biol. Technol.* **2013**, *85*, 30–38. [CrossRef]
28. Mishra, P.; Woltering, E.; Brouwer, B.; Hogeveen-van Echtelt, E. Improving moisture and soluble solids content prediction in pear fruit using near-infrared spectroscopy with variable selection and model updating approach. *Postharvest Biol. Technol.* **2021**, *171*, 111348. [CrossRef]
29. Sun, X.; Subedi, P.; Walsh, K.B. Achieving robustness to temperature change of a NIRS-PLSR model for intact mango fruit dry matter content. *Postharvest Biol. Technol.* **2020**, *162*, 111117. [CrossRef]
30. Anderson, N.T.; Walsh, K.B.; Subedi, P.P.; Hayes, C.H. Achieving robustness across season, location and cultivar for a NIRS model for intact mango fruit dry matter content. *Postharvest Biol. Technol.* **2020**, *168*, 111202. [CrossRef]

31. Mishra, P.; Nikzad-Langerodi, R. Partial least square regression versus domain invariant partial least square regression with application to near-infrared spectroscopy of fresh fruit. *Infrared Phys. Technol.* **2020**, *111*, 103547. [CrossRef]
32. Kennard, R.; Stone, L.; Walsh, K.B. Computer aided design of experiments. *Technometrics* **1969**, *11*, 137–148. [CrossRef]
33. Suykens, J.; Vandewalle, J.; Moor, B.D. Least squares support vector machine classifiers. *Neural Process. Lett.* **1999**, *9*, 293–300. [CrossRef]
34. Suykens, J.; Vandewalle, J.; Moor, B.D. Optimal control by least squares support vector machines. *Neural Process. Lett.* **2001**, *14*, 23–35. [CrossRef]
35. Meng, Z.; Liu, H.; An, X.; Yin, Y.; Jin, C.; Zhang, A. Prediction model of wheat straw moisture content based on SPA-SSA-BP. *Trans. Chin. Soc. Agric. Mach.* **2022**, *53*, 231–238. [CrossRef]
36. Liu, H.; Wang, Y.; An, X.; Wei, Y.; Luo, L.; Chen, X. Study on detection method of wheat unsound kernel based on near-infrared hyperspectral imaging technology. *Spectrosc. Spectr. Anal.* **2019**, *39*, 223–229. [CrossRef]
37. Yun, Y.; Wang, W.; Tan, M.; Liang, Y.; Li, H.; Cao, D.; Lu, H.; Xu, Q. A strategy that iteratively retains informative variables for selecting optimal variable subset in multivariate calibration. *Anal. Chim. Acta* **2014**, *807*, 36–43. [CrossRef]
38. Luck, W. *Structure of Water and Aqueous Solutions*; Verlag Chemie: Weinheim, Germany, 1974; pp. 248–284.
39. Arroume, M.; Idrissi, A.; Turrell, G. Infrared spectroscopic investigation of the effect of temperature on the ν3 band shapes of CO_2 and N_2O in some liquid alkanes. *J. Mol. Struct.* **1992**, *266*, 153–158. [CrossRef]
40. Delwiche, S.; Mekwatanakarn, W.; Wang, C. Soluble solids and simple sugars measurement in intact mango using near infrared spectroscopy. *Horttechnology* **2008**, *18*, 410–416. [CrossRef]

foods

MDPI

Article

Intelligent Evaluation of Stone Cell Content of Korla Fragrant Pears by Vis/NIR Reflection Spectroscopy

Tongzhao Wang [1,2], Yixiao Zhang [1,2], Yuanyuan Liu [1,2,*], Zhijuan Zhang [1,2] and Tongbin Yan [1,2]

1 Agricultural Engineering Key Laboratory, Department of Xinjiang Uygur Autonomous Region, University of Education, Alar 843300, China
2 College of Mechanical and Electrical Engineering, Tarim University, Alar 843300, China
* Correspondence: lyylvjin@taru.edu.cn

Abstract: Stone cells are a distinctive characteristic of pears and their formation negatively affects the quality of the fruit. To evaluate the stone cell content (SCC) of Korla fragrant pears, we developed a Vis/NIR spectroscopy system that allowed for the adjustment of the illuminating angle. The successive projective algorithm (SPA) and the Monte Carlo uninformative variable elimination (MCUVE) based on the sampling algorithm were used to select characteristic wavelengths. The particle swarm optimization (PSO) algorithm was used to optimize the combination of penalty factor C and kernel function parameter g. Support vector regression (SVR) was used to construct the evaluation model of the SCC. The SCC of the calibration set ranged from 0.240% to 0.657% and that of the validation set ranged from 0.315% to 0.652%. The SPA and MCUVE were used to optimize 57 and 83 characteristic wavelengths, respectively. The combinations of C and g were (6.2561, 0.2643) and (2.5133, 0.1128), respectively, when different characteristic wavelengths were used as inputs of SVR, indicating that the first combination had good generalization ability. The correlation coefficients of the SPA-SVR model after pre-processing the standardized normal variate (SNV) for both sets were 0.966 and 0.951, respectively. These results show that the SNV-SPA-SVR model satisfied the requirements of intelligent evaluation of SCC in Korla fragrant pears.

Keywords: successive projective algorithm; uninformative variable elimination; support vector regression; Korla fragrant pear; stone cell content; intelligent evaluation

Citation: Wang, T.; Zhang, Y.; Liu, Y.; Zhang, Z.; Yan, T. Intelligent Evaluation of Stone Cell Content of Korla Fragrant Pears by Vis/NIR Reflection Spectroscopy. *Foods* **2022**, *11*, 2391. https://doi.org/10.3390/foods11162391

Academic Editor: Arun K. Bhunia

Received: 2 July 2022
Accepted: 6 August 2022
Published: 9 August 2022

1. Introduction

Korla fragrant pears are popular with consumers worldwide due to their beautiful skin color, sweet and crisp flesh, and rich fragrance. Consumers pay close attention to appearance quality and edible quality when they purchase Korla fragrant pears. A special characteristic of pears is that various stress factors induce the formation of stone cells, composed of large amounts of lignin and cellulose, which negatively affects the edibility and quality of the fruits [1]. There is a large number of soluble cells in some Korla fragrant pears of inferior quality, such as 'green top fruit' and rough skin fruit [2,3]. The taste is more delicate if there is lower stone cell content (SCC) in fresh Korla fragrant pears. Up to now, research on stone cells has mostly been concerned with their content [4] and structural characterization [5–7] in different germplasms of pears. Moreover, traditional methods of evaluation of the quality of edible fruit are destructive and require long processing times.

Visible and near-infrared (Vis/NIR) spectroscopy is fast, safe, and contactless technology that has been used to evaluate some qualities of fruits, including soluble solids content (SSC) [8–12], firmness [13–16], titratable acidity [17], dry matter [18], and polyphenols to amino ratio [19]. Intelligent evaluating research about SCC has not been conducted. In addition, samples are needed to change detecting places manually, and the illuminating angles cannot be changed automatically when the Vis/NIR spectroscopy systems are used.

While Vis/NIR spectrometry usually yields thousands of response signals per detection point, not all wavelengths are associated with corresponding chemical bonds as a quality parameter of one kind of agro-product. Furthermore, the spectra at certain wavelengths do not make significant contributions to the evaluation models. Therefore, it is essential to extract characteristic wavelengths to establish intelligent models that are robust, require less computing time, and are highly predictive. Successive projective algorithm (SPA) [20,21] and uninformative variable elimination (UVE) [22,23] are two filter-type [24] algorithms that have been widely used to extract characteristic wavelengths. The SPA and UVE were evaluated using the results of correlation coefficient (R), root mean square error (RMSE), multiple linear regression (MLR), and partial least square regression (PLSR) as bases to estimate the combinations of the wavelengths. MLR [25,26] and PLSR [27,28] have different principles of establishing evaluating models. It is unreasonable to use different evaluating models to estimate the combinations of characteristic wavelengths.

A support vector machine (SVM) is a generalized linear classifier that categorizes data into two classes based on supervised learning. It has several advantages, including stability, sparsity, and simplicity. Previous applications of this algorithm mainly included qualitative analyses [29–32]. Recently, support vector regression (SVR) has been applied in quantitative evaluations as part of the development of SVM theories [33–35].

Taking all these factors into account, this study was conducted with three aims: (1) to set up an intelligent spectra acquisition system for Korla fragrant pears in which the optical subsystem has an adjustable irradiation angle and samples can be rotated at a specific angle; (2) to choose characteristic wavelengths to simplify the detection of SCC in Korla fragrant pears by SPA and UVE; and (3) to establish an SVR evaluating model after optimizing penalty factor C and kernel function parameter gamma (g).

2. Materials and Methods

2.1. Korla Fragrant Pears and Pretreatment

Korla fragrant pears were collected from a plantation (Alar, Xinjiang, China) from 15 to 20 September 2021. A total of 120 fruit samples were selected with uniform spindle shape (diameter 61–85 mm) and weight (110–130 g), and without visual damage on the surface.

Samples were soaked in a mixture of water and a special fruit cleaning agent (Almawin, Germany), with chemical compositions of plant sugar surfactant, citric acid, organic lemon extract, glycerin, and lactic acid, for about 30 s, and then rinsed twice with distilled water. The cleaned pears were air-dried at room temperature (20 °C) and then stored in a preservation box at 4 °C. Prior to Vis/NIR spectra acquisition, samples were placed at room temperature for 30 min. Each Korla fragrant pear was coded by a labeled paper (24 × 12 mm) which was attached to the end of the calyx.

2.2. Vis/NIR Spectroscopy System and Diffuse Reflectance Spectra Acquisition

The Vis/NIR spectroscopy system is shown in Figure 1. The system is composed of a spectra acquisition unit, a light source, a sample rotating unit, and a computer. The spectroscopy system is placed in the dark room.

A shortwave spectrometer (USB2000+, Ocean Optics Inc., Dunedin, FL, USA) and an optical fiber component (QP400-2-VIS-BX, Ocean Optics Inc., Dunedin, FL, USA) make up the spectra acquisition unit. The range of wavelengths is between 468 nm and 1155 nm, and the bits of A/D conversion is 12. The spectrometer sends spectra to the computer by serial communication. There is 1 receiving optical fiber in the center of the optical fiber component, and the numerical aperture is 0.22 ± 0.02 mm.

The light source unit contains 2 Halogen tungsten lamp beads (20 W, Philips). Each lamp bead is fixed in a cup-shaped container which has two connecting through-holes along the center axis and two positioning blind holes perpendicular to the central axis. Each group has two mounting plates, as shown in the partial enlargement in Figure 1, with one connecting through-hole and five positioning through-holes. The distances between each of the five positioning holes and the connecting through-hole are the same, and the angle between the two adjacent connecting holes is 15°. The angle of the light beams can be adjusted via the cup-shaped container and the mounting plate group.

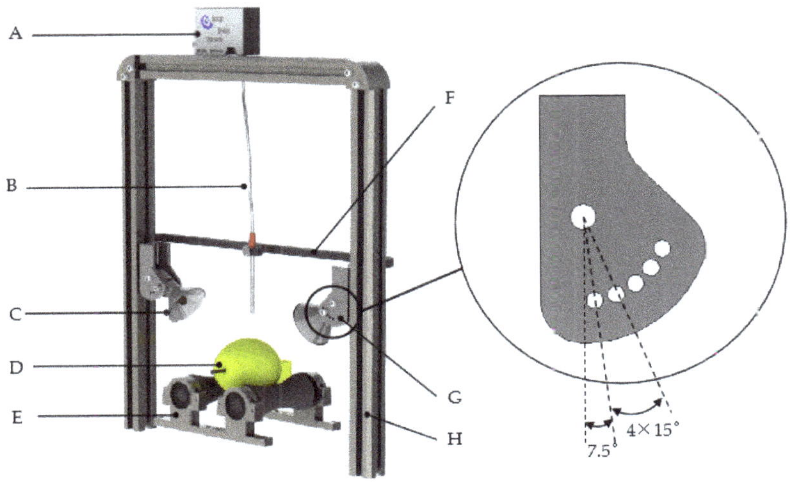

Figure 1. Vis/NIR spectra acquisition system for Korla fragrant pears. A: spectrometer; B: optical fiber; C: halogen lamp; D: sample; E: rotating stage; F: optical fiber bracket; G: lamp mounting plate; H: system mounting rack.

The sample rotating unit consists of two rubber rollers that rotate in the same direction. The distance between the centers of the rollers is 5 mm larger than the diameter of each roller. The maximum rotation of the Korla fragrant pear samples is 120° ± 0.5°.

The distance between the circular lamp rack and the sample stage is 80 mm, and the angle of each lamp bead in the vertical direction is 37.5°. In this set-up, a circle of incident light, with a diameter of 70 mm, irradiates the upper part of the Korla fragrant pears.

Prior to the acquisition of spectral data, a Korla fragrant pear weighing 120 g ± 1 g was placed on the sample rotating unit, after which the parameters of the spectrometer were adjusted. The optical fiber probe was placed in the upper center of the Korla fragrant pear at a vertical distance of 10 mm, where the detecting radius of the sample was about 2.2 mm. The integration time, scanning times, and smoothing were set to 20 ms, 5, and 2, respectively, when the maximum reflection intensity of a standard diffuse reflection whiteboard (DR300-WS-PTFE, Ocean Optics Inc., Dunedin, FL, USA) was about 55,705 counts (about 85% of the maximum value). A white reference and black reference were obtained by turning the halogen lamp on and off, respectively, when the standard diffuse reflection whiteboard was placed 10 mm away from the fiber probe.

Each sample was placed on the sample rotating stage so that the surface near the maximum diameter along the minor axis was under the optical fiber probe. The first group of spectral data for the first point was obtained at the vertically downward position of the number-marked side, and the other two groups were obtained at a rotation of 120° and 240°, respectively, along the long axis. The mean spectral values of the three points were taken as spectral data for each sample.

2.3. Measurement of SCC

The SCC of Korla fragrant pears was measured using a modified gravimetric method [36]. After removing the peel and core, three pieces of pulp (11× 11× 8.3 mm) were cut around the spectral data collecting points and weighed with a precision electronic balance (FA3004, Shanghai Liangping Instrument Co., Ltd., Shanghai, China). The total weight of the three pieces was recorded.

The three pieces of pulp were sealed together with a self-sealing bag and cooled for 24 h at −18 °C. Next, the pulp was thawed and homogenized in 50 mL of distilled water using a small-sized tissue smasher (FL1902, Ningbo Kajafa Electrical Technology Co., Ltd., Ningbo, China) at 22,000 r/min for 1 min. The homogenate was poured into a 1000 mL beaker (beaker A). The inner wall of the plastic container of the smasher was rinsed 2–3 times with distilled water, and the cleaning solution was added to beaker A. Next, 600 mL of distilled water was added to the mixture in beaker A, stirred with a glass rod for 1 min, and allowed to stand for about 30 s. The upper suspension in breaker A was poured out into beaker B. This process was repeated 2–3 times until there was no suspended substance. The same operation was conducted for the mixture in beaker B.

The precipitated stone cells in beaker A and beaker B were filtered through filter paper. Next, the filter paper was dried at 60–65 °C in a drying oven until the weight remained unchanged. Then, the dried stone cells were collected and weighed. The SCC of sample i was calculated as:

$$X_i = \frac{m_{itotal} - m_{ifilter}}{m_i} \times 100\% \quad (1)$$

where i is the serial number for samples; X_i refers to the SCC; m_{itotal} refers to the total weight of filter paper and stone cells; $m_{ifilter}$ refers to the weight of filter paper; and m_i refers to the weight of the selected pulp.

2.4. Spectral Preprocessing and Sample Set Division

The original spectra inevitably shifted and displayed background noise due to the influences of the data acquisition environment, sample size, instrument, and other factors. The stability of spectral data and the signal-to-noise ratio could effectively be improved through the reasonable use of preprocessing methods. A multiple scattering correction (MSC) was used to eliminate baseline drift [37]. The standardized normal variate (SNV) was used for the centering and calibration of the spectral data in each wavelength [38]. Several Savitzky–Golay (S-G) filters [39], with frame sizes of 3–9 and fitting orders of 1–7, were used to improve the smoothing effect. The optimal combination of frame size and fitting order was chosen according to R and RMSE values. Each spectral preprocessing method and the combinations of S-G and MSC, or S-G and SNV, were used. The preprocessed spectra were used to construct different PLSR models of SCC. The best method or combination was selected according to corresponding R and RMSE values.

The sample-set partitioning method based on the joint x-y distance algorithm (SPXY) divided the samples into a calibration set (Cs) and a validation set (Vs) where spectral data and SCC were taken as the input data. The proportion of Cs:Vs was 3:1 in this study.

2.5. Algorithms of Selecting Characteristic Wavelengths

2.5.1. SPA

The SPA is a forward variable selection method that uses simple operations to minimize the collinearity of variables in vector space [40]. Three phases are required to select characteristic wavelengths which have the least collinearities.

First, K chains with N_max variables are created by using QR decomposition of spectral matrix $Spec_{Ncal \times K}$. The number of N_max should be between the minimum value defined by the data processor and the smaller of Ncal and K. Here, Ncal and K represent the number of samples in Cs and wavelengths, respectively.

Second, K × N_maxsets of characteristic wavelengths were selected according to the root mean square error of Vs (RMSEV). Each regression coefficient vector B of the PLSR model was calculated according to Equation (2). The RMSEV of the corresponding PLSR model was calculated according to Equation (3). The set of characteristic wavelengths with the minimum RMSEV was selected.

$$Specc \times B = Refc \quad (2)$$

$$\mathrm{RMSEV(j)} = \sqrt{\frac{1}{\mathrm{N_{val}}} \sum_{i=1}^{\mathrm{N_{val}}} \left(\mathrm{Ref_v(i)} - \hat{\mathrm{Ref}}_{\mathrm{v}}(\mathrm{i})\right)^2} \quad (3)$$

where Specc refers to the set of preprocessed spectral data, which has Nrows ($0 < N < N\_max$) and S columns ($0 < S < K$); Refc refers to the measured values of SCC corresponding to the selected N samples in Cs; $\mathrm{Ref_v(i)}$ refers to the measured value of SCC of sample i in Vs; $\hat{\mathrm{Ref}}_{\mathrm{v}}(\mathrm{i})$ refers to the predicted SCC value calculated by selected spectral data and B.

Third, uninformative wavelengths were further eliminated according to the F-test. A correlation index was defined for each selected wavelength at the end of phase 2. The index was the absolute value of the arithmetic product of the regression coefficient and the standard deviation. The originally selected characteristic wavelengths were rearranged in descending order according to the correlation indexes. Another set of PLSR models was established with the spectral data of the first j wavelengths and SCC. Corresponding RMSEVs were calculated. The critical value, t_{RMSEV}, was calculated by the inverse function of the sum distribution function for the F distribution, as shown by Equation (4) for which the significance value α was 0.25 and the degrees of freedom were the same. The wavelengths whose RMSEVs were less than t_{RMSEV} were chosen as the final characteristic ones.

$$t_{RMSEV} = \frac{\mathrm{RMSEV(j)}}{\min(\mathrm{RMSEV(j)})} \quad (4)$$

2.5.2. UVE Combined with Monte Carlo Sampling (MCUVE) and PLSR

The informative wavelengths were selected by UVE based on the regression coefficients of PLSR models. The Monte Carlo sampling method was used to randomly select *N* kinds of sample groups. The PLSR regression coefficient vector $\beta(j,:)$ was obtained from the spectra and corresponding SCC vector of the jth group. The stability value C(k) at the kth wavelength was calculated by Equation (5). Wavelengths were sorted according to the values of vector C from the largest to the smallest. Evaluating models were established by adding new spectra of one wavelength, which had a smaller stability value. The wavelengths were selected as characteristic wavelengths with the minimum value of RMSEV.

$$\mathrm{C(k)} = \frac{\mathrm{mean}(\beta\,(k,:))}{\mathrm{std}(\beta(k,:))} \quad (5)$$

where mean($\beta(k,:)$) and std($\beta(k,:)$) refer to the mean coefficient and standard deviation at the kth wavelength, respectively.

2.6. Modeling Algorithm

The radial basis function (RBF) is a good generalization of the kernel function of SVR. The particle swarm optimization algorithm (PSO) was used to determine the optimal combination of C and g [41] in order to obtain a model with good performance. The model was evaluated by RMSEV and R.

3. Results

3.1. Statistics of SCC Measured Values

There were 90 samples in the Cs and 30 samples in the Vs. The SCC values of both sets are shown in Table 1. The SCC ranged from 0.240% to 0.657% in the Cs and from 0.315% to 0.652% in the Vs. The combinations of mean value and standard deviation (SD) were (0.486%, 0.100%) and (0.481%, 0.083%), respectively. The range of SCC in Cs covered that in Vs, which ensured the feasibility of the evaluating model. An ANOVA test was taken to check SCC values in Cs and Vs through SPSS software (Version 23, International Business Machines Corporation, Armonk, NY, USA). The *p* value was 0.008, indicating that there was a significant difference between Cs and Vs.

Table 1. Statistics of SCC in Cs and Vs.

Sample Set	Numbers	Min (%)	Max (%)	Mean (%)	SD (%)	p
Cs	90	0.240	0.657	0.486	0.100	0.008
Vs	30	0.315	0.652	0.481	0.083	

Max: the maximum value of the dataset; Min: the minimum value of the dataset; Cs: calibration sets; Vs: validation sets.

3.2. Spectral Characteristics and Different Preprocessing Methods

Spectra in the range of 498–1020 nm were considered effective, owing to the large amount of noise at both ends of the original spectrum. The effective original spectral curves of Korla fragrant pears are shown in Figure 2a. There were two reflective valleys near 680 nm and 980 nm and two reflective peaks near 550 nm and 750 nm. The spectra near 750 nm and 980 nm were related to carbohydrate content [42], and O-H [43] in the flesh of fruits. SCC had a negative correlation with carbohydrate content; therefore, the spectra at these wavelengths were indirectly related to stone cells. Spectra near 550 nm and 680 nm were related to anthocyanins and chlorophyll in the sample epidermis, respectively [44]. Korla fragrant pears with high levels of stone cells usually have green skin; therefore, SCC also had some relationship with spectra near 550 nm and 680 nm.

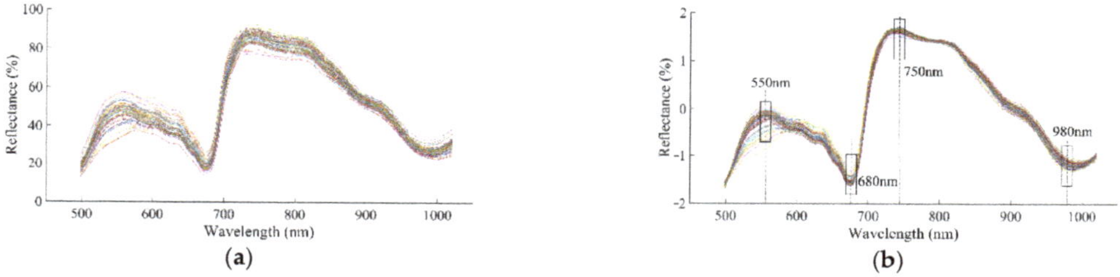

Figure 2. Reflective spectral curves. (**a**) Raw spectrum; (**b**) spectrum after SNV pretreating.

Using preprocessing algorithms could improve the evaluation accuracy compared with not using them. The principal component numbers of PLSR models were all 10 after different preprocessing algorithms. The optimal combination of frame size and fitting order was (7, 5) where R of the calibration set and validation set grew the largest, according to Table 2. The evaluation results of PLSR models based on different spectral preprocessing algorithms are shown in Table 3. Evaluation models based on MSC and SNV had higher Rs and lower RMSEs, while those based on S-$G_{(7, 5)}$ had lower Rs and higher RMSEs, in Cs and Vs. The robustness of the PLSR model based on SNV was better than that of MSC according to the different values of Rs between Cs and Vs. The addition of S-$G_{(7, 5)}$ did not improve the ability of evaluation because the combination of two-point smoothing and S-$G_{(7, 5)}$ eliminated some effective spectral information. The model established on the basis of the SNV preprocessing algorithm achieved the best results, with R and RMSE of 0.9189 and 0.0277% in the Cs, and 0.8935 and 0.0315% in the Vs. Spectral curves based on SNV are shown in Figure 2b.

Table 2. Correlation coefficients of Cs and Vs with different S-G parameters.

Fitting Order \ Frame Size	None	3	5	7	9
none	0.8613 0.8214				
1		0.8276 0.8007	0.7867 0.7616	0.7403 0.7150	0.7012 0.6710
2			0.8306 0.8035	0.7928 0.7710	0.7789 0.7458
3			0.8414 0.8137	0.8227 0.8006	0.8023 0.7853
4				0.8527 0.8195	0.8419 0.8059
5				0.8926 0.8210	0.8647 0.8100
6					0.8589 0.8128
7					0.8527 0.8026

Correlation coefficients on the top and the bottom of different combinations of frame size and fitting order refer to correlation coefficients of validation set and calibration set, respectively.

Table 3. Evaluation of PLSR based on different spectral preprocessing algorithms.

Parameter	Preprocessing Algorithm	Factor Number	R_C	$RMSE_C$ (%)	R_V	$RMSE_V$ (%)
Stone cell content (%)	None	9	0.8613	0.0360	0.8214	0.0412
	MSC	10	0.9191	0.0277	0.8879	0.0325
	SNV	10	0.9189	0.0277	0.8935	0.0315
	S-G$_{(7,5)}$	10	0.8926	0.0319	0.8210	0.0409
	S-G$_{(7,5)}$& MSC	10	0.9001	0.0308	0.8614	0.0361
	S-G$_{(7,5)}$& SNV	10	0.8999	0.0308	0.8641	0.0356

R_C: the correlation coefficient of the calibration set; $RMSE_C$: root mean square error of the calibration set; R_V: the correlation coefficient of the validation set; $RMSE_V$: root mean square error of the validation set; S-G(7,5): Savitzky–Golay filter with a frame size of 7 and fitting order of 5.

3.3. Characteristic Wavelengths

Fewer than 100 wavelengths were required to simplify the SCC evaluating models. The changing processes of RMSEV with different wavelength candidate subsets, which were chosen by SPA or MCUVE, are shown in Figure 3a,b, respectively. Using the combination of (SPA, PLSR) or (MCUVE, PLSR), characteristic wavelengths were selected (57 and 83, respectively). The minimum RMSEVs of SPA and MCUVE were 0.0692% and 0.0685%, respectively.

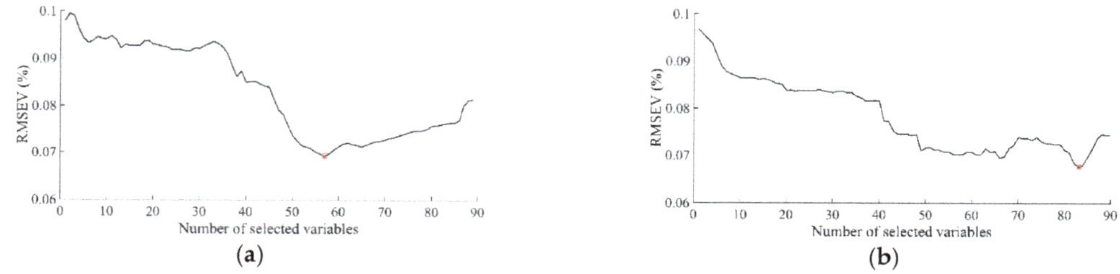

Figure 3. Changing processes of RMSEV with different wavelengths. (**a**) SPA; (**b**) MCUVE.

As shown in Figure 4a, 46 characteristic wavelengths distributed densely between 846.8 and 940.6 nm were obtained for the first selection method, while 10 characteristic wavelengths were scattered between 498 and 750 nm and 1 characteristic wavelength was located at 997.1 nm. Characteristic wavelengths selected by the second method were distributed mainly in the ranges of 757.7–796.7 nm, 828.3–847.5 nm, 866.8–910.3 nm, and 952.9–1006.9 nm, as shown in Figure 4b. The two methods both selected characteristic wavelengths in the range of 828.3–910.3 nm, which were correlated with the third overtone stretch of O-H and C-H functional groups.

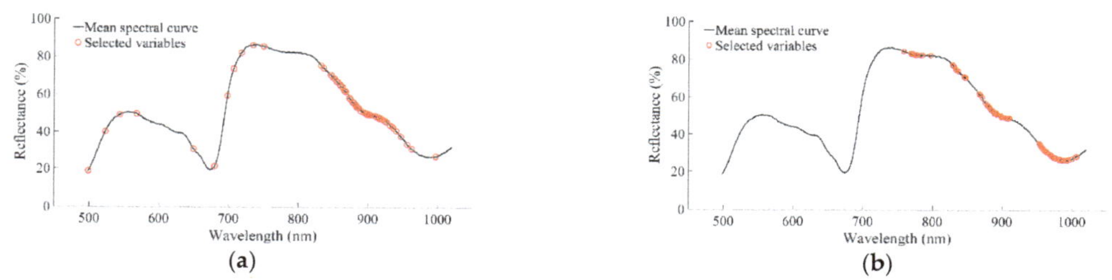

Figure 4. Distribution of characteristic wavelengths. (**a**) SPA; (**b**) MCUVE.

3.4. SCC Evaluation Based on PSO-SVR

When the selected 57 wavelengths or 83 wavelengths were used as inputs, the optimal values of C were 2.5133 and 6.2561, respectively, as shown in Figure 5a,b. This indicated that the error tolerability produced by the first wavelength group was stricter than that produced by the second group and might result in evaluation model overfitting. The optimal values of g were 0.1128 and 0.2643 for the first and second group, respectively, indicating that the support vector number of the first wavelength group was less than that of the second. The optimal combination of C and g was obtained in the 36th iteration for MCUVE where the fitness was 0.01394%, and for SPA in the 100th iteration where the fitness was 0.01404%.

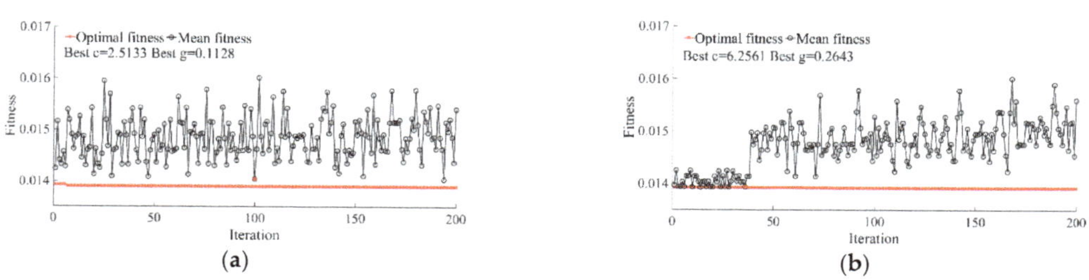

Figure 5. Optimization of SVR parameters. (**a**) SPA; (**b**) MCUVE.

The evaluation of SCC based on SPA-POS-SVR is shown in Figure 6a. The correlation coefficients of the Cs and Vs were 0.949 and 0.928, respectively, and the RMSEs of both sets were 0.0253% and 0.0297%, respectively. The evaluation ability in the SCC range of 0.2000–0.4000% and 0.6000–0.7000% was better than that in the SCC range of 0.4000–0.7000%, with samples of Cs and Vs evenly distributed in the interval with smaller deviations. This showed that the robustness of SPA-PSO-SVR was poor and that the adaptability of the global evaluation was low.

Figure 6. Scatter plot of the calibration set (×) and verification set (o) of stone cell content. (**a**) SPA; (**b**) MCUVE.

The evaluation results of SCC based on MCUVE-POS-SVR are shown in Figure 6b. The correlation coefficients of the Cs and Vs were 0.966 and 0.951, respectively, and the RMSEs of both sets were 0.0209% and 0.0239%, respectively. The evaluation ability in the SCC range of 0.2000–0.7000% was better. Moreover, several scattering points of the Cs and Vs had relatively large deviations in the SCC ranges of 0.3000–0.4000% and 0.6000–0.7000%. Overall, the evaluation accuracy and robustness of MCUVE-PSO-SVR were better than those of SPA-PSO-SVR. The MCUVE-PSO-SVR model of SCC can be applied in online systems or portable equipment for evaluating the qualities of Korla fragrant pear.

The parameters of our spectroscopy system were set according to Korla fragrant pear. The SCC evaluating model and parameters of spectroscopy system would not be suitable for other kinds of pear, such as 'Yali' pear, 'Dangsha' pear, and so on. The minimum SCC of all samples was 0.240% in Cs and 0.315% in Vs according to Table 1. The largest evaluating errors were 8.711% and 10.845% in Cs and Vs. The evaluating precision met the requirements of application. Therefore, the minimum limit of SCC that could be detected was 0.240% by the MCUVE-PSO-SVR model.

4. Discussion

It has been shown previously that spectroscopy technology can be applied in quantificational evaluations of the quality parameters of agricultural products. This was a new exploration to intelligently evaluate SCC for Korla fragrant pears by Vis/NIR reflection spectroscopy. An intelligent evaluating system was introduced. Its illuminating angle could be adjusted and Korla fragrant pears could be rotated along the long axis with a set angle.

Several processes were undertaken in order to obtain an accurate and stable evaluating model of SCC. The SNV algorithm had the best evaluation effect among the spectral pretreatment methods, according to evaluating accuracy and robustness. A total of 83 characteristic wavelengths were selected by using Vis/NIR spectra combined with MCUVE algorithm. Finally, the MCUVE-PSO-SVR model was established. The correlation coefficients of the Cs and Vs were 0.966 and 0.951, and the root mean square errors of both sets were 0.0209% and 0.0239%. The results demonstrate that the model could achieve the quantitative detection of SCC in Korla fragrant pear.

The nondestructive and rapid evaluation system combined with the MCUVE-PSO-SVR model for the SCC of Korla fragrant pears could meet the requirements of detection time and evaluation precision. It could be rendered more suitable for industry work requirements through the adjustment of key parameters and the development of supporting equipment. Its transformation and applications in industry are more meaningful, and present an attractive target for future research.

Author Contributions: Resources, Y.L.; data curation, Y.Z.; writing—original draft preparation, T.W.; writing—review and editing, Y.L.; visualization, Z.Z.; supervision, T.Y.; project administration, Y.L. All authors have read and agreed to the published version of the manuscript.

Funding: This work was supported by the National Natural Science Foundation of China, grant number 31960498, the Open Project Program of the Key Laboratory of Colleges & Universities under the Department of Education of Xinjiang Uygur Autonomous Region, grant number TDNG20221044 and TDNG20200102.

Institutional Review Board Statement: Not applicable.

Informed Consent Statement: Not applicable.

Data Availability Statement: The data presented in this study are available on request from the corresponding author. The data are not publicly available due to the request of funding scientific research projects.

Conflicts of Interest: The authors declare that they have no known competing financial interests or personal relationships that could have appeared to influence the work reported in this paper.

References

1. Xue, Y.S.; Xu, S.Z.; Xue, C.; Wang, R.Z.; Zhang, M.Y.; Li, J.M.; Zhang, S.L.; Wu, J. Pear-process: A new phenotypic tool for stone cell trait evaluation in pear fruit. *J. Integr. Agric.* **2020**, *19*, 1625–1634. [CrossRef]
2. Li, D.H.; Meng, X.T.; Xu, M.Q.; Che, F.B.; Pan, Y.; Zhao, D.Y. Research on identification features of quality classification of Korla fragrantpear (*Pyrus sinkiangensis* Yu). *Storage Process* **2018**, *18*, 5–15.
3. Xu, H.B.; Wu, J.; Wang, Z.P.; Gao, Y.M.; Wang, Z.P.; Zhao, Z.Q. Characteristics of the vibration spectral response of rough peel pear among Korla pear. *Sci. Technol. Food Ind.* **2015**, *36*, 57–66.
4. Zhang, J.Y.; Li, J.M.; Xue, C.; Wang, R.Z.; Zhang, M.Y.; Qi, K.J.; Fan, J.; Hu, H.J.; Zhang, S.L.; Wu, j. The variation of stone cell content in 236 germplasms of sand pear (*Pyrus pyrifolia*) and identification of related candidate genes. *Hortic. Plant J.* **2021**, *7*, 108–116. [CrossRef]
5. Tao, S.T.; Khanizadeh, S.; Zhang, H.; Zhang, S.L. Anatomy, ultrastructure and lignin distribution of stone cells in two Pyrus species. *Plant Sci.* **2009**, *176*, 413–419. [CrossRef]
6. Jin, Q.; Yan, C.C.; Qiu, J.X.; Zhang, N.; Lin, Y.; Cai, Y.P. Structural characterization and deposition of stone cell lignin in Dangshan Su pear. *Sci. Hortic.* **2013**, *155*, 123–130. [CrossRef]
7. Yan, C.C.; Yin, M.; Zhang, N.; Jin, Q.; Fang, Z.; Lin, Y.; Cai, Y.P. Stone cell distribution and lignin structure in various pear varieties. *Sci. Hortic.* **2014**, *174*, 142–150. [CrossRef]
8. Fan, S.X.; Wang, Q.Y.; Tian, X.; Yang, G.Y.; Xia, Y.; Li, J.B.; Huang, W.Q. Non-destructive evaluation of soluble solids content of apples using a developed portable Vis/NIR device. *Biosyst. Eng.* **2020**, *193*, 138–148. [CrossRef]
9. Yuan, L.M.; Mao, F.; Chen, X.J.; Li, L.M.; Huang, G.Z. Non-invasive measurements of 'Yunhe' pears by vis/NIRS technology coupled with deviation fusion modeling approach. *Postharvest Biol. Technol.* **2020**, *160*, 111067. [CrossRef]
10. João, P.S.N.; Mateua, W.D.A.; Izabella, P.C.; Luis, C.C.J.; Gustavo, H.A.T. Determination of 'Palmer' mango maturity indices using portable near infrared (VIS/NIR) spectrometer. *Postharvest Biol. Technol.* **2017**, *130*, 75–80.
11. Wang, J.Y.; Guo, Z.M.; Zou, C.X.; Jiang, S.Q.; El-Seedi, H.R.; Zou, X.B. General Model of Multi-Quality Detection for Apple from Different Origins by Vis/NIR Transmittance Spectroscopy. *J. Food Meas. Charact.* 2022, *accepted*. [CrossRef]
12. Guo, Z.M.; Wang, M.M.; Agyekum, A.A.; Wu, J.Z.; Chen, X.B.; Zou, M.; El-Seedi, H.R.; Tao, F.F.; Qin, O.Y.; Zou, X.B. Quantitative detection of apple watercore and soluble solids content by near infrared transmittance spectroscopy. *J. Food Eng.* **2020**, *279*, 109955. [CrossRef]
13. Li, J.B.; Zhang, H.L.; Zhan, B.S.; Zhang, Y.F.; Li, R.L.; Li, J.B. Nondestructive firmness measurement of the multiple cultivars of pears by Vis/NIR spectroscopy coupled with multivariate calibration analysis and MC-UVE-SPA method. *Infrared Phys. Technol.* **2020**, *104*, 103154. [CrossRef]
14. Nur, F.M.K.; Puneet, M.; Rob, E.S.; Ernst, J.W.; Martin, P.B. Assessing firmness in mango comparing broadband and miniature spectrophotometers. *Infrared Phys. Technol.* **2021**, *115*, 103733.
15. René, M.; Carolina, C.; Magnólia, L.S.N.; Emanuel, J.N.M.; Juan, P.Z.; Sergio, T.F. Non-destructive evaluation and detection of internal physiological disorders in 'Keitt' mango using a hand-held Vis/NIR spectrometer. *Postharvest Biol. Technol.* **2020**, *167*, 111251.
16. Bizzani, M.; Flores, D.W.M.; Colnago, L.A.; Ferreira, M.D. Non-invasive spectroscopic methods to estimate orange firmness, peel thickness, and total pectin content. *Microchem. J.* **2017**, *133*, 168–174. [CrossRef]

17. Annelisa, A.B.; Fernanda, C.; Abadia, R.N.; Clarissa, D.; Flávio, A.S.; Gustavo, H.A.T.; Luis, C.C.J. Non-destructive determination of color, titratable acidity, and dry matter in intact tomatoes using a portable Vis/NIR spectrometer. *J. Food Compos. Anal.* **2022**, *107*, 104288.
18. Sylvia, T.; Marianne, G.B.; Karen, K.P.; Sergey, V.K. Evaluating pear (cv. Clara Frijs) dry matter and soluble solids content with near infrared spectroscopy. *LWT-Food Sci. Technol.* **2014**, *59*, 1107–1113.
19. Guo, Z.M.; Barimah, A.O.; Yin, L.M.; Chen, Q.S.; Shi, J.Y.; EI-Seedi, H.R.; Zou, X.B. Intelligent evaluation of taste constituents and polyphenols-to-amino acids ratio in matcha tea powder using near infrared spectroscopy. *Food Chem.* **2021**, *353*, 129372. [CrossRef]
20. Valber, E.A.; Adriano, A.G.; David, D.S.F.; Héctor, C.G.; Roberto, K.H.G.; Mario, C.U.A. Vis/NIR spectrometric determination of Brix and sucrose in sugar production samples using kernel partial least squares with interval selection based on the successive projections algorithm. *Talanta* **2018**, *181*, 38–43.
21. Sun, Y.; Gu, X.Z.; Sun, K.; Hu, H.J.; Xu, M.; Wang, Z.J.; Tu, K.; Pan, L.Q. Hyperspectral reflectance imaging combined with chemometrics and successive projections algorithm for chilling injury classification in peaches. *LWT-Food Sci. Technol.* **2017**, *75*, 557–564. [CrossRef]
22. Jan, P.M.A.; Yvan, V.H.; Lutgarde, M.C.B. Improved variable reduction in partial least squares modelling by Global-Minimum Error Uninformative-Variable Elimination. *Anal. Chim. Acta* **2017**, *982*, 37–47.
23. Li, C.; Zhao, T.L.; Li, C.; Mei, L.; Yu, E.; Dong, Y.T.; Chen, J.H.; Zhu, S.J. Determination of gossypol content in cottonseeds by near infrared spectroscopy based on Monte Carlo uninformative variable elimination and nonlinear calibration methods. *Food Chem.* **2017**, *221*, 990–996. [CrossRef] [PubMed]
24. Li, L.J.; Huang, W.Q.; Wang, Z.L.; Liu, S.Q.; He, X.; Fan, S.X. Calibration transfer between developed portable Vis/NIR devices for detection of soluble solids contents in apple. *Postharvest Biol. Technol.* **2022**, *183*, 111720. [CrossRef]
25. Pedro, S.S.; Ana, S.A.; Carla, M.B. Use of artificial neural network model for rice quality evaluation based on grain physical parameters. *Foods* **2021**, *10*, 3016.
26. Wang, H.W.; Shangguan, L.Y.; Wu, J.J.; Guan, R. Multiple linear regression modeling for compositional data. *Neurocomputing* **2013**, *122*, 490–500. [CrossRef]
27. Nadya, V.; Claudia, M.; Jimy, O.; Tony, C.; Himer, A.; Wilson, C. Comparison between artificial neural network and partial least squares regression models for hardness modeling during the ripening process of Swiss-type cheese using spectral profiles. *J. Food Eng.* **2018**, *219*, 8–15.
28. Sahar, R.; Hossein, M.; Bahareh, J.; Aslan, A.; Mohammad, S. Achieving a robust Vis/NIR model for microbial contamination detection of Persian leek by spectral analysis based on genetic, iPLS algorithms and VIP scores. *Postharvest Biol. Technol.* **2021**, *175*, 111413.
29. Jair, C.; Farid, G.L.; Lisbeth, R.M.; Asdrubal, L. A comprehensive survey on support vector machine classification:Applications, challenges and trends. *Neurocomputing* **2020**, *408*, 189–215.
30. Onuwa, O.; Christopher, E.N. Deep support vector machine for hyperspectral image classification. *Pattern Recognit.* **2020**, *103*, 107298.
31. Ji, Y.M.; Sun, L.J.; Li, Y.S.; Li, J.; Liu, S.C.; Xie, X.; Xu, Y.T. Non-destructive classification of defective potatoes based on hyperspectral imaging and support vector machine. *Infrared Phys. Technol.* **2019**, *99*, 71–79. [CrossRef]
32. Camilo, L.M.M.; Kássio, M.G.L.; Francis, L.M. Uncertainty estimation and misclassification probability for classification models based on discriminant analysis and support vector machines. *Anal. Chim. Acta* **2019**, *1063*, 40–46.
33. Xu, S.X.; Zhao, Y.C.; Wang, M.Y.; Shi, X.Z. Determination of rice root density from Vis–NIR spectroscopy by support vector machine regression and spectral variable selection techniques. *Catena* **2017**, *157*, 12–23. [CrossRef]
34. Alireza, S.; Adel, B.; Miguel, G. Evaluation of banana quality indices from color features using support vector regression. *Talanta* **2016**, *148*, 54–61.
35. Li, P.; Ma, J.C.; Zhong, N. Raman spectroscopy combined with support vector regression and variable selection method for accurately evaluating salmon fillets storage time. *Optik* **2021**, *247*, 167879. [CrossRef]
36. Nie, J.Y.; Li, J.; Yang, Z.F.; Zhang, H.J.; Li, M.Q. Study on the conditions for measuring stone cell content in pear flesh by freezing method. *J. Fruit Sci.* **2006**, *01*, 133–135.
37. Liu, Q.; Chen, S.X.; Zhou, D.D.; Chao, D.; Wang, J.H.; Zhou, H.S.; Tu, K.; Pan, L.Q.; Li, P.X. Nondestructive detection of weight loss rate, surface color, vitamin C content, and firmness in mini-Chinese cabbage with nanopackaging by fourier transform-near infrared spectroscopy. *Foods* **2021**, *10*, 2309. [CrossRef]
38. Wang, F.; Zhao, C.J.; Yang, G.J. Development of a non-destructive method for detection of the juiciness of pear via VIS/NIR spectroscopy combined with chemometric methods. *Foods* **2020**, *9*, 1778. [CrossRef]
39. Fu, X.S.; Hong, X.Z.; Liao, J.Y.; Ji, Q.G.; Li, C.F.; Zhang, M.Z.; Ye, Z.H.; Yu, X.P. Fingerprint approaches coupled with chemometrics to discriminate geographic origin of imported salmon in China's consumer market. *Foods* **2021**, *10*, 2986. [CrossRef]
40. Mario, C.U.A.; Teresa, C.B.S.; Roberto, K.H.G.; Takashi, Y.; Henrique, C.C.; Valeria, V. The successive projections algorithm for variable selection in spectroscopic multicomponent analysis. *Chemom. Intell. Lab. Syst.* **2001**, *57*, 65–73.
41. Esfandiarpour-Boroujeni, I.; Karimi, E.; Shirani, H.; Esmaeilizadeh, M.; Mosleh, Z. Yield evaluation of apricot using a hybrid particle swarm optimizationimperialist competitive algorithm- support vector regression (PSO-ICA-SVR) method. *Sci. Hortic.* **2019**, *257*, 108756. [CrossRef]

42. Cen, H.; He, Y.; Huang, M. Combination and comparison of multivariate analysis for the identification of orange varieties using visible and near infrared reflectance spectroscopy. *Eur. Food Res. Technol.* **2007**, *225*, 699–705. [CrossRef]
43. Jamshidi, M.; Hamdami, N.; Dohkani, S.; Keramat, J. Single-and multi-objective optimization of low fat ice-cream formulation, based on genetic algorithms. *J. Agric. Sci. Technol.* **2012**, *14*, 1285–1296.
44. Tian, X.; Wang, Q.Y.; Li, J.B.; Peng, F.; Huang, W.Q. Non-destructive evaluation of soluble solids content of pear based on fruit surface feature classification and multivariate regression analysis. *Infrared Phys. Technol.* **2018**, *92*, 336–344. [CrossRef]

foods

MDPI

Article

Predicting Mandarin Fruit Acceptability: From High-Field to Benchtop NMR Spectroscopy

Ignacio Migues [1,2], Fernando Rivas [3], Guillermo Moyna [4], Simon D. Kelly [5] and Horacio Heinzen [1,*]

1 Laboratorio de Farmacognosia y Productos Naturales, Departamento de Química Orgánica, Facultad de Química, Universidad de la República, Montevideo 11800, Uruguay
2 Programa de Posgrados de la Facultad de Química, Universidad de la República, Montevideo 11800, Uruguay
3 Programa Nacional de Investigación en Producción Citrícola, Instituto Nacional de Investigación Agropecuaria (INIA), Salto 50000, Uruguay
4 Laboratorio de Espectroscopía y Fisicoquímica Orgánica, Departamento de Química del Litoral, CENUR Litoral Norte, Universidad de la República, Paysandú 60000, Uruguay
5 Food Safety Control Laboratory, Joint FAO/IAEA Centre of Nuclear Techniques in Food and Agriculture, Department of Nuclear Sciences and Applications, International Atomic Energy Agency, 1400 Vienna, Austria
* Correspondence: heinzen@fq.edu.uy; Tel.: +598-2924-4068

Abstract: Recent advances in nuclear magnetic resonance (NMR) have led to the development of low-field benchtop NMR systems with improved sensitivity and resolution suitable for use in research and quality-control laboratories. Compared to their high-resolution counterparts, their lower purchase and running costs make them a good alternative for routine use. In this article, we show the adaptation of a method for predicting the consumer acceptability of mandarins, originally reported using a high-field 400 MHz NMR spectrometer, to benchtop 60 MHz NMR systems. Our findings reveal that both instruments yield comparable results regarding sugar and citric acid levels, leading to the development of virtually identical predictive linear models. However, the lower cost of benchtop NMR systems would allow cultivators to implement this chemometric-based method as an additional tool for the selection of new cultivars.

Keywords: acceptability; benchtop NMR; mandarins; NMR

Citation: Migues, I.; Rivas, F.; Moyna, G.; Kelly, S.D.; Heinzen, H. Predicting Mandarin Fruit Acceptability: From High-Field to Benchtop NMR Spectroscopy. *Foods* **2022**, *11*, 2384. https://doi.org/10.3390/foods11162384

Academic Editors: Zhiming Guo, Zhao Zhang and Dong Hu

Received: 24 June 2022
Accepted: 1 August 2022
Published: 4 August 2022

1. Introduction

Since the early days of nuclear magnetic resonance (NMR), considerable efforts have been invested to increase sensitivity and spectral resolution through the use of magnets with stronger fields. These endeavors have gone hand in hand with the development of novel superconducting materials and cryogenic technologies [1]. However, these systems are generally expensive and have high running and maintenance costs, driving many NMR spectrometer manufacturers to develop smaller and more accessible systems based on cryogen-free permanent magnets. These low-field instruments have magnetic fields below 2.3 T (i.e., ^{1}H resonance frequencies under 100 MHz), fit on a regular laboratory benchtop, and are even suitable for use in field experiments [2,3]. The basis of these instruments is the use of rare-earth ring-shaped magnets that produce relatively strong and homogeneous fields [3]. Their lower sensitivity can sometimes be offset by concentrating the samples or using a variety of methodologies for the enhancement of Boltzmann polarization [4]. Similarly, issues with chemical shift resolution can be addressed through the application of different signal acquisition and processing techniques, including solvent suppression and gradient-based pulse sequences [3,5].

Although low-field benchtop NMR spectrometers may not be suitable for natural product research due to their lower sensitivity and resolution, they have been used successfully in the quality control of phytopharmaceuticals and in food analysis, to mention a few examples [6]. In academia, the use of benchtop NMR is increasing progressively.

The low operating costs and ease of use of these instruments allow students not only to control their research products, but to follow chemical reactions in real time or even perform quantitative analyses [6–9]. Indeed, the quantitation of natural products using benchtop NMR has been employed in the quality control of drugs [10] and for the detection of adulterations in pharmaceutical products [11,12]. The use of low-field NMR in routine quality control of foods has also been demonstrated [6]. Examples of such applications include the determination of alcohol content in beverages [13] and the study of food authenticity and food fraud by targeted and untargeted analysis, where wine, coffee, oils, or even meat are examples [14–18]. For certain products, subdisciplines have been developed to study metabolomic profiles. For example, the term "MEATabolomics" refers to the application of metabolomic analysis to correlate the composition of meat with its sensory attributes [19,20].

In food analysis, untargeted approaches are preferred when trying to discover flavor-related compounds, which are followed with targeted analyses to measure the content of specific compounds or study metabolic pathways of interest [21,22]. Citrus metabolomics has been emerging in the last few years to control industrial processes or to evaluate flavor traits that influence consumer preferences [23–25]. However, little research has been conducted to adapt high-field NMR techniques to low-field systems. As stated by Castaing-Cordier and coworkers [26], benchtop instruments can be used in many applications due to recent advances in terms of sensitivity and resolution. Recently, we proved the usefulness of high-field NMR to predict consumer preferences in mandarins. Although interesting from an academic point of view, the high cost of the 400 MHz spectrometer employed in the study hampers its application by the local citrus industry [25]. The aim of the present work is to show an updated protocol for the analysis of mandarin consumer preferences using benchtop NMR systems that could be accessible to citrus fruit cultivators. As shown herein, our results indicate that chemometric-based consumer acceptability models of identical quality can be obtained regardless of magnetic field.

2. Materials and Methods

The samples used in the comparisons were a selection of aqueous mandarin extracts obtained during the development of the original method at 400 MHz [25]. Five extract replicates for each mandarin variety were lyophilized and stored under nitrogen in sealed containers until analysis. They were then dissolved in 600 μL of deuterium oxide (MagniSolv™, 99.9% D, Merck, Darmstadt, Germany), transferred to 5 mm NMR tubes (Norell® Standard Series™ Sigma-Aldrich, Darmstadt, Germany) and analyzed immediately.

A Bruker Avance III 400 spectrometer (Bruker, Ettlingen, Germany) was used to perform the high-field NMR experiments, while a Magritek Spinsolve 60 benchtop NMR spectrometer (Magritek GmbH, Aachen, Germany) was used to obtain the data at the low field. The 400 MHz spectra were obtained at a ^{1}H frequency of 400.13 MHz using a z-gradient BBFO-Plus probe (298 K). Spectra were recorded using a spectral width of 8 KHz, a data size of 32 K, and using a 30° excitation pulse. A total 64 scans with a relaxation delay of 1 s between scans were averaged, leading to an analysis time of 4.1 min per sample. The 60 MHz data were obtained at room temperature using a ^{1}H frequency of 62.32 MHz, a spectral width of 5 KHz, a data size of 32 K, and using a 90° excitation pulse. A total of 256 scans with a relaxation delay of 1 s between scans were averaged in this case, resulting in a total analysis time of 64.0 min per sample.

All spectra were processed using MNova (version 11.0, MestreLab Research, S.L., Santiago de Compostela, Spain) following an identical protocol, which included zero filling to 64 K and apodization with a 0.3 Hz exponential window function prior to Fourier transformation, manual phase and baseline correction, and referencing to the signal of the anomeric proton of α-glucose at 5.22 ppm. The spectra were then aligned using the derivative method and the average spectrum as a reference [27].

Once all spectra were aligned, the integral of the signal belonging to the sucrose glucosyl anomeric proton at 5.40 ppm was given an arbitrary value of 1.00. Then, the areas of the signals corresponding to the anomeric protons of α-glucose at 5.22 ppm, β-glucose at 4.63 ppm, and the multiplet arising from the H-3 and H-4 protons of the β-furanose form of fructose at 4.09 ppm, together with the four citric acid methylene protons centered at approximately 2.8 ppm, were scaled to that of the sucrose signal. The integration ranges for the sugar signals mentioned above were, respectively, 5.54 to 5.32, 5.29 to 5 16, 4.63 to 4.53, and 4.10 to 4.07 ppm in both instruments. Due to slight differences in the temperature of the experiments, the citrate signals were integrated from 3.02 to 2.73 ppm in the high-field spectrometer, and between 2.81 and 2.54 ppm on the benchtop instrument

The relative area values were corrected using the sweetness scale of Schiffman and coworkers [28,29], being 1.0 for sucrose, 1.3 for fructose and 0.6 for α- and β-glucose. The ratio sweetening power/citric acid was calculated as follows, where n represents each of the sugars considered:

$$\frac{\sum_n (Sugar\ sweetness \times Sugar\ content)_n}{Citric\ acid\ content} \quad (1)$$

The correlation between the mandarin acceptability and the sweetening power/citric acid ratio was determined using the same mandarin varieties for both spectrometer systems, the R^2 of the regressions was determined and the root mean square error (RMSE) of each model was calculated.

3. Results and Discussion

Figure 1 shows spectra obtained at 400 and 60 MHz for the same aqueous extract, respectively. Given its higher resolution, the spectrum obtained at 400 MHz allows for the identification of most protons from the species of interest. On the other hand, several of these signals appear overlapped at 60 MHz, making the initial assignment of resonances a harder task that requires technical know-how.

However, the signals corresponding to the sugar anomeric protons and citric acid methylene protons of interest are in relatively uncluttered regions of the spectrum, and therefore their identification and quantitation is achievable. Indeed, if the selection of the integration ranges is rigorous and consistent with those employed at the high field, the integration of the signals corresponding to anomeric protons of sucrose, glucose, and fructose, as well as the citric acid methylene protons, allows us to apply the methodology developed previously [25] to predict the acceptability of the mandarin samples (Table 1).

Table 1. Results of the sensory evaluation (acceptability), sweetening power/citric acid ratio, predicted acceptability using the model and RMSE of the prediction of each model (60 and 400 MHz data).

Variety	Acceptability	60 MHz			400 MHz		
		Sweetening Power/Citric Acid *	Predicted Acceptability	RMSE	Sweetening Power/Citric Acid *	Predicted Acceptability	RMSE
B475B	7.4	10.85	6.88	0.29	9.31	6.76	0.35
F7P3	6.9	11.53	7.21		10.12	7.27	
B475A	5.5	8.45	5.72		7.70	5.75	
B79	4.8	6.75	4.89		6.34	4.90	
M16	3.6	3.88	3.50		4.16	3.53	

* The reported values correspond to the average of 5 replicates. The RSD was less than 20% for all the varieties analyzed.

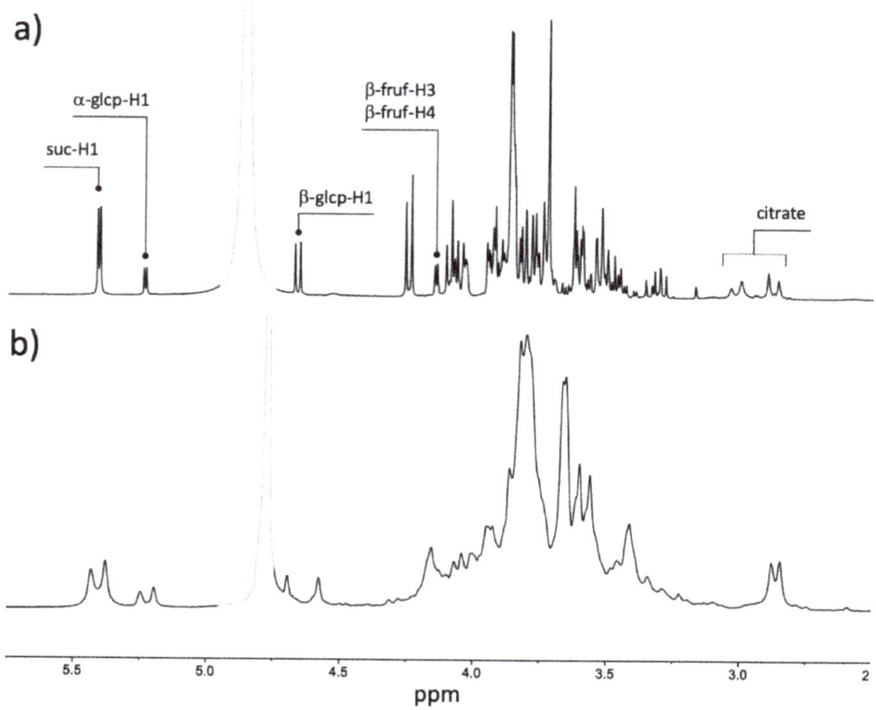

Figure 1. Comparison of ^{1}H spectra of the aqueous extract of mandarin variety B475B obtained with 400 and 60 MHz spectrometers ((**a**) and (**b**), respectively). Resonances employed in the estimations are annotated in the 400 MHz spectrum. The grayed-out region in both spectra corresponds to the residual HDO peak.

The sweetness/citric acid ratio of the samples determined at the two frequencies considered had high correlation ($R^2 > 0.99$, Figure 2), showing the equivalence of both systems and their fitness for the intended purpose of the method.

It is then possible to study the correlation between the acceptability of the mandarin samples determined by consumers and the sweetening power/citric acid ratio obtained using the 60 and 400 MHz systems (Figure 3). Using this set of data, a linear regression model with an R^2 of 0.94 and an RMSE of 0.35 was obtained using data recorded at 400 MHz. The corresponding regression parameters of the linear model derived using sugar and citric acid concentrations determined with the 60 MHz instrument were 0.96 and 0.29, indicating that acceptability prediction models of similar quality were obtained regardless of the instrument employed in their development.

In addition, the correlation between the predicted acceptability using both models was very high ($R^2 > 0.99$), further proving the equivalence of the models derived from the two instrumental systems (Figure 4).

It is worth pointing out that although models derived from data at 400 and 60 MHz are of the same predictive quality, special attention is needed when identifying and integrating data in the low-field instrument. As stated earlier and shown in Figure 1, there is considerable signal overlap in the 3.00 to 4.30 ppm region and expertise is required to assign peaks and process these spectra accurately.

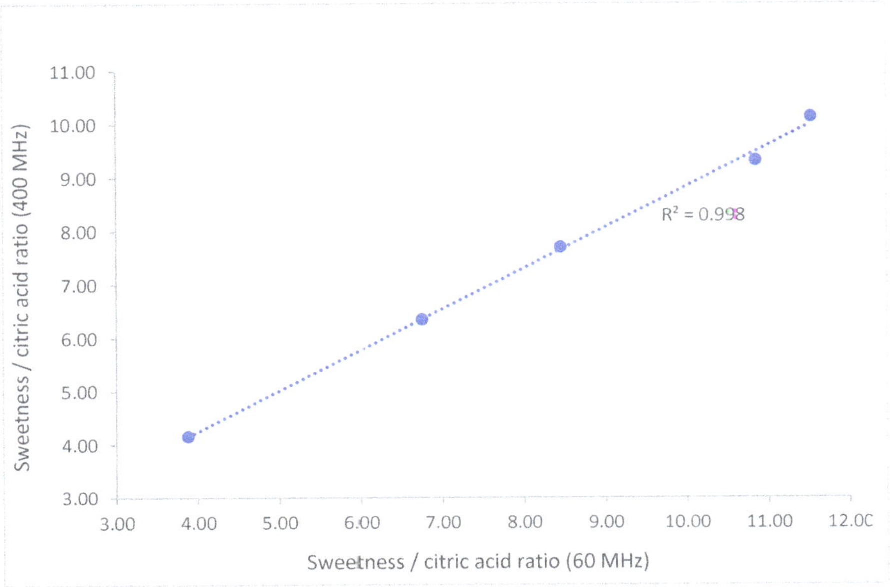

Figure 2. Correlation between the measurement of the sweetness/citric acid ratio of the samples obtained at 400 and 60 MHz.

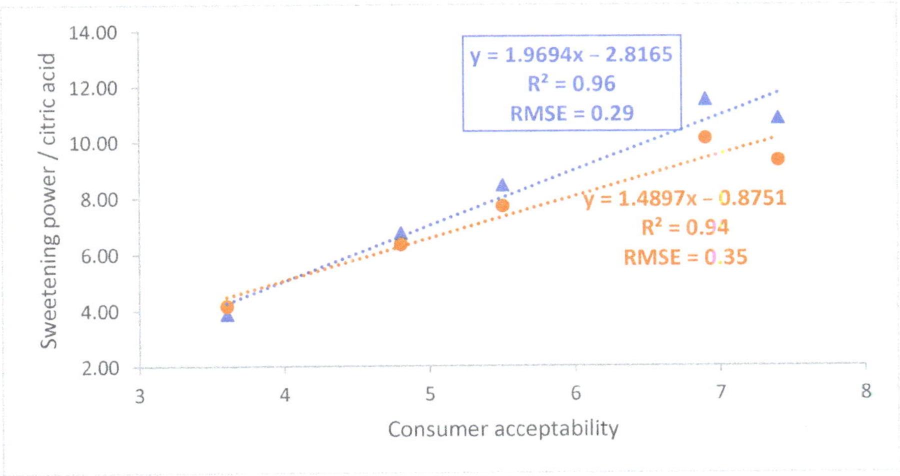

Figure 3. Correlation between consumer acceptability and sweetening power/citric acid determined using 60 and 400 MHz data (blue triangles and orange circles, respectively).

Conversely, and due to the lower resolution of low-field instruments, the variations in the chemical shifts of sugar signals with pH have less impact on spectra recorded at 60 MHz [30]. This makes spectral referencing and alignment simpler in these instruments.

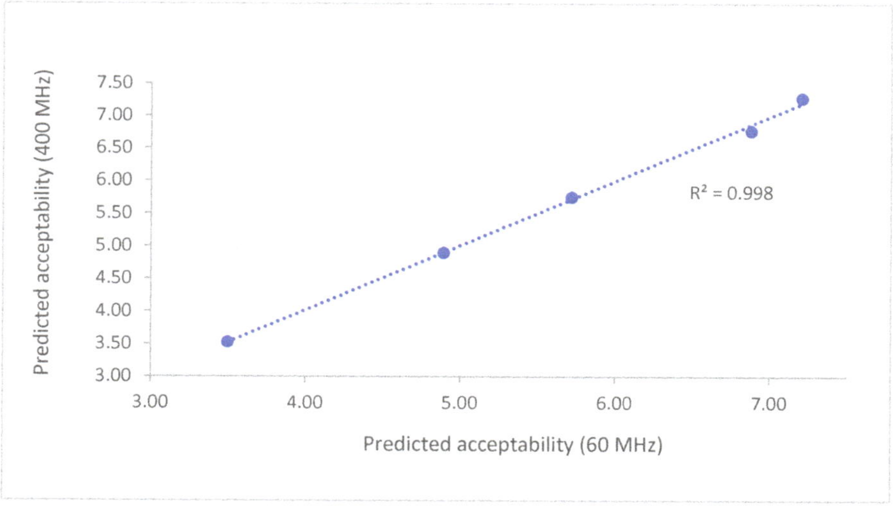

Figure 4. Correlation of the predicted acceptability using the models obtained with both instruments.

4. Conclusions

As demonstrated above, low-field NMR systems can be employed in the development of consumer acceptability prediction models that have identical quality to those derived from high-field NMR data. The lower purchase and running costs of benchtop spectrometers makes these chemometric-based tools more accessible for routine inclusion in fruit breeding programs, such as the Uruguayan Programa Nacional de Investigación en Producción Citrícola. Furthermore, the continuing advances in benchtop NMR instruments, which include the implementation pure shift pulse sequences, solvent suppression techniques, and multidimensional and multinuclear methods, will facilitate their application to other fields of food analysis and metabolomics.

Author Contributions: Conceptualization, S.D.K. and H.H.; methodology, I.M.; formal analysis, I.M.; investigation, I.M.; resources, F.R. and S.D.K.; data curation, I.M.; writing—original draft preparation, I.M.; writing—review and editing, G.M., S.D.K. and H.H.; visualization, I.M.; supervision, G.M., S.D.K., H.H.; project administration, S.D.K. and H.H.; funding acquisition, S.D.K. All authors have read and agreed to the published version of the manuscript.

Funding: This research was funded by the International Atomic Energy Agency (IAEA, award CRP D52042, "Implementation of Nuclear Techniques for Authentication of Foods with High-Value Labelling Claims") and the Agencia Nacional de Investigación e Innovación (ANII, award POS_NAC_2020_1_164240, "Estudio de Perfiles Metabólicos como Herramienta para la Caracterización de Mezclas de Compuestos Bioactivos"). The APC was covered by IAEA and the Programa de Desarrollo de las Ciencias Básicas (PEDECIBA).

Institutional Review Board Statement: Not applicable.

Informed Consent Statement: Not applicable.

Data Availability Statement: The data presented in this study are available on request from the corresponding author.

Acknowledgments: The authors would like to thank the staff at INIA Salto Grande and Área de Evaluación Sensorial (Departamento de Ciencia y Tecnología de Alimentos (CYTAL), Facultad de Química, Universidad de la República).

Conflicts of Interest: The authors declare no conflict of interest.

References

1. Freeman, R.; Robert, J.B. *A Brief History of High Resolution NMR BT—NMR at Very High Field*; Robert, J.B., Ed.; Springer: Berlin/Heidelberg, Germany, 1991; pp. 1–16.
2. Blümich, B. Low-field and benchtop NMR. *J. Magn. Reson.* **2019**, *306*, 27–35. [CrossRef]
3. Giberson, J.; Scicluna, J.; Legge, N.; Longstaffe, J. Chapter Three—Developments in benchtop NMR spectroscopy 2015–2020. *Annu. Rep. NMR Spectrosc.* **2021**, *102*, 153–246.
4. Halse, M.E. Perspectives for hyperpolarisation in compact NMR. *TrAC Trends Anal. Chem.* **2016**, *83*, 76–83. [CrossRef]
5. Kunjir, S.; Rodriguez-Zubiri, M.; Coeffard, V.; Felpin, F.-X.; Giraudeau, P.; Farjon, J. Merging Gradient-Based Methods to Improve Benchtop NMR Spectroscopy: A New Tool for Flow Reaction Optimization. *ChemPhysChem* **2020**, *21*, 2311–2319. [CrossRef]
6. van Beek, T.A. Low-field benchtop NMR spectroscopy: Status and prospects in natural product analysis. *Phytochem. Anal.* **2021**, *32*, 24–37. [CrossRef] [PubMed]
7. Kent, J.E.; Bell, N.G.A. Molecular Properties of Caffeine Explored by NMR: A Benchtop NMR Experiment for Undergraduate Physical-Chemistry Laboratories. *J. Chem. Educ.* **2019**, *96*, 786–791. [CrossRef]
8. Lee, Y.; Matviychuk, Y.; Holland, D.J. Quantitative analysis using external standards with a benchtop NMR spectrometer. *J. Magn. Reson.* **2020**, *320*, 106826. [CrossRef] [PubMed]
9. Araneda, J.F.; Mendonça Barbosa, T.; Hui, P.; Leclerc, M.C.; Ma, J.; Maier, A.F.G.; Riegel, S.D. Incorporating Benchtop NMR Spectrometers in the Undergraduate Lab: Understanding Resolution and Circumventing Second-Order Effects. *J. Chem. Educ.* **2021**, *98*, 1227–1232. [CrossRef]
10. Araneda, J.F.; Chu, T.; Leclerc, M.C.; Riegel, S.D.; Spingarn, N. Quantitative analysis of cannabinoids using benchtop NMR instruments. *Anal. Methods* **2020**, *12*, 4853–4857. [CrossRef]
11. Assemat, G.; Balayssac, S.; Gerdova, A.; Gilard, V.; Caillet, C.; Williamson, D.; Malet-Martino, M. Benchtop low-field 1H Nuclear Magnetic Resonance for detecting falsified medicines. *Talanta* **2019**, *196*, 163–173. [CrossRef] [PubMed]
12. Keizers, P.H.J.; Bakker, F.; Ferreira, J.; Wackers, P.F.K.; van Kollenburg, D.; van der Aa, E.; van Beers, A. Benchtop NMR spectroscopy in the analysis of substandard and falsified medicines as well as illegal drugs. *J. Pharm. Biomed. Anal.* **2020**, *178*, 112939. [CrossRef]
13. Isaac-Lam, M.F. Determination of alcohol content in alcoholic beverages using 45 MHz benchtop NMR spectrometer. *Int. J. Spectrosc.* **2016**, *2016*, 2526946. [CrossRef] [PubMed]
14. Defernez, M.; Wren, E.; Watson, A.D.; Gunning, Y.; Colquhoun, I.J.; Le Gall, G.; Williamson, D.; Kemsley, E.K. Low-field 1H NMR spectroscopy for distinguishing between arabica and robusta ground roast coffees. *Food Chem.* **2017**, *216*, 106–113. [CrossRef] [PubMed]
15. Sobolev, P.A.; Mannina, L.; Proietti, N.; Carradori, S.; Daglia, M.; Giusti, M.A.; Antiochia, R.; Capitani, D. Untargeted NMR-Based Methodology in the Study of Fruit Metabolites. *Molecules* **2015**, *20*, 4088–4108. [CrossRef] [PubMed]
16. Gunning, Y.; Jackson, A.J.; Colmer, J.; Taous, F.; Philo, M.; Brignall, R.M.; El Ghali, T.; Defernez, M.; Kemsley, E.K. High-throughput screening of argan oil composition and authenticity using benchtop 1H NMR. *Magn. Reson. Chem.* **2020**, *58*, 1177–1186. [PubMed]
17. Gunning, Y.; Taous, F.; El Ghali, T.; Gibbon, J.D.; Wilson, E.; Brignall, R.M.; Kemsley, E.K. Mitigating instrument effects in 60 MHz 1H NMR spectroscopy for authenticity screening of edible oils. *Food Chem.* **2022**, *370*, 131333. [CrossRef]
18. Sobolev, A.P.; Thomas, F.; Donarski, J.; Ingallina, C.; Circi, S.; Marincola, F.C.; Capitani, D.; Mannina, L. Use of NMR applications to tackle future food fraud issues. *Trends Food Sci. Technol.* **2019**, *91*, 347–353. [CrossRef]
19. Antonelo, D.S.; Cônsolo, N.R.B.; Gómez, J.F.M.; Beline, M.; Pavan, B.; Souza, C.; Goulart, R.S.; Colnago, L.A.; Silva, S.L. NMR-based metabolomics to assess metabolites correlated with beef sensory properties. In Proceedings of the 65th International Congress of Meat Science and Technology, Berlin, Germany, 4–9 August 2019; pp. 801–803.
20. Muroya, S.; Ueda, S.; Komatsu, T.; Miyakawa, T.; Ertbjerg, P. MEATabolomics: Muscle and meat metabolomics in domestic animals. *Metabolites* **2020**, *10*, 188. [CrossRef] [PubMed]
21. Pavagadhi, S.; Swarup, S. Metabolomics for evaluating flavor-associated metabolites in plant-based products. *Metabolites* **2020**, *10*, 197. [CrossRef] [PubMed]
22. Hatzakis, E. Nuclear Magnetic Resonance (NMR) Spectroscopy in Food Science: A Comprehensive Review. *Compr. Rev. Food Sci. Food Saf.* **2019**, *18*, 189–220. [CrossRef]
23. Salvino, R.A.; Colella, M.F.; De Luca, G. NMR-based metabolomics analysis of Calabrian citrus fruit juices and its application to industrial process quality control. *Food Control* **2021**, *121*, 107619. [CrossRef]
24. Cirillo, A.; Magri, A.; Scognamiglio, M.; D'Abrosca, B.; Fiorentino, A.; Petriccione, M.; Di Vaio, C. Evaluation of Morphological, Qualitative, and Metabolomic Traits during Fruit Ripening in Pomegranate (*Punica granatum* L.). *Horticulturae* **2022**, *8*, 384. [CrossRef]
25. Migues, I.; Hodos, N.; Moltini, A.I.; Gámbaro, A.; Rivas, F.; Moyna, G.; Heinzen, H. 1H NMR metabolic profiles as selection tools of new mandarin cultivars based on fruit acceptability. *Sci. Hortic.* **2021**, *287*, 110262. [CrossRef]
26. Castaing-Cordier, T.; Ladroue, V.; Besacier, F.; Bulete, A.; Jacquemin, D.; Giraudeau, P.; Farjon, J. High-field and benchtop NMR spectroscopy for the characterization of new psychoactive substances. *Forensic Sci. Int.* **2021**, *321*, 110718. [CrossRef] [PubMed]
27. Alamprese, C. Food Authentication. In *Spectroscopic Methods in Food Analysis*; Franca, A.S., Nollet, L.M.L., Eds.; CRC Press: Miami, FL, USA, 2018; pp. 327–352.

28. Schiffman, S.S.; Booth, B.J.; Losee, M.L.; Pecore, S.D.; Warwick, Z.S. Bitterness of sweeteners as a function of concentration. *Brain Res. Bull.* **1995**, *36*, 505–513. [CrossRef]
29. Stampanoni, C.R. Influence of acid and sugar content on sweetness, sourness and the flavour profile of beverages and sherbets. *Food Qual. Prefer.* **1993**, *4*, 169–176. [CrossRef]
30. Matviychuk, Y.; Yeo, J.; Holland, D.J. A field-invariant method for quantitative analysis with benchtop NMR. *J. Magn. Reson.* **2019**, *298*, 35–47. [CrossRef]

foods

MDPI

Article

Calibration of Near Infrared Spectroscopy of Apples with Different Fruit Sizes to Improve Soluble Solids Content Model Performance

Xiaogang Jiang [1,2], Mingwang Zhu [1,2], Jinliang Yao [1,2], Yuxiang Zhang [3] and Yande Liu [1,2,*]

1 School of Mechatronics & Vehicle Engineering, East China Jiaotong University, Nanchang 330013, China; jxg_ecjtu@163.com (X.J.); zmw15256864150@163.com (M.Z.); a15755567391@163.com (J.Y.)
2 Institute of Intelligent Mechanical and Electrical Equipment Innovation, Nanchang 330013, China
3 Huaiyin College of Technology, Huaiyin 223003, China; yxzhang@hyit.edu.cn
* Correspondence: jxliuyd@163.com

Abstract: The transmission spectrum of apples is affected by the fruit's size, which leads to poor prediction performance of the soluble solids content (SSC) models built for their different apple sizes. In this paper, three sets of near infrared (NIR) spectra of apples with various apple diameters were collected by applying NIR spectroscopy detection equipment to compare the spectra differences among various apple diameter groups. The NIR spectra of apples were corrected by studying the extinction rates within different apples. The corrected spectra were used to develop a partial least squares prediction model for their soluble solids content. Compared with the prediction model of the soluble solids content of apples without size correction, the R_p of PLSR improved from 0.769 to 0.869 and RMSEP declined from 0.990 to 0.721 in the small fruit diameter group; the R_p of PLSR improved from 0.787 to 0.932 and RMSEP declined from 0.878 to 0.531 in the large fruit diameter group. The proposed apple spectra correction method is effective and can be used to reduce the influence of sample diameter on NIR spectra.

Keywords: apple; NIR; size correction; extinction coefficient; fruit diameter difference

Citation: Jiang, X.; Zhu, M.; Yao, J.; Zhang, Y.; Liu, Y. Calibration of Near Infrared Spectroscopy of Apples with Different Fruit Sizes to Improve Soluble Solids Content Model Performance. *Foods* **2022**, *11*, 1923. https://doi.org/10.3390/foods11131923

Academic Editors: Zhiming Guo, Zhao Zhang and Dong Hu

Received: 4 June 2022
Accepted: 24 June 2022
Published: 28 June 2022

1. Introduction

Apples are known for their texture, flavor, visual effect, and nutritional value [1,2]. NIR spectroscopy has become the representative and main development direction of modern non-destructive testing with its unique advantages of simplicity, efficiency, and non-destructiveness, and is an effective way to solve the classification of agricultural products [3–6]. The application of NIR spectroscopy in fruit and vegetable quality inspection has been reported, mainly focusing on citrus, apple, pear, tomato, and other species [7–10].

Many scholars have studied the application of NIR spectroscopy in the internal quality of fruits. In terms of algorithms, Travers et al. [11] developed partial least squares (PLS) models based on spectra in the wavelength ranges of 680–1000 nm and 1100–2350 nm, respectively, after extracting the characteristic wavelengths using the competitive adaptive re-weighted sampling algorithm (CARS) for predicting the dry matter (DM) and SSC of pears. The feature wavelengths selected by CARS successfully highlighted the differences between the prediction models based on the two different spectral ranges. In near infrared spectroscopy research, the relationship between near-infrared reflectance and transmittance spectra of kiwifruit and soluble solids was investigated by Schaare et al. [12]. The analysis showed that modeling using transmittance spectra was better than reflectance, with a test set correlation coefficient of 0.961 and a test set root mean square error of 0.8%. Tian et al. [13] used NIR spectra to predict nuclear mold in apples of different fruit sizes, correcting the NIR spectra of apples with different degrees of disease. The accuracy of apple disease degree prediction established by the corrected spectra was up to 90%. In terms of actual testing, Arana et al. [14] examined the soluble solids of white grapes and the method chosen was NIR spectroscopy, which achieved a good prediction of the soluble solids of white grapes. Jha et al. [15] examined the internal quality of seven Indian mangoes by NIR spectroscopy between 1200 nm and 2200 nm for both soluble solids and acidity and established PLS with R_p of only 0.715 and 0.703. Liu et al. [16] developed a generalized UVE-PLS model for apple brix by collecting diffuse transmission spectra of red Fuji apples from three different locations, namely Qixia, Luochuan, and Huining, highlighting the potential of spectroscopic techniques for fruit quality detection in different origins. Antonucci et al. [17] conducted a study on the internal quality of oranges by spectroscopic techniques and achieved good results in predicting their acidity and soluble solids by regression analysis using the PLS model, with correlation coefficients of 0.843 and 0.812 for soluble solids and acidity of oranges, respectively. Ni et al. [18] performed the NIR spectral model transfer of different instruments by filtering the wavelength information of different NIR instruments. Two datasets of maize and scutellaria samples measured by different NIR instruments were used to test the performance of the method, where the overall prediction performance of the SWCSS-PLS model for the secondary measurement samples was much better than that of the full-wavelength PLS model. Meng Qinglong et al. [19] collected the reflection spectra of fresh "Fuji" apples from 400 to 1000 nm, and used different pretreatment and different characteristic wavelength screening methods to establish various models to predict the SSC content of apples. It can be better used for the detection of apple SSC. The above studies did not consider the sample size. Ideally, the samples for NIR modeling should include all of the variables affecting the NIR spectra, but this is very difficult for the detection of complex variables in the internal quality of apples. If more variability samples are included in the model, the prediction accuracy of the model decreases and further confirmation is needed to meet the requirements. In this paper, it was found that the light intensity of the transmittance spectra of the internal pulp of apples showed a log-linear relationship with their fruit diameter. Therefore, a size-correction method for apples is proposed, based on which the NIR spectra of all apples with different fruit diameters are transformed into a single spectrum that is used to eliminate the effect of size on the performance of the apple SSC prediction model. Compared with the apple soluble solids content prediction model without size correction, the proposed apple size correction method effectively solves the problem of poor prediction accuracy of apple SSC model due to apple size.

2. Materials and Methods

2.1. Test Materials

Apples were harvested from a red Fuji apple orchard with 480 apples divided into three fruit size groups (65–75 mm, 75–85 mm, and 85–95 mm), with a total of 160 values under every fruit size set. Spectra information and SSC were collected at the markers.

2.2. Spectral Acquisition

The spectra of apples were collected by near-infrared online inspection equipment [20], and the structure of the equipment is shown in Figure 1. The spectrometer was a high-precision spectrometer (QE65Pro, Ocean, Manhattan, NY, USA). The light source system uses 100 W Osram halogen lamps evenly distributed along both sides of the main drive chain, with five lamps on each side. To achieve a stable value of light source attenuation, a white Teflon sphere was used as a reference to calibrate the NIR spectroscopy online detection device after 30 min of warm-up each time the power was turned on. The current value of the regulated power supply was adjusted and the range of energy spectrum change of the transmission spectrum was observed until the standard deviation of the adjacent energy spectrum intensity was within 1% and the the NIR spectra are reproducible, and the apple sample spectrum acquisition was started. The parameters of the spectrum acquisition were set as follows: duration time of 100 ms, motion speed of 5 m/s, and spectral wavelength range of 370–1150 nm.

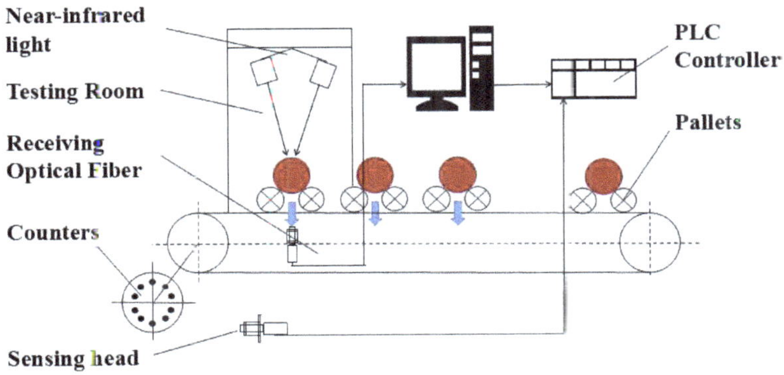

Figure 1. Near-infrared spectra acquisition device.

2.3. SSC Measurement of Samples

After collecting NIR spectra of apple samples, 1-cm-thick slices were cut along the equator of the apples and subsequently divided into 4 equal portions according to the label. The SSC of the extracted apple juice was determined with a saccharimeter (PR-101a, ATAGO, Nagasaki, Japan). The measurements were repeated three times to take the average value as the final SSC value.

2.4. Data Processing

The 160 sample spectra under each fruit diameter group were divided into a calibration set (120) and a prediction set (40) using the Kennard-Stone (K-S) algorithm. Since the NIR spectral data matrix of each fruit diameter group is 160 × 1044, to reduce the errors caused by non-experimental factors, this study used Unscrambler (Version 9.7, CAMO, City of Oslo, Norway) software to process the spectra using different pretreatment methods (Multiple scattering correction, MSC; standard normal variable transformation, SNV; Savitzky-Golay smoothing, S-G smothing). The partial least squares (PLS) method was then used to establish the apple SSC detection model.

Partial least squares regression (PLSR) is widely used in NIR spectral analysis to decompose the spectral array X and the concentration array Y simultaneously to strengthen the corresponding computational relationship and ensure the best model is obtained. The PLS regression model is shown in Equation (1):

$$Y = bX + e \tag{1}$$

where b denotes the vector of regression coefficients and e denotes the model residuals.

The performance of the model is judged by the correlation coefficient R_p and the root mean square error value (RMSEP). Equations of R_p can be found in Equation (2) and RMSEP can be found in Equation (3).

$$R_p = \sqrt{1 - \frac{\sum_{i=1}^{n}(y_i - \hat{y}_i)}{\sum_{i=1}^{n}(y_i - \overline{y})}} \tag{2}$$

$$\text{RMSEP} = \sqrt{\frac{1}{n-1}\sum_{i=1}^{n}(y_i - \hat{y}_i)^2} \tag{3}$$

where n is the number of experimental samples, y_i is the actual value of the i-th sample in the prediction set measured by the standard method, $\hat{y}_i$ is the predicted value of the i-th sample in the prediction set measured by NIR spectroscopy and mathematical model, and $\overline{y}_i$ is the mean value of the SSC of all apples in the prediction set.

3. Results and Analysis

3.1. Sample Chemical Index Statistics Results

The 160 apple spectra under the calibration set (120) and the prediction set (40) by the K-S algorithm were sorted and the sorted apple SSC values are presented in Table 1. The SSC range of the modeling set under each fruit size group was larger than the SSC range of the prediction set, which allows for improved forecasting of apple SSC.

Table 1. SSC values for apples of different fruit sizes.

Fruit Size Range	Data Type	Number of Samples	Max/Brix	Min/Brix	Mean/Brix	Deviation
65–75 mm	Calibration Set	120	17.2	9.8	13.8	1.62
	Prediction set	40	16.7	10.9	13.8	1.24
75–85 mm	Calibration Set	120	17.2	8.3	13.2	1.60
	Prediction set	40	16.5	8.7	12.73	1.78
85–95 mm	Calibration Set	120	15.3	10.9	13.3	1.73
	Prediction set	40	15.3	11.3	13.3	1.92

3.2. Near-Infrared Spectra of Three Groups of Fruit Size Apples

The mean spectra of apples under each fruit size group are shown in Figure 2. With the increase of the fruit diameter, the corresponding spectra energy of apples is smaller. The strongest spectra energy was collected from 65–75 mm apples, and the weakest spectra energy was collected from 85–95 mm apples. The effective wavelength range was set from 350 to 850 nm due to the weak signal and little effective information at both ends of the spectra. The spectral trends of apples with different fruit sizes were the same, with differences in absorption intensity. The spectral curves showed prominent absorption peaks near 645, 710, and 810 nm, and troughs near 675, 758, and 830 nm, respectively. The absorption peak at 645 nm was mainly influenced by the color of the epidermis [21], the absorption of chlorophyll near 675 nm might be the absorption of the chlorophyll [22], and the trough near 758 nm was related to the O-H triplet stretching vibration [23], and the weaker trough near 830 nm was related to the N-H triplet stretching vibration [24].

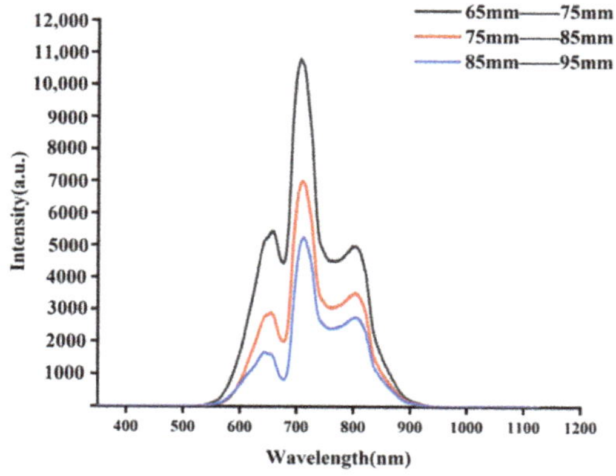

Figure 2. Raw spectra of three sets of fruit diameter samples.

As can be seen in Figure 2, the intensity of the NIR spectra is decreasing as the fruit diameter increases. This is due to the fact that there is an attenuation of the NIR light intensity due to the flesh of the fruit when the light is transmitted inside the apple.

The degree of attenuation of the NIR light intensity increases with the increase of the light range. The degree of attenuation of NIR light intensity inside the apple shows a logarithmic relationship with the apple fruit diameter [25], which can be fitted as a function of Equation (4).

$$I = I_0 \exp(-u_e d) \quad (4)$$

where I_0 is the light intensity emitted by the NIR source, I is the received NIR spectral intensity, d is the light transmission length, and u_e is the attenuation degree factor of NIR spectra.

In the spectral acquisition device shown in Figure 1, d is the fruit diameter of the apple. As d increases, the NIR light intensity becomes more attenuated during the propagation inside the apple, and the intensity of the obtained apple NIR spectra becomes smaller. It can be seen that the difference in apple fruit diameter will affect the light intensity of its NIR spectra.

3.3. PLSR Results of SSC for Mixed Apple Size

To verify that apple size differences affect their NIR spectra and lead to poor prediction performance of the developed apple SSC prediction model The calibration and prediction sets of fruit diameter groups 65–75 mm, 75–85 mm, and 85–95 mm were used as the calibration and prediction sets of the mixed fruit diameter apple prediction model to establish the PLSR of SSC for different apple sizes, and the model effects are shown in Table 2.

Table 2. PLSR findings for SSC of mixed apple size groups.

Number of Calibration Set	Number of Prediction Set	R_c	RMSEC	R_p	RMSEP
360	120	0.733	1.011	0.722	1.086

As can be seen from Table 2, the SSC model built with mixed apple fruit diameter has a poor prediction performance with an R_p of 0.722 and an RMSEP of 1.086. When there is a large difference in apple fruit diameter in the model, it will cause the problem of poor prediction performance of the established model, so size correction of apple fruit diameter is needed to improve the prediction performance of its SSC model.

3.4. PLSR Results of SSC for Each Fruit Size Set

The number of LVs in the PLSR model was set from 1 to 20 to prevent the overfitting or underfitting of the model. Table 3 shows the PLSR results established for individual fruit size sets after several pretreatment methods.

From Table 3, it can be seen that the PLSR prediction performance of the SSC established by SNV pretreatment of apple NIR spectra for the three sets of fruit sizes is the best and the correlation coefficients R_p of the models are 0.863, 0.947, and 0.917, respectively, and the root mean square error values RMSEP of the prediction sets are 0.771, 0.622, and 0.752, respectively. The scatter plot of PLSR prediction is shown in Figure 3. It can be seen that the pretreatment method SNV can eliminate the effect of sample particle size on NIR spectra [26], thus solving the influence of spectral dispersion due to the unequal sample dimensions.

3.5. Individual Fruit Size Groups Predicted Other Fruit Size Groups

It can be seen from Table 3, the PLSR prediction performance of SSC established when the apple size was 75–85 mm was better. To investigate whether the prediction performance of the model could be achieved when more variance samples were included in the model, the PLSR results for the remaining two fruit size groups using the medium apple size group to predict the SSC are shown in Table 4.

Table 3. PLSR findings for SSC of various apple size groups after spectra pretreatment.

Fruit Size Range	Pretreatment	LVs	R_c	RMSEC	R_p	RMSEP
65–75 mm	Original	11	0.931	0.592	0.853	0.786
	MSC	8	0.902	0.700	0.857	0.785
	SNV	12	0.972	0.376	0.863	0.771
	S-G smoothing	12	0.907	0.683	0.854	0.794
75–85 mm	Original	10	0.951	0.534	0.941	0.654
	MSC	9	0.964	0.462	0.941	0.654
	SNV	11	0.976	0.373	0.947	0.622
	S-G smoothing	11	0.950	0.540	0.937	0.677
85–95 mm	Original	10	0.916	0.389	0.898	0.827
	MSC	9	0.908	0.371	0.814	0.817
	SNV	10	0.936	0.295	0.917	0.752
	S-G smoothing	11	0.909	0.369	0.854	0.852

MSC: multivariate scattering correction; SNV: standard normal variables transformation; S-G smoothing: Savitzky-Golay smoothing; Lvs, latent variable individual.

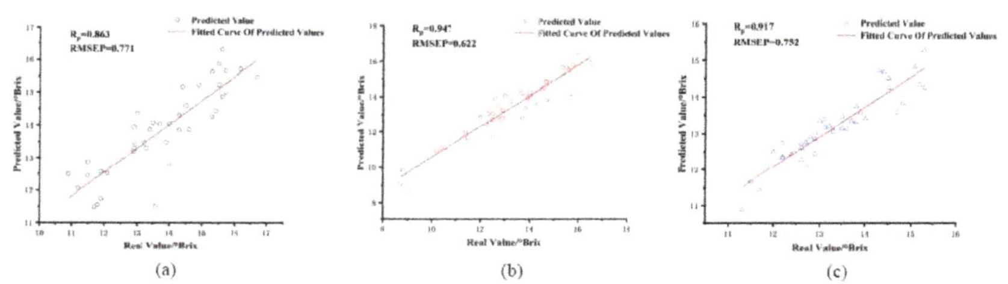

Figure 3. Scatter plot of PLSR prediction of apple SSC for three apple diameter sets. (**a**) apple size 65–75 mm, (**b**) apple size 75–85 mm, and (**c**) apple size 85–95 mm.

Table 4. The fruit size group alone predicted the SSC results of other fruit size groups.

Calibration Set	Prediction Set	R_c	RMSEC	R_p	RMSEP
75–85 mm	65–75 mm	0.951	0.534	0.769	0.990
	85–95 mm	0.958	0.412	0.787	0.878

As can be seen from Table 4, the PLSR prediction performance of soluble solids content built from the modeling and prediction sets was poor when the difference in apple fruit size was significant between them. Compared with Table 3, the correlation coefficient R_p decreased from 0.863 to 0.769 and the root means square error value RMSEP increased from 0.771 to 0.990 for the PLSR of the small fruit size group. The correlation coefficient R_p decreased from 0.917 to 0.787 and the root means square error value RMSEP increased from 0.752 to 878 for the PLSR of the large fruit size group. The scatter plot of its PLSR is shown in Figure 4. Apple size differences significantly impacted the accuracy of the SSC model. With the same variety of apples, there will be differences in volume size, and the size differences will affect the detection performance when performing NIR spectroscopy, so it is necessary to correct the NIR spectra of apples of different sizes to improve the detection performance of NIR detection equipment.

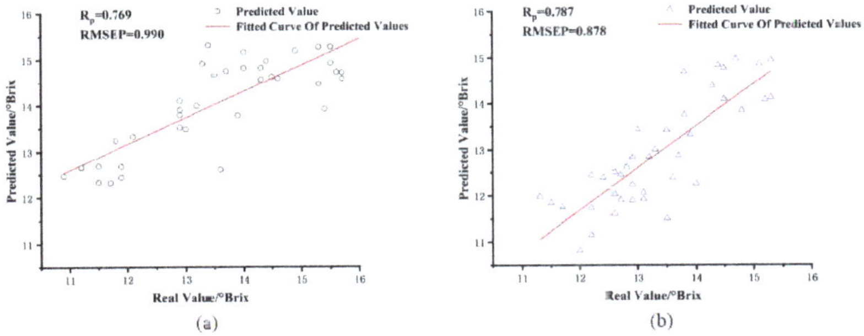

Figure 4. Scatter plot of PLSR with different fruit diameter of modeling set and prediction set. (**a**) 75–85 mm apple size group predicted 65–75 mm apple size group, (**b**) 75–85 mm apple size group predicted 85–95 mm apple size group.

3.6. Correction of Near Infrared Spectroscopy for Apples of Various Diameters

From Formula (4), apple size affects the prediction model of soluble solids content, and the light intensity of the apple and its fruit diameter are logarithmic functions. The deformation of Formula (4) can be obtained as Formula (5).

$$\ln(I) = \ln(I_0) - u_e d \tag{5}$$

If the light intensity of the apple at its two internal depths d_1 and d_2 are I_1 and I_2, respectively, Formula (5) can be deformed as:

$$-u_e = \frac{\ln(I_1) - \ln(I_2)}{d_1 - d_2} \tag{6}$$

From Formula (5), we can find the extinction coefficient of apples or the collection method in Figure 1, the light range d is the fruit diameter at the equator of apples, and I is the light intensity of the transmission spectra of apples collected by the fiber optic probe. From Table 3, it can be seen that the PLSR performance of SSC with medium apple size is better, so the average spectra of the medium fruit size group are taken as IR and the average fruit size of the medium fruit size group is taken as d_1, and the average extinction coefficients of all samples can be obtained as shown in the following Formula (7).

$$-u_e = \frac{\sum_{i=1}^{n}\left(\frac{\ln(I_R)-\ln(I_i)}{d_R-d_i}\right)}{n} \tag{7}$$

where I_R is the reference spectra, I_i is the spectra of the apple sample i-th, d_R is the reference fruit diameter, d_i is the average fruit size of the apple sample i, n is the number of samples, and the extinction coefficient applicable to all apple samples can be obtained from Formula (7). The inverse operation of Formula (7) leads to Formula (8).

$$I_i^* = \exp(-u_e(d_R - d_i) + \ln(I_i)) \tag{8}$$

where I_i^* is the size-corrected spectra of apple sample i according to its fruit diameter. The size-corrected spectra of all samples can be obtained according to Formula (8), and their average size-corrected spectra of different apple fruit diameter groups are shown in Figure 5.

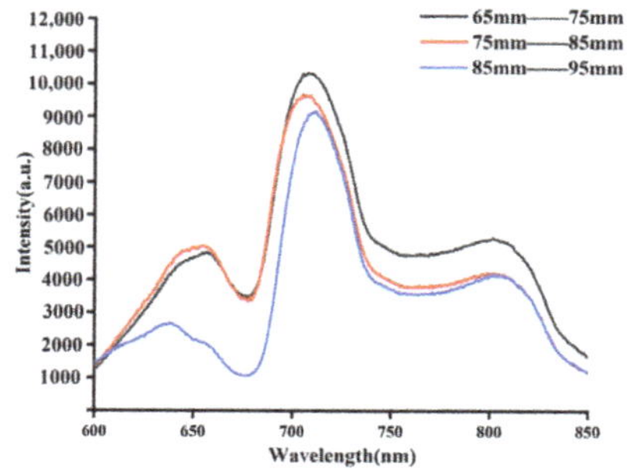

Figure 5. Apple spectra after size correction.

Compared with the uncorrected apple spectra, the size-corrected apple spectra have a cross-over phenomenon in the spectra of each fruit diameter group, and the spacing between the vertical aspects of each fruit diameter group in the spectra is reduced compared with Figure 2. This spacing exists as a result of the differences in apple fruit diameter. Our proposed size correction approach was used to correct the near-infrared spectra for apples of various dimensions.

The corrected NIR spectra of apples were used to build the PLSR of SSC with different fruit sizes. The large and small fruit size groups were predicted using the medium fruit size group, and the predicted results are shown in Table 5.

Table 5. PLSR results were established after the spectra correction.

Calibration Set	Prediction Set	R_c	RMSEC	R_p	RMSEP
75–85 mm	65–75 mm	0.951	0.570	0.869	0.721
	85–95 mm	0.969	0.459	0.932	0.531

As can be seen from Tables 4 and 5, the model prediction performance of the corrected NIR spectra compared to the PLSR built for the original apple spectra was significantly improved. Among them, the correlation coefficient R_p of the PLSR established for the small fruit size group improved from 0.769 to 0.869, and RMSEP decreased from 0.990 to 0.721. the correlation coefficient R_p of the PLSR established for the large fruit size group improved from 0.787 to 0.932. the RMSEP decreased from 0.878 to 0.531. the PLSR of the two fruit size groups scatters plots are shown in Figure 6. The results show that after the spectral correction of Formula (7), the spectra of apples of different sizes can be converted into a standard spectrum to correct the NIR spectra of apples of different sizes, which is used to improve the performance of its prediction model. Similarly, this spectral correction method can be applied to other fruits such as pear, citrus, and watermelon.

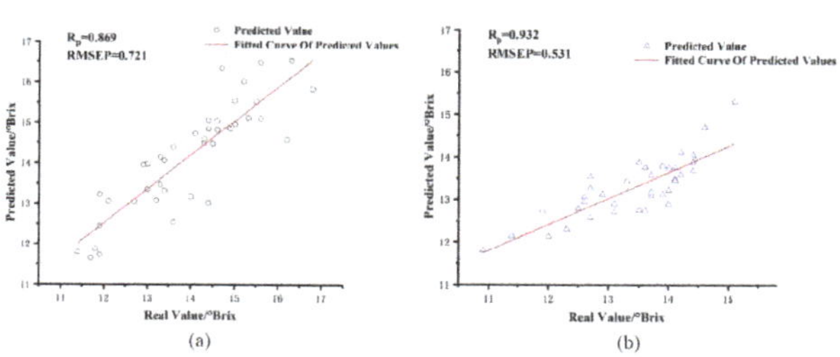

Figure 6. The PLSR scatter plot was created by the corrected spectra. (**a**) 75–85 mm fruit size group predicted 65–75 mm fruit size group, (**b**) 75–85 mm fruit size group predicted 85–95 mm fruit size group.

4. Conclusions

This paper presents a method that can correct the near-infrared spectra of apples with various dimensions to enhance the performance of SSC prediction models. The transmission spectra of apples of different sizes were converted to the same standard to obtain the extinction coefficients of transmitted light. The transmission spectra were corrected according to the average extinction coefficients of apples, and the corrected spectra were employed to model the soluble solids content of apples. Compared with the model of the soluble solids content of apples without size correction, the correlation coefficient R_p of PLSR for the small fruit diameter group increased from 0.769 to 0.869 and RMSEP decreased from 0.990 to 0.721. The correlation coefficient R_p of PLSR for the large fruit size group increased from 0.787 to 0.932 and RMSEP decreased from 0.878 to 0.531. The apple size correction method proposed in this paper is reliable. By improving the model algorithm in the application of NIR online inspection device, the standard spectra are set to size correct the apple spectra of different fruit diameters and reduce the influence of apple fruit size variation on its SSC model.

Author Contributions: X.J.: research, writing, theoretical guidance. M.Z.: preparation, methodology. J.Y.: grammar check. Y.Z.: format detection. Y.L.: format detection. All authors have read and agreed to the published version of the manuscript.

Funding: This research was funded by the National Natural Science Foundation of China (grant number: 31760344); the Science and Technology Research Project of Education, Department of Jiangxi Province (grant number: GJJ200615), and the APC was funded by the National Natural Science Foundation of China.

Data Availability Statement: The data presented in this study are available on request from the corresponding author.

Conflicts of Interest: The authors declare no conflict of interest.

References

1. Gao, J.F.; Zhang, H.L.; Kong, W.W.; He, Y. Nondestructive Discrimination of Waxed Apples Based on Hyperspectral Imaging Technology. *Spectrosc. Spectr. Anal.* **2013**, *33*, 1922–1926.
2. Harker, F.R.; Gunson, F.A.; Jaeger, S.R. The case for fruit quality: An interpretive review of consumer attitudes and preferences for apples. *Postharvest Biol. Technol.* **2003**, *28*, 333–347. [CrossRef]
3. Zhao, J.; Kuang, L.X.; Xu, F.X.; Su, Y.; Feng, X.Q. Different placement methods of oranges on near red Effect of the variability of the effective acidity model of external NDEs. *Chin. J. Food* **2014**, *14*, 246–250.
4. Caramês, E.T.S.; Alamar, P.D.; Poppi, R.J.; Pallone, J.A.L. Quality control of cashew apple and guava nectar by near infrared spectroscopy. *J. Food Compos. Anal.* **2017**, *56*, 41–46. [CrossRef]
5. Clark, C.J.; Mcglone, A.V.; Jordan, R.B. Detection of Brownheart in 'Braeburn' apple by transmission NIR spectroscopy. *Postharvest Biol. Technol.* **2003**, *28*, 87–96. [CrossRef]
6. Gong, Y.J.; Pei, J.Q.; Li, H.B.; Feng, Y.L.; Ning, X.F. Design of Portable Rapid Non-Destructive Testing System for Apple Quality. *J. Shenyang Agric. Univ.* **2017**, *48*, 238–243.
7. McGlone, V.A.; Martinsen, P.J.; Clark, C.J.; Jordan, R.B. On-line detection of Brownheart in Braeburn apples using near infrared transmission measurements. *Postharvest Biol. Technol.* **2005**, *37*, 142–151. [CrossRef]
8. Li, L.S.; Liu, Y.D.; Hu, J.; Song, Y.; Zhu, X.R.; Huang, M.; Zhou, X.Q. Application of near-infrared nondestructive testing technology in fruit ripeness discrimination. *J. East China Jiaotong Univ.* **2021**, *38*, 95–105.
9. McGlone, V.A.; Martinsen, P.J. Transmission Measurements on Intact Apples Moving at High Speed. *J. Near Infrared Spectrosc.* **2004**, *12*, 37–43. [CrossRef]
10. Sun, T.; Mo, X.X.; Liu, M.H. Effect of fruit peel on the accuracy of visible/near infrared detection of soluble solids in navel oranges. *Spectrosc. Spect. Anal.* **2018**, *38*, 1406–1412.
11. Travers, S.; Bertelsen, M.G.; Petersen, K.K. Predicting pear (cv. Clara Frijs) dry matter and soluble solids content with near infrared spectroscopy. *LWT-Food Sci. Technol.* **2014**, *59*, 1107–1113. [CrossRef]
12. Schaare, P.N.; Fraser, D.G. Comparison of reflectance, interactance and transmission modes of visible-near infrared spectroscopy for measuring internal properties of kiwifruit (*Actinidia chinensis*). *Postharvest Biol. Technol.* **2000**, *20*, 175–184. [CrossRef]
13. Tian, S.; Zhang, J.; Zhao, J.; Zhang, Z.; Zhang, H. Effective modification through transmission Vis/NIR spectra affected by fruit size to improve the prediction of moldy apple core. *Infrared Phys. Technol.* **2019**, *100*, 117–124. [CrossRef]
14. Arana, I.; Jarén, C.; Arazuri, S. Maturity, variety and origin determination in white grapes (*Vitis Vinifera* L.) using near infrared reflectance technology. *J. Near Infrared Spec.* **2005**, *13*, 349–359. [CrossRef]

15. Jha, S.N.; Jaiswal, P.; Narsaiah, K. Non-destructive prediction of sweetness of intact mango using near infrared spectroscopy. *Sci. Hortic.* **2012**, *138*, 171–175. [CrossRef]
16. Liu, Y.D.; Xu, H.; Sun, X.D.; Jiang, X.G.; Rao, Y.; Xu, J.; Wang, J.Z. Visible near-infrared spectroscopy for online detection of sugar content of apples of different origins. *Chin. J. Optics.* **2020**, *13*, 482–491.
17. Antonucci, F.; Pallottino, F.; Paglia, G.; Palma, A.; D€Aquino, S.; Menesatti, P. Non-destructive estimation of mandarin maturity status through portable VIS-NIR spectrophotometer. *Food Bioprocess Technol.* **2011**, *4*, 809–813. [CrossRef]
18. Ni, L.; Han, M.; Luan, S. Screening wavelengths with consistent and stable signals to realize calibration model transfer of near infrared spectra. *Spectrochim. Acta Part A Mol. Biomol. Spectrosc.* **2019**, *206*, 350–358. [CrossRef]
19. Meng, Q.L.; Sang, J.; Huang, R.S.; Chen, L.T.; Zhang, Y. Visible/NIR NDT of soluble solids in apples. *Food Ferment. Ind.* **2020**, *46*, 205–209.
20. Liu, Y.D.; Zhang, Y.; Jiang, X.G. Determination of hardness and soluble solids content of peach in different storage periods. *Spectrosc. Spectr. Anal.* **2021**, *41*, 306–312.
21. Murray, I.; Williams, P. *Near Infrared Technology in the Agricultural and Food Industries*; American Association of Cereal Chemists, Inc.: St. Paul, MN, USA, 1987; pp. 17–34.
22. McDevitt, R.M.; Gavin, A.J.; Andrés, S. The ability of visible and near-infrared reflectance spectroscopy (NIRS) to predict the chemical composition of ground chicken carcasses and to discriminate between carcasses from different enotypes. *J. Near Infrared Spectroscopy* **2005**, *13*, 109–117. [CrossRef]
23. Zhou, L.J.; Wu, H.J.; Li, T.; Wang, Z.Y. Determination of fatty acids in broiler breast meat by near-infrared reflectance spectroscopy. *Meat Sci.* **2012**, *90*, 658–664. [CrossRef] [PubMed]
24. Prieto, N.; Ross, D.W.; Navajas, R.I. Online prediction of fatty acid profiles in crossbred Limousin and Aberdeen Angus beef cattle using near infrared reflectance spectroscopy. *Animal* **2011**, *5*, 155–165. [CrossRef]
25. Fraser, D.G.; Kunnemeyer, R.; McGlone, V.A.; Jordan, R.B. Near infrared (NIR) light penetration into an apple. *Postharvest Biol. Technol.* **2001**, *22*, 191–194. [CrossRef]
26. Li, J.; Sun, S.K.; Jiang, L.W. Study on nondestructive identification method of green tea based on near infrared spectroscopy and chemometrics. *J. Anal. Test.* **2020**, *39*, 1344–1350.

foods

MDPI

Article

Assembled Reduced Graphene Oxide/Tungsten Diselenide/Pd Heterojunction with Matching Energy Bands for Quick Banana Ripeness Detection

Xian Li [1], Chengcheng Xu [2], Xiaosong Du [2], Zhen Wang [1], Wenjun Huang [2], Jie Sun [2], Yang Wang [2,*] and Zhemin Li [1,3,*]

[1] Agricultural Information Institute, Chinese Academy of Agricultural Sciences, Beijing 100081, China; lixian@caas.cn (X.L.); zhenskar@163.com (Z.W.)
[2] State Key Laboratory of Electronic Thin Films and Integrated Devices, School of Optoelectronic Science and Engineering, University of Electronic Science and Technology of China, Chengdu 610054, China; xcc_uestc@163.com (C.X.); xsdu@uestc.edu.cn (X.D.); 202022050434@std.edu.cn (W.H.); sjjya@163.com (J.S.)
[3] Graduate School of Chinese Academy of Agricultural Sciences, Beijing 100081, China
* Correspondence: landlord@uestc.edu.cn (Y.W.); lizhemin@caas.cn (Z.L.)

Abstract: The monitoring of ethylene is of great importance to fruit and vegetable quality, yet routine techniques rely on manual and complex operation. Herein, a chemiresistive ethylene sensor based on reduced graphene oxide (rGO)/tungsten diselenide (WSe_2)/Pd heterojunctions was designed for room-temperature (RT) ethylene detection. The sensor exhibited high sensitivity and quick p-type response/recovery (33/13 s) to 10–100 ppm ethylene at RT, and full reversibility and excellent selectivity to ethylene were also achieved. Such excellent ethylene sensing behaviors could be attributed to the synergistic effects of ethylene adsorption abilities derived from the negative adsorption energy and the promoted electron transfer across the WSe_2/Pd and rGO/WSe_2 interfaces through band energy alignment. Furthermore, its application feasibility to banana ripeness detection was verified by comparison with routine technique through simulation experiments. This work provides a feasible methodology toward designing and fabricating RT ethylene sensors, and may greatly push forward the development of modernized intelligent agriculture.

Keywords: fruit quality monitoring; room-temperature ethylene sensor; density functional theory; adsorption energy; band energy alignment

Citation: Li, X.; Xu, C.; Du, X.; Wang, Z.; Huang, W.; Sun, J.; Wang, Y.; Li, Z. Assembled Reduced Graphene Oxide/Tungsten Diselenide/Pd Heterojunction with Matching Energy Bands for Quick Banana Ripeness Detection. *Foods* **2022**, *11*, 1879. https://doi.org/10.3390/foods11131879

Academic Editor: Seung-Hyun Kim

Received: 20 May 2022
Accepted: 23 June 2022
Published: 24 June 2022

1. Introduction

Ethylene is an important plant hormone that regulates the physiological and biochemical changes in climacteric fruits and vegetables to control their maturity, freshness, softness, and deterioration [1,2]. The released ethylene can potentially express the flavor quality of fruits coupled with sugar content [3,4]. The accumulated ethylene molecules inside a fresh fruit package could also stimulate physiological activity and consequently, accelerate fruit deterioration, which limits their storage life and leads to product losses [5]. Moreover, the wounding and spoilage of fruits also induces the biosynthesis of ethylene [6]. According to a report from the Food and Agriculture Organization (FAO), 1.3 billion tons of food loss per year was reported, which represented 33% of total food production, among which fruits and vegetables held the highest loss rate (45%) [7]. Fruit abnormalities in early stages can be discovered promptly through ethylene monitoring and thus most of these losses could be avoided. Moreover, for the timely export of fresh climacteric fruits, the production as well as respiration of ethylene should also be of concern during long-supply chains [8]. Therefore, the continuous and accurate detection/monitoring of ethylene released from fruits is vital for managing and controlling the harvesting, storage, package, transportation, and selling process of climacteric fruits, which is much more prominent especially in today's intelligent agriculture era.

So far, various techniques including chromatography [9], spectroscopy [10,11], electrochemical sensor [12], chemical sensor [13], and fluorescence probe [14] have been reported for ethylene detection, among which chemical sensor stands out, owing to its real-time response, high sensitivity, manpower operation, and low cost. Recently, many attempts have been made to develop high-performance chemiresistive ethylene sensors, in which the gas concentration is translated into a resistive electrical signal for detection. Bulk or nanostructured metal oxide semiconductors (MOSs, WO_3 [15], SnO_2 [16], and ZnO [17]) have been reported to show high sensitivity to ethylene. However, in metal oxide-based ethylene sensors, high temperature (typically 170–500 °C) or light irradiation is usually required to activate the reaction between adsorbed oxygen and ethylene, which results in high system complexity and high power consumption and thus greatly hinders its practical applications. Therefore, how to achieve high sensitivity to ethylene without additional activation energy is of great urgency and importance for practical agricultural applications due to its merits in reducing power consumption and system complexity. For MOSs, doping with noble metals such as Au, Pt, Pd [18–20], and transition metal halides ($CuCl_2$, $NiCl_2$ [21]) has been demonstrated to greatly lower the operating temperature. However, how to achieve excellent ethylene sensing performances at room temperature is still challenging.

The construction of a sensing material system for room-temperature ethylene detection should be considered from the following two aspects. Firstly, the gas sensing response arises from the physical adsorption of ethylene molecules onto the sensing films, and thus the adsorption capabilities of the sensing films should be optimized to achieve high sensitivity toward ethylene. It has been theoretically proven that the negative adsorption energy of the target analyte-sensing film system is beneficial for the adsorption of target analyte molecules, which also has been proven experimentally [22,23]. Secondly, when the target analyte molecules were adsorbed onto the sensing film, the electron transfer takes place between the target analyte and sensing film [24,25]. How to translate this electron transfer process into electrical resistance change greatly depends on the energy level alignment of the sensing material system, where suitable energy alignment promotes the electron transfer and thus results in a larger sensing response.

Following this regard, the ternary reduced graphene oxide (rGO)/tungsten diselenide (WSe_2)/Pd heterojunctions were designed and fabricated toward room-temperature ethylene detection. The negative adsorption energy of the ternary heterojunctions provides enough active sites for ethylene molecules adsorption, and the electron transfer across the rGO/WSe_2 and WSe_2/Pd interfaces through band energy alignment greatly promotes the sensing response. Compared to the solely one or two components-based heterojunctions, the ternary heterojunction-based ethylene sensor exhibits higher sensitivity and quicker p-type response to the ppm level of ethylene at room temperature, and the sensitivity to 10 ppm of ethylene was 0.001%, with the response and recovery time being 33 and 13 s. Moreover, the sensor exhibits full reversibility and excellent selectivity at room temperature. Furthermore, the application feasibility of the sensor to fruit quality monitoring was verified by comparison with routine techniques through banana ripeness detection simulation experiments. This work provides a feasible methodology for designing and fabricating a sensing material system toward room-temperature ethylene detection, pushing forward the development of modernized intelligent agriculture.

2. Materials and Methods

2.1. Preparation

GO aqueous solution (0.5 mg/mL, Hangzhou Gaoxi Tech Co., Ltd., Hangzhou, China), WSe_2 nanosheets dispersion (0.5 mg/mL, Nanjing MKNANO Tec Co., Ltd., Nanjing, China), and Pd nanoparticles (NPs, 99.9% metals basis, ≤1 μm, Shanghai Aladdin Biochemical Tech Co., Ltd., Shanghai, China) were prepared. Polyelectrolytes poly (diallyldimethylammonium chloride) (PDDA, 200,000–350,000, 20 wt. % aqueous solution, polycation), and poly (sodium-pstyrenesulfonate) (PSS, 70,000, polyanion) were purchased from Sigma-Aldrich. PDDA and PSS were diluted and dissolved in deionized (DI) water to 1 wt. % and

2 wt. %, respectively. Pd NPs were then dispersed in PSS solution with a concentration of 0.5 mg/mL. All reagents were directly used without further purification.

2.2. Sensor Fabrication

Ethylene sensitive films were deposited on SiO_2/Si substrates pre-patterned with Ti/Au (20 nm/50 nm) interdigital electrodes (IDEs) by the layer-by-layer self-assembling method. The cleaned substrate was vertically immersed in PDDA solution for 15 min and dried with nitrogen, followed by immersing in DI water for removal of excessive PDDA molecules. Therefore, positive charges were fixed on the surface of substrates. Since the GO, WSe_2, and PSS-Pd solutions were negatively charged, the positive charged substrate was then vertically immersed in negatively charged solution for 15 min for self-assembling of nanostructures on IDEs due to the electrostatic adsorption effects. Finally, the sensing films were annealed at 200 °C for 1 h to reduce GO.

2.3. Instruments and Measurements

The surface morphology and microstructure of sensing materials and films were observed by scanning emission microscopy (SEM, Zeiss Gemini, Oberkochen, Germany) and transmission electron microscope (TEM, FEI G2F20, Hillsboro, OR, USA). The chemical compositions and surface states of the samples were examined by X-ray photoelectron spectroscopy (XPS, Thermo Sacalab 250Xi, Waltham, MA, USA). The Raman spectra were recorded with He-Ne laser excitation at 532 nm using a Raman spectrometer (Renishaw inVia, Wotton-under-Edge, UK).

The ethylene-sensing properties of the sensors were measured at room temperature by a homemade dynamic test system (Figure S1) [26]. The ethylene standard gas (10–100 ppm in dry air), main interference gas (CO_2, 30,000 ppm in dry air), and carrier gas dry air were supplied by Chengdu Xuyuan Chemical Co., Ltd., Chengdu, China. The concentration of tested ethylene and CO_2 was controlled by the mass flow control (MFC300, Wuxi Aitoly Electronics Co., Ltd., Wuxi, China) with the dry air as carrier gas. The fabricated ethylene sensors were put into the test chamber and their resistances were recorded by a real-time multimeter resistance acquisition system (Keithley 2700, Cleveland, OH, USA). The sensing response of the sensor was defined as $(R_g - R_0)/R_0$, where R_g and R_0 represent the steady-state resistance value of the sensor in the tested gas atmosphere and dry air, respectively. The response/recovery time was defined as the time required for 90% change of the resistance during the adsorption/desorption process.

2.4. Theoretical Calculation

DFT (Density Functional Theory) calculations were carried out using the Vienna Ab-initio Simulation Package (VASP) with the frozen-core all-electron projector-augment-wave method. The Perdew—Burke—Ernzerhof (PBE) of Generalized Gradient Approximation (GGA) was adopted to describe the exchange, correlation potential, and structure optimization. Van der Waals interactions were considered by the DFT-D2 method of Grimme. The plane wave basis set cut-off energy was set to 500 eV, and the Monkhorst-Pack k-point sampling was set to 2 × 2 × 1. The geometry optimizations were performed until the forces on each ion was reduced below 0.01 eV/Å.

The adsorption energy (E_{ads}) was calculated as:

$$E_{ads} = E_{adsorbent+gas} - (E_{adsorbent} + E_{gas}) \quad (1)$$

where $E_{adsorbent}$ was the energy of the sensitive materials, E_{gas} represented the energy of the adsorbed ethylene molecule, and $E_{adsorbent+gas}$ was the total energy of the adsorbed system. Energetically, the negative adsorption energy values are desirable for the adsorption process.

2.5. Banana Ripeness Detection Experiments

The performance verification of the as-fabricated ethylene sensor was conducted by applications in banana ripeness detection. Yellowish green bananas were obtained from the local market and stored at room temperature (20 °C). During storage, the color of the samples changed with the increase of storage time, and the four typical color stages that are yellowish green, all yellow, yellow with brown speckles, and brown were chosen for detection of released ethylene. The bananas at the target sampling stage were put into a sealed quartz glass container. The containers were sealed for a period of time for collection of ethylene, and then the sensor was placed into the container as well to record its in-time resistance values for the detection of released ethylene. After sensing for 100 s, the sensor was taken out from the container, and placed in ambient air to recover to the initial resistance state. The above steps were repeated four times to obtain the sensing response to bananas at different ripeness stages. At the same time, the gas in the container was also extracted by a gas syringe (25 μL, VICI Precision Sampling, Inc., Baton Rouge, LA, USA) for gas analysis by gas chromatography-mass spectrometry (Agilent GC-MS 7890B-5977A, Santa Clara, CA, USA). The parameters of GC-MS were set as: 250 °C inlet temperature, 40 °C column temperature for 3 min, 40 °C /min speed to 100 °C, 1 mL/min flow rate, splitless, 230 °C ion source, and 150 °C MS quadrupole temperature. Identification of ethylene was confirmed by comparing the collected mass spectra with the spectra in the National Institute for Standards and Technology (NIST 14) data bank. The relative content of the ethylene was determined using the area normalization method. Three sampling analysis was one replicate and the average results were used.

3. Results

3.1. Theoretical Design of the Ethylene Sensitive Material System

A model of graphene consisting of 55 atoms was constructed. The length of the C-C bond in basal plane was 1.42540 Å, which was close to the C-C bond length of the graphite planar structure [27]. The geometric structure of the 16 monolayer WSe_2 was also discussed [28,29] and the length of the W-Se bond varied from 2.51006 to 2.51244 Å. The 4 Pd atoms with each volume of 62.01 Å^3 were adopted. Lattice constants of graphene and WSe_2 were calculated as 12.34 and 13.26, respectively. After compounding, the lattice constant of graphene/WSe_2 was 12.64, indicating graphene was stretched while WSe_2 was compressed. The length between ethylene and graphene, WSe_2, and Pd, were 2.86 Å, 2.88 Å, and 2.98 Å, respectively, forming no chemical bonds and declaring van der Waals adsorption, as shown in Figure 1. While in ternary structure, there was a chemical bond (1.74 Å) linked between Pd and WSe_2, and the distance between ethylene and WSe_2 was shortened to 2.2 Å, indicating that the adsorption had been changed.

Energies of the above structures and the ethylene adsorbed on the structures were obtained (Table 1). After assemble with WSe_2, the energy of graphene almost doubled. While after metal Pd modification, the energy of the graphene became slightly more powerful. Adsorption energies were calculated and are shown in Table 2. The adsorption of graphene to ethylene was only −0.14; a negative value indicated the adsorption occurred energetically and a weak value declared the van der Waals adsorption. Compounded Pd or WSe_2 helped to enhance the adsorption capacity of graphene to ethylene. It was noticed that capacity drastically increased to 5.5 times through ternary compounding.

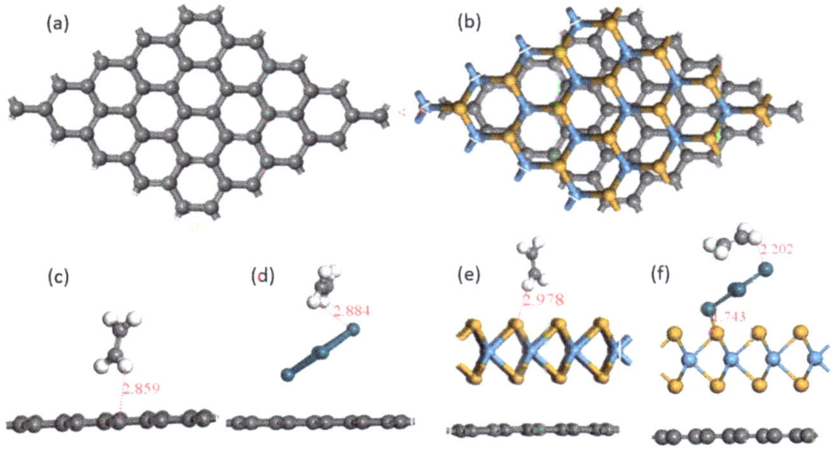

Figure 1. Optimized geometric models of (**a**) graphene and (**b**) graphene/WSe_2, and adsorption structures of ethylene molecules on (**c**) graphene, (**d**) graphene/Pd, (**e**) graphene/WSe_2, and (**f**) graphene/WSe_2/Pd.

Table 1. The calculated single point energies of the adsorbed systems.

Structure	Energy (eV)
Ethylene	−31.75
Graphene	−458.56
Graphene/WSe_2	−817.16
Graphene/Pd	−471.35
Graphene/WSe_2/Pd	−805.71
Ethylene adsorbed on graphene surface	−490.45
Ethylene adsorbed on graphene/Pd surface	−503.27
Ethylene adsorbed on graphene/WSe_2 surface	−849.10
Ethylene adsorbed on graphene/WSe_2/Pd surface	−838.24

Table 2. Adsorption energies of the ethylene adsorption system.

Adsorbed System	Adsorption Energy (eV)
Ethylene-graphene	−0.14
Ethylene-graphene/Pd	−0.17
Ethylene-graphene/WSe_2	−0.19
Ethylene-graphene/WSe_2/Pd	−0.78

3.2. Fabrication and Characterization of Ethylene Sensitive Films

Four sensitive material systems were fabricated including rGO, rGO/WSe_2, rGO/Pd, and rGO/WSe_2/Pd to achieve the most promising candidates for room-temperature ethylene sensing. Firstly, the self-assembled films were characterized by XPS to investigate their chemical composition and surface states. Figure 2a,b shows the C 1s spectra of GO and rGO films, respectively. For GO films, C-C, C-O, and C=O are the three main types of carbon bonds, corresponding to the characteristic peaks at 284.8 eV, 287.0 eV, and 288.4 eV. After annealing, the peak of the C-O bond decreases greatly and the C-C bond becomes the main peak, indicating that most of the oxygen-containing functional groups in GO are removed and thus GO is reduced. After deposition of WSe_2 nanosheets, the chemical composition of the rGO/WSe_2 sensitive films was examined by XPS as well, shown in Figure 2c. Two obvious peaks appeared at 34.2 and 32.1 eV and were attributed to $W4f_{7/2}$ and $W4f_{5/2}$ of W^{4+}, respectively, while the peaks at 54.2 eV and 55 eV correspond to the

Se3d$_{3/2}$ and Se3d$_{5/2}$ of Se^{2-}. Then, the Raman spectrum was also recorded to exhibit the structural properties of the rGO/WSe_2 sensitive films, as displayed in Figure 2d. The characteristic peak of WSe_2 at 251 cm^{-1} is clearly observed, which could be attributed to the overlapped peaks of the interlayer mode A_{1g} (253 cm^{-1}) and in-plane vibration mode $E^1{}_{2g}$ (250 cm^{-1}). The peaks at 1349 cm^{-1} and 1590 cm^{-1} correspond to the D and G band of rGO, respectively, and its intensity ratio is almost 1.0, indicating the reduction of GO into rGO with some defects. Moreover, the broad 2D band at 2470–3000 cm^{-1} shows that the fabricated rGO film is composed of some layers of the rGO sheets.

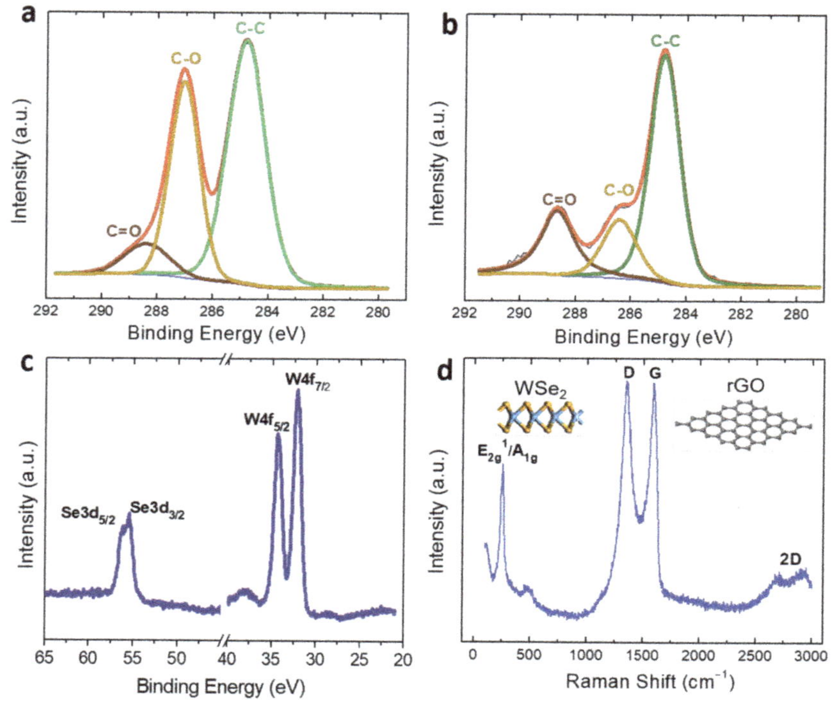

Figure 2. XPS C 1s spectrum of (**a**) GO and (**b**) rGO. (**c**) XPS spectra of W 4f and Se 3d of WSe_2. (**d**) Raman spectrum of rGO/WSe_2 sensitive films.

The surface morphology of rGO and its heterojunction films were observed through SEM. The pure rGO films show uniform morphology with a few wrinkles (Figure 3a). After deposition of lamellar WSe_2 nanosheets, the rGO/WSe_2 composite films remain uniform with excellent interlayer coupling between rGO and WSe_2 nanosheets (Figure 3b,c). As shown in Figure S2, the EDS element mapping images of W, Se, C, and O confirm that these four elements are homogeneously present in the whole film region, which further proves formation of rGO/WSe_2 heterojunction. Pd NPs are evenly distributed on rGO and rGO/WSe_2 films (Figure 3c,d), which was beneficial for the sensing enhancement effects of Pd NPs. The corresponding high-resolution TEM (HRTEM) images exhibit a specific lattice spacing of the (102) facet for WSe_2, (100) facet for WSe_2, and (111) facet for Pd, which is measured to be 0.260 nm, 0.282 nm, and 0.225 nm, respectively, as shown in Figure 3d,e.

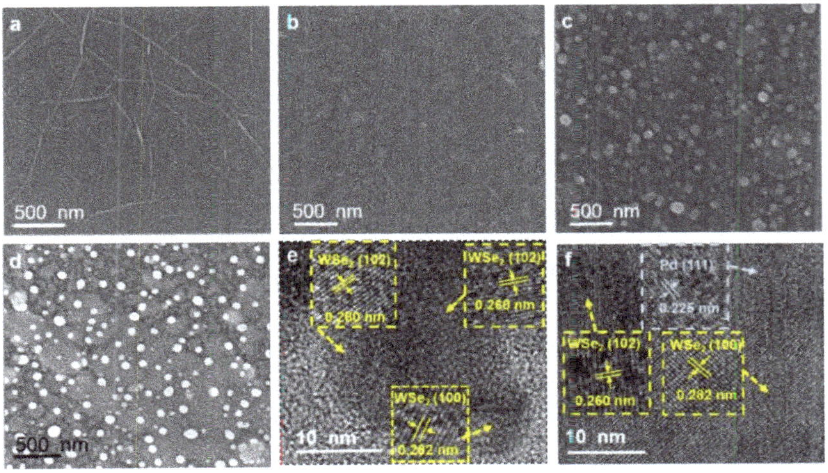

Figure 3. SEM images of the self-assembled (**a**) rGO, (**b**) rGO/WSe_2, (**c**) rGO/Pd, and (**d**) rGO/WSe_2/Pd, high-resolution TEM images of (**e**) rGO/WSe_2 and (**f**) rGO/WSe_2/Pd.

3.3. Ethylene-Sensing Characteristics

Figure 4a shows the real-time sensing response versus time curves of four sensitive films (rGO, rGO/WSe_2, rGO/Pd, and rGO/WSe_2/Pd) when exposed to 10–100 ppm ethylene at room temperature. When exposed to ethylene, all resistance values decline rapidly and then approximately reach the saturation states after almost the same periods of time. After ethylene was purged by dry air, the resistance values gradually recovered to their initial states, which indicates the p-type semiconducting behaviors of all fabricated sensing films. The sensing response of all sensors increases with the increase of ethylene concentration ranging from 10 to 100 ppm, and it could be clearly seen that both of the rGO/WSe_2 heterostructures and Pd NPs could greatly promote the ethylene sensing characteristics. Among them, pure rGO film shows the lowest sensing response, and the response to 100 ppm ethylene is nearly twice and four times larger after assembling with Pd and WSe_2, respectively, and particularly, ten times for rGO/WSe_2/Pd heterojunctions, compared with pure rGO based sensors. These results are consistent with the theoretical analysis results shown above, which further demonstrates the enhancement effects of rGO/WSe_2 heterostructures and Pd NPs for ethylene sensing.

The repeatability of all kinds of sensitive films to 50 and 100 ppm ethylene was measured and shown in Figure 4b. It could be clearly seen that except for the case of rGO films in 50 ppm ethylene, almost similar response curves including response value, response/recovery time, and also stable baselines are obtained in five successive cycles, indicating the excellent repeatability properties of sensitive films. Moreover, to further prove its practical application in the agricultural environment, the selectivity properties of rGO/WSe_2/Pd composite films were investigated. The typical interfering gas when used for fruit ripeness detection is CO_2, which is produced by the respiration of fruits. The sensing response of the composite sensor to 50 ppm ethylene is significantly larger than that of 3000 ppm CO_2 (Figure 4c), verifying the excellent selectivity of the rGO/WSe_2/Pd sensitive films in fruit ripeness detection application scenarios. In addition, the response and recovery time (90% change in sensor resistance) were extracted and shown in Figure 4d,e. Among four kinds of sensitive films, the rGO/WSe_2/Pd sensitive film exhibits the shortest response time (33 s) and recovery time (13 s) to 10 ppm ethylene. It could be attributed to the 2D structures of the rGO and WSe_2 nanosheets, Se vacancies existing in the surface of WSe_2 films, and also high activity of Pd NPs.

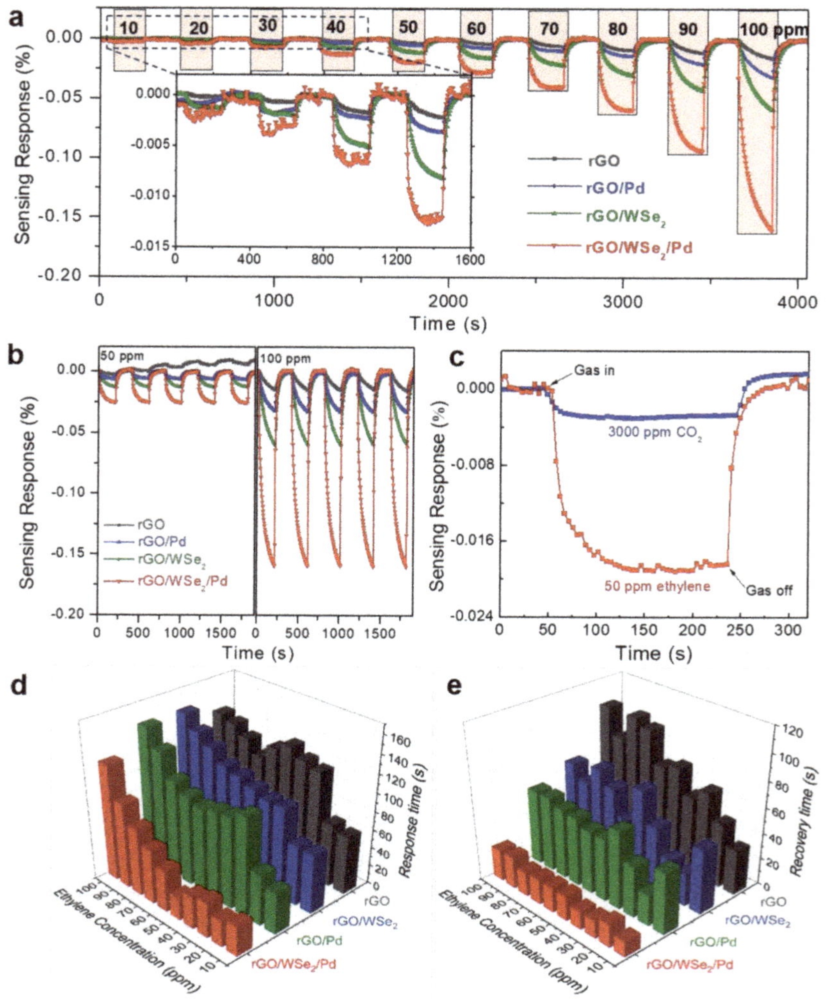

Figure 4. (**a**) Real-time sensing response of rGO-based self-assembled sensitive film to 10–100 ppm ethylene at room temperature, (**b**) repeatability, and (**c**) selectivity properties of self-assembled rGO/WSe_2/Pd sensitive film when exposed to ethylene. (**d**) Response time and (**e**) recovery time of the ethylene sensor based on rGO/WSe_2/Pd sensitive films.

3.4. Banana Ripeness Detection Applications

The fruit ripeness detection experiments were then designed and conducted in the laboratory to verify the application feasibility of our as-fabricated ethylene sensors by comparison with traditional GC-MS technology. In our experiments, the bananas, a typical climacteric fruit, were adopted, and it had been demonstrated that ethylene was released during the ripeness process of bananas. The level of banana ripeness can be commonly divided into four stages: unripe, slightly ripe, ripe, and overripe [30], accompanying the gradual change of banana color and released ethylene concentration. The unripe bananas exhibit solid light green color with some light greenish-yellow, and show no significant aroma. Mostly yellow with very faint green at tips and along edges is defined as slightly ripe, which usually exhibits some faint banana aroma. After that, bananas step into the ripe stage, and the color become solid yellow with no green, but sporadic small brown spots. At this stage, the released aroma is the strongest. Afterwards, the bananas enter into the overripe stage with the color of dark brown and even black, during which the released aroma gradually drops into a relatively low level. According to the characteristics of the above four ripeness stages, the color changes of the obtained banana samples could be clearly observed (Figure 5a), and the bananas at different ripeness stages were placed into different contained. The containers were sealed for collection of ethylene, and after about 20 min, the sensors were placed into the containers for ethylene sensing toward banana ripeness detection. The sensing response to bananas at different ripeness stages is shown in Figure 5b, from which it could be clearly seen that the ethylene concentration increases gradually until the bananas reached the ripe stage. The sensing response is calculated to be 16.8%, 25.4%, 33.8%, and 30.2%, corresponding to the unripe, slightly ripe, ripe, and overripe bananas, respectively. The normalized peak area of the released ethylene measured by GC-MS exhibits the same changing trend with the banana ripeness stage i.e., the ethylene concentration increases until the bananas are ripe, and further declines when they become overripe, as shown in Figure 5c.

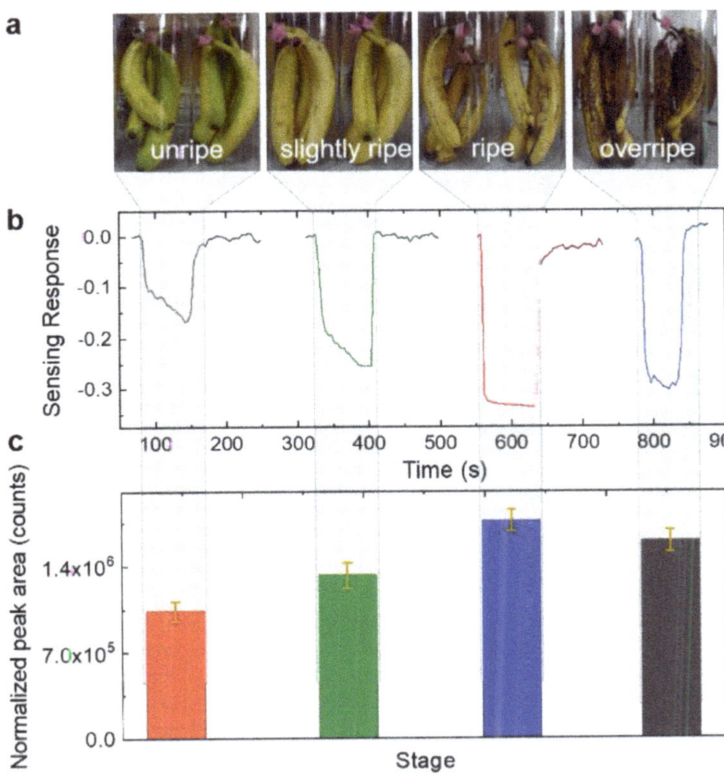

Figure 5. (**a**) Photos of banana samples at different storage stages: yellowish green (1st stage, unripe), all yellow (2nd stage, slightly ripe), all yellow with brown speckles (3rd stage, ripe), and dark brown (4th stage, overripe). (**b**) The real-time resistance versus time curves of the rGO/WSe_2/Pd heterojunction-based chemiresistive sensor when placed into the sealed container with bananas at different ripeness stages. (**c**) The normalized peak area of released ethylene from banana samples at different ripeness stages by GC-MS technology.

4. Discussion

According to the calculated adsorption energies, all the negative values manifested the advantage of room temperature operation of the fabricated ternary sensitive materials system with energy-level alignment, which may be caused by the formed chemical bond, further changing the energy band structure. The calculated adsorption energies were consistent with the sensing response of each adsorbed system to ethylene. Previous studies show that rGO is p-type materials. When ethylene molecules interact with p-type rGO sensitive films, electrons will transfer from rGO to ethylene, owing to higher Fermi levels of rGO than ethylene, resulting in a decrease of electron density of the rGO films and thus a negative sensing response to ethylene gas. When composited with Pd NPs, the sensing response to ethylene gas becomes more negative in rGO/Pd composites, which is inconsistent with the positive sensing response arising from the catalytic effects of Pd NPs toward ethylene molecules and thus excludes the role of catalytic properties on sensing response enhancements. This ethylene sensing enhancement could be attributed to the local doping effects of Pd NPs on rGO when the device is exposed to ethylene [31]. The large work function difference between high work function Pd NPs and rGO results in the local hole-doping of the rGO at the Pd NPs/rGO interfaces and thus lowers the Fermi energy levels of the rGO/Pd composites. Combined with the higher adsorption energy, both contributed to more electrons transferred from rGO/Pd to ethylene, and leads to an enhanced negative sensing response to ethylene.

In the case of rGO/WSe_2 bilayer heterojunctions, the device resistance is dominated by conductive rGO films due to the bilayer device structure deposited on IDEs and much higher electrical conductivity of rGO than WSe_2. When rGO and WSe_2 are brought into contact, it is expected that electrons would pass from rGO to WSe_2 until the equilibrium of the Fermi level is achieved. As a result, a Schottky-type junction is formed across the rGO/WSe_2 interface with a downward band bending and a hole depletion region in WSe_2 near the surface [32,33]. When the rGO/WSe_2 heterojunction is exposed to ethylene, the electron transfer from WSe_2 to the adsorbed ethylene molecules would lead to an increase of hole concentration in WSe_2. The increased hole concentration in WSe_2 causes a larger Fermi level difference between rGO and WSe_2, leading to more holes transferred from WSe_2 to rGO, and thus significantly increased sensitivity to ethylene. Furthermore, when Pd NPs are introduced to the rGO/WSe_2 bilayer film, the Schottky junction effects are further enhanced by the hole doping effects in WSe_2 induced by Pd NPs [34]. Combined with more negative adsorption energy of the rGO/WSe_2/Pd composites, more holes will transfer from WSe_2 to rGO, resulting in a further increased sensitivity to ethylene. Moreover, the unique structures of 2D materials and the self-assembled sensitive films could increase the adsorption sites for gas molecules, accelerate the charge transfer between rGO and WSe_2 nanosheets, and promote the synergistic effects between them.

For climacteric fruit such as the banana, when it changes from ripe to overripe, the released ethylene concentration decreases. This indicates a direct relationship between released ethylene concentration and the banana ripeness stage, which still needs further quantitative research. Moreover, the released aroma of climacteric fruits has a positive correlation with the released inner ethylene during the storage process [3]. Therefore, intelligent ethylene sensor techniques cannot only reflect the ripeness information of climacteric fruits, but also can be potentially adopted to evaluate their flavor quality. The consistent results from our fabricated ethylene sensor and GC-MS confirm the reliability and feasibility of our proposed ethylene sensor methodology toward banana ripeness detection. Furthermore, the proposed ethylene sensor exhibits distinct advantages of low costs, fast and accurate detection, simple operation, and in situ monitoring. The ethylene sensors can also further be developed into electronic noses and integrated into picking robots, grading equipment, packaged boxes, and shelves, which could push forward the development of intelligent agriculture.

5. Conclusions

In conclusion, a room-temperature ethylene sensor based on a ternary rGO/WSe_2/Pd heterojunction and its banana ripeness detection applications were demonstrated. The ethylene sensor shows excellent figures of merit including high sensitivity, rapid response/recovery, full repeatability, and high selectivity, making it promising for practical banana ripeness detection applications. Such excellent room-temperature ethylene sensing behaviors could be attributed to the following two aspects: (1) The negative adsorption energy of the ternary heterojunctions provides enough active sites for ethylene molecules adsorption, and (2) The electron transfer across the rGO/WSe_2 and WSe_2/Pd interfaces is greatly promoted through band energy alignment. Furthermore, its reliability and feasibility toward fruit quality monitoring applications were confirmed and validated through banana ripeness detection simulation experiments by comparison with traditional GC-MS technology. This work provides a feasible methodology for construction of a room-temperature ethylene sensing material system, and also may innovatively push forward the applications of sensor technology in the intelligent agriculture field to acquire the ripeness and quality information of climacteric fruit quickly and in-time.

Supplementary Materials: The following supporting information can be downloaded at: https://www.mdpi.com/article/10.3390/foods11131879/s1, Figure S1: Schematic diagram of the gas sensing setup; Figure S2: EDS mapping results of the rGO/WSe_2 heterojunction films.

Author Contributions: Conceptualization, funding acquisition, writing—original draft preparation, review and editing, X.L.; investigation, data curation, C.X. and J.S.; writing—review and editing, X.D.; software, visualization, Z.W. and W.H.; supervision, Y.W. and Z.L. All authors have read and agreed to the published version of the manuscript.

Funding: This research was funded by Central Public-interest Scientific Institution Basal Research Fund, China (Grand No.Y2019XK18), and the Innovation Project of Chinese Academy of Agricultural Sciences, China (Grand No.CAAS-ASTIP-2016-AII).

Institutional Review Board Statement: Not applicable.

Informed Consent Statement: Not applicable.

Data Availability Statement: Data are available from the authors.

Conflicts of Interest: The authors declare no conflict of interest.

References

1. Eccher, G.; Begheldo, M.; Boschetti, A.; Ruperti, B.; Botton, A. Roles of ethylene production and ethylene receptor expression in regulating apple fruitlet abscission. *Plant Physiol.* **2015**, *169*, 125–137. [CrossRef] [PubMed]
2. Hwang, S.I.; Star, A. Picking flowers with carbon nanotube sensors. *ACS Cent. Sci.* **2020**, *6*, 451–463. [CrossRef] [PubMed]
3. Qi, W.Y.; Wang, H.J.; Zhou, Z.; Yang, P.; Wu, W.B.; Li, Z.M.; Li, X. Ethylene emission as a potential indicator of Fuji apple flavor quality evaluation under low temperature. *Hortic. Plant J.* **2020**, *6*, 231–239. [CrossRef]
4. Cao, Y.D.; Wang, H.J.; Wu, W.B.; Yang, P.; Zhou, Z.; Wang, Z.; Li, Z.M.; Li, X. Simple and effective characterization of Fuji apple flavor quality by ethylene and sugar content. *Food Anal. Methods* **2021**, *14*, 2576–2584. [CrossRef]
5. East, A.R.; Samarakoon, H.C.; Pranamornkith, T.; Bronlund, J.E. A review of ethylene permeability of films. *Packag. Technol. Sci.* **2015**, *28*, 732–740. [CrossRef]
6. Owino, W.O.; Nakano, R.; Kubo, Y.; Inaba, A. Differential regulation of genes encoding ethylene biosynthesis enzymes and ethylene response sensor ortholog during ripening and in response to wounding in avocados. *J. Am. Soc. Hortic. Sci.* **2002**, *127*, 520–527. [CrossRef]
7. Afreen, H.; Bajwa, I.S. An IoT-based real-time intelligent monitoring and notification system of cold storage. *IEEE Access* **2021**, *9*, 38236–38253. [CrossRef]
8. Nagle, M.; Yasunaga, E.; Mahayothee, B.; Muller, J. Potenial for sensor systems to monitor fruit physiology of mango during long-distance transport. *Environ. Control Biol.* **2018**, *56*, 33–38. [CrossRef]
9. Pereira, L.; Pujol, M.; Garcia-Mas, J.; Phillips, M.A. Non-invasive quantification of ethylene in attached fruit headspace at 1 p.p.b. by gas chromatography-mass spectrometry. *Plant J.* **2017**, *91*, 172–183. [CrossRef]
10. Eberhardta, A.; Schmitt, K.; Rademacher, S.; Huber, J.; Bauersfeld, M.L.; Wöllenstein, J. Nondispersive infrared photometer based on a rotating interference filter for investigation of climacteric fruit ripening. *Procedia Eng.* **2016**, *168*, 1223–1226. [CrossRef]

11. Tabassum, S.; Kumar, D.P.; Kumar, R. Copper complex-coated nanopatterned fiber-tip guided mode resonance fevice for selective detection of ethylene. *IEEE Sens.* **2021**, *21*, 17420–17429. [CrossRef]
12. Valente, J.; Almeida, R.; Kooistra, L. A comprehensive study of the potential application of flying ethylene-sensitive sensors for ripeness detection in apple orchards. *Sensors* **2019**, *19*, 372. [CrossRef] [PubMed]
13. Wilson, D. Chemical sensors for farm-to-table monitoring of fruit quality. *Sensors* **2021**, *21*, 1634. [CrossRef]
14. Sun, M.; Yang, X.; Zhang, Y.; Wang, S.; Wong, M.W.; Ni, R.; Huang, D. Rapid and visual detection and quantitation of ethylene released from ripening fruits: The new use of grubbs catalyst. *J. Agric. Food Chem.* **2019**, *67*, 507–513. [CrossRef]
15. Nimittrakoolchai, O.U.; Supothina, S. High-yield precipitation synthesis of tungsten oxide platelet particle and its ethylene gas-sensing characteristic. *Mater. Chem. Phys.* **2008**, *112*, 270–274. [CrossRef]
16. Ahn, H.; Hoh, J.H.; Kim, S.B.; Overfelt, R.A.; Yoon, Y.S.; Kim, D.J. Effect of annealing and argon-to-oxygen ratio on sputtered SnO_2 thin film sensor for ethylene gas detection. *Mater. Chem. Phys.* **2010**, *124*, 563–568. [CrossRef]
17. Sholehah, A.; Faroz, D.F.; Huda, N.; Utari, L.; Septiani, N.L.W.; Yuliarto, B. Synthesis of ZnO flakes on flexible substrate and its application on ethylene sensing at room temperature. *Chemosensors* **2020**, *8*, 2. [CrossRef]
18. Pimtong-Ngam, Y.; Jiemsirilers, S.; Supothina, S. Preparation of tungsten oxide-tin oxide nanocomposites and their ethylene sensing characteristics. *Sens. Actuators A-Phys.* **2007**, *139*, 7–11. [CrossRef]
19. Kathirvelan, J.; Vijayaraghavan, R.; Thomas, A. Ethylene detection using TiO_2-WO_3 composite sensor for fruit ripening applications. *Sens. Rev.* **2017**, *37*, 147–154. [CrossRef]
20. Ivanov, P.; Llobet, E.; Vergara, A.; Stankova, M.; Vilanova, X.; Hubalek, J.; Gracia, I.; Cane, C.; Correig, X. Towards a micro-system for monitoring ethylene in warehouses. *Sens. Actuators B-Chem.* **2005**, *111–112*, 63–70. [CrossRef]
21. Zhang, J.N.; Zhang, L.Z.; Leng, D.Y.; Ma, F.; Zhang, Z.Y.; Zhang, Y.Y.; Wang, W.; Liang, Q.F.; Gao, J.Z.; Lu, H.B. Nanoscale Pd catalysts decorated WO_3-SnO_2 heterojunction nanotubes for highly sensitive and selective acetone sensing. *Sens. Actuators B-Chem.* **2020**, *306*, 127575. [CrossRef]
22. Liang, X.Y.; Ding, N.; Ng, S.P.; Wu, C.-M.L. Adsorption of gas molecules on Ga-doped graphene and effect of applied electric field: A DFT study. *Appl. Surf. Sci.* **2017**, *411*, 11–17. [CrossRef]
23. Tabari, L.; Farmanzadeh, D. Yttrium doped graphene oxide as a new adsorbent for H_2O, CO, and ethylene molecules: Dispersion-corrected DFT calculations. *Appl. Surf. Sci.* **2020**, *500*, 144029. [CrossRef]
24. Liu, C.H.; Tai, H.L.; Zhang, P.; Ye, Z.B.; Su, Y.J.; Jiang, Y.D. Enhanced ammonia-sensing properties of PANI-TiO_2-Au ternary self-assembly nanocomposite thin film at room temperature. *Sens. Actuators B-Chem.* **2017**, *246*, 85–95. [CrossRef]
25. Zhang, D.Z.; Pan, W.J.; Zhou, L.J.; Yu, S.J. Room-temperature benzene sensing with Au-doped ZnO nanorods/exfoliated WSe_2 nanosheets and Density Functional Theory simulations. *ACS Appl. Mater. Interfaces* **2021**, *13*, 33392–33403. [CrossRef]
26. Li, X.; Xu, J.L.; Jiang, Y.D.; He, Z.Z.; Liu, B.H.; Xie, H.K.; Li, H.; Li, Z.M.; Wang, Y.; Tai, H.L. Toward agricultural ammonia volatilization monitoring: A flexible polyaniline/$Ti_3C_2T_x$ hybrid sensitive films based gas sensor. *Sens. Actuators B-Chem.* **2020**, *316*, 128144. [CrossRef]
27. Dreyer, D.R.; Park, S.; Bielawski, C.W.; Ruoff, R.S. The chemistry of graphene oxide. *Chem. Soc. Rev.* **2010**, *39*, 228–240. [CrossRef]
28. Kim, J.H.; Mirzaei, A.; Kim, H.W.; Kim, S.S. Realization of Au-decorated WS_2 nanosheets as low power-consumption and selective gas sensors. *Sens. Actuators B-Chem.* **2019**, *296*, 126659. [CrossRef]
29. Neri, G. Thin 2D: The New Dimensionality in Gas Sensing. *Chemosensors* **2017**, *5*, 21. [CrossRef]
30. Phillips, K.M.; McGinty, R.C.; Couture, G.; Pehrsson, P.R.; McKillop, K.; Fukagawa, N.K. Dietary fiber, starch, and sugars in bananas at different stages of ripeness in the retail market. *PLoS ONE* **2021**, *16*, e0253366. [CrossRef]
31. Wang, J.W.; Rathi, S.; Singh, B.; Lee, I.; Maeng, S.; Joh, H.-I.; Kim, G.-H. Dielectrophoretic assembly of Pt nanoparticle-reduced graphene oxide nanohybrid for highly-sensitive multiple gas sensor. *Sens. Actuators B-Chem.* **2015**, *220*, 755–761. [CrossRef]
32. Huang, Q.W.; Zeng, D.W.; Li, H.Y.; Xie, C.S. Room temperature formaldehyde sensors with enhanced performance, fast response and recovery based on zinc oxide quantum dots/graphene nanocomposites. *Nanoscale* **2012**, *4*, 5651–5658. [CrossRef] [PubMed]
33. Liu, Z.Q.; Zhao, H.Y.; Li, N.; Zhang, Y.; Zhang, X.Y.; Du, Y.P. Assembled 3D electrocatalysts for efficient hydrogen evolution: WSe_2 layers anchored on graphene sheets. *Inorg. Chem. Front.* **2016**, *3*, 313. [CrossRef]
34. Chen, W.Y.Y.; Yerembetova, A.; Washer, B.M.; Jiang, X.F.; Shuvo, S.N.; Peroulis, D.; Wei, A.; Stanciu, L.A. Selective detection of ethylene by MoS_2-carbon nanotube networks coated with Cu(I)-Pincer complexes. *ACS Sens.* **2020**, *5*, 1699–1706. [CrossRef]

foods

Article

Identification of Maize with Different Moldy Levels Based on Catalase Activity and Data Fusion of Hyperspectral Images

Wenchao Wang [1,2], Wenqian Huang [2], Huishan Yu [1,*] and Xi Tian [2,3,*]

1 College of Physical Science and Information Engineering, Liaocheng University, Liaocheng 252000, China; wangwc0501@163.com
2 Beijing Research Center of Intelligent Equipment for Agriculture, Beijing 100097, China; huangwq@nercita.org.cn
3 College of Engineering, China Agricultural University, Beijing 100083, China
* Correspondence: 13906350692@163.com (H.Y.); tianx@nercita.org.cn (X.T.)

Abstract: Maize is susceptible to mold infection during growth and storage due to its large embryo and high moisture content. Therefore, it is essential to distinguish the moldy sample from healthy groups to prevent the spread of mold and avoid huger economic losses. Catalase is a metabolite in the growth of microorganisms; hence, all maize samples were accurately divided into four moldy grades (health, mild, moderate, and severe levels) by determining their catalase activity. The visible and shortwave near-infrared (Vis-SWNIR) and longwave near-infrared (LWNIR) hyperspectral images were investigated to jointly identify the moldy levels of maize. Spectra and texture information of each maize sample were extracted and used to build the classification models of maize with different moldy levels in pixel-level fusion and feature-level fusion. The result showed that the feature-level fusion of spectral and texture within Vis-SWNIR and LWNIR regions achieved the best results, overall prediction accuracy reached 95.00% for each moldy level, all healthy maize was correctly classified, and none of the moldy samples were misclassified as healthy level. This study illustrated that two hyperspectral image systems, with complementary spectral ranges, combined with feature selection and data fusion strategies, could be used synergistically to improve the classification accuracy of maize with different moldy levels.

Keywords: maize; moldy level; catalase activity; hyperspectral image; data fusion; feature selection

Citation: Wang, W.; Huang, W.; Yu, H.; Tian, X. Identification of Maize with Different Moldy Levels Based on Catalase Activity and Data Fusion of Hyperspectral Images. *Foods* **2022**, *11*, 1727. https://doi.org/10.3390/foods11121727

Academic Editor: Corrado Costa

Received: 6 May 2022
Accepted: 10 June 2022
Published: 13 June 2022

1. Introduction

Maize is an important food crop, feed crop, and cash crop [1]. Compared with wheat and rice, maize has larger embryos, and its moisture content at harvest reaches about 30%, higher than the 25% and 22.5% of wheat and rice, respectively, which makes it more susceptible to mold infection during growth and storage [2]. Among them, *Aspergillus flavus* is the strain that most easily and commonly infects maize. aflatoxin (AFB1), produced by *Aspergillus flavus*, is extremely carcinogenic and toxic, and is the most toxic mold secondary metabolite in contaminated food [3,4]. Humans and animals eating food contaminated with AFB1 is a serious threat to life and health safety. Therefore, the early real-time detection of moldy maize has very important research significance.

Traditional detection methods for moldy maize include sensory evaluation and physical and chemical component detection [5]. Sensory evaluation is simple, time-saving, and low cost, but the evaluation results are easily disrupted by the external environment and the subjective emotions of the inspectors. Additionally, the toxic substances will also pose a threat to the health of the inspectors. The physical and chemical component determination is generally detected by high-performance liquid chromatography (HPLC), polymerase chain reaction (PCR), and enzyme-linked immunosorbent assay (ELISA) [6–8]. Although these methods can achieve more accurate measurement and qualitative analysis, they require expensive testing equipment and professional technicians, and the testing process is complex and time-consuming [9]. Hence, using traditional methods, it is difficult to achieve simple, rapid, and non-destructive detection, which cannot meet the actual needs for identification of kernel maize.

In recent years, non-destructive detection technologies, such as electronic nose, machine vision, near-infrared, and hyperspectral imaging, have been successfully applied to the classification of moldy maize [10,11]. Electronic nose technology [12] is mainly used specific sensors to identify the level of mold based on the change of volatile organic compounds (VOCs) information. Leggieri et al. [13] used electronic nose technology to determine the concentration of AFB1 and fumonisins (FBs) in maize; the prediction model of AFB1 and FBs built by the artificial neural networks were 78% and 77%, respectively. However, the concentration of gas is easy to change in the flow state, which affected the discrimination accuracy. Machine vision detection mainly adopts machine learning algorithms to extract features from kernel images and then establish the classification model based on extracted features [14]. Shi Ying [15] extracted R-channel eigenvalues of RGB images (red, green and blue three-channel color image) and classified maize kernel samples with different levels of mold using the Back Propagation (BP) neural network. Visible-near infrared (Vis-NIR) spectroscopy technology connects the spectral information with the internal content of substances and uses the spectral curve to analyze the changes of internal components of seeds in the process of mold growth [16]. Therefore, the machine vision and Vis-NIR spectroscopy can express the external and internal changes of the target samples respectively. However, both internal quality and external characteristics of maize will change during the moldy process; neither machine vision nor Vis-NIR spectroscopy can obtain internal and external quality information at the same time. Hyperspectral imaging technology combines spectral analysis technology with image processing technology, which can simultaneously obtain the spectral data with internal component information and the image data with appearance feature information, realizing the rapid, pollution-free and non-destructive detection [17–19].

In terms of using hyperspectral imaging technology to identify moldy maize, Tao et al. [20] used random frog (RF) combined with partial least-squares discriminant analysis (PLS-DA) to qualitatively analyze the healthy maize and polluted maize inoculated with aflatoxigenic fungus at different culture days based on long wave near-infrared (LWNIR) hyperspectral images; the classification accuracy of the calibration set and verification set was 82.3% and 94.9%, respectively. Williams et al. [21] evaluated the fungal development in maize kernels using LWNIR; principal component analysis (PCA) was firstly used to remove the interference of noise, such as background, bad pixels, and shadows, from the hyperspectral images. Three distinct clusters related to the degree of infection were found in the scoring plots of PC4 and PC5. Dai et al. [22] established a classification model of moldy maize with different culture days (0 days, 2 days, 4 days, 6 days, and 8 days) based on 9 characteristic wavelengths selected from visible and short wave near-infrared (Vis-SWNIR) hyperspectral imaging using fisher discriminant analysis (FDA); the classification accuracy of the calibration set and validation set were 100% and 98.67%, respectively, illustrating that the characteristic wavelengths could represent the main information about moldy levels of maize samples. Del Fiore et al. [23] used Vis-SWNIR hyperspectral imaging combined with multivariate statistical analysis to identify maize kernels infected with fungi under different growths. The results showed that hyperspectral imaging was able to quickly distinguish between healthy and infected maize, i.e., 48 h after inoculation with mycorrhizal fungi. Previous studies have shown that both the spectral ranges of Vis-SWNIR and LWNIR can be used to distinguish the moldy level of maize; however, it has not been found that fusing the spectral information of different hyperspectral systems can construct a classification model of maize with different moldy levels. Yu et al. [24] studied the influence of Vis-SWNIR and LWNIR hyperspectral imaging systems on the prediction ability of total volatile basic nitrogen (TVB-N) content in tilapia fillets during refrigeration; the results showed that the fused spectral data of both sensors achieved a better prediction result than that of individual sensor. Meanwhile, the study of fusing spectral data with texture data to discriminate the moldy level of maize is less extensive. Ma et al. [25] developed the classification model of fresh and frozen meats based on the spectral and texture information extracted from Vis-SWNIR hyperspectral images; the research showed that the classification model built by the feature fusion of spectra and texture was better than that of spectra and texture alone. Therefore, fusing the information of spectra and texture obtained from different hyperspectral image systems would be a new idea for constructing an accurate classification model of maize with different moldy levels.

Mold growth is uncontrollable. Hence, moldy levels may not be uniform among different samples at the same culture time and a small number of samples were not consistent with the designed moldy levels. Catalase (CAT) is a metabolite in the growth of *Aspergillus flavus* and other microorganisms [26]. Zhang et al. [27] found that the correlation coefficient between the number of mold colonies and the activity value of CAT reached more than 0.9 in various grains such as wheat, rice, and maize. In addition, CAT is the precursor product of AFB1 produced by *Aspergillus flavus*. Zhang et al. [28] analyzed the correlation between CAT activity value and AFB1 content of moldy maize and found that both of them had the same change curves. The above research showed that CAT activity value could reflect the moldy levels of maize. However, at present, it is not found that the model for discrimination of maize with different moldy levels was established based on the feature fusion of hyperspectral imaging information and CAT activity value of moldy maize.

In this study, we proposed a new method to better divide the maize with different moldy levels by monitoring the CAT activity value of maize samples infected with *Aspergillus flavus* under different culture days. The objective of this study is to examine the potential of using multi-levels data fusion of hyperspectral images to identify the maize with different moldy levels. The specific objectives of this study were to: (1) analyze the difference of spectra and texture of Vis-SWNIR and LWNIR hyperspectral images of maize samples with different moldy levels; (2) examine the ability of different pretreatment methods and classifier for identification of maize samples with different moldy levels; (3) compare the classification ability of the models based on the pixel-level fusion of spectra and different texture parameters; (4) evaluate the effects of features selected by different variable selection methods on the classification models of feature-level fusion; and (5) establish the best classification model of maize with different moldy levels by integrating the spectrum and texture data with Vis-SWNIR and LWNIR regions.

2. Materials and Methods

2.1. Maize Sample Preparation

"Zhengdan 958" is widely planted in China due to its advantages of high and stable yield; hence, "Zhengdan 958" was selected as the experimental maize sample of this study. To reduce the influence of the bacteria carried by the maize itself, maize kernels with the same size and appearance were selected manually. All maize kernels were surface sterilized by soaking in 2% sodium hypochlorite solution for 5 min and then rinsed three times with distilled water. The conidia suspension of *Aspergillus flavus* (BNCC142801 purchased from BeNa Biotechnology Research Center, Xinyang, Henan province, China) was diluted to 10^{-3} in sterile water and inoculated into maize kernels. Simulation of the maize mold process by inoculation with molds has been widely used in laboratory studies. Tao et al. [20] used two Aspergillus flavus, AF13 (aflatoxin-producing Aspergillus flavus) and AF36 (non-aflatoxin-producing Aspergillus flavus), for artificial laboratory inoculation to study the changes in maize. In this study, all inoculated maize kernels were divided into 240 groups and placed in petri dishes with the embryo side facing upward. Each group contained about 30 kernels, weighing 10 ± 0.5 g. All samples were cultured in a constant temperature and humidity incubator with a temperature of 30 °C and relative humidity of 80%.

2.2. Hyperspectral Image Acquisition System

The Vis-SWNIR and LWNIR hyperspectral reflectance imaging system (Figure 1) built in the Intelligent Detection Laboratory of China Agricultural Intelligent Equipment Technology Center, was used to acquire the hyperspectral images of moldy maize samples in the wavelength range 327–1098 (nm) and 930–2548 (nm). The Vis-SWNIR hyperspectral imaging system consists of an imaging spectrometer (ImSpector V10EQE, Spectra Imaging Ltd, Oulu, Finland), an electron multi-plying charge-coupled device (EMCCD) camera (Andor Luca EMCCD DL-604 M, Andor Technology plc., Belfast, UK) with a resolution of 502 × 500 and a camera lens (OLE23-f/2.4, Spectral Imaging Ltd., Oulu, Finland), and a spectraCube data acquisition software (Isuzu Optics Corp., Xinzhu, Taiwan, China) controls the operation of mobile platform and acquisition of hyperspectral images. The LWNIR hyperspectral imaging system consists of an imaging spectrometer (ImSpector N25E, Spectral Imaging Ltd., Oulu, Finland), a charge-coupled device (CCD) camera (Xeva-2.5-320, Xenics Ltd., Leuven, Belgium) with a resolution of 320 × 256 and a camera lens (HSIA-OLE22, Spectral Imaging Ltd., Oulu, Finland), spectral acquisition software (Isuzu Optics Corp., Xinzhu, Taiwan, China). The two hyperspectral acquisition systems shared two 300 w halogen lampsadjusted at an angle of about 45 to provide a stable light source, a motorized displacement stage (EZHR17EN, AllMotion, Inc., Union City, CA, USA) for sample placement, and a computer (Dell, Intel (R) Core (TM) i5-2400 CPU @ 3.10 GHz) with two types of hyperspectral acquisition software.

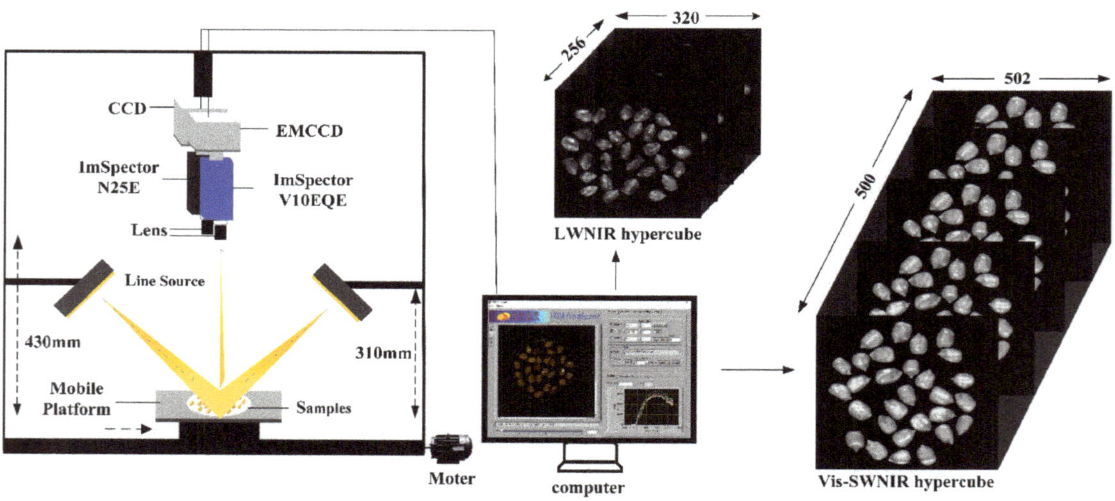

Figure 1. Hyperspectral image acquisition system.

The moldy levels of maize varied with the culture time. In this study, to artificially cultivate the maize samples with different moldy grades, sixty samples were taken out from a constant temperature and humidity incubator on day 0, day 2, day 4, and day 6 as the moldy samples at healthy, mild, moderate, and severe levels, respectively. In order to collect the high-quality images without saturation and distortion, hyperspectral images of samples were acquired by line scanning method, and the distance between the lens to the moving platform, the exposure time and the mobile platform speed were set to 430 mm, 3 ms, 2.6 mm/s for Vis-SWNIR hyperspectral systems and 310 mm, 5 ms, 40 mm/s for LWNIR hyperspectral systems, respectively.

2.3. Determination of CAT Activity

CAT activity value could reflect the activity strength of mold. Therefore, CAT activity value of the maize samples at four moldy levels was determined using the potassium permanganate titration method [29] after hyperspectral image collection. The specific steps of CAT activity determination were as follows:

Step 1: Weigh the sample and place it in a conical flask;

Step 2: Add 40 mL distilled water and 5 mL 0.3% hydrogen peroxide, and set another control group (40 mL distilled water and 5 mL 0.3% hydrogen peroxide in an empty conical bottle), and shake them in a shaker for 20 min;

Step 3: Add 5 mL 3 mol/L sulfuric acid, shake for 5 min, take it out, filter with filter paper, take out 10 mL filtrate, titrate with 0.005 mol/L potassium permanganate solution until light pink, and do not change color for 30 s.

The value of CAT activity (Equation (1)) is expressed as the volume (mL) of 0.005 mol/L potassium permanganate consumed in unit weight (g) and time (h). Where V_1 and V_2 were the titrated potassium permanganate volume (mL) of the control group and sample group, M was the mass (g) of the sample, and T was the unit time (h).

$$Value(\frac{mL}{g \times h}) = \frac{(V_1 - V_2)}{M \times T} \tag{1}$$

2.4. Hyperspectral Image Processing and Information Extraction

The original hyperspectral image needs to be corrected to eliminate the influence of light source and camera dark current changes [30]. The standard white reference image was acquired using a white Teflon plate (99% reflectivity) under the same sampling environment as the sample. Turn off the light source and cover the lens to obtain a black reference image (0% reflectivity). The corrected image is calculated using the black and white reference image by Equation (2):

$$I_C = \frac{I_O - I_B}{I_W - I_B} \tag{2}$$

where I_O was the original hyperspectral image, I_W and I_B represented the white reference and black reference images, respectively, and I_C was the corrected hyperspectral image. In this study, the hyperspectral image correction and subsequent data processing were performed in MATLAB 2019B (The MathWorks, Inc., Natick, MA, USA).

To extract the information of region of interest (ROI), the mask method was used to segment the target and background of the corrected hyperspectral images. The gray images at 849 and 1098 nm were used to construct a binary mask by setting appropriate thresholds, because the spectral intensity difference between the gray image background and maize was largest at 849 and 1098 nm wavelength images for Vis-SWNIR and LWNIR hyperspectral images, respectively. Then, the corresponding hyperspectral image was multiplied by the filtered mask to remove the background information. The original and denoised RGB images of maize with different mold levels were shown in Figure 2. After acquiring the ROI region, the average spectrum was extracted from all pixels of ROI region for each wavelength of the hyperspectral image. Due to the noise and useless information in the beginning and end bands, a total of 389 spectral variables within 399–1001 nm and 112 spectral variables within 1005–1701 nm were obtained from the hyperspectral images of Vis-SWNIR and LWNIR regions, respectively.

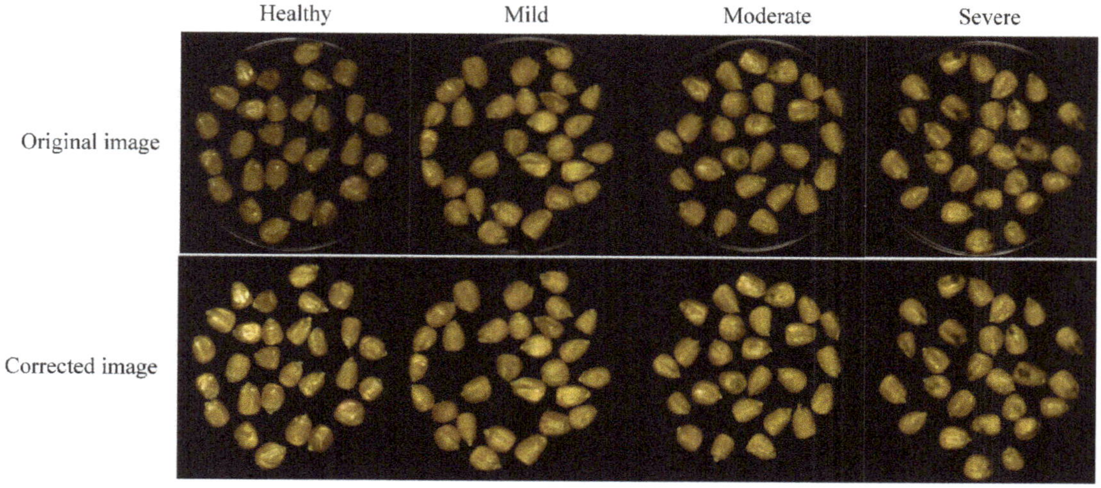

Figure 2. Original and denoised RGB images (red, green and blue three-channel color image) of maize with different mold levels.

The extraction of texture features was realized by the gray-level co-occurrence matrix (GLCM). The GLCM described the probability of occurrence of two pixels with different distances and directions in a gray image [31]. The data of four texture parameters (contrast, correlation, energy, and homogeneity) in each ROI band was extracted from the GLCM, by Equation (3)–(6). In this study, the pixel distance was set as 1, and only the GLCM in the four directions of 0°, 45°, 90°, and 135° was considered. The average value in the four directions was used to describe each texture parameter characteristic. After removing the noise bands, four texture parameters feature matrices with sizes of 240 × 389 (240: number of samples; 389: number of variables) and 240 × 112 were obtained in the Vis-SWNIR and LWNIR bands, respectively.

$$contrast = \left(\sum_{i=1}^{N}\sum_{i=1}^{N}(i-j)^2P(i,j)\right) \tag{3}$$

$$correction = \frac{\sum_{i=1}^{N}\sum_{i=1}^{N}(ij)P(i,j) - \mu_i\mu_j}{\sigma_i\sigma_j} \tag{4}$$

$$energy = \sum_{i=1}^{N}\sum_{i=1}^{N}P(i,j)^2 \tag{5}$$

$$homogeneity = \sum_{i=1}^{N}\sum_{i=1}^{N}\frac{P(i,j)}{1+(i-j)^2} \tag{6}$$

where (i,j) was the pixel coordinate, $P(i,j)$ was the joint probability with two neighboring pixels, and N (set N = 8 in this research) was the number of gray-levels. μ_i, μ_j, σ_i, and σ_j represented the mean and standard deviation of the row and columns in the GLCM, respectively.

2.5. Spectral Data Preprocessing

The hyperspectral data were easily interfered by random noise, stray light, background, and equipment in the hyperspectral images acquisition. To eliminate the influence of environmental factors and improve the correlation between spectral data and chemical composition, it was necessary to preprocess the raw spectrum [32]. Hence, moving smooth, multiple scattering correction (msc), detrend, and mean centralization (center) were used in this study. Studies have indicated that the smooth was used to remove noise interference in the spectrum and improve the signal-to-noise ratio. Its basic idea is to smooth the raw data through the "averaging" or "fitting" of several points in a finite size spectral window. The spectral window size must be an odd number, and the wider the window, the lower the spectral resolution. Msc used the method of least squares to fit the linear relationship between each spectrum and the average spectrum. This means that msc could eliminate scattering bias. Detrend is an approach to eliminate the baseline drift in the spectrum and the influence of different sampling batches on the spectrum. Firstly, a trend line was derived from spectral values and wavelengths through least squares fitting, and then the trend line was subtracted from the original spectrum. The center was effective in enhancing the differences between data, its basic idea is to remove the column, row, or overall average from each column, row, or both separately [33–35]. In this study, smooth was firstly used to reduce the noise and interference existed in original spectra, and then msc, detrend, and center were employed secondly to process the spectra on the basis of smooth. The best spectral preprocessing method was determined by comparing the effects of different pretreatment methods on classification accuracy, then the spectral data processed by the best method were fused with texture information for further analysis.

2.6. Data Fusion

Data fusion was a process of combining information from different independent information sources, which could express the described objects or processes in more detail and complete than using a single information source alone. Generally, data fusion was divided into pixel-level fusion, feature-level fusion, and decision-level fusion according to the fusion level from low to high [36]. In the study, the spectral data and texture data obtained from hyperspectral images within Vis-SWNIR and LWNIR regions were fused at pixel-level and feature-level respectively for developing a high accuracy and robust classification model of moldy maize.

Pixel-level fusion was simply merging the data information of different sources [37], so the fused features contained more variables, which was conducive to further data processing. However, it could also input the irrelevant and redundant variables into the model. In this study, pixel-level fusion models were built by fusing the spectra matrices with texture parameters, for Vis-SWNIR and LWNIR regions. For the new matrix formed after data fusion, one row represented the characteristic information of the same sample, and one column represented the eigenvalues at a specific wavelength. The optimal combination of spectrum and texture parameters was obtained by evaluating the classification accuracy of the developed models.

Feature-level fusion was to extract features from a single data block using the variable selection method and then integrate the processed feature matrix [38,39]. Compared with pixel-level fusion, feature-level fusion could adjust the number of features from different data, especially when there were large differences between single data blocks. In this study, both spectral matrix and texture matrix obtained from Vis-SWNIR and LWNIR ranges had 389 and 112 variables, respectively, these data were often multicollinearity and redundant, especially between adjacent bands. Therefore, feature wavelength selection was commonly adopted to select the key wavelengths from full-band data, which could reduce redundant and noisy information, as well as simplify the model. The optimal combination of spectral data and texture parameters in pixel-level fusion was used as the data source of feature-level fusion. Three kinds of variable selection methods including variable combination population analysis (VCPA) [40], iteratively retains informative variables (IRIV) [41], and hybrid method mVCPA-IRIV [42] were used to select the features that carry the information of moldy maize from the spectral and texture parameters data, and then combined them into a new data matrix to build the feature-level model. The number of columns in the new data matrix was the number of features obtained from the two data.

2.7. Discriminant Model and Evaluation

In order to obtain an accurate and reliable classifier, the original spectral data were used to establish a classification model with SVM, Random Forest (RF), and K-nearest neighbors (KNN). The performance of different classifiers was compared, and the optimal classifier method was selected to be used in the subsequent data processing classification algorithm.

The basic idea of SVM was to divide the segmentation hyperplane with the maximum classification interval according to the training samples in the feature space. When facing the nonlinear problem, the kernel function was introduced and transformed into a linear problem in high-dimensional space through nonlinear transformation. SVM was often used for problems with a small sample set or linear indivisibility [43]. The radial basis function (RBF) kernel function had more advantages in dealing with the nonlinear relationship between feature information and categories [44], hence RBF was selected as the kernel function of SVM in this study. The optimal loss parameter and kernel parameter was searched by the cross-verification grid optimization method.

RF was an ensemble learning method based on the Bagging algorithm, which could be used to solve classification and regression problems. RF had the advantages of processing high-dimension data, strong adaptability to data sets, and fast training speed [45]. In this study, when the RF classifier was trained, the number of decision trees was set to 50 to store the observation results of each tree.

KNN was a commonly used classification algorithm. Its core idea was to select k nearest neighbor samples in the feature space. In these K samples, if most samples belong to a certain category, the test samples also belong to this category [46]. In this study, when the KNN algorithm was used for training, parameters were automatically optimized to obtain the optimal nearest neighbor number and distance measurement parameters.

The rationality of data set division affects the prediction performance of the classification model. To avoid the influence of artificially selected calibration prediction sets on the results, in this study, all 240 samples were sequentially divided into 4 moldy levels based on the determined CAT activity value, so there were 60 samples in healthy, mild, moderate and severe levels, respectively. Then, the 60 samples of each category were randomly divided into calibration and prediction sets with a proportion of 3:1. Hence, 180 samples were selected as the calibration set to build the calibration model, and the 60 remaining samples were selected as the prediction set for evaluating the performance of the established model.

The performance of the model was evaluated from four aspects: classification accuracy of the calibration set and prediction set, and overfitting. Generally, a good model should have higher classification accuracy and lower differences between calibration and prediction sets. The main key steps of this study were shown in Figure 3.

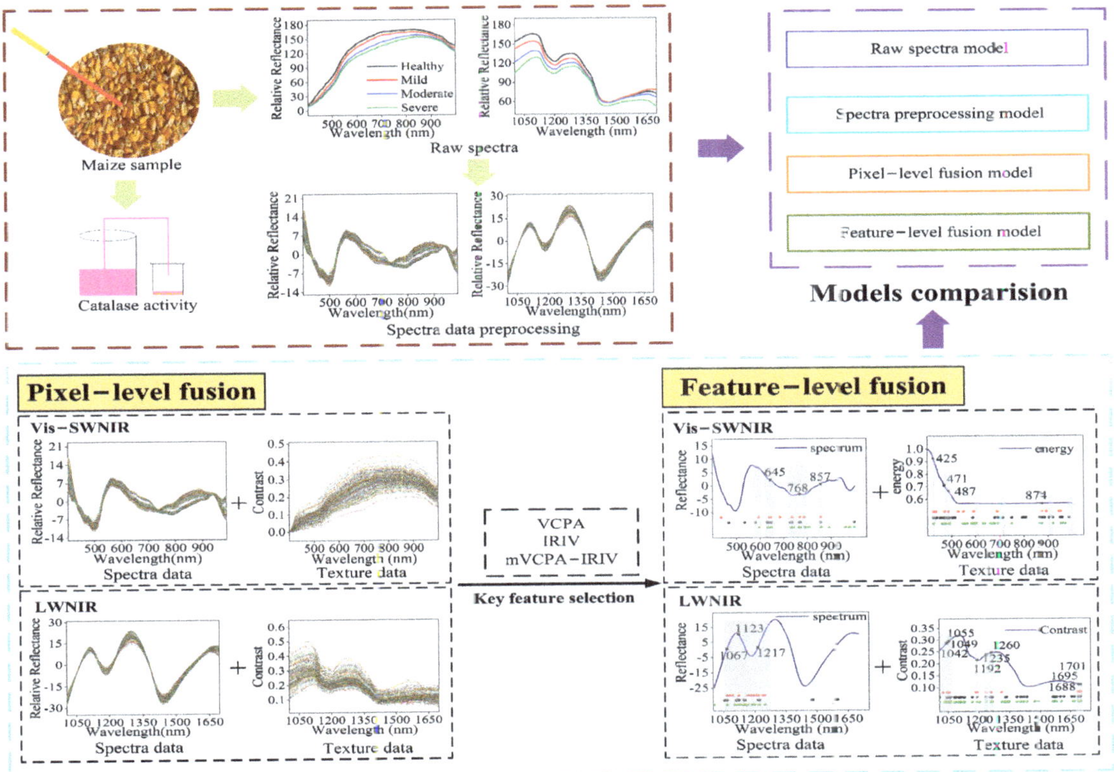

Figure 3. The experimental scheme of the data fusion model for identification of maize with different moldy levels.

3. Results and Analysis

3.1. CAT Activity Analysis of Maize with Different Moldy Levels

Table 1 shows the range, mean, and standard deviation of CAT activity values under different moldy levels. The maize kernels used in this experiment were sterilized, and the CAT activity value of the sample was 0 under the healthy level. The results showed that CAT activity values increased with the aggravation of maize moldy levels, and CAT activity values increased rapidly in the early stages of mold and slowly in the late stages, which was related to the growth pattern of the mold. The CAT activity values of different moldy levels had obvious gradient differences, indicating that CAT activity values could be used to determine the level of moldy maize.

Table 1. Catalase (CAT) activity value of maize with different moldy levels.

Moldy Level	Mean mL/(h × g)	Standard Deviation mL/(h × g)
Healthy level	0	0
Mild level	1.57	0.13
Moderate level	1.91	0.09
Severe level	2.24	0.12

3.2. Spectral and Texture Characterization

The curves of original spectra and texture data extracted from the hyperspectral images of Vis-SWNIR and LWNIR regions were shown in Figure 4. The solid lines and the shaded part of Figure 4a,b are the average spectra and the standard deviation of maize with different moldy levels, respectively. Figure 4c–j shows the texture data of contrast, correlation, energy, and homogeneity extracted from the ROI of all samples. It was clear that both spectra data and texture data of maize with different moldy levels had similar trends, but their reflectance intensity were significantly different, which may be related to the decomposition of chemical substances in the process of maize mold.

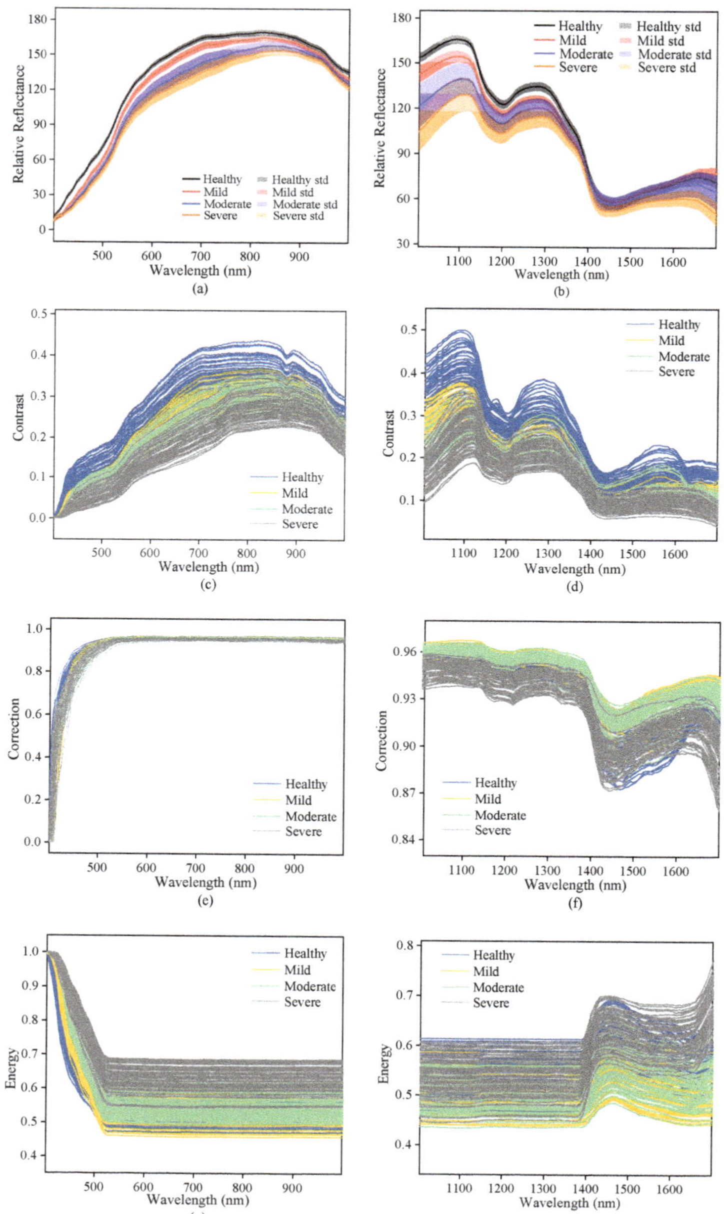

Figure 4. *Cont.*

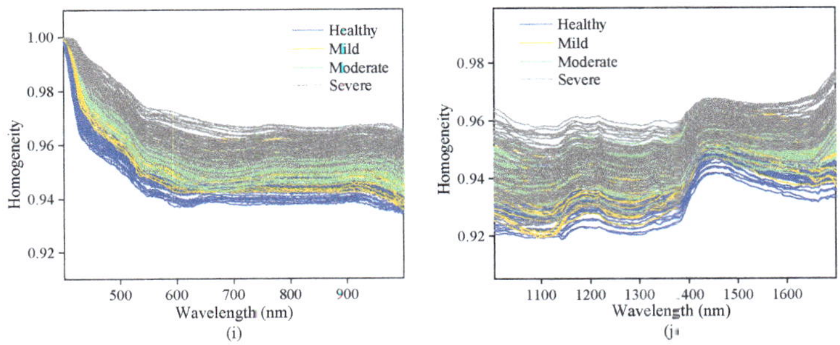

Figure 4. Spectra and texture curves of maize with different moldy levels in the Vis-SWNIR region (left) and LWNIR region (right): (**a**) original spectra in Vis-SWNIR region; (**b**) original spectra in LWNIR region; (**c**) contrast parameter in Vis-SWNIR region; (**d**) contrast parameter in LWNIR region; (**e**) correction parameter in Vis-SWNIR region; (**f**) correction parameter in LWNIR region; (**g**) energy parameter in Vis-SWNIR region; (**h**) energy parameter in LWNIR region; (**i**) homogeneity parameter in Vis-SWNIR region; (**j**) homogeneity parameter in LWNIR region.

By analyzing the spectral curve characteristics, it can be easily found that the more serious maize mold, the lower the spectral reflection intensity in both wavelength regions, indicating that the light absorption capacity of mold tissue was stronger than that of maize tissue. The spectral curve is monotonous in the Vis-SWNIR region, the average spectral curve gradually increased in the region of 399–820 nm and then decreases slowly. However, the spectral curve was complicated and varied in the LWNIR region. Two obvious reflectance peaks were captured around 1100 nm and 1300 nm, respectively. The former may be related to C-H in lipids [47], and the latter can be designated as a combination between the first overtone of N-H stretching with the fundamental N-H in-plane bending and C-N stretching with N-H in-plane bending vibrations [48]. In addition, there were two obvious absorption peaks at 1192 nm and 1445 nm. The peak at 1192 nm may be associated with the second overtone of C-H stretching in carbohydrates [49] and at 1445 nm may be related to the O-H bond in water and the first overtone of C-H in protein [50]. There were significant differences in the reflectance spectra and texture intensity between different moldy levels. These differences may provide the possibility of classifying the maize with different moldy levels. However, the spectra of maize samples with different moldy levels crossed in some wavelength intervals (1400–1701 nm), and there was no significant correlation between the reflectance spectra and the moldy levels. Hence, the spectra and textures should be fused to research the classification ability of their latent information.

3.3. Comparison and Optimization of Different Classifiers and Preprocessing Methods

In non-destructive detection technology based on visible and near-infrared hyperspectral images, many spectral preprocessing methods and classifiers can be used to construct the classification model. In order to improve the development efficiency of the classification models, the classifier and spectral preprocessing method were firstly determined in this study based on the original spectral data. Three classification algorithms, including SVM, RF, and KNN, were used to build the classification model based on the original spectra extracted from different hyperspectral systems. Table 2 shows the CAT activity value of the calibration and prediction datasets. It should be pointed out that the samples in the calibration and prediction sets were kept unchanged and no single sample was used in calibration set and prediction set at the same time. The results of different classification algorithms were shown in Figure 5. Compared the overall classification accuracy yielded by different classifiers, SVM was the most robust classifier with the accuracy of calibration and prediction sets of 93.89% and 86.67%, and 86.67% and 85% for Vis-SWNIR and LWNIR regions, respectively. Hence, SVM was used as the only classifier to build the classification models in the subsequent data processing. In terms of the classification accuracy built by the SVM models, the LWNIR region was poorer than the Vis-SWNIR region, which may be due to the weak correlation between the spectral intensity and the level of mold in the 1400–1701 wavelength range, as can be seen by the spectral characteristic curve.

Table 2. Division of maize with different moldy levels in the calibration and prediction sets.

Moldy Level	Calibration Set			Prediction Set		
	Num	Mean	Standard Deviation	Num	Mean	Standard Deviation
Healthy level	35	0	0	15	0	0
Mild level	35	1.58	0.17	15	1.58	0.14
Moderate level	35	1.94	0.18	15	1.96	0.20
Severe level	35	2.20	0.14	15	2.13	0.24

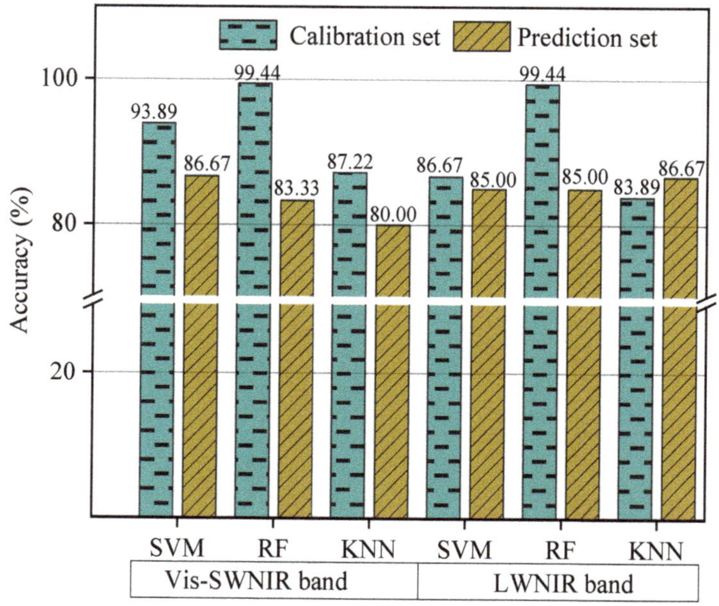

Figure 5. The classification results of different classifiers based on the original spectra of Vis-SWNIR and LWNIR hyperspectral systems.

In terms of spectral preprocessing methods, to reduce the noise and interference in the original spectral information, the 9-point smooth method was firstly used to eliminate the noise existed in the original spectra. Then msc, detrend, and center methods were carried out on the basis of smoothed spectra to further optimize the spectral data, and the influence of different preprocessing methods on classification accuracy was compared. The average spectra preprocessed by different method were shown in Figure 6a–f. It can be seen that these preprocessing methods effectively eliminate signal offset and light scattering. The classification results of different preprocessing methods were shown in Table 3. For the case of the Vis-SWNIR region, the best preprocessing method was smooth-detrend with the classification accuracy of 86.67% and 88.33% for the calibration set and prediction set, respectively, because the accuracy gap between the calibration set and prediction set was the smallest. For the case of the LWNIR region, the model developed by smooth-detrend spectra achieved better prediction performance with the classification accuracy of 90.56% and 88.33% for the calibration set and prediction set, respectively. All classification models developed by the spectra preprocessed by smooth-detrend were superior to the model developed by original spectra, proving that spectral processing could greatly improve the reliability of classification models. The detrend could eliminate the influence of different sampling batches on the spectrum and improve the robustness and accuracy of the classification model. Sanchez et al. [35] improved the prediction ability of strawberry quality parameters by using detrend spectral pretreatment method. Paz et al. [51] found that detrend pretreatment had better prediction effect on sugar content and hardness in plums. Furthermore, the method of spectral pretreatment depends largely on the analyte being modeled and must be based on the judgment of the analyst [52]. In this study, smooth-detrend was selected as the most optimal spectral preprocessing method, and the spectra pretreated by smooth-detrend preprocessing were used for subsequent analysis instead of the original spectra in both Vis-SWNIR and LWNIR regions.

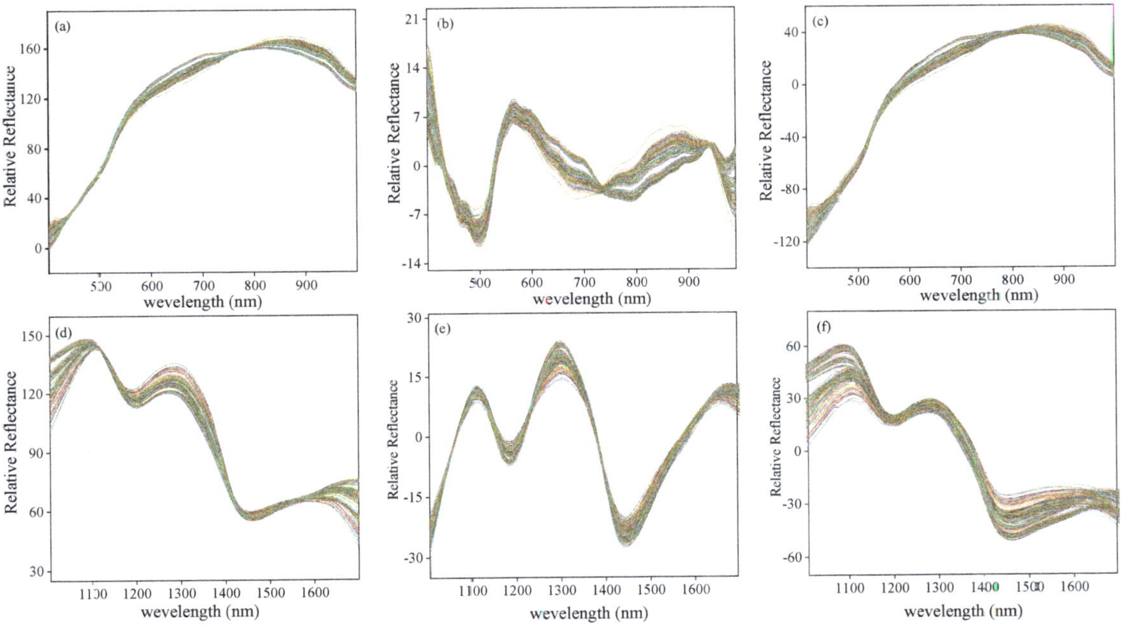

Figure 6. The preprocessed spectra by smooth-msc (**a**,**d**), smooth-detrend (**b**,**e**), and smooth-center (**c**,**f**) methods for the Vis-SWNIR and LWNIR regions, respectively.

Table 3. The classification performance of the SVM models established by different preprocessing methods.

Classifier	Sensor	Spectral Preprocessing Method	Calibration Set Accuracy (%)	Prediction Set Accuracy (%)
SVM	Vis-SWNIR	smooth-msc	84.44	88.33
		smooth-detrend	86.67	88.33
		smooth-center	85.56	90.00
	LWNIR	smooth-msc	90.56	86.67
		smooth-detrend	90.56	88.33
		smooth-center	91.11	85.00

3.4. Pixel-Level Fusion Based on Full Wavelengths Spectra and Texture Data

Spectra and texture data carried the component content and component distribution information of the target sample, respectively. In order to develop a higher accuracy classification model, four texture parameters, including contrast, correction, energy, and homogeneity, were extracted and used to create a new fusion matrix by fusing with the spectral data at the pixel-level level. Then the most optimal combination of spectra and texture was determined by establishing classification models based on the fused data.

The classification results of the pixel-level fusion of spectral and texture information were shown in Table 4. Compared with the results obtained by individual spectral data, the classification ability of fusion data varied with the participation of different texture features. The texture feature of energy and contrast had a positive effect on improving the classification models, with the accuracy of prediction sets of 90% and 90% for Vis-SWNIR and LWNIR regions, respectively. In general, the contrast parameter reflects the clarity of the image according to the depth of the texture groove, and the energy parameter reflects the randomness of the image texture. The amount of mold increased with the increase of cultured time, and the mold mainly concentrated in the embryo region of the maize, which may be the reason why both texture features were more conducive to the classification of maize with different moldy levels. It should be pointed out that, other combinations of spectra and texture features had not yielded the desired results, suggesting that the prediction ability of pixel-level fusion was not the accumulation of data quantity. Although the pixel-level fusion of spectra and textures directly merged the data of different sources, this method could input valuable information to the model, but it can also add a large number of uncorrelated and noisy variables, resulting in the fused data could not significantly improve the predictive power of the model. Similar results were obtained using NIR and ATR-FTIR data blocks to detection of adulteration in honey [39], models based on a data matrix generated by pixel-level data fusion show no significant improvement in accuracy.

Table 4. The classification results of the pixel-level fusion of spectral and texture information.

Sensor	Data Source		Calibration Set Accuracy (%)	Prediction Set Accuracy (%)
	Spectra	Texture		
Vis-SWNIR	smooth-detrend	contrast	85.56	86.67
		correction	85.56	86.67
		energy	92.22	90.00
		homogeneity	85.56	86.67
LWNIR	smooth-detrend	contrast	92.22	90.00
		correction	92.22	88.33
		energy	97.78	85.00
		homogeneity	92.22	88.33

3.5. Classification Model Built by Feature-Level Fusion of Spectra and Texture Data

The classification of maize with different moldy levels based on hyperspectral imaging technology involved the rapid collection of a substantial number of hyperspectral images, which were composed of two spatial dimensions and one spectral dimension data. Then, the spectral or texture information was extracted from these hyperspectral images and used to predict the categories of each maize sample. A large number of spectral or texture variables in the full wavelength range often contained noise from the environment and instrumental sources, leading to complexity and poor predicting ability of a calibration model. In addition, when used for online or at-line purposes, the complex calibration models developed with the whole spectrum will not be applicable. To resolve these issues, the features of spectra and textures were extracted alone using variable selection algorithms including VCPA, IRIV, and mVCPA-IRIV; then, the classification model of feature-level was established based on the extracted features for Vis-SWNIR and LWNIR regions, respectively.

In order to analyze the features selected from spectral and texture data, characteristic wavelengths selected by the same variable selection method were concatenated together and their distribution maps were shown in Figure 7. Comparing the number of selected characteristic bands, the VCPA selected fewer variables than IRIV and mVCPA-IRIV for spectral and texture data in both Vis-SWNIR and LWNIR regions. The number of selected characteristic bands was greater in texture data than that of spectral data for all three variable selection methods, illustrating that the texture had more information about the moldy maize than that of spectral data. Comparing the distribution of characteristic bands selected by three variable selection methods, it could be found that many common regions were determined in both spectral and texture data. The shared regions were concentrated at 629–649, 728–743, 764–772, 855–860, and 1055–1248 nm for the spectral data (Figure 6a,c), while the shared regions were observed to be concentrated at 410–490, 584–592, 679–693, 866–876, 953–963, 1029–1060, 1167–1192, 1235–1267, and 1688–1701 nm for the texture data (Figure 6b,d). According to previous studies, Stasiewicz et al. [53] classified different levels of Aspergillus flavus in maize at 850 nm near 857 nm. Moreover, 768 nm and 853 nm were used to differentiate the fungal contaminated maize from healthy samples in Chu's study [54]. Furthermore, 1029–1267 nm belonged to the second overtone of N-H stretching of proteins, as well as C-H stretching in lipids [47,48]; 1688–1701 nm was attributed to the second overtone of S-H, these were associated with protein, fat and starch [55]. Mold growth broke down the fat, protein, and starch of the maize kernels, which would change the reflection spectra and textures features of maize kernels, and resulting in the selection of the above characteristic bands. Except for those shared regions, there were some differences in the selected characteristic bands, which may be caused by the different principles of variable selection methods.

The classification results of the feature-level fusion of spectral and texture information were shown in Table 5. It could be found that the feature-level fusion models achieved better accuracy and reliability in Vis-SWNIR range than that of LWNIR region. In particular, the feature-level fusion models of VCPA, IRIV, and mVCPA-IRIV increased by 3.33%, 5.00%, and 1.67%, respectively, compared to the model based on pixel-level fusion for Vis-LWNIR region (Table 4). In detail, the prediction accuracy of the model based on the features selected by VCPA, IRIV, and mVCPA-IRIV was 93.33%, 95%, and 91.67% for Vis-SWNIR, and 90%, 83.33%, and 91.97% for LWNIR region. Although the IRIV method achieved the best prediction results in Vis-SWNIR region, the prediction ability was very poor in LWNIR region. In addition, the IRIV was time-consuming in variable selection and the number of selected variables used for modeling was much larger than that of the VCPA approach. However, the VCPA method achieved the classification accuracy of 93.33% and 90% for Vis-SWNIR and LWNIR regions, which was more robust than IRIV and mVCPA-IRIV. It is worth noting that, as a hybrid variable selection method, mVCPA-IRIV did not yield the best prediction results either in the Vis-LWNIR or in the LWNIR region. This result was consistent with the results of employing two-step hybrid methods to determine TVB-N contents in tilapia fillet for single Vis-NIR and NIR data blocks [24]. This may be because feature selection was based on a single data block with a relatively small number of variables, which was not suitable for the hybrid variable selection methods.

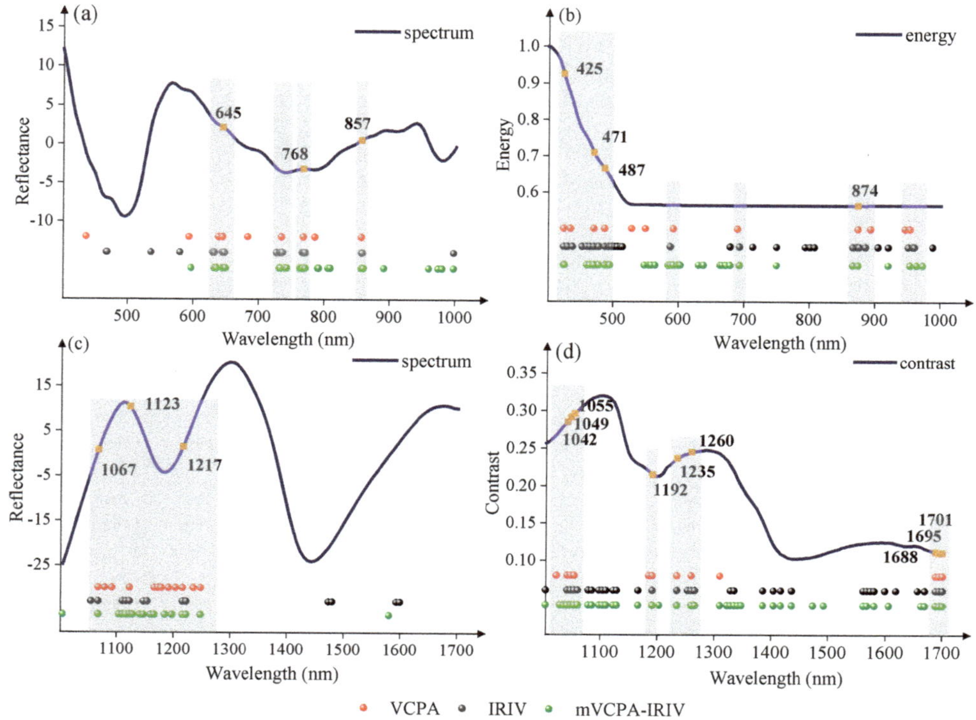

Figure 7. The distribution of key wavelengths selected by different variable selection algorithms from spectra and texture data of Vis-SWNIR and LWNIR regions, respectively. (**a**) Spectrum in Vis-SWNIR region, (**b**) energy parameters in Vis-SWNIR region, (**c**) spectrum in LWNIR region, and (**d**) contrast parameters in LWNIR region. Note: red points, black points, and green points represent the variables retained by VCPA, IRIV, and mVCPA-IRIV in the spectrum or texture parameters, respectively. Orange points represent the bands jointly selected by all three algorithms.

Table 5. The classification results of the feature-level fusion of spectral and texture information.

Integration Method	Sensor	Data Source	Variable Selection Algorithm	Characteristic Number		Calibration Set Accuracy (%)	Prediction Set Accuracy (%)
				Spectra	Texture		
Feature-level fusion	Vis-SWNIR	smooth-detrend energy	VCPA	9	12	94.44	93.33
			IRIV	21	28	97.78	95.00
			mVCPA-IRIV	28	39	93.89	91.67
	LWNIR	smooth-detrend contrast	VCPA	12	12	96.67	90.00
			IRIV	13	35	100.00	83.33
			mVCPA-IRIV	17	41	99.44	91.97

In conclusion, feature-level fusion model on the basis of variable selection had a large potential for distinguishing the maize with different moldy levels than the model of pixel-level fusion with full-band in both Vis-SWNIR and LWNIR regions. Remarkably, in the study by Anting et al. [37], the feature-level-PCA strategy using sample image and the spectra to evaluate the fermentation degree of black tea had a similar result with our proposed strategy.

3.6. Determination of the Optimal Feature-Level Fusion Model

Currently, the above results were based on two separate hyperspectral systems; meanwhile, typical hyperspectral imaging systems rarely extended the wavelength range of 399–1701 nm. Generally, more detailed and comprehensive feature information can be acquired from a wider spectral range, which makes sense given that integrating spectral and corresponding texture parameter data from Vis-SWNIR and LWNIR systems. The feature-level fusion model built by characteristic variables selected by VCPA obtained the most reliable classification result, hence, these variables (9 spectral features and 12 energy features for the Vis-SWNIR region, and 12 spectral features and 12 energy features for the LWNIR region) were integrated to build the classification model. The overall prediction accuracy was 96.11% and 95% for calibration and prediction sets, respectively, which was greater than the pixel-level fusion model and feature-fusion model of independent sensor. Although the number of variables used for modeling increased, it was far lower than the number of variables for full wavelength data. Similar results were also obtained in the internal bruising detection of blueberry by combining two hyperspectral systems with feature fusion strategies [56]. Figure 8 shows that the overall predicted results for the maize with different moldy levels. Except for the moderate level, all the moldy maize groups reached a high accuracy of more than 95%. In particular, all healthy maize was correctly classified. Some moderate levels samples were misclassified as mild or severe levels, resulting in a classification ability of only 90% for moderate levels, which agreed with the result of Yao et al. [57]. The moldy maize at moderate levels was difficult to accurately identify, which may be caused by the reduced variation between different moldy levels. This phenomenon also could be found through the determination of the CAT activity; with the aggravation of moldy level, the increase of CAT activity value among different categories decreased. However, it was worth emphasizing that none of the moldy samples were misclassified as healthy level, illustrating that the classification model had a certain practicality and objectivity.

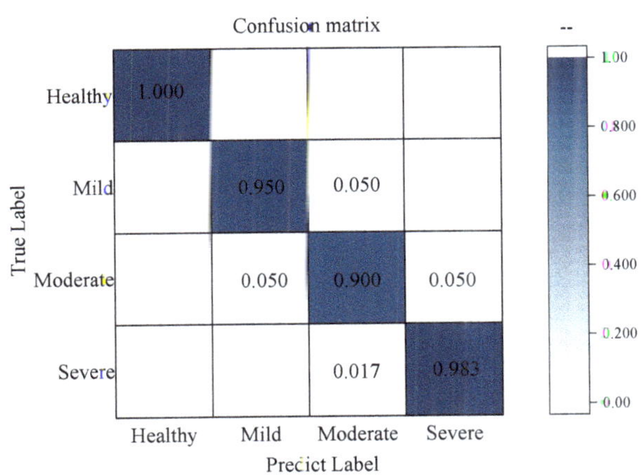

Figure 8. Confusion matrix of overall prediction results for all samples.

By comparing previous studies, some non-destructive testing techniques to identify grain mold have been studied extensively. These single technologies, such electronic nose [13], machine vision [15], Vis-SWNIR hyperspectral systems [22,23], and LWNIR hyperspectral systems [20,21] have been used to monitor the health condition of maize during storage. Remarkably, these studies obtained satisfactory results. Due to the growth and multiplication of mold, both the internal quality and external characteristics of maize change during the moldy process. Hence, the strategy of using single technologies to evaluate the quality of maize was limited. In our study, the spectral and different texture parameter data were extracted based on the collected Vis-SWNIR and LWNIR hyperspectral images. The data fusion strategy significantly improved the richness of information, which was helpful for building a robust classification model. Therefore, it can be concluded that feature-level fusion model based on spectral and texture information of two hyperspectral systems could be used to improve the classification accuracy of maize with different moldy levels.

4. Conclusions

As an important grain crop, maize is susceptible to mold infection during growth and storage due to its large embryo area and high moisture content. Therefore, it is essential to distinguish the moldy sample from healthy groups to prevent the spread of mold and avoid huger economic losses. Hyperspectral imaging technology combines spectral analysis technology with image processing technology, which can simultaneously obtain the spectral data with internal component information and the image data with appearance feature information, realizing the rapid, pollution-free, and non-destructive detection. In this study, the hyperspectral images of maize with different moldy levels were collected within Vis-SWNIR and LWNIR regions, and the spectra and texture information were extracted and used to establish the classification model with the methods of pixel-level and feature-level fusions. The results showed that data fusion strategies at both levels achieved better classification results than spectra alone. For pixel-level data fusion of spectral and texture information, the energy and contrast achieved positive effect on improving the classification model, with prediction accuracy of 90% and 90% for Vis-SWNIR and LWNIR regions, respectively. The improvement in model detection accuracy is not very apparent, as some irrelevant variables are introduced along with useful information. For feature-level data fusion of spectral and texture information, the variables selected by VCPA significantly increase the classification accuracy, with prediction accuracy of 93.33% and 90% for Vis-SWNIR and LWNIR regions, respectively. Feature-level fusion models based on the key variable combination of two hyperspectral systems were best for the classification of maize with different moldy levels, with an overall prediction accuracy of 95.00% for each moldy level.

This paper mainly focused on the identify of moldy maize; it should be noted that the data fusion strategies presented in this study are generally suitable to the quality detection of other grain crops such as wheat, rice, and peanut. Although this study had shown that the great feasibility of using hyperspectral imaging technology and multi-source data fusion method to discriminate the maize with different moldy levels, there must be some differences between naturally and artificially moldy maize samples, and we will use the data fusion strategies to classify moldy maize under natural growth in our future work. Additionally, AFB1 is a metabolite of mold with high toxicity; therefore, the growth monitoring of mold plays an important role in the early warning of AFB1 pollution in maize. CAT is a precursor product synthesized by AFB1, its dynamic activity reflects the level of mold activity, and it has a significant relationship with the content of AFB1. Therefore, we will develop a model for warning the AFB1 contamination based on the relationship between CAT of mold and AFB1 under different moldy conditions, which provides theoretical basis and technical guarantee for safe storage of maize.

Author Contributions: Conceptualization, W.W. and X.T.; software, W.H.; validation, W.W. and W.H.; data curation, W.W. and X.T.; writing—original draft preparation, W.W.; writing—review and editing, W.W., H.Y. and X.T.; project administration, H.Y. and X.T.; funding acquisition, H.Y. and X.T. All authors have read and agreed to the published version of the manuscript.

Funding: This study was supported by National Natural Science Foundation of China (Grant No. 31901402) and Young Elite Scientists Sponsorship Program by CAST (Grant No. 2019QNRC001).

Institutional Review Board Statement: Not applicable.

Informed Consent Statement: Not applicable.

Data Availability Statement: The data that support the findings of this study are available upon request from the authors.

Conflicts of Interest: The authors declare no conflict of interest.

References

1. Ambrose, A.; Kandpal, L.M.; Kim, M.S.; Lee, W.H.; Cho, B.K. High speed measurement of corn seed viability using hyperspectral imaging. *Infrared Phys. Technol.* **2016**, *75*, 173–179. [CrossRef]
2. Hesseltine, C.W. Natural occurrence of mycotoxins in cereals. *Mycopathol. Mycol. Appl.* **1974**, *53*, 141–153. [CrossRef] [PubMed]
3. Pearson, T.C.; Wicklow, D.T.; Pasikatan, M.C. Reduction of Aflatoxin and Fumonisin Contamination in Yellow Corn by High-Speed Dual-Wavelength Sorting. *Cereal Chem. J.* **2004**, *81*, 490–498. [CrossRef]
4. Kimuli, D.; Wang, W.; Wang, W.; Jiang, H.Z.; Zhao, X.; Chu, X. Application of SWIR hyperspectral imaging and chemometrics for identification of aflatoxin B1 contaminated maize kernels. *Infrared Phys. Technol.* **2018**, *89*, 351–362. [CrossRef]
5. Wang, W.; Ni, X.Z.; Lawrence, K.C.; Yoon, S.C.; Heitschmidt, G.W.; Feldner, P. Feasibility of detecting Aflatoxin B1 in single maize kernels using hyperspectral imaging. *J. Food Eng.* **2015**, *166*, 182–192. [CrossRef]
6. Campone, L.; Piccinelli, A.L.; Celano, R.; Rastrelli, L. Application of dispersive liquid-liquid microextraction for the determination of aflatoxins B1, B2, G1 and G2 in cereal products. *J. Chromatogr. A* **2011**, *1218*, 7648–7654. [CrossRef]
7. Maragos, C.M.; Busman, M. Rapid and advanced tools for mycotoxin analysis: A review. *Food Addit. Contam. Part A Chem. Anal. Control. Expo. Risk Assess.* **2010**, *27*, 688–700. [CrossRef]
8. Krnjaja, V.; Stanojkovic, A.; Stankovic, S.; Lukic, M.; Bijelic, Z.; Mandic, V.; Micic, N. Fungal contamination of maize grain samples with a special focus on toxigenic genera. *Biotechnol. Anim. Husb.* **2017**, *33*, 233–241. [CrossRef]
9. Guo, Z.M.; Wang, M.M.; Wu, J.Z.; Tao, F.F.; Chen, Q.S.; Wang, Q.Y.; Ouyang, Q.; Shi, J.Y.; Zou, X.B. Quantitative assessment of zearalenone in maize using multivariate algorithms coupled to Raman spectroscopy. *Food Chem.* **2019**, *286*, 282–288. [CrossRef]
10. Huang, K.Y.; Chien, M.C. A Novel Method of Identifying Paddy Seed Varieties. *Sensors* **2017**, *17*, 809. [CrossRef]
11. Singh, C.B.; Jayas, D.S.; Paliwal, J.; White, N.D.G. Detection of Sprouted and Midge-Damaged Wheat Kernels Using Near-Infrared Hyperspectral Imaging. *Cereal Chem. J.* **2009**, *86*, 256–260. [CrossRef]
12. Gu, S.; Chen, W.; Wang, Z.; Wang, J.; Huo, Y. Rapid detection of Aspergillus spp. infection levels on milled rice by headspace-gas chromatography ion-mobility spectrometry (HS-GC-IMS) and E-nose. *LWT* **2020**, *132*, 109758. [CrossRef]
13. Leggieri, M.C.; Mazzoni, M.; Fodil, S.; Moschini, M.; Bertuzzi, T.; Prandini, A.; Battilani, P. An electronic nose supported by an artificial neural network for the rapid detection of aflatoxin B1 and fumonisins in maize. *Food Control.* **2021**, *123*, 107722. [CrossRef]
14. Ng, H.F.; Wilcke, W.F.; Morey, R.V.; Lang, J.P. Machine Vision Color Calibration in Assessing Corn Kernel Damage. *Trans. ASAE* **1998**, *41*, 727–732. [CrossRef]
15. Shi, Y. Research on the Visual Detection Method of Corn Seed Grain Mold and Mold Degree. Master's Thesis, Yangzhou University, Yangzhou, China, 2021.
16. Fan, Y.M.; Ma, S.C.; Wu, T.T. Individual wheat kernels vigor assessment based on NIR spectroscopy coupled with machine learning methodologies. *Infrared Phys. Technol.* **2020**, *105*, 103213. [CrossRef]
17. Zhang, B.H.; Huang, W.Q.; Li, J.B.; Zhao, Z.J.; Fan, S.X.; Wu, J.T.; Liu, C.L. Principles, developments and applications of computer vision for external quality inspection of fruits and vegetables: A review. *Food Res. Int.* **2014**, *62*, 326–343. [CrossRef]
18. Singh, C.B.; Jayas, D.S.; Paliwal, J.; White, N.D.G. Identification of insect-damaged wheat kernels using short-wave near-infrared hyperspectral and digital colour imaging. *Comput. Electron. Agric.* **2010**, *73*, 118–125. [CrossRef]
19. Zhou, Q.; Huang, W.Q.; Fan, S.X.; Zhao, F.; Liang, D.; Tian, X. Non-destructive discrimination of the variety of sweet maize seeds based on hyperspectral image coupled with wavelength selection algorithm. *Infrared Phys. Technol.* **2020**, *109*, 103418. [CrossRef]
20. Tao, F.F.; Yao, H.B.; Hruska, Z.; Kincaid, R.; Rajasekaran, K.; Bhatnagar, D. A novel hyperspectral-based approach for identification of maize kernels infected with diverse *Aspergillus flavus* fungi. *Biosyst. Eng.* **2020**, *200*, 415–430. [CrossRef]
21. Williams, P.J.; Geladi, P.; Britz, T.J.; Manley, M. Investigation of fungal development in maize kernels using NIR hyperspectral imaging and multivariate data analysis. *J. Cereal Sci.* **2012**, *55*, 272–278. [CrossRef]
22. Dai, S.S.; Yong, Y. Fisher discriminant analysis for moldy degrees of maize samples based on the feature selection of hyperspectral data. *Food Mach.* **2018**, *34*, 68–72. [CrossRef]
23. Del Fiore, A.; Reverberi, M.; Ricelli, A.; Pinzari, F.; Serranti, S.; Fabbri, A.A.; Fanelli, C. Early detection of toxigenic fungi on maize by hyperspectral imaging analysis. *Int. J. Food Microbiol.* **2010**, *144*, 64–71. [CrossRef] [PubMed]
24. Yu, H.D.; Qing, L.W.; Yan, D.T.; Xia, G.; Zhang, C.; Yun, Y.H.; Zhang, W. Hyperspectral imaging in combination with data fusion for rapid evaluation of tilapia fillet freshness. *Food Chem.* **2021**, *348*, 129129. [CrossRef] [PubMed]
25. Ma, J.; Pu, H.; Sun, D.W.; Gao, W.; Qu, J.H.; Ma, K.Y. Application of Vis–NIR hyperspectral imaging in classification between fresh and frozen-thawed pork Longissimus Dorsi muscles. *Int. J. Refrig.* **2015**, *50*, 10–18. [CrossRef]
26. Sippy, N.; Luxton, R.; Lewis, R.J.; Cowell, D.C. Rapid electrochemical detection and identification of catalase positive micro-organisms. *Biosens. Bioelectron.* **2003**, *18*, 741–749. [CrossRef]
27. Zhang, S.B.; Zhai, H.C.; Hu, Y.S.; Wang, L.; Yu, G.H.; Huang, S.X.; Cai, J.P. A rapid detection method for microbial spoilage of agro-products based on catalase activity. *Food Control.* **2014**, *42*, 220–224. [CrossRef]

28. Zhang, Y.L.; Zai, H.C.; Zhang, S.B.; Cai, J.P. Early warning of AFB_1 contamination in stored maize based on monitoring of peroxidase activity. *Food Mach.* **2017**, *33*, 110–113. [CrossRef]
29. Zhang, Q.B.; Li, T.; Wang, C.Q.; Cai, Y.; Yang, J.W.; Zhang, R.P.; Li, B. Microbial Agents: Effects on Activities of Urease and Catalase in Flue-cured Tobacco Rhizosphere Soil. *Chin. Agric. Sci. Bull.* **2016**, *32*, 46–50. [CrossRef]
30. Li, J.B.; Huang, W.Q.; Chen, L.P.; Fan, S.X.; Zhang, B.H.; Guo, Z.M.; Zhao, C.J. Variable Selection in Visible and Near-Infrared Spectral Analysis for Noninvasive Determination of Soluble Solids Content of 'Ya' Pear. *Food Anal. Methods* **2014**, *7*, 1891–1902. [CrossRef]
31. Haralick, R.M.; Shanmugam, K.S. Combined spectral and spatial processing of ERTS imagery data. *Remote Sens. Environ.* **1974**, *3*, 3–13. [CrossRef]
32. Guo, Z.M.; Wang, M.M.; Agyekum, A.A.; Wu, J.Z.; Chen, Q.S.; Zuo, M.; El-Seedi, H.R.; Tao, F.F.; Shi, J.Y.; Ouyang, Q.; et al. Quantitative detection of apple watercore and soluble solids content by near infrared transmittance spectroscopy. *J. Food Eng.* **2020**, *279*, 109955. [CrossRef]
33. Wang, H.L.; Peng, J.Y.; Xie, C.Q.; Bao, Y.D.; He, Y. Fruit quality evaluation using spectroscopy technology: A review. *Sensors* **2015**, *15*, 11889–11927. [CrossRef]
34. Ilari, J.L.; Martens, H.; Isaksson, T. Determination of Particle Size in Powders by Scatter Correction in Diffuse Near-Infrared Reflectance. *Appl. Spectrosc.* **1988**, *42*, 722–728. [CrossRef]
35. Sánchez, M.T.; Haba la Haba, M.J.; Benítez-López, M.; Fernández-Novales, J.; Garrido-Varo, A.; Pérez-Marín, D. Non-destructive characterization and quality control of intact strawberries based on NIR spectral data. *J. Food Eng.* **2012**, *110*, 102–108. [CrossRef]
36. Borras, E.; Ferre, J.; Boque, R.; Mestres, M.; Acena, L.; Calvo, A.; Busto, O. Olive oil sensory defects classification with data fusion of instrumental techniques and multivariate analysis (PLS-DA). *Food Chem.* **2016**, *203*, 314–322. [CrossRef]
37. An, T.; Huang, W.Q.; Tian, X.; Fan, S.X.; Duan, D.D.; Dong, C.W.; Zhao, C.J.; Li, G.L. Hyperspectral imaging technology coupled with human sensory information to evaluate the fermentation degree of black tea. *Sens. Actuators B. Chem.* **2022**, *366*, 131994. [CrossRef]
38. Borràs, E.; Ferré, J.; Boqué, R.; Mestres, M.; Aceña, L.; Busto, O. Data fusion methodologies for food and beverage authentication and quality assessment—A review. *Anal. Chim. Acta* **2015**, *891*, 1–14. [CrossRef]
39. Huang, F.R.; Song, H.; Guo, L.; Guang, P.W.; Yang, X.H.; Li, L.Q.; Yang, M.X. Detection of adulteration in Chinese honey using NIR and ATR-FTIR spectral data fusion. *Spectrochim. Acta Part. A Mol. Biomol. Spectrosc.* **2020**, *235*, 118297. [CrossRef]
40. Yun, Y.H.; Wang, W.T.; Deng, B.C.; Lai, G.B.; Liu, X.B.; Ren, D.B.; Liang, Y.Z.; Fan, W.; Xu, Q.S. Using variable combination population analysis for variable selection in multivariate calibration. *Anal. Chim. Acta* **2015**, *862*, 14–23. [CrossRef]
41. Yun, Y.H.; Wang, W.T.; Tan, M.L.; Liang, Y.Z.; Li, H.D.; Cao, D.S.; Lu, H.M.; Xu, Q.S. A strategy that iteratively retains informative variables for selecting optimal variable subset in multivariate calibration. *Anal. Chim. Acta* **2014**, *807*, 36–43. [CrossRef]
42. Yun, Y.H.; Li, H.D.; Deng, B.C.; Cao, D.S. An overview of variable selection methods in multivariate analysis of near-infrared spectra. *TrAC Trends Anal. Chem.* **2019**, *113*, 102–115. [CrossRef]
43. Sánchez, A.V.D. Advanced support vector machines and kernel methods. *Neurocomputing* **2003**, *55*, 5–20. [CrossRef]
44. Li, J.B.; Huang, W.Q.; Zhao, C.J.; Zhang, B.H. A comparative study for the quantitative determination of soluble solids content, pH and firmness of pears by Vis/NIR spectroscopy. *J. Food Eng.* **2013**, *116*, 324–332. [CrossRef]
45. Nunes Miranda, E.; Henrique Groenner Barbosa, B.; Henrique Godinho Silva, S.; Augusto Ussi Monti, C.; Yue Phin Tng, D.; Rezende Gomide, L. Variable selection for estimating individual tree height using genetic algorithm and random forest. *For. Ecol. Manag.* **2022**, *504*, 119828. [CrossRef]
46. Zhao, D.D.; Hu, X.Y.; Xiong, S.W.; Tian, J.; Xiang, J.W.; Zhou, J.; Li, H.H. k-means clustering and kNN classification based on negative databases. *Appl. Soft Comput.* **2021**, *110*, 107732. [CrossRef]
47. Marques, A.S.; Castro, J.N.F.; Costa, F.J.M.D.; Neto, R.M.; Lima, K.M.G. Near-infrared spectroscopy and variable selection techniques to discriminate Pseudomonas aeruginosa strains in clinical samples. *Microchem. J.* **2016**, *124*, 306–310. [CrossRef]
48. Daszykowski, M.; Wrobel, M.S.; Czarnik-Matusewicz, M.; Walczak, B. Near-infrared reflectance spectroscopy and multivariate calibration techniques applied to modelling the crude protein, fibre and fat content in rapeseed meal. *Analyst* **2008**, *133*, 1523–1531. [CrossRef]
49. Alhamdan, A.M.; Atia, A. Non-destructive method to predict Barhi dates quality at different stages of maturity utilising near-infrared (NIR) spectroscopy. *Int. J. Food Prop.* **2017**, *20*, 2950–2959. [CrossRef]
50. Li, J.B.; Luo, W.; Wang, Z.L.; Fan, S.X. Early detection of decay on apples using hyperspectral reflectance imaging combining both principal component analysis and improved watershed segmentation method. *Postharvest Biol. Technol.* **2019**, *149*, 235–246. [CrossRef]
51. Paz, P.; Sanchez, M.T.; Perez-Marin, D.; Guerrero, J.E.; Garrido-Varo, A. Nondestructive determination of total soluble solid content and firmness in plums using near-infrared reflectance spectroscopy. *J. Agric. Food Chem.* **2008**, *56*, 2565–2570. [CrossRef]
52. Delwiche, S.R.; Reeves, J.B., III. The effect of spectral pre-treatments on the partial least squares modeling of agricultural products. *J. Near Infrared Spectrosc.* **2004**, *12*, 177–182. [CrossRef]
53. Stasiewicz, M.J.; Falade, T.D.O.; Mutuma, M.; Mutiga, S.K.; Harvey, J.J.W.; Fox, G.; Nelson, R.J. Multi-spectral kernel sorting to reduce aflatoxins and fumonisins in Kenyan maize. *Food Control* **2017**, *78*, 202–214. [CrossRef]
54. Chu, X. Hyperspectral Imaging Identifications for Cereal Fungi and Detection Methods for Moldy Maize Kernels. Ph.D. Thesis, China Agriculture University, Beijing, China, 2018.

55. Wang, Z.L.; Tian, X.; Fan, S.X.; Zhang, C.; Li, J.B. Maturity determination of single maize seed by using near-infrared hyperspectral imaging coupled with comparative analysis of multiple classification models. *Infrared Phys. Technol.* **2021**, *112*, 103596. [CrossRef]
56. Fan, S.X.; Li, J.B.; Huang, W.Q.; Chen, L.P. Data Fusion of Two Hyperspectral Imaging Systems with Complementary Spectral Sensing Ranges for Blueberry Bruising Detection. *Sensors* **2018**, *18*, 4463. [CrossRef]
57. Yao, H.; Hruska, Z.; Kincaid, R.; Brown, R.; Cleveland, T.; Bhatnagar, D. Correlation and classification of single kernel fluorescence hyperspectral data with aflatoxin concentration in corn kernels inoculated with *Aspergillus flavus* spores. *Food Addit. Contam. Part A Chem. Anal. Control Expo. Risk Assess.* **2010**, *27*, 701–709. [CrossRef]

MDPI
St. Alban-Anlage 66
4052 Basel
Switzerland
www.mdpi.com

Foods Editorial Office
E-mail: foods@mdpi.com
www.mdpi.com/journal/foods

www.ingramcontent.com/pod-product-compliance
Lightning Source LLC
Chambersburg PA
CBHW061606220926
51850CB00005B/438
9783036593241